MULTIVARIABLE CALCULUS

MULTIVARIABLE CALCULUS

Produced by the Consortium based at Harvard and funded by a National Science Foundation Grant.

William G. McCallum
University of Arizona

Deborah Hughes-Hallett
Harvard University

Daniel Flath
University of South Alabama

Douglas Quinney
University of Keele

Andrew M. Gleason
Harvard University

Wayne Raskind
University of Southern California

Sheldon P. Gordon
Suffolk County Community College

Jeff Tecosky-Feldman
Haverford College

David Mumford
Harvard University

Joe B. Thrash
University of Southern Mississippi

Brad G. Osgood
Stanford University

Thomas W. Tucker
Colgate University

with the assistance of

Adrian Iovita
*Centre Interuniversitaire en
Calcul Mathématique Algébrique*

John Wiley & Sons, Inc.
New York   Chichester   Weinheim   Brisbane   Singapore   Toronto

Dedicated to Amy, Nell, Abby, and Sally.

Cover Photo by Greg Pease

This material is based upon work supported by the National Science Foundation under Grant No. DUE-9352905.

This book was set in Times Roman by the Consortium based at Harvard using LATEX, Mathematica, and the package AsTeX, which was written by Alex Kasman for this project.

ISBN: 0-471-31151-0

Printed in the United States of America

10 9 8 7 6 5 4 3

PREFACE

Calculus is one of the greatest achievements of the human intellect. Inspired by problems in astronomy, Newton and Leibniz developed the ideas of calculus 300 years ago. Since then, each century has demonstrated the power of calculus to illuminate questions in mathematics, the physical sciences, engineering, and the social and biological sciences.

Calculus has been so successful because of its extraordinary power to reduce complicated problems to simple rules and procedures. Therein lies the danger in teaching calculus: it is possible to teach the subject as nothing but the rules and procedures – thereby losing sight of both the mathematics and of its practical value. With the generous support of the National Science Foundation, our group set out to create a new calculus curriculum that would restore that insight. This book is the second stage in that endeavor. The first stage is our single variable text.

Basic Principles

The two principles that guided our efforts in developing the single variable book remain valid. The first is our prescription for restoring the mathematical content to calculus:

> **The Rule of Three:** *Every topic should be presented geometrically, numerically and algebraically.*

We continually encourage students to think and write about the geometrical and numerical meaning of what they are doing. It is not our intention to undermine the purely algebraic aspect of calculus, but rather to reinforce it by giving meaning to the symbols. In the homework problems dealing with applications, we continually ask students what their answers mean in practical terms.

The second principle, inspired by Archimedes, is our prescription for restoring practical understanding:

> **The Way of Archimedes:** *Formal definitions and procedures evolve from the investigation of practical problems.*

Archimedes believed that insight into mathematical problems is gained by investigating mechanical or physical problems first.[1] For the same reason, our text is problem driven. Whenever possible, we start with a practical problem and derive the general results from it. By practical problems we usually, but not always, mean real world applications. These two principles have led to a dramatically new curriculum – more so than a cursory glance at the table of contents might indicate.

---

[1] . . . I thought fit to write out for you and explain in detail . . . the peculiarity of a certain method, by which it will be possible for you to get a start to enable you to investigate some of the problems in mathematics by means of mechanics. This procedure is, I am persuaded, no less useful even for the proof of the theorems themselves; for certain things first became clear to me by a mechanical method, although they had to be demonstrated by geometry afterwards because their investigation by the said method did not furnish an actual demonstration. But it is of course easier, when we have previously acquired, by the method, some knowledge of the questions, to supply the proof than it is to find it without any previous knowledge. From *The Method*, in *The Works of Archimedes* edited and translated by Sir Thomas L. Heath (Dover, NY)

Technology

In multivariable calculus, even more so than in single variable calculus, computer technology can be put to great advantage to help students learn to think mathematically. For example, looking at surface graphs and contour diagrams is enormously helpful in understanding functions of many variables. Furthermore, the ability to use technology effectively as a tool in itself is of the greatest importance. Students are expected to use their judgment to determine where technology is useful.

However, the book does not require any specific software or technology, and we have accommodated those without access to sufficiently powerful technology by providing supplementary master copies for overhead slides, showing surface graphs, contour diagrams, parametrized curves, and vector fields. Ideally, students should have access to technology with the ability to draw surface graphs, contour diagrams, and vector fields, and to calculate multiple integrals and line integrals numerically. Failing that, however, the combination of hand held graphing calculators and the overhead transparencies is quite satisfactory, and has been used successfully by test sites.

What Student Background is Expected?

Students using this book should have successfully completed a course in single variable calculus. It is not necessary for them to have used the single variable book from the same consortium in order for them to learn from this book.

The book is thought-provoking for well-prepared students while still accessible to students with weaker backgrounds. Providing numerical and graphical approaches as well as the algebraic gives students another way of mastering the material. This approach encourages students to persist, thereby lowering failure rates.

Content

Our approach to designing this curriculum was the same as the one we took in our single variable book: we started with a clean slate, and compiled a list of topics that we thought were fundamental to the subject, after discussions with mathematicians, engineers, physicists, chemists, biologists, and economists. In order to meet individual needs or course requirements, topics can easily be added or deleted, or the order changed.

We assume throughout that functions of two or more variables are defined on regions with piecewise smooth boundaries.

Chapter 11: Functions of Many Variables

We introduce functions of many variables from several points of view, using surface graphs, contour diagrams, and tables. This chapter plays the same role for this course as Chapter 1 did for the single variable course; it gives students the skills to read graphs and contour diagrams and think graphically, to read tables and think numerically, and to apply these skills, along with their algebraic skills, to modeling the real world. We pay particular attention to the idea of a cross-section of a function, obtained by varying one variable independently of the others. We have found that it is useful for the student to study this notion before going on to partial derivatives and gradients. We study linear functions in detail, in preparation for the notion of local linearity. We conclude with a section on continuity.

Chapter 12: A Fundamental Tool: Vectors

We define vectors as geometric objects having direction and magnitude, with displacement vectors as the model, and then introduce the representation of vectors in terms of coordinates. We give equivalent geometric and algebraic definitions of the dot and cross product.

Chapter 13: Differentiating Functions of Many Variables

We introduce the basic notions of partial derivative, directional derivative, gradient, and differential. In keeping with the spirit of the single variable book, we put these notions in the framework of local linearity. We also use local linearity to introduce the notion of differentiability, and in the discussion of the multivariable chain rule. We discuss higher order partial derivatives, their interpretation in partial differential equations, and their application to quadratic Taylor approximations. We conclude with a section on differentiability.

Chapter 14: Optimization

We apply the ideas of the previous chapter to optimization problems, both constrained and unconstrained. We derive the second derivative test for local extrema by first considering the case of quadratic polynomials, and then appealing to the quadratic Taylor approximation. We discuss the existence of global extrema for continuous functions on closed and bounded regions. In the section on constrained optimization, we discuss Lagrange multipliers, equality and inequality constraints, problems with more than one constraint, and the Lagrangian.

Chapter 15: Integrating Functions of Many Variables

We motivate the multivariable definite integral graphically by considering the problem of estimating total population from a contour diagram for population density, using finer and finer grids. We continue with numerical examples using tables, and then give two methods of calculating multiple integrals: analytically, by means of iterated integrals, and numerically, by the Monte Carlo method. We discuss both double and triple integrals in Cartesian, polar, spherical, and cylindrical coordinates. We also discuss applications to multivariate probability.

Chapter 16: Parametric Curves and Surfaces

We start with the problem of representing curves parametrically, then use parameterized curves to represent motion. We define velocity and acceleration geometrically, then give the formulas in terms of components. We continue with a section on parameterized surfaces, and then discuss the connection between implicit, explicit, and parametric representations of surfaces using the implicit function theorem. The final section discusses one of the original applications of calculus: Newton's explanation Kepler's Laws of planetary motion.

Chapter 17: Vector Fields

In this brief chapter we introduce vector-valued functions of many variables, or vector fields. This chapter lays the foundation for the geometric approach in the next three chapters to line integrals, flux integrals, divergence, and curl. We start with physical examples, such as velocity vector fields and force fields, and include many sketches of vector fields to help build geometric intuition. We also discuss flow lines of vector fields and their relation with systems of differential equations.

Chapter 18: Line Integrals

We present the concept of integrating a vector field along a path with a coordinate-free definition. We spend some time building intuition using sketches of vector fields with paths superimposed, before introducing the method of calculating line integrals using parametrizations. We then discuss conservative fields, gradient fields, and the Fundamental Theorem of Calculus for Line Integrals. We continue with a discussion of non-conservative vector fields and Green's Theorem, and give the curl test for a conservative vector field. We conclude with a proof of Green's Theorem using the change of variables formula.

Chapter 19: Flux Integrals

We introduce the flux integral of a vector field through a parameterized surface in the same way as we introduced line integrals. First we give a coordinate-free definition, then we discuss examples where the flux integral (or at least its sign) can be calculated geometrically. Then we show how to calculate flux integrals over surface graphs, portions of cylinders, and portions of spheres. We conclude with a section on flux integrals over arbitrary parameterized surfaces.

Chapter 20: Calculus of Vector Fields

We introduce divergence and curl in a coordinate-free way; the divergence in terms of flux density, and curl in terms of circulation density. We then give the formulas in Cartesian coordinates. In the single variable book we derived the Fundamental Theorem of Calculus by pointing out that the integral of the rate of change is the total change. In much the same way, we derive the Divergence Theorem by showing that the integral of flux density over a volume is the total flux out of the volume and Stokes' theorem by showing that the integral of circulation density over a surface is the total circulation around its boundary. We discuss the three fundamental theorems of multivariable calculus, and show how they lead to the three-dimensional curl test for a conservative vector field. We conclude with a section proving the Divergence Theorem and Stokes' Theorem using the change of variables formula.

Changes from the Preliminary Edition

We have incorporated suggestions from users of the Preliminary Edition, who have helped us make the exposition as clear and concise as possible. Most of the figures have been redrawn, particularly the three-dimensional figures.

- *Chapter 11*. We have added a section on Limits and Continuity.

- *Chapter 12*. The geometric and algebraic definitions of the dot and cross products are now given in tandem at the beginning of each section. Each section gives an argument explaining why the two definitions are equivalent.

- *Chapter 13*. The material on directional derivatives and gradients has been substantially reorganized. We introduce both directional derivatives and gradient vectors in Section 13.4, but only in the two-dimensional case. We have replaced geometric definition of the gradient with the algebraic definition, motivated by the formula for computing directional derivatives; the geometric properties are then derived from this formula. Section 13.5 contains new material on the relation between 2-dimensional and 3-dimensional gradients, and on situations where the gradient doesn't have a geometric interpretation. In the section on the chain rule we have added a new example from physical chemistry. We have added a new section at the end on differentiability from a graphical and intuitive point of view, which discusses the relation between differentiability, partial derivatives, and continuity.

- *Chapter 14*. The material about global extrema from the old Section 14.1 has been moved into Section 14.2, so that Section 14.1 now focuses exclusively on critical points and their classification. The theoretical material about closed and bounded sets has been moved to the end of Section 14.2. This results in a greater emphasis on the main ideas.

- *Chapter 15*. The introductory examples have been made shorter so the definition is reached more quickly.

- *Chapter 16*. We have substantially reorganized the material in Sections 16.1–16.3. The new Section 16.1 concentrates on the geometric idea of representing a curve parametrically. The new Section 16.2 develops the idea of a parametric curve as representing a motion, and introduces the velocity and acceleration vectors. The material on implicit, explicit, and parametric representations has been moved to the new Section 16.4 on the Implicit Function Theorem. We have added Section 16.5 on Newton's explanation of Kepler's Laws.

- *Chapter 18.* We have added material on the three-dimensional curl test, foreshadowing Chapter 20. We have added a new section giving a proof of Green's Theorem.

- *Chapter 19.* We have reorganized this chapter into three sections, with some new material. Section 19.1 contains new material on calculating flux integrals without parameterizations by using simple geometric arguments to reduce the integral to a double integral. Section 19.2 has new material on calculating flux integrals over portions of cylinders and spheres.

- *Chapter 20.* Reflecting the changes made in Chapter 12, the geometric and algebraic definitions of divergence and curl are now presented together, with an intuitive argument explaining why they give the same result. We have added some more challenging examples. and material on divergence-free and curl-free vector fields. A new section on the three fundamental theorems discusses the three-dimensional curl test and the divergence test for curl fields. We have replaced the old final section with a new one giving proofs of the Divergence Theorem and Stokes' Theorem based on the idea of parameterizing a region and using change of variables to reduce the proof to the case of rectangular regions, in tune with the standard modern proof using differential forms. This is a challenging section for strong students which serves as an excellent capstone section for students taking this course at the honors level.

- *Answers to Odd Numbered Problems.* We have incorporated the Student Answer Manual into the book. This gives short answers to those odd numbered problems which have short answers.

Options for a One-Semester Course

Instructors using the text in a one semester course have the following two choices: They can stop at the end of Chapter 18, allowing time for a thorough treatment of parameterized curves and surfaces, line integrals, and Green's theorem; or they can continue to Chapter 20, covering only the earlier sections in Chapters 16, 17, 18, and 19, which will yield a brief treatment of line and flux integrals from a geometric point of view and give students enough background to understand the Divergence theorem and Stokes' theorem.

Supplementary Materials

- **Instructor's Manual** with teaching tips, calculator programs, some overhead transparency masters and sample exams and quizzes.
- **Instructor's Solution Manual** with complete solutions to all problems.
- **Student's Solution Manual** with complete solutions to every other odd-numbered problem.
- **MultiGraph** for Windows surface plotting software.

Our Experiences

In the process of developing the ideas incorporated in this book, we have been conscious of the need to test the materials thoroughly in a wide variety of institutions serving many different types of students. Consortium members have used previous versions of the book at a broad range of institutions. During the 1995–1996 academic year we were assisted by colleagues at over 100 schools who class-tested the Preliminary Edition and reported their experiences and those of their students. This diverse group of schools used the book in semester and quarter systems, in computer labs, small groups, and traditional settings, and with a number of different technologies. We appreciate the valuable suggestions they made, which we have tried to incorporate into the First Edition of the text.

Acknowledgements

Thanks to Ruby Aguirre, Ed Alexander, Carole Anderson, Leonid Andreev, Ralph Baierlein, Paul Balister, Frank Beatrous, Jerrie Beiberstein, Melanie Bell, Ebo Bentil, Yoav Bergner, Shelina Bhojani, Thomas Bird, Paul Blanchard, Melkana Brakalova, John Bravman, David Bressoud, R. Campbell, Phil Cheifetz, Oksana Cherniavskaya, C. K. Cheung, Dave Chua, Dean Chung, Robert Condon, Eric Connally, Radu Constantinescu, Pat Corn, Josh Cowley, Jie Cui, Caspar Curjel, Bill Dunn, Mike Esposito, Pavel Etingof, Bill Faris, Hermann Flaschka, Leonid Friedlander, Leonid Fridman, Greg Fung, Deborah Gaines, Amanda Galtman, Avijit Gangopadhyay, Howard Georgi, Scott Gilbert, Marty Greenlee, David Grenda, Benedict Gross, John Hagood, David G. Harris, Angus Hendrick, John Huth, Robert Indik, Raj Jesudason, Qin Jing, Jerry Johnson, Millie Johnson, Joe Kanapka, Alex Kasman, Matthias Kawski, David Kazhdan, Misha Kazhdan, Thomas Kerler, Charlie Kerr, Mike Klucznik, Sandy Koonce, Matt Kruse, Ted Laetsch, Sylvain Laroche, Janny Leung, Dave Levermore, Lei Li, Weiye Li, Li Liu, Carlos Lizzaraga, Patti Frazer Lock, John Lucas, Alex Mallozzi, Brad Mann, Elliot Marks, Ricardo Martinez, Eric Mazur, Mark McConnell, Dan McGee, Tom McMahon, Georgia Mederer, Andrew Metrick, Michal Mlejnek, Jean Morris, Don Myers, Bridget Neale, Alan Newell, James Osterburg, Myles Paige, Ed Park, Ted Pyne, Howard Penn, Tony Phillips, Laura Piscitelli, Ago Pisztora, Steve Prothero, Rebecca Rapoport, David Richards, Ann Ryu, Walter Seaman, Russ Shachter, Barbara Shipman, Mary Sibayan, Jeff Silver, Chris Sinclair, Yum-Tong Siu, Keith Stroyan, Noah Syroid, Francis Su, Suds Sudholz, Mike Tabor, Cliff Taubes, Ralph Teixeira, Denise Todd, Elias Toubassi, Jerry Uhl, Doug Ulmer, Adrian Vajiac, Bill Velez, Faye Villalobos, Jianmei Wang, Joseph Watkins, Xianbao Xu, and Bruce Yoshiwara.

In particular, we would like to thank Paul Feehan for his remarkable contributions.

William G. McCallum   Sheldon P. Gordon   Wayne Raskind

Deborah Hughes-Hallett   David Mumford   Jeff Tecosky-Feldman

Daniel E. Flath   Brad G. Osgood   Joe B. Thrash

Andrew M. Gleason   Douglas Quinney   Thomas W. Tucker

To Students: How to Learn from this Book

- This book may be different from other math textbooks that you have used, so it may be helpful to know about some of the differences in advance. This book emphasizes at every stage the *meaning* (in practical, graphical or numerical terms) of the symbols you are using. There is much less emphasis on "plug-and-chug" and using formulas, and much more emphasis on the interpretation of these formulas than you may expect. You will often be asked to explain your ideas in words or to explain an answer using graphs.

- The book contains the main ideas of multivariable calculus in plain English. Your success in using this book will depend on your reading, questioning, and thinking hard about the ideas presented. Although you may not have done this with other books, you should plan on reading the text in detail, not just the worked examples.

- There are very few examples in the text that are exactly like the homework problems. This means that you can't just look at a homework problem and search for a similar–looking "worked out" example. Success with the homework will come by grappling with the ideas of calculus.

- Many of the problems that we have included in the book are open-ended. This means that there may be more than one approach and more than one solution, depending on your analysis. Many times, solving a problem relies on common sense ideas that are not stated in the problem but which you will know from everyday life.

- This book assumes that you have access to a graphing calculator or computer; preferably one that can draw surface graphs, contour diagrams, and vector fields, and can compute multivariable integrals and line integrals numerically. There are many situations where you may not be able to find an exact solution to a problem, but you can use a calculator or computer to get a reasonable approximation. An answer obtained this way is usually just as useful as an exact one. However, the problem does not always state that a calculator is required, so use your judgement.

- This book attempts to give equal weight to three methods for describing functions: graphical (a picture), numerical (a table of values) and algebraic (a formula). Sometimes you may find it easier to translate a problem given in one form into another. For example, if you have to find the maximum of a function, you might use a contour diagram to estimate its approximate position, use its formula to find equations that give the exact position, then use a numerical method to solve the equations. The best idea is to be flexible about your approach: if one way of looking at a problem doesn't work, try another.

- Students using this book have found discussing these problems in small groups very helpful. There are a great many problems which are not cut-and-dried; it can help to attack them with the other perspectives your colleagues can provide. If group work is not feasible, see if your instructor can organize a discussion session in which additional problems can be worked on.

- You are probably wondering what you'll get from the book. The answer is, if you put in a solid effort, you will get a real understanding of one of the most important accomplishments of the millennium – calculus – as well as a real sense of the power of mathematics in the age of technology.

CONTENTS

14 OPTIMIZATION: LOCAL AND GLOBAL EXTREMA   175

15 INTEGRATING FUNCTIONS OF MANY VARIABLES   215

20 CALCULUS OF VECTOR FIELDS

APPENDICES

ANSWERS TO ODD NUMBERED PROBLEMS

INDEX

CHAPTER ELEVEN

FUNCTIONS OF SEVERAL VARIABLES

Many quantities depend on more than one variable: the amount of food grown depends on the amount of rain and the amount of fertilizer used; the rate of a chemical reaction depends on the temperature and the pressure of the environment in which it proceeds; the strength of the gravitational attraction between two bodies depends on their masses and their distance apart; and the rate of fallout from a volcanic explosion depends on the distance from the volcano and the time since the explosion. Each example involves a function of two or more variables. In this chapter, we will see many different ways of looking at functions of several variables.

11.1 FUNCTIONS OF TWO VARIABLES

Function Notation

Suppose you are planning to take out a five-year loan to buy a car and you need to calculate what your monthly payment will be; this depends on both the amount of money you borrow and the interest rate. These quantities can vary separately: the loan amount can change while the interest rate remains the same, or the interest rate can change while the loan amount remains the same. To calculate your monthly payment you need to know both. If the monthly payment is m, the loan amount is L, and the interest rate is $r\%$, then we express the fact that m is a function of L and r by writing:

$$m = f(L, r).$$

This is just like the function notation of one-variable calculus. The variable m is called the dependent variable, and the variables L and r are called the independent variables. The letter f stands for the *function* or rule that gives the value of m corresponding to given values of L and r.

A function of two variables can be represented graphically, numerically by a table of values, or algebraically by a formula. In this section we will give examples of each of these three ways of viewing a function.

Graphical Example: A Weather Map

Figure 11.1 shows a weather map from a newspaper. What information does it convey? It displays the predicted high temperature, T, in degrees Fahrenheit (°F), throughout the US on that day. The curves on the map, called *isotherms*, separate the country into zones, according to whether T is in the 60s, 70s, 80s, 90s, or 100s. (*Iso* means same and *therm* means heat.) Notice that the isotherm separating the 80s and 90s zones connects all the points where the temperature is exactly 90°F.

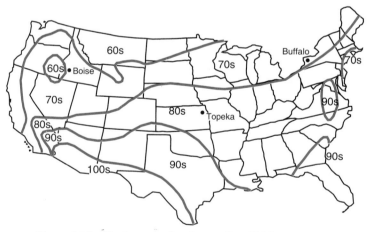

Figure 11.1: Weather map showing predicted high temperatures, T, on a summer day

Example 1   Estimate the predicted value of T in Boise, Idaho; Topeka, Kansas; and Buffalo, New York.

Solution   Boise and Buffalo are in the 70s region, and Topeka is in the 80s region. Thus, the predicted temperature in Boise and Buffalo is between 70 and 80 while the predicted temperature in Topeka is between 80 and 90.

In fact, we can say more. Although both Boise and Buffalo are in the 70s, Boise is quite close to the $T = 70$ isotherm, whereas Buffalo is quite close to the $T = 80$ isotherm. So we estimate that the temperature will be in the low 70s in Boise and the high 70s in Buffalo. Topeka is about halfway between the $T = 80$ isotherm and the $T = 90$ isotherm. Thus, we guess that the temperature in Topeka will be in the mid 80s. In fact, the actual high temperatures for that day were 71°F for Boise, 79°F for Buffalo, and 86°F for Topeka.

The predicted high temperature, T, illustrated by the weather map is a function of (that is, depends on) two variables, often longitude and latitude, or miles east-west and miles north-south of a fixed point, say, Topeka. The weather map in Figure 11.1 is called a *contour map* or *contour diagram* of that function. Section 11.3 shows another way of visualizing functions of two variables using surfaces; Section 11.4 looks at contour maps in detail.

Numerical Example: Beef Consumption

Suppose you are a beef producer and you want to know how much beef people will buy. This depends on how much money people have and on the price of beef. The consumption of beef, C (in pounds per week per household) is a function of household income, I (in thousands of dollars per year), and the price of beef, p (in dollars per pound). In function notation, we write:

$$C = f(I, p).$$

Table 11.1 contains values of this function. Values of p are shown across the top, values of I are down the left side, and corresponding values of $f(I, p)$ are given in the table.[1] For example, to find the value of $f(40, 3.50)$, we look in the row corresponding to $I = 40$ under $p = 3.50$, where we find the number 4.05. Thus,

$$f(40, 3.50) = 4.05.$$

This means that, on average, if a household's income is $40,000 a year and the price of beef is $3.50/lb, the family will buy 4.05 lbs of beef per week.

TABLE 11.1   *Quantity of beef bought (pounds/household/week)*

		Price of beef, p ($/lb)			
		3.00	3.50	4.00	4.50
Household income per year, I ($1000)	20	2.65	2.59	2.51	2.43
	40	4.14	4.05	3.94	3.88
	60	5.11	5.00	4.97	4.84
	80	5.35	5.29	5.19	5.07
	100	5.79	5.77	5.60	5.53

Notice how this differs from the table of values of a one-variable function, where one row or one column is enough to list the values of the function. Here many rows and columns are needed because the function has a value for every *pair* of values of the independent variables.

[1] Adapted from Richard G. Lipsey, *An Introduction to Positive Economics 3rd Ed.*, Weidenfeld and Nicolson, London, 1971

Algebraic Examples: Formulas

In both the weather map and beef consumption examples, there is no formula for the underlying function. That is usually the case for functions representing real-life data. On the other hand, for many idealized models in physics, engineering, or economics, there are exact formulas.

Example 2   Give a formula for the function $M = f(B, t)$ where M is the amount of money in a bank account t years after an initial investment of B dollars, if interest is accrued at a rate of 5% per year compounded (a)   annually   (b)   continuously.

Solution   (a)   Annual compounding means that M increases by a factor of 1.05 every year, so

$$M = f(B, t) = B(1.05)^t.$$

(b)   Continuous compounding means that M grows according to the exponential function e^{kt}, with $k = 0.05$, so

$$M = f(B, t) = Be^{0.05t}.$$

Example 3   A cylinder with closed ends has a radius r and a height h. If its volume is V and its surface area is A, find formulas for the functions $V = f(r, h)$ and $A = g(r, h)$.

Solution   Since the area of the circular base is πr^2, we have

$$V = f(r, h) = \text{Area of base} \cdot \text{Height} = \pi r^2 h.$$

The surface area of the side is the circumference of the bottom, $2\pi r$, times the height h, giving $2\pi rh$. Thus,

$$A = g(r, h) = 2 \cdot \text{Area of base} + \text{Area of side} = 2\pi r^2 + 2\pi rh.$$

Strategy to Investigate Functions of Two Variables: Vary One Variable at a Time

We can learn a great deal about functions of two or more variables by letting one variable vary at a time while holding the others fixed, thus obtaining a function of one variable.

The Wave

Suppose you are in a stadium where the audience is doing the wave. This is a ritual in which members of the audience stand up and sit down in such a way as to create a wave that moves around the stadium. Normally a single wave travels all the way around the stadium, but we assume there is a continuous sequence of waves. What sort of function will describe the motion of the audience? To keep things simple, we consider just one row of spectators. We consider the function which describes the motion of each individual in the row. This is a function of two variables: x (the seat number) and t (the time in seconds). For each value of x and t, we write $h(x, t)$ for the height (in feet) above the ground of the head of the spectator in seat number x at time t seconds. Suppose we are told that

$$h(x, t) = 5 + \cos(0.5x - t).$$

Example 4   (a)   Explain the significance of $h(x, 5)$ in terms of the wave. Find the period of $h(x, 5)$. What does this period represent?

(b)   Explain the significance of $h(2, t)$ in terms of the wave. Find the period of $h(2, t)$. What does this period represent?

Solution (a) Fixing $t = 5$ means we are taking a particular moment in time; letting x vary means we are looking along the whole row at that instant. Thus, the function $h(x, 5) = 5 + \cos(0.5x - 5)$ gives the heights along the row at the instant $t = 5$. Figure 11.2 shows the graph of $h(x, 5)$ which is a snapshot of the row at $t = 5$. The heights form a wave of period 4π, or about 12.6 seats. This period tells us that the length of the wave is about 13 seats.

(b) Fixing $x = 2$ means we are concentrating on the spectator in seat number 2; letting t vary means we are watching the motion of that spectator as time passes. Thus, the function $h(2, t)$ describes the motion of the spectator in seat 2 as a function of time. Figure 11.3 shows the graph of $h(2, t) = 5 + \cos(1 - t)$. Notice that the value of h varies between 4 feet and 6 feet as the spectator sits down and stands up. The period is 2π, or about 6.3 seconds. This period represents the time it takes for the spectator to stand up and sit down once.

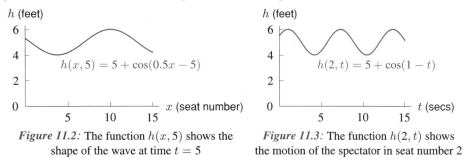

Figure 11.2: The function $h(x, 5)$ shows the shape of the wave at time $t = 5$

Figure 11.3: The function $h(2, t)$ shows the motion of the spectator in seat number 2

In general, the one-variable function $h(a, t)$ gives the motion of the spectator in seat a. Figure 11.3 shows the motion of the person in seat number 2. If we pick someone in another seat, we get a similar function, except that the graph may be shifted to the right or to the left.

Example 5 Show that the graph of $h(7, t)$ has the same shape as the graph of $h(2, t)$.

Solution The motion of the person in seat 7 is described by

$$h(7, t) = 5 + \cos(0.5(7) - t) = 5 + \cos(3.5 - t).$$

Since $h(2, t) = 5 + \cos(1 - t)$, we can rewrite $h(7, t)$ as

$$h(7, t) = 5 + \cos(1 + 2.5 - t) = 5 + \cos(1 - (t - 2.5)) = h(2, t - 2.5).$$

We see that $h(7, t)$ is the same function as $h(2, t)$, except that the t has been replaced by $t - 2.5$. Thus, the graph of $h(7, t)$ is the graph of $h(2, t)$ shifted 2.5 seconds to the right, which means 2.5 seconds later. This means the spectator in seat 7 stands up 2.5 seconds later than the person in seat 2. (See Figure 11.4.) This lag is what makes the wave travel around the stadium. If all the spectators stood up and sat down at the same time, there would be no wave.

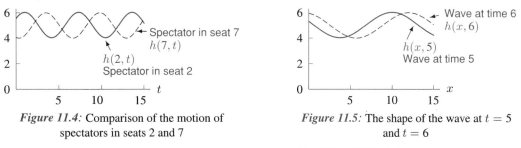

Figure 11.4: Comparison of the motion of spectators in seats 2 and 7

Figure 11.5: The shape of the wave at $t = 5$ and $t = 6$

Example 6   Use the result of Example 5 to find the speed of the wave.

Solution   Since the spectator in seat 7 does the same thing as the spectator in seat 2 but 2.5 seconds later, the wave has moved 5 seats in 2.5 seconds. Thus, the speed is $5/2.5 = 2$ seats per second.

Example 7   Use the functions $h(x, 5)$ and $h(x, 6)$ to show that the speed of the wave is 2 seats per second.

Solution   Figure 11.5 shows graphs of $h(x, 5) = 5 + \cos(0.5x - 5)$ and $h(x, 6) = 5 + \cos(0.5x - 6)$, that is, snapshots of the wave at $t = 5$ and $t = 6$. The shape of the wave at $t = 6$ is the same as the shape at $t = 5$, only shifted to the right by about two seats. Thus, the wave is moving at a rate of about 2 seats per second. To confirm that the speed is exactly 2 seats per second, we must use algebra. When $t = 5$, the equation of the wave is

$$h(x, 5) = 5 + \cos(0.5x - 5),$$

which has a peak where

$$0.5x - 5 = 0,$$

so, at

$$x = 10,$$

that is, in the 10th seat. One second later, at $t = 6$, the equation of the wave is

$$h = 5 + \cos(0.5x - 6),$$

which has a peak where

$$0.5x - 6 = 0,$$

so, at

$$x = 12,$$

that is, in the 12th seat. Thus, the wave moved 2 seats in one second.

The Beef Data

For a function given by a table of values, such as the beef consumption data, we allow one variable to vary at a time by looking at one row or one column. For example, to fix the income at 40, we look at the row $I = 40$. This row gives values of the function $f(40, p)$. Since I is fixed, we now have a function of one variable that shows how much beef is bought at various prices by people who earn $40,000 a year. Table 11.2 shows that $f(40, p)$ decreases as p increases. The other rows tell the same story; for each income, I, the consumption of beef goes down as the price, p, increases.

TABLE 11.2   *Beef consumption by households making $40,000*

p	3.00	3.50	4.00	4.50
$f(40, p)$	4.14	4.05	3.94	3.88

Weather Map

What happens on the weather map in Figure 11.1 when we allow only one variable at a time to vary? Suppose x represents miles west-east of Topeka and y represents miles north-south, and suppose we move along the west-east line through Topeka. We keep y fixed at 0 and let x vary. Along this line, the high temperature T goes from the 60s along the west coast, to the 70s in Nevada and Utah, to the 80s in Topeka, to the 90s just before the east coast, then returns to the 80s. A possible graph is shown in Figure 11.6. Other graphs are possible because we don't know for sure how the temperature varies between contours.

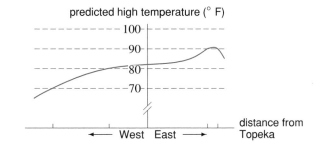

Figure 11.6: Predicted high temperature on an east-west line through Topeka

Problems for Section 11.1

Problems 1–3 refer to the weather map in Figure 11.1 on page 2.

1. Give the range of daily high temperatures for
   (a) Pennsylvania   (b) North Dakota   (c) California.

2. Sketch the graph of the predicted high temperature T on a line north-south through Topeka.

3. Sketch the graphs of the predicted high temperature on a north-south line and an east-west line through Boise.

For Problems 4–8 refer to Table 11.1 on page 3, where p is the price of beef and I is annual household income.

4. Make a table showing the amount of money, M, that the average household spends on beef (in dollars per household per week) as a function of the price of beef and household income.

5. Give tables for beef consumption as a function of p, with I fixed at $I = 20$ and $I = 100$. Give tables for beef consumption as a function of I, with p fixed at $p = 3.00$ and $p = 4.00$. Comment on what you see in the tables.

6. How does beef consumption vary as a function of household income if the price of beef is held constant?

7. Make a table of the proportion, P, of household income spent on beef per week as a function of price and income. (Note that P is the fraction of income spent on beef.)

8. Express P, the proportion of household income spent on beef per week, in terms of the original function $f(I, p)$ which gave consumption as a function of p and I.

9. Sketch the graph of the bank account function f in Example 2(a) on page 4, holding B fixed at three different values and letting only t vary. Then sketch the graph of f, holding t fixed at three different values and letting only B vary. Explain what you see.

10. You are planning a long driving trip and your principal cost will be gasoline.
    (a) Make a table showing how the daily fuel cost varies as a function of the price of gasoline (in dollars per gallon) and the number of gallons you buy each day.
    (b) If your car goes 30 miles on each gallon of gasoline, make a table showing how your daily fuel cost varies as a function of your daily travel distance and the price of gas.

11. Consider the acceleration due to gravity, g, at a height h above the surface of a planet of mass m.

    (a) If m is held constant, is g an increasing or decreasing function of h? Why?
    (b) If h is held constant, is g an increasing or decreasing function of m? Why?

12. The *temperature adjusted for wind-chill* is a temperature which tells you how cold it feels, as a result of the combination of wind and temperature. Table 11.3 shows the temperature adjusted for wind-chill as a function of wind speed and temperature.

TABLE 11.3   *Temperature adjusted for wind-chill (°F)*

Wind Speed (mph)	Temperature (°F)							
	35	30	25	20	15	10	5	0
5	33	27	21	16	12	7	0	−5
10	22	16	10	3	−3	−9	−15	−22
15	16	9	2	−5	−11	−18	−25	−31
20	12	4	−3	−10	−17	−24	−31	−39
25	8	1	−7	−15	−22	−29	−36	−44

    (a) If the temperature is 0°F and the wind speed is 15 mph, how cold does it feel?
    (b) If the temperature is 35°F, what wind speed makes it feel like 22°F?
    (c) If the temperature is 25°F, what wind speed makes it feel like 20°F?
    (d) If the wind is blowing at 15 mph, what temperature feels like 0°F?

13. Using Table 11.3, make tables of the temperature adjusted for wind-chill as a function of wind speed for temperatures of 20°F and 0°F.

14. Using Table 11.3, make tables of the temperature adjusted for wind-chill as a function of temperature for wind speeds of 5 mph and 20 mph.

15. Suppose the function for the stadium wave on page 4 was given by $h(x,t) = 5 + \cos(x - 2t)$. How does this wave compare with the original wave? What is the speed of this wave (in seats per second)?

16. Suppose the stadium wave on page 4 was moving in the opposite direction, right-to-left instead of left-to-right. Give a possible formula for h.

Problems 17–20 concern a vibrating guitar string. Suppose you pluck a guitar string and watch it vibrate. Snapshots of the guitar string at millisecond intervals are shown in Figure 11.7.

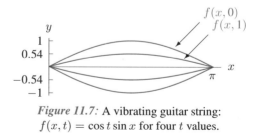

Figure 11.7: A vibrating guitar string:
$f(x,t) = \cos t \sin x$ for four t values.

Think of the guitar string stretched tight along the x-axis from $x = 0$ to $x = \pi$. Each point on the string has an x-value, $0 \le x \le \pi$. As the string vibrates, each point on the string moves back and forth on either side of the x-axis. Let $y = f(x,t)$ be the displacement at time t of the point on the string located x units from the left end. Then a possible formula for $y = f(x,t)$ is

$$y = f(x,t) = \cos t \sin x, \quad 0 \le x \le \pi, \quad t \text{ in milliseconds.}$$

17. (a) Sketch graphs of y versus x for fixed t values, $t = 0, \pi/4, \pi/2, 3\pi/4, \pi$.
     (b) Use your graphs to explain why this function could represent a vibrating guitar string.

18. Explain what the functions $f(x,0)$ and $f(x,1)$ represent in terms of the vibrating string.

19. Explain what the functions $f(0,t)$ and $f(1,t)$ represent in terms of the vibrating string.

20. Describe the motion of the guitar strings whose displacements are given by the following:
     (a) $y = g(x,t) = \cos 2t \sin x$ (b) $y = h(x,t) = \cos t \sin 2x$

11.2 A TOUR OF THREE-DIMENSIONAL SPACE

Cartesian Coordinates in Three-Space

Imagine three coordinate axes meeting at the *origin*: a vertical axis, and two horizontal axes at right angles to each other. (See Figure 11.8.) Think of the xy-plane as being horizontal, while the z-axis extends vertically above and below the plane. The labels x, y, and z show which part of each axis is positive; the other side is negative. We generally use *right-handed axes* in which looking down the positive z-axis gives the usual view of the xy-plane. We specify a point in 3-space by giving its coordinates (x, y, z) with respect to these axes. Think of the coordinates as instructions telling you how to get to the point; start at the origin, go x units along the x-axis, then y units in the direction parallel to the y-axis and finally z units in the direction parallel to the z-axis. The coordinates can be positive, zero or negative; a zero coordinate means "don't move in this direction," and a negative coordinate means "go in the negative direction parallel to this axis." For example, the origin has coordinates $(0,0,0)$, since we get there from the origin by doing nothing at all.

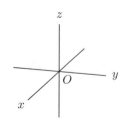

Figure 11.8: Coordinate axes in three-dimensional space

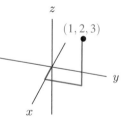

Figure 11.9: The point $(1,2,3)$ in 3-space

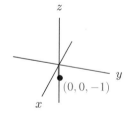

Figure 11.10: The point $(0,0,-1)$ in 3-space

Example 1   Describe the position of the points with coordinates $(1,2,3)$ and $(0,0,-1)$.

Solution   We get to the point $(1,2,3)$ by starting at the origin, going 1 unit along the x-axis, 2 units in the direction parallel to the y-axis, and 3 units up in the direction parallel to the z-axis. (See Figure 11.9.)

To get to $(0,0,-1)$, we don't move at all in the x and y directions, but move 1 unit in the negative z direction. So the point is on the negative z-axis. (See Figure 11.10.) You can check that the position of the point is independent of the order of the x, y, and z displacements.

Example 2   You start at the origin, go along the y-axis a distance of 2 units in the positive direction, and then move vertically upward a distance of 1 unit. What are the coordinates of your final position?

Solution   You started at the point $(0, 0, 0)$. When you went along the y-axis, your y-coordinate increased to 2. Moving vertically increased your z-coordinate to 1; your x-coordinate didn't change because you did not move in the x direction. So your final coordinates are $(0, 2, 1)$. (See Figure 11.11.)

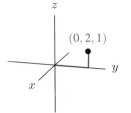

Figure 11.11: The point $(0, 2, 1)$ is reached by moving 2 along the y-axis and 1 upward

It is often helpful to picture a three dimensional coordinate system in terms of a room. The origin is a corner at floor level where two walls meet the floor. The z-axis is the vertical intersection of the two walls; the x- and y-axes are the intersections of each wall with the floor. Points with negative coordinates lie behind a wall in the next room or below the floor.

Graphing Equations in Three-Dimensional Space

We can graph equations involving the variables x, y, and z in three-dimensional space.

Example 3   What do the graphs of the equations $z = 0$, $z = 3$, and $z = -1$ look like?

Solution   Graphing an equation means drawing the set of all points in space whose coordinates satisfy the equation. So to graph $z = 0$ we need to visualize the set of points whose z-coordinate is zero. If the z-coordinate is 0, then we must be at the same vertical level as the origin, that is, we are in the horizontal plane containing the origin. So the graph of $z = 0$ is the middle plane in Figure 11.12. The graph of $z = 3$ is a plane parallel to the graph of $z = 0$, but three units above it. The graph of $z = -1$ is a plane parallel to the graph of $z = 0$, but one unit below it.

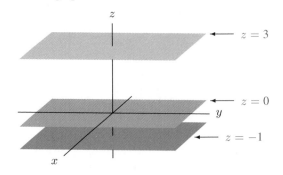

Figure 11.12: The planes $z = -1$, $z = 0$, and $z = 3$

The plane $z = 0$ contains the x- and y-coordinate axes, and hence is called the xy-coordinate plane, or xy-plane for short. There are two other coordinate planes. The yz-plane contains both the y- and the z-axes, and the xz-plane contains the x- and z-axes. (See Figure 11.13.)

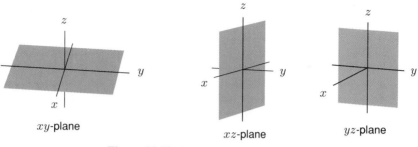

xy-plane xz-plane yz-plane

Figure 11.13: The three coordinate planes

Example 4   Which of the points $A = (1, -1, 0)$, $B = (0, 3, 4)$, $C = (2, 2, 1)$, and $D = (0, -4, 0)$ lies closest to the xz-plane? Which point lies on the y-axis?

Solution   The magnitude of the y-coordinate gives the distance to the xz-plane. The point A lies closest to that plane, because it has the smallest y-coordinate in magnitude. To get to a point on the y-axis, we move along the y-axis, but we don't move at all in the x or z directions. Thus, a point on the y-axis has both its x- and z-coordinates equal to zero. The only point of the four that satisfies this is D. (See Figure 11.14.)

In general, if a point has one of its coordinates equal to zero, it lies in one of the coordinate planes. If a point has two of its coordinates equal to zero, it lies on one of the coordinate axes.

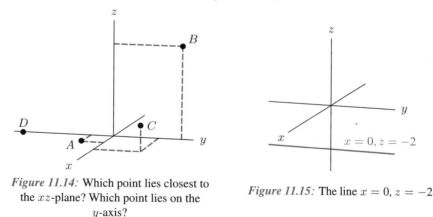

Figure 11.14: Which point lies closest to the xz-plane? Which point lies on the y-axis?

Figure 11.15: The line $x = 0$, $z = -2$

Example 5   You are 2 units below the xy-plane and in the yz-plane. What are your coordinates?

Solution   Since you are 2 units below the xy-plane, your z-coordinate is -2. Since you are in the yz-plane, your x-coordinate is 0; your y-coordinate can be anything. Thus, you are at the point $(0, y, -2)$. The set of all such points forms a line parallel to the y-axis, 2 units below the xy-plane, and in the yz-plane. (See Figure 11.15.)

Example 6   You are standing at the point $(4, 5, 2)$, looking at the point $(0.5, 0, 3)$. Are you looking up or down?

Solution   The point you are standing at has z-coordinate 2, whereas the point you are looking at has z-coordinate 3; hence you are looking up.

Example 7   Imagine that the yz-plane in Figure 11.15 is a page of this book. Describe the region behind the page.

Solution   The positive part of the x-axis pokes out of the page; moving in the positive x direction brings you out in front of the page. The region behind the page corresponds to negative values of x, and so it is the set of all points in three-dimensional space satisfying the inequality $x < 0$.

Distance

In 2-space, the formula for the distance between two points (x, y) and (a, b) is given by

$$\text{Distance} = \sqrt{(x - a)^2 + (y - b)^2}.$$

The distance between two points (x, y, z) and (a, b, c) in 3-space is represented by PG in Figure 11.16. The side PE is parallel to the x-axis, EF is parallel to the y-axis, and FG is parallel to the z-axis.

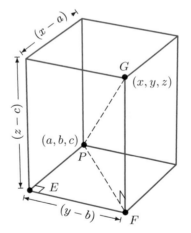

Figure 11.16: The diagonal PG gives the distance between the points (x, y, z) and (a, b, c)

Using Pythagoras' theorem twice gives

$$(PG)^2 = (PF)^2 + (FG)^2 = (PE)^2 + (EF)^2 + (FG)^2 = (x - a)^2 + (y - b)^2 + (z - c)^2.$$

Thus, a formula for the distance between the points (x, y, z) and (a, b, c) in 3-space is

$$\boxed{\text{Distance} = \sqrt{(x - a)^2 + (y - b)^2 + (z - c)^2}.}$$

Example 8   Find the distance between $(1, 2, 1)$ and $(-3, 1, 2)$.

Solution   The formula gives

$$\text{Distance} = \sqrt{(-3 - 1)^2 + (1 - 2)^2 + (2 - 1)^2} = \sqrt{18} = 4.24.$$

Example 9   Find an expression for the distance from the origin to the point (x, y, z).

Solution   The origin has coordinates $(0, 0, 0)$, so the distance from the origin to (x, y, z) is given by

$$\text{Distance} = \sqrt{(x - 0)^2 + (y - 0)^2 + (z - 0)^2} = \sqrt{x^2 + y^2 + z^2}.$$

Example 10   Find the equation for a sphere of radius 1 with center at the origin.

Solution   The sphere consists of all points (x, y, z) whose distance from the origin is 1, that is, which satisfy the equation

$$\sqrt{x^2 + y^2 + z^2} = 1.$$

This is an equation for the sphere. If we square both sides we get the equation in the form

$$x^2 + y^2 + z^2 = 1.$$

Note that this equation represents the *surface* of the sphere. The solid ball enclosed by the sphere is represented by the inequality $x^2 + y^2 + z^2 \leq 1$.

Problems for Section 11.2

1. Which of the points $A = (1.3, -2.7, 0)$, $B = (0.9, 0, 3.2)$, $C = (2.5, 0.1, -0.3)$ is closest to the yz-plane? Which one lies on the xz-plane? Which one is farthest from the xy-plane?

2. Which of the points $A = (23, 92, 48)$, $B = (-60, 0, 0)$, $C = (60, 1, -92)$ is closest to the yz-plane? Which one lies on the xz-plane? Which one is farthest from the xy-plane?

3. You are at the point $(-1, -3, -3)$, standing upright and facing the yz-plane. You walk 2 units forward, turn left, and walk for another 2 units. What is your final position? From the point of view of an observer looking at the coordinate system in Figure 11.8 on page 9, are you in front of or behind the yz-plane? Are you to the left or to the right of the xz-plane? Are you above or below the xy-plane?

4. You are at the point $(3, 1, 1)$ facing the yz-plane. Assume you are standing upright. You walk 2 units forward, turn left, and walk for another 2 units. What is your final position? From the point of view of an observer looking at the coordinate system in Figure 11.8 on page 9, are you in front of or behind the yz-plane? Are you to the left of or to the right of the xz-plane? Are you above or below the xy-plane?

Sketch graphs of the equations in Problems 5–7 in 3-space.

5. $x = -3$ 6. $y = 1$ 7. $z = 2$ and $y = 4$

8. Find a formula for the shortest distance between a point (a, b, c) and the y-axis.

9. Describe the set of points whose distance from the x-axis is 2.

10. Describe the set of points whose distance from the x-axis equals the distance from the yz-plane.

11. Which of the points $P = (1, 2, 1)$ and $Q = (2, 0, 0)$ is closest to the origin?

12. Which two of the three points $P_1 = (1, 2, 3)$, $P_2 = (3, 2, 1)$ and $P_3 = (1, 1, 0)$ are closest to each other?

13. A cube is located such that its top four corners have the coordinates $(-1, -2, 2)$, $(-1, 3, 2)$, $(4, -2, 2)$ and $(4, 3, 2)$. Give the coordinates of the center of the cube.

14. A rectangular solid lies with its length parallel to the y-axis, and its top and bottom faces parallel to the plane $z = 0$. If the center of the object is at $(1, 1, -2)$ and it has a length of 13, a height of 5 and a width of 6, give the coordinates of all eight corners and draw the figure labeling the eight corners.

15. Which of the points $P_1 = (-3, 2, 15)$, $P_2 = (0, -10, 0)$, $P_3 = (-6, 5, 3)$ and $P_4 = (-4, 2, 7)$ is closest to $P = (6, 0, 4)$?

16. On a set of x, y, and z axes oriented as in Figure 11.8 on page 9, draw a straight line through the origin, lying in the xz-plane and such that if you move along the line with your x-coordinate increasing, your z-coordinate is decreasing.

17. On a set of x, y and z axes oriented as in Figure 11.8 on page 9, draw a straight line through the origin, lying in the yz-plane and such that if you move along the line with your y-coordinate increasing, your z-coordinate is increasing.

18. Find the equation of the sphere of radius 5 centered at the origin.

19. Find the equation of the sphere of radius 5 centered at $(1, 2, 3)$.

20. Given the sphere

$$(x - 1)^2 + (y + 3)^2 + (z - 2)^2 = 4,$$

(a) Find the equations of the circles (if any) where the sphere intersects each coordinate plane.

(b) Find the points (if any) where the sphere intersects each coordinate axis.

11.3 GRAPHS OF FUNCTIONS OF TWO VARIABLES

How do You Visualize a Function of Two Variables?

The weather map on page 2 is one way of visualizing a function of two variables. In this section we see how to visualize a function of two variables in another way, using a surface in 3-space.

The Graph of a Function and How to Plot One

For a function of one variable, $y = f(x)$, the graph of f is the set of all points (x, y) in 2-space such that $y = f(x)$. In general, these points lie on a curve in the plane. When a computer or calculator graphs f, it approximates by plotting points in the xy-plane and joining consecutive points by line segments. The more points, the better the approximation.

Now consider a function of two variables.

The **graph** of a function of two variables, f, is the set of all points (x, y, z) such that $z = f(x, y)$. In general, the graph of a function of two variables is a surface in 3-space.

Plotting the Graph of the Function $f(x, y) = x^2 + y^2$

To sketch the graph of f we connect points as for a function of one variable. We first make a table of values of f, such as in Table 11.4.

TABLE 11.4   *Table of values of $f(x, y) = x^2 + y^2$*

		-3	-2	-1	0	1	2	3
	-3	18	13	10	9	10	13	18
	-2	13	8	5	4	5	8	13
	-1	10	5	2	1	2	5	10
x	0	9	4	1	0	1	4	9
	1	10	5	2	1	2	5	10
	2	13	8	5	4	5	8	13
	3	18	13	10	9	10	13	18

Now we plot points. For example, we plot $(1, 2, 5)$ because $f(1, 2) = 5$ and we plot $(0, 2, 4)$ because $f(0, 2) = 4$. Then, we connect the points corresponding to the rows and columns in the table. The result is called a *wire-frame* picture of the graph. Filling in between the wires gives a surface. That is the way a computer drew the graphs in Figure 11.17 and 11.18. The more points that are plotted, the more the picture looks like the surface in Figure 11.19.

You should check to see if the sketches make sense. Notice that the graph goes through the origin since $(x, y, z) = (0, 0, 0)$ satisfies $z = x^2 + y^2$. Observe that if x is held fixed and y is allowed to vary, the graph dips down and then goes back up, just like the entries in the rows of Table 11.4. Similarly, if y is held fixed and x is allowed to vary, the graph dips down and then goes back up, just like the columns of Table 11.4.

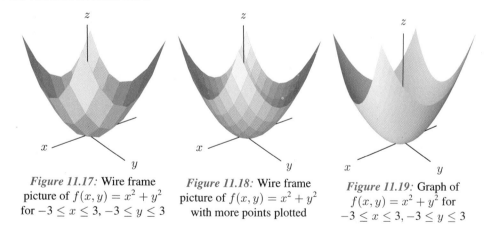

Figure 11.17: Wire frame picture of $f(x, y) = x^2 + y^2$ for $-3 \le x \le 3, -3 \le y \le 3$

Figure 11.18: Wire frame picture of $f(x, y) = x^2 + y^2$ with more points plotted

Figure 11.19: Graph of $f(x, y) = x^2 + y^2$ for $-3 \le x \le 3, -3 \le y \le 3$

New Graphs from Old

We can use the graph of a function to visualize the graphs of related functions.

Example 1   Let $f(x, y) = x^2 + y^2$. Describe in words the graphs of the following functions:
   (a) $g(x, y) = x^2 + y^2 + 3$,   (b) $h(x, y) = 5 - x^2 - y^2$,   (c) $k(x, y) = x^2 + (y - 1)^2$.

Solution   We know from Figure 11.19 that the graph of f is a bowl with its vertex at the origin. From this we can work out what the graphs of g, h, and k will look like.

(a)   The function $g(x, y) = x^2 + y^2 + 3 = f(x, y) + 3$, so the graph of g is the graph of f, but raised by 3 units. See Figure 11.20.

(b)   Since $-x^2 - y^2$ is the negative of $x^2 + y^2$, the graph of $-x^2 - y^2$ is an upside down bowl. Thus, the graph of $h(x, y) = 5 - x^2 - y^2 = 5 - f(x, y)$ looks like an upside down bowl with vertex at $(0, 0, 5)$, as in Figure 11.21.

(c)   The graph of $k(x, y) = x^2 + (y - 1)^2 = f(x, y - 1)$ is a bowl with vertex at $x = 0$, $y = 1$, since that is where $k(x, y) = 0$, as in Figure 11.22.

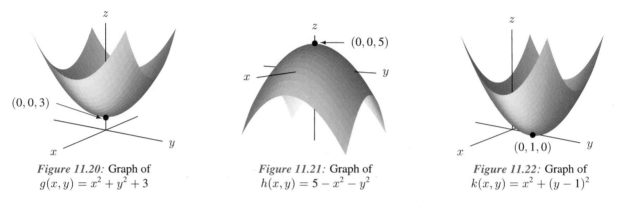

Figure 11.20: Graph of
$g(x, y) = x^2 + y^2 + 3$

Figure 11.21: Graph of
$h(x, y) = 5 - x^2 - y^2$

Figure 11.22: Graph of
$k(x, y) = x^2 + (y - 1)^2$

Example 2   Describe the graph of $G(x, y) = e^{-(x^2 + y^2)}$. What symmetry does it have?

Solution   Since the exponential function is always positive, the graph lies entirely above the xy-plane. From the graph of $x^2 + y^2$ we see that $x^2 + y^2$ is zero at the origin and gets larger as we move farther from the origin in any direction. Thus, $e^{-(x^2 + y^2)}$ is 1 at the origin, and gets smaller as we move away from the origin in any direction. It can't go below the xy-plane; instead it flattens out, getting closer and closer to the plane. We say the surface is *asymptotic* to the xy-plane. (See Figure 11.23.)

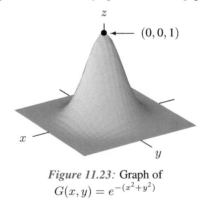

Figure 11.23: Graph of
$G(x, y) = e^{-(x^2 + y^2)}$

Now consider a point (x, y) on the circle $x^2 + y^2 = r^2$. Since

$$G(x, y) = e^{-(x^2 + y^2)} = e^{-r^2},$$

the value of the function G is the same at all points on this circle. Thus, we say the graph of G has *circular symmetry*.

Cross-Sections and the Graph of a Function

We have seen that a good way to analyze a function of two variables is to let one variable vary at a time while the other is kept fixed.

> For a function $f(x, y)$, the function we get by holding x fixed and letting y vary is called a **cross-section** of f with x fixed. The graph of the cross-section of $f(x, y)$ with $x = c$ is the curve, or cross-section, we get by intersecting the graph of f with the plane $x = c$. We define a cross-section of f with y fixed similarly.

For example, the cross-section of $f(x, y) = x^2 + y^2$ with $x = 2$ is $f(2, y) = 4 + y^2$. The graph of this cross-section is the curve we get by intersecting the graph of f with the plane perpendicular to the x-axis at $x = 2$. (See Figure 11.24.)

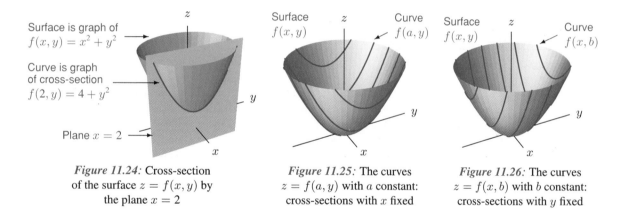

Figure 11.24: Cross-section of the surface $z = f(x, y)$ by the plane $x = 2$

Figure 11.25: The curves $z = f(a, y)$ with a constant: cross-sections with x fixed

Figure 11.26: The curves $z = f(x, b)$ with b constant: cross-sections with y fixed

Figure 11.25 shows graphs of other cross-sections of f with x fixed; Figure 11.26 shows graphs of cross-sections with y fixed.

Example 3   Describe the cross-sections of the function $g(x, y) = x^2 - y^2$ with y fixed and then with x fixed. Use these cross-sections to describe the shape of the graph of g.

Solution   The cross-sections with y fixed at $y = b$ are given by

$$z = g(x, b) = x^2 - b^2.$$

Thus, each cross-section with y fixed gives a parabola opening upwards, with minimum $z = -b^2$. The cross-sections with x fixed are of the form

$$z = g(a, y) = a^2 - y^2,$$

which are parabolas opening downwards with a maximum of $z = a^2$. (See Figures 11.27 and 11.28) The graph of g is shown in Figure 11.29. Notice the upward opening parabolas in the x-direction and the downward opening parabolas in the y-direction. We say that the surface is *saddle-shaped*.

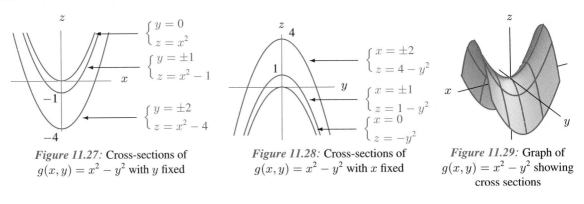

Figure 11.27: Cross-sections of
$g(x, y) = x^2 - y^2$ with y fixed

Figure 11.28: Cross-sections of
$g(x, y) = x^2 - y^2$ with x fixed

Figure 11.29: Graph of
$g(x, y) = x^2 - y^2$ showing
cross sections

Linear Functions

Linear functions are central to single variable calculus; they are equally important in multivariable calculus. You may be able to guess the shape of the graph of a linear function of two variables. (It's a plane.) Let's look at an example.

Example 4   Describe the graph of $f(x, y) = 1 + x - y$.

Solution   The plane $x = a$ is vertical and parallel to the yz-plane. Thus, the cross-section with $x = a$ is the line $z = 1 + a - y$ which slopes downward in the y-direction. Similarly, the plane $y = b$ is parallel to the xz-plane. Thus, the cross-section with $y = b$ is the line $z = 1 + x - b$ which slopes upward in the x-direction. Since all the cross-sections are lines, you might expect the graph to be a flat plane, sloping down in the y-direction and up in the x-direction. This is indeed the case. (See Figure 11.30.)

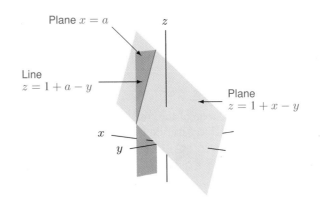

Figure 11.30: Graph of the plane $z = 1 + x - y$ showing
cross-section with $x = a$

When One Variable is Missing: Cylinders

Suppose we graph an equation like $z = x^2$ which has one variable missing. What does the surface look like? Since y is missing from the equation, the cross-sections with y fixed are all the same parabola, $z = x^2$. Letting y vary up and down the y-axis, this parabola sweeps out the trough-shaped surface shown in Figure 11.31. The cross-sections with x fixed are horizontal lines, obtained by cutting the surface by a plane perpendicular to the x-axis.

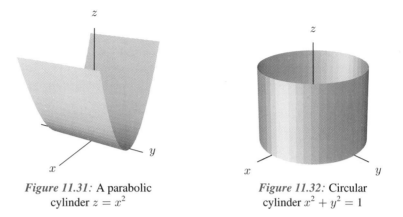

Figure 11.31: A parabolic
cylinder $z = x^2$

Figure 11.32: Circular
cylinder $x^2 + y^2 = 1$

This surface is called a *parabolic cylinder*, because it is formed from a parabola in the same way that an ordinary cylinder is formed from a circle; it has a parabolic cross-section instead of a circular one.

Example 5   Graph the equation $x^2 + y^2 = 1$ in 3-space.

Solution   Although the equation $x^2 + y^2 = 1$ does not represent a function, the surface representing it can be graphed by the method used for $z = x^2$. The graph of $x^2 + y^2 = 1$ in the xy-plane is a circle. Since z does not appear in the equation, the intersection of the surface with any horizontal plane will be the same circle $x^2 + y^2 = 1$. Thus, the surface is the cylinder shown in Figure 11.32.

Problems for Section 11.3

1. The surface in Figure 11.33 is the graph of the function $z = f(x, y)$ for positive x and y.

    (a) Suppose y is fixed and positive. Does z increase or decrease as x increases? Sketch a graph of z against x.

    (b) Suppose x is fixed and positive. Does z increase or decrease as y increases? Sketch a graph of z against y.

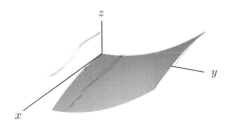

Figure 11.33

2. Match the following descriptions of a company's success with the graphs in Figure 11.34.

    (a) Our success is measured in dollars, plain and simple. More hard work won't hurt, but it also won't help.

    (b) No matter how much money or hard work we put into the company, we just couldn't make a go of it.

    (c) Although we aren't always totally successful, it seems that the amount of money invested doesn't matter. As long as we put hard work into the company our success will increase.

    (d) The company's success is based on both hard work and investment.

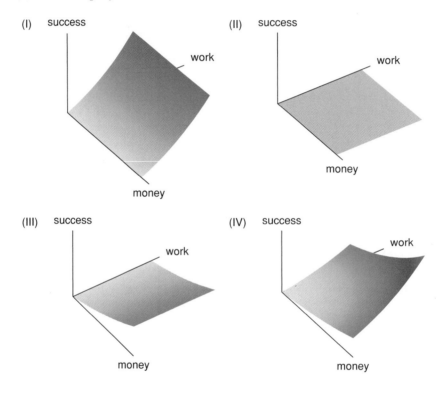

Figure 11.34

3. For each of the following functions, decide whether it could be a bowl, a plate, or neither. Consider a plate to be any fairly flat surface and a bowl to be anything that could hold water, assuming the positive z-axis is up.

    (a) $z = x^2 + y^2$ (b) $z = 1 - x^2 - y^2$ (c) $x + y + z = 1$
    (d) $z = -\sqrt{5 - x^2 - y^2}$ (e) $z = 3$

4. For each function in Problem 3 sketch cross-sections:

    (i) With x fixed at $x = 0$ and $x = 1$.
    (ii) With y fixed at $y = 0$ and $y = 1$.

5. Match the following functions with their graphs in Figure 11.35.

    (a) $z = \dfrac{1}{x^2 + y^2}$ (b) $z = -e^{-x^2 - y^2}$ (c) $z = x + 2y + 3$
    (d) $z = -y^2$ (e) $z = x^3 - \sin y$.

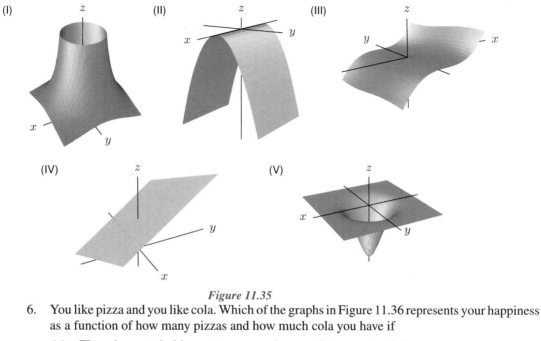

Figure 11.35

6. You like pizza and you like cola. Which of the graphs in Figure 11.36 represents your happiness as a function of how many pizzas and how much cola you have if

(a) There is no such thing as too many pizzas and too much cola?
(b) There is such a thing as too many pizzas or too much cola?
(c) There is such a thing as too much cola but no such thing as too many pizzas?

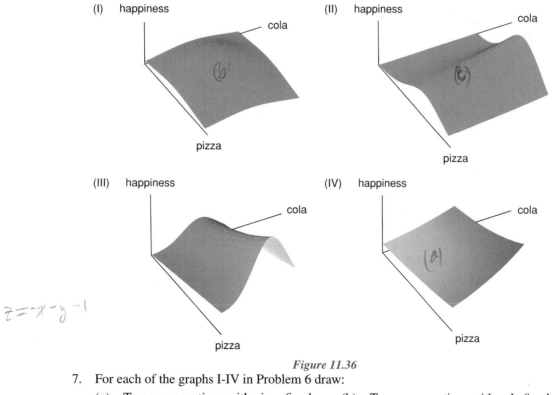

$z = -x - y - 1$

Figure 11.36

7. For each of the graphs I-IV in Problem 6 draw:

(a) Two cross-sections with pizza fixed   (b) Two cross-sections with cola fixed.

8. Figure 11.37 contains graphs of the parabolas $z = f(x, b)$ for $b = -2, -1, 0, 1, 2$. Which of the graphs of $z = f(x, y)$ in Figure 11.38 best fits this information?

Figure 11.37

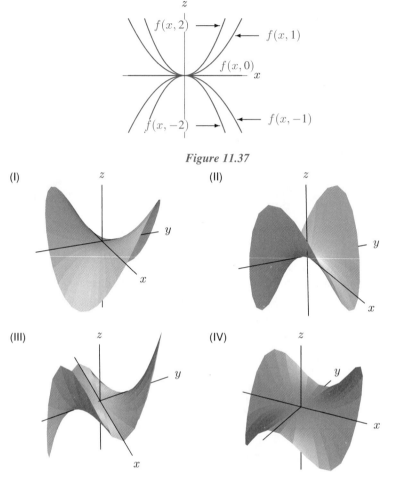

Figure 11.38

9. Imagine a single wave traveling along a canal. Suppose x is the distance along the canal from the middle, t is the time, and z is the height of the water above the equilibrium level. The graph of z as a function of x and t is shown in Figure 11.39.

   (a) Draw the profile of the wave for $t = -1, 0, 1, 2$. (Show the x-axis to the right and the z-axis vertically.)

   (b) Is the wave traveling in the direction of increasing or decreasing x?

   (c) Sketch a surface representing a wave traveling in the opposite direction.

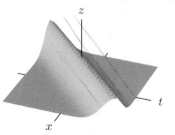

Figure 11.39

10. Describe in words the cross-sections with t fixed and the cross-sections with x fixed of the vibrating guitar string function

$$f(x, t) = \cos t \sin x, \quad 0 \le x \le \pi, \quad 0 \le t \le 2\pi$$

on page 8. Explain the relation of these cross-sections to the graph of f.

11. Use a computer or calculator to draw the graph of the vibrating guitar string function:

$$g(x, t) = \cos t \sin 2x, \quad 0 \le x \le \pi, \quad 0 \le t \le 2\pi.$$

Relate the shape of the graph to the cross-sections with t fixed and those with x fixed.

12. Use a computer or a calculator to draw the graph of the traveling wave function:

$$h(x, t) = 3 + \cos(x - 0.5t), \quad 0 \le x \le 2\pi, \quad 0 \le t \le 2\pi.$$

Relate the shape of the graph to the cross-sections with t fixed and those with x fixed.

13. Consider the function f given by $f(x, y) = y^3 + xy$. Draw graphs of cross-sections with:
    (a) x fixed at $x = -1$, $x = 0$, and $x = 1$.   (b) y fixed at $y = -1$, $y = 0$, and $y = 1$.

14. A swinging pendulum consists of a mass at the end of a string. At one moment the string makes an angle x with the vertical and the mass has speed y. At that time, the energy, E, of the pendulum is given by the expression[2]

$$E = 1 - \cos x + \frac{y^2}{2}.$$

    (a) Consider the surface representing the energy. Sketch a cross-section of the surface:
        (i)   Perpendicular to the x-axis at $x = c$.
        (ii) Perpendicular to the y-axis at $y = c$.
    (b) For each of the graphs in Figures 11.40 and 11.41 use your answer to part (a) to decide which is the x-axis and which is the y-axis and to put reasonable units on each one.

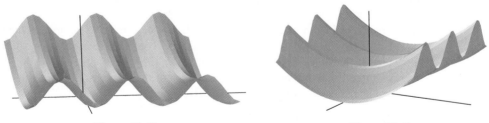

Figure 11.40   *Figure 11.41*

11.4 CONTOUR DIAGRAMS

The surface which represents a function of two variables often gives a good idea of the function's general behavior—for example, whether it is increasing or decreasing as one of the variables increases. However it is difficult to read numerical values off a surface and it can be hard to see all of the function's behavior from a surface. Thus, functions of two variables are often represented by contour diagrams like the weather map on page 2. Contour diagrams have the additional advantage that they can be extended to functions of three variables.

[2]Adapted from *Calculus in Context*, by James Callahan, Kenneth Hoffman, (New York: W.H. Freeman, 1995)

Topographical Maps

One of the most common examples of a contour diagram is a topographical map like that shown in Figure 11.42. It gives the elevation in the region and is a good way of getting an overall picture of the terrain: where the mountains are, where the flat areas are. Such topographical maps are frequently colored green at the lower elevations and brown, red, or white at the higher elevations.

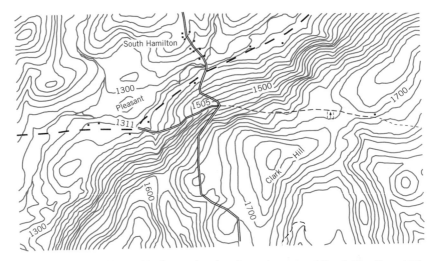

Figure 11.42: A topographical map showing the region around South Hamilton, NY

The curves on a topographical map that separate lower elevations from higher elevations are called *contour lines* because they outline the contour or shape of the land.[3] Because every point along the same contour has the same elevation, contour lines are also called *level curves* or *level sets*. The more closely spaced the contours, the steeper the terrain; the more widely spaced the contours, the flatter the terrain. (Provided, of course, that the elevation between contours varies by a constant amount.) Certain features have distinctive characteristics. A mountain peak is typically surrounded by contour lines like those in Figure 11.43. A pass in a range of mountains may have contours that look like Figure 11.44. A long valley has parallel contour lines indicating the rising elevations on both sides of the valley (see Figure 11.45); a long ridge of mountains has the same type of contour lines, only the elevations decrease on both sides of the ridge. Notice that the elevation numbers on the contour lines are as important as the curves themselves.

Figure 11.43: Mountain peak

Figure 11.44: Pass between two mountains

Figure 11.45: Long valley

Figure 11.46: Impossible contour lines

There are some things contour lines cannot do. Two contours corresponding to different elevations cannot cross each other as shown in Figure 11.46. If they did, the point of intersection of the two curves would have two different elevations, which is impossible (assuming the terrain has no overhangs). We will often follow the convention of drawing contours for equally spaced values of z.

---

[3]In fact they are usually not straight lines, but curves. They may also be in disconnected pieces.

Corn Production

Contour maps can also be useful to display information about a function of two variables without reference to a surface. Consider how to represent the effect of different weather conditions on US corn production. What would happen if the average temperature were to increase (due to global warming, for example)? What would happen if the rainfall were to decrease (due to a drought)? One way of estimating the effect of these climatic changes is to use Figure 11.47. This is a contour diagram giving the corn production $f(R, T)$ in the United States as a function of the total rainfall, R, in inches, and average temperature, T, in degrees Fahrenheit, during the growing season.[4] Suppose at the present time, $R = 15$ inches and $T = 76°$F. Production is measured as a percentage of the present production; thus, the contour through $R = 15$, $T = 76$ has value 100, that is, $f(15, 76) = 100$.

Example 1   Use Figure 11.47 to estimate $f(18, 78)$ and $f(12, 76)$ and explain the answer in terms of corn production.

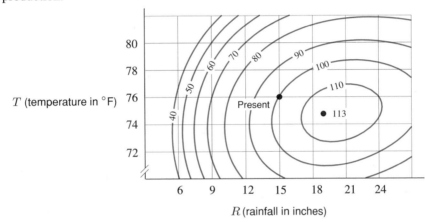

Figure 11.47: Corn production as a function of rainfall and temperature

Solution   The point with R-coordinate 18 and T-coordinate 78 is on the contour $C = 100$, so $f(18, 78) = 100$. This means that if the annual rainfall were 18 inches and the temperature were 78°F, the country would produce about the same amount of corn as at present, although it would be wetter and warmer than it is now.

   The point with R-coordinate 12 and T-coordinate 76 is about halfway between the $C = 80$ and the $C = 90$ contours, so $f(12, 76) \approx 85$. This means that if the rainfall fell to 12 inches and the temperature stayed at 76°, then corn production would drop to about 85% of what it is now.

Example 2   Describe in words the cross-sections with T and R constant through the point representing present conditions. Give a common sense explanation of your answer.

Solution   To see what happens to corn production if the temperature stays fixed at 76°F but the rainfall changes, look along the horizontal line $T = 76$. Starting from the present and moving left along the line $T = 76$, the values on the contours decrease. In other words, if there is a drought, corn production decreases. Conversely, as rainfall increases, that is, as we move from the present to the right along the line $T = 76$, corn production increases, reaching a maximum of more than 110% when $R = 21$, and then decreases (too much rainfall floods the fields).

[4]Adapted from S. Beaty and R. Healy, *The Future of American Agriculture*, Scientific American, Vol. 248, No.2, February, 1983

If, instead, rainfall remains at the present value and temperature increases, we move up the vertical line $R = 15$. Under these circumstances corn production decreases; a $2°$ increase causes a 10% drop in production. This makes sense since hotter temperatures lead to greater evaporation and hence drier conditions, even with rainfall constant at 15 inches. Similarly, a decrease in temperature leads to a very slight increase in production, reaching a maximum of around 102% when $T = 74$, followed by a decrease (the corn won't grow if it is too cold).

Contour Diagrams and Graphs

Contour diagrams and graphs are two different ways of representing a function of two variables. How do we go from one to the other? In the case of the topographical map, the contour diagram was created by joining all the points at the same height on the surface and dropping the curve into the xy-plane.

How do we go the other way? Suppose we wanted to plot the surface representing the corn production function $C = f(R, T)$ given by the contour diagram in Figure 11.47. Along each contour the function has a constant value; if we take each contour and lift it above the plane to a height equal to this value, we get the surface in Figure 11.48.

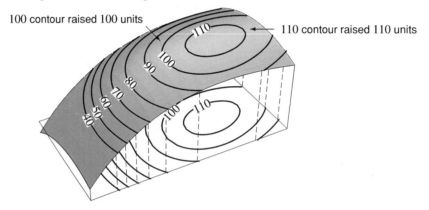

Figure 11.48: Getting the graph of the corn yield function from the contour diagram

Notice that the raised contours are the curves we get by slicing the surface horizontally. In general, we have the following result:

> Contour lines, or level curves, are obtained from a surface by slicing it with horizontal planes.

Finding Contours Algebraically

Algebraic equations for the contours of a function f are easy to find if we have a formula for $f(x, y)$. Suppose the surface has equation

$$z = f(x, y).$$

A contour is obtained by slicing the surface with a horizontal plane with equation $z = c$. Thus, the equation for the contour at height c is given by:

$$f(x, y) = c.$$

Example 3   Find equations for the contours of $f(x, y) = x^2 + y^2$ and draw a contour diagram for f. Relate the contour diagram to the graph of f.

Solution The contour at height c is given by

$$f(x,y) = x^2 + y^2 = c.$$

This is a contour only for $c \geq 0$, For $c > 0$ it is a circle of radius $\sqrt{c}$. For $c = 0$, it is a single point (the origin). Thus, the contours at an elevation of $c = 1, 2, 3, 4, \ldots$ are all circles centered at the origin of radius 1, $\sqrt{2}$, $\sqrt{3}$, 2, ... The contour diagram is shown in Figure 11.49. The bowl–shaped graph of f is shown in Figure 11.50. Notice that the graph of f gets steeper as we move further away from the origin. This is reflected in the fact that the contours become more closely packed as we move further from the origin; for example, the contours for $c = 6$ and $c = 8$ are closer together than the contours for $c = 2$ and $c = 4$.

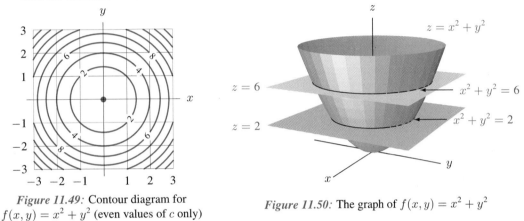

Figure 11.49: Contour diagram for
$f(x,y) = x^2 + y^2$ (even values of c only)

Figure 11.50: The graph of $f(x,y) = x^2 + y^2$

Example 4 Draw a contour diagram for $f(x,y) = \sqrt{x^2 + y^2}$ and relate it to the graph of f.

Solution The contour at level c is given by

$$f(x,y) = \sqrt{x^2 + y^2} = c$$

For $c > 0$ this is a circle, just as in the previous example, but here the radius is c instead of $\sqrt{c}$. For $c = 0$, it is the origin. Thus, if the level c increases by 1, the radius of the contour increases by 1. This means the contours are equally spaced concentric circles (see Figure 11.51) which do not become more closely packed further from the origin. Thus, the graph of f has the same constant slope as we move away from the origin (see Figure 11.52), making it a cone rather than a bowl.

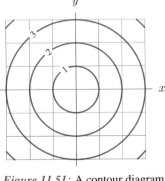

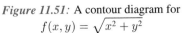

Figure 11.51: A contour diagram for
$f(x,y) = \sqrt{x^2 + y^2}$

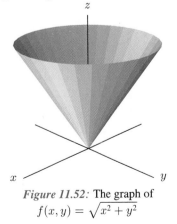

Figure 11.52: The graph of
$f(x,y) = \sqrt{x^2 + y^2}$

In both of the previous examples the level curves are concentric circles because the surfaces have circular symmetry. Any function of two variables which depends only on the quantity $(x^2 + y^2)$ has such symmetry: for example, $G(x, y) = e^{-(x^2+y^2)}$ or $H(x, y) = \sin(\sqrt{x^2 + y^2})$.

Example 5   Draw a contour diagram for $f(x, y) = 2x + 3y + 1$.

Solution   The contour at level c has equation $2x + 3y + 1 = c$. Rewriting this as $y = -(2/3)x + (c-1)/3$, we see that the contours are parallel lines with slope $-2/3$. The y-intercept for the contour at level c is $(c-1)/3$; each time c increases by 3, the y-intercept moves up by 1. The contour diagram is shown in Figure 11.53.

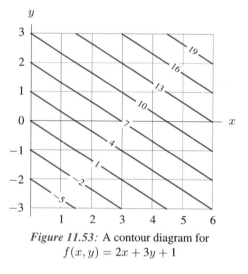

Figure 11.53: A contour diagram for
$f(x, y) = 2x + 3y + 1$

Contour Diagrams and Tables

Sometimes we can get an idea of what the contour diagram of a function looks like from its table.

Example 6   Relate the values of $f(x, y) = x^2 - y^2$ in Table 11.5 to its contour diagram in Figure 11.54.

TABLE 11.5   *Table of values of* $f(x, y) = x^2 - y^2$

3	0	−5	−8	−9	−8	−5	0
2	5	0	−3	−4	−3	0	5
1	8	3	0	−1	0	3	8
0	9	4	1	0	1	4	9
−1	8	3	0	−1	0	3	8
−2	5	0	−3	−4	−3	0	5
−3	0	−5	−8	−9	−8	−5	0
	−3	−2	−1	0	1	2	3

(left column labeled y, bottom row labeled x)

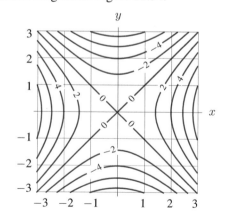

Figure 11.54: Contour map of $f(x, y) = x^2 - y^2$

Solution   One striking feature of the values in Table 11.5 is the zeros along the diagonals. This occurs because $x^2 - y^2 = 0$ along the lines $y = x$ and $y = -x$. So the $z = 0$ contour consists of these two lines. In the triangular region of the table that lies to the right of both diagonals, the entries are positive. To the left of both diagonals, the entries are also positive. Thus, in the contour diagram, the positive contours will lie in the triangular regions to the right and left of the lines $y = x$ and $y = -x$. Further, the table shows that the numbers on the left are the same as the numbers on the right; thus, each contour will have two pieces, one on the left and one on the right. See Figure 11.54. As we move away from the origin along the x-axis, we cross contours corresponding to successively larger values. On the saddle-shaped graph of $f(x, y) = x^2 - y^2$ shown in Figure 11.55, this corresponds to climbing out of the saddle along one of the ridges. Similarly, the negative contours occur in pairs in the top and bottom triangular regions; the values get more and more negative as we go out along the y-axis. This corresponds to descending from the saddle along the valleys that are submerged below the xy-plane in Figure 11.55. Notice that we could also get the contour diagram by graphing the family of hyperbolas $x^2 - y^2 = 0, \pm 2, \pm 4, \ldots$.

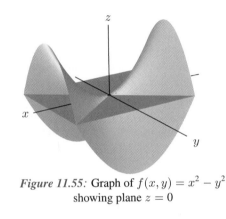

Figure 11.55: Graph of $f(x, y) = x^2 - y^2$
showing plane $z = 0$

Using Contour Diagrams: The Cobb-Douglas Production Function

Suppose you are running a small printing business, and decide to expand because you have more orders than you can handle. How should you expand? Should you start a night-shift and hire more workers? Should you buy more expensive but faster computers which will enable the current staff to keep up with the work? Or should you do some combination of the two?

Obviously, the way such a decision is made in practice involves many other considerations— such as whether you could get a suitably trained night shift, or whether there are any faster computers available. Nevertheless, you might model the quantity, P, of work produced by your business as a function of two variables: your total number, N, of workers, and the total value, V, of your equipment.

How would you expect such a production function to behave? In general, having more equipment and more workers enables you to produce more. However, increasing equipment without increasing the number of workers will increase production a bit, but not beyond a point. (If equipment is already lying idle, having more of it won't help.) Similarly, increasing the number of workers without increasing equipment will increase production, but not past the point where the equipment is fully utilized, as any new workers would have no equipment available to them.

Example 7   Explain why the contour diagram in Figure 11.56 does not model the behavior expected of the production function, whereas the contour diagram in Figure 11.57 does.

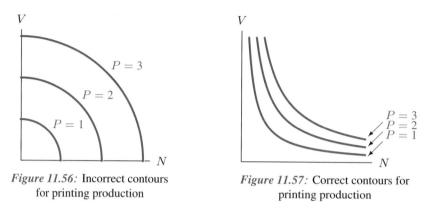

Figure 11.56: Incorrect contours for printing production

Figure 11.57: Correct contours for printing production

Solution   Look at the contour diagram in Figure 11.56. Fixing V at a particular value and letting N increase means moving to the right on the contour diagram. As you do so, you cross contours with larger and larger P values, meaning that production increases indefinitely. On the other hand, in Figure 11.57, as you move in the same direction you find yourself moving nearly parallel to the contours, crossing them less and less frequently. Therefore, production increases more and more slowly as N increases while V is held fixed. Similarly, if you hold N fixed and let V increase, the contour diagram in Figure 11.56 shows production increasing at a steady rate, whereas Figure 11.57 shows production increasing, but at a decreasing rate. Thus, Figure 11.57 fits the expected behavior of the production function best.

Formula for a Production Function

Production functions with the qualitative behavior we want are often approximated by formulas of the form

$$P = f(N, V) = cN^\alpha V^\beta$$

where P is the total quantity produced and c, α, and β are positive constants with $0 < \alpha < 1$ and $0 < \beta < 1$.

Example 8   Show that the contours of the function $P = cN^\alpha V^\beta$ have approximately the shape of the contours in Figure 11.57.

Solution   The contours are the curves where P is equal to a constant value, say P_0, that is, where

$$cN^\alpha V^\beta = P_0.$$

Solving for V we get

$$V = \left(\frac{P_0}{c}\right)^{1/\beta} N^{-\alpha/\beta}.$$

Thus, V is a power function of N with a negative exponent, and hence its graph has the shape shown in Figure 11.57.

The Cobb-Douglas Production Model

In 1928, Cobb and Douglas used a similar function to model the production of the entire US economy in the first quarter of this century. Using government estimates of P, the total yearly production between 1899 and 1922, of K, the total capital investment over the same period, and of L, the total labor force, they found that P was well approximated by the *Cobb-Douglas production function*

$$P = 1.01 L^{0.75} K^{0.25}.$$

This function turned out to model the US economy surprisingly well, both for the period on which it was based, and for some time afterwards.

Problems for Section 11.4

For the functions in Problems 1–9, sketch a contour diagram with at least four labeled contours. Describe in words the contours and how they are spaced.

1. $f(x, y) = x + y$
2. $f(x, y) = xy$
3. $f(x, y) = x^2 + y^2$
4. $f(x, y) = 3x + 3y$
5. $f(x, y) = -x^2 - y^2 + 1$
6. $f(x, y) = x^2 + 2y^2$
7. $f(x, y) = \sqrt{x^2 + 2y^2}$
8. $f(x, y) = y - x^2$
9. $f(x, y) = \cos \sqrt{x^2 + y^2}$

10. Figure 11.58 is a contour diagram of the monthly payment on a 5-year car loan as a function of the interest rate and the amount you borrow. Suppose that the interest rate is 13% and that you decide to borrow $6,000.

    (a) What is your monthly payment?
    (b) If interest rates drop to 11%, how much more can you borrow without increasing your monthly payment?
    (c) Make a table of how much you can borrow, without increasing your monthly payment, as a function of the interest rate.

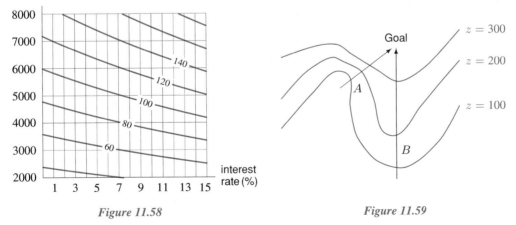

Figure 11.58   Figure 11.59

11. Figure 11.59 shows a contour map of a hill with two paths, A and B.

    (a) On which path, A or B, will you have to climb more steeply?
    (b) On which path, A or B, will you probably have a better view of the surrounding countryside? (Assuming trees do not block your view.)
    (c) Alongside which path is there more likely to be a stream?

12. Each of the contour diagrams in Figure 11.60 shows population density in a certain region. Choose the contour diagram that best corresponds to each of the following situations. Many different matchings are possible. Pick any reasonable one and justify your choice.

    (a) The center of the diagram is a city.
    (b) The center of the diagram is a lake.
    (c) The center of the diagram is a power plant.

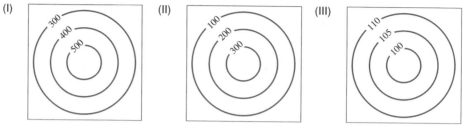

Figure 11.60

For each of the surfaces in Problems 13–15, sketch a possible contour diagram, marked with reasonable z-values. (Note: There are many possible answers.)

13.   14.   15.

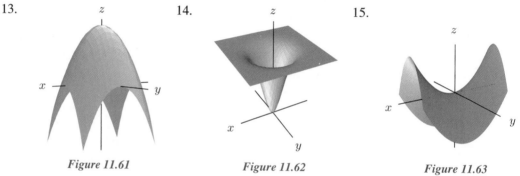

Figure 11.61   *Figure 11.62*   *Figure 11.63*

16. Figure 11.64 shows the density of the fox population P (in foxes per square kilometer) for southern England. Draw two different cross-sections along a north-south line and two different cross-sections along an east-west line of the population density P.

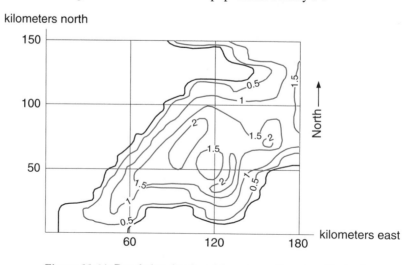

Figure 11.64: Population density of foxes in southwestern England

17. Use a computer or a calculator to sketch a contour diagram for the vibrating string function

$$f(x, t) = \cos t \sin 2x, \qquad 0 \le x \le \pi, \ 0 \le t \le \pi.$$

Use $c = -2/3, -1/3, 0, 1/3, 2/3$. (You will not be able to do this algebraically.)

18. On page 4 we introduced the traveling wave function

$$h(x, t) = 5 + \cos(0.5x - t).$$

Draw a contour diagram using $c = 4, 4.5, 5, 5.5, 6$ for this function. Explain how your contour diagram relates to the cross-sections of h discussed on page 4. Where are the contours most closely spaced? Most widely spaced?

19. Draw contour diagrams for each of the pizza-cola-happiness graphs given in Problem 6 on page 21.

20. A manufacturer sells two goods, one at a price of $3000 a unit and the other at a price of $12000 a unit. Suppose a quantity q_1 of the first good and q_2 of the second good are sold at a total cost of $4000 to the manufacturer.

    (a) Express the manufacturer's profit, π, as a function of q_1 and q_2.
    (b) Sketch curves of constant profit in the q_1q_2-plane for $\pi = 10000$, $\pi = 20000$, and $\pi = 30000$ and the break-even curve $\pi = 0$.

21. The cornea is the front surface of the eye. Corneal specialists use a TMS, or Topographical Modeling System, to produce a "map" of the curvature of the eye's surface. A computer analyzes light reflected off the eye and draws level curves joining points of constant curvature. The regions between these curves are colored different colors.

    The first two pictures in Figure 11.65 are cross-sections of eyes with constant curvature, the smaller being about 38 units and the larger about 50 units. For contrast, the third eye has varying curvature.

    (a) Describe in words how the TMS map of an eye of constant curvature will look.
    (b) Draw the TMS map of an eye with the cross-section in Figure 11.66. Assume the eye is circular when viewed from the front, and the cross-section is the same in every direction. Put reasonable numeric labels on your level curves.

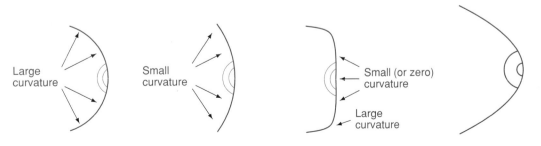

Large curvature

Small curvature

Small (or zero) curvature

Large curvature

Figure 11.65: Pictures of eyes with different curvature   *Figure 11.66*

22.   Match Tables 11.6–11.9 with the contour diagrams (I) - (IV) in Figure 11.67.

TABLE 11.6

$y \backslash x$	-1	0	1
-1	2	1	2
0	1	0	1
1	2	1	2

TABLE 11.7

$y \backslash x$	-1	0	1
-1	0	1	0
0	1	2	1
1	0	1	0

TABLE 11.8

$y \backslash x$	-1	0	1
-1	2	0	2
0	2	0	2
1	2	0	2

TABLE 11.9

$y \backslash x$	-1	0	1
-1	2	2	2
0	0	0	0
1	2	2	2

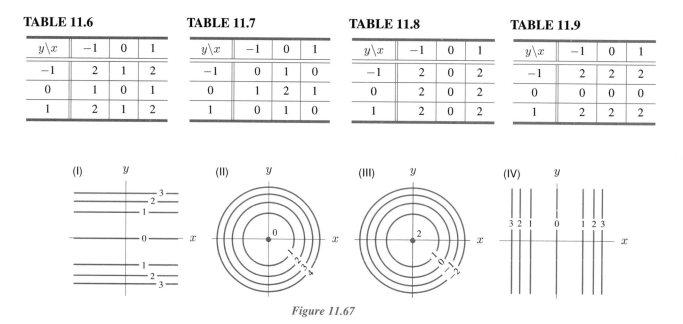

Figure 11.67

23.   Match the surfaces (a)–(e) in Figure 11.68 with the contour diagrams (I)–(V) in Figure 11.69.

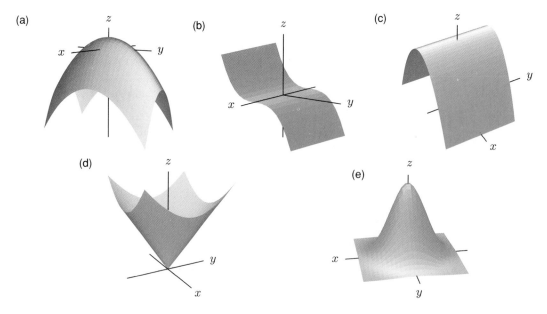

Figure 11.68

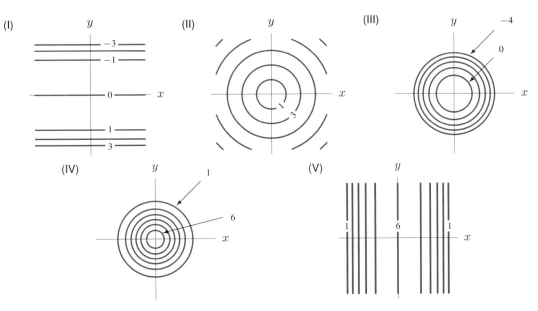

(I)

(II)

(III)

(IV)

(V)

Figure 11.69

24.   The map in Figure 11.70 is from the undergraduate senior thesis of Professor Robert Cook, Director of Harvard's Arnold Arboretum. It shows level curves of the function giving the species density of breeding birds at each point in the US, Canada, and Mexico.

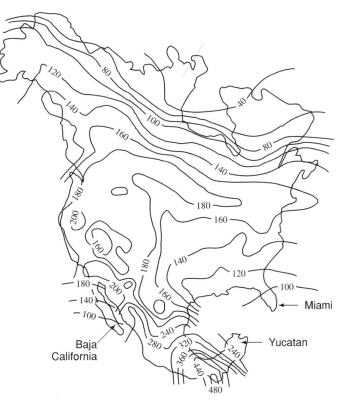

Figure 11.70

Using the map in Figure 11.70, decide if the the following statements are true or false. Explain your answers.

(a) Moving from south to north across Canada, the species density increases.
(b) The species density in the area around Miami is over 100.
(c) In general, peninsulas (for example, Florida, Baja California, the Yucatan) have lower species densities than the areas around them.
(d) The greatest rate of change in species density with distance is in Mexico. If you think this is true, mark the point and direction which give the maximum rate of change and explain why you picked the point and direction you did.

25. The temperature T (in °C) at any point in the region $-10 \leq x \leq 10$, $-10 \leq y \leq 10$ is given by the function

$$T(x,y) = 100 - x^2 - y^2.$$

(a) Sketch isothermal curves (curves of constant temperature) for $T = 100°C$, $T = 75°C$, $T = 50°C$, $T = 25°C$, and $T = 0°C$.
(b) Suppose a heat-seeking bug is put down at any point on the xy-plane. In which direction should it move to increase its temperature fastest? How is that direction related to the level curve through that point?

26.

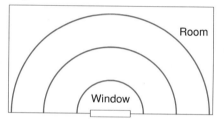

Figure 11.71

Figure 11.71 shows the level curves of the temperature H in a room near a recently opened window. Label the three level curves with reasonable values of H if the house is in the following locations.

(a) Minnesota in winter (where winters are harsh).
(b) San Francisco in winter (where winters are mild).
(c) Houston in summer (where summers are hot).
(d) Oregon in summer (where summers are mild).

27. Match each Cobb-Douglas production function with the correct graph in Figure 11.72 and with the correct statements.

(a) $F(L, K) = L^{0.25} K^{0.25}$
(b) $F(L, K) = L^{0.5} K^{0.5}$
(c) $F(L, K) = L^{0.75} K^{0.75}$

(D) Tripling each input triples output.
(E) Quadrupling each input doubles output.
(G) Doubling each input almost triples output.

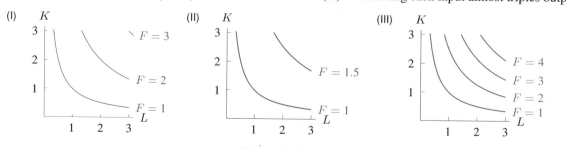

Figure 11.72

28. Consider the Cobb-Douglas production function $P = f(L, K) = 1.01L^{0.75}K^{0.25}$. What is the effect on production of doubling both labor and capital?

29. A general Cobb-Douglas production function has the form

$$P = cL^{\alpha}K^{\beta}.$$

An important economic question concerns what happens to production if labor and capital are both scaled up? For example, does production double if both labor and capital are doubled?

Economists talk about

- *increasing returns to scale* if doubling L and K more than doubles P,
- *constant returns to scale* if doubling L and K exactly doubles P,
- *decreasing returns to scale* if doubling L and K less than doubles P.

What conditions on α and β lead to increasing, constant, or decreasing returns to scale?

30. Figure 11.73 shows the contours of the temperature along one wall of a heated room through one winter day, with time indicated as on a 24-hour clock. The room has a heater located at the left-most corner of the wall and one window in the wall. The heater is controlled by a thermostat about 2 feet from the window.

    (a) Where is the window?
    (b) When is the window open?
    (c) When is the heat on?
    (d) Draw graphs of the temperature along the wall of the room at 6 am, at 11 am, at 3 pm (15 hours) and at 5 pm (17 hours).
    (e) Draw a graph of the temperature as a function of time at the heater, at the window and midway between them.
    (f) The temperature at the window at 5 pm (17 hours) is less than at 11 am. Why do you think this might be?
    (g) To what temperature do you think the thermostat is set? How do you know?
    (h) Where is the thermostat?

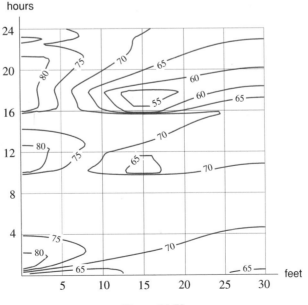

Figure 11.73

11.5 LINEAR FUNCTIONS

What is a Linear Function of Two Variables?

Linear functions played a central role in one-variable calculus because many one-variable functions are locally linear. (That is, their graphs look like a line when we zoom in.) In two-variable calculus, *linear functions* are those whose graph is a plane. In Chapter 13, we will see that most of the two-variable functions that we work with have graphs which look like planes when we zoom in.

What Makes a Plane Flat?

What is it about a function $f(x, y)$ that makes its graph $z = f(x, y)$ a plane? Linear functions of *one* variable have straight line graphs because they have constant slope. No matter where we are on the graph, the slope is the same: a given increase in the x-coordinate always gives the same increase in the y-coordinate. In other words, the ratio $\Delta y / \Delta x$ is constant. On a plane, the situation is a bit more complicated. If we walk around on a tilted plane, the slope is not always the same: it depends on the direction in which we walk. However, it is still true that at every point on the plane, the slope is the same as long as we choose the same direction. If we walk in the direction parallel to the x-axis, we always find ourselves walking up or down with the same slope; the same thing is true if we walk in the direction parallel to the y-axis. In other words, the slope ratios $\Delta z / \Delta x$ (with y fixed) and $\Delta z / \Delta y$ (with x fixed) are both constant.

Example 1   We are on a plane that cuts the z-axis at $z = 5$, has slope 2 in the x direction and slope -1 in the y direction. What is the equation of the plane?

Solution   Finding the equation of the plane means constructing a formula for the z-coordinate of the point on the plane directly above the point (x, y) in the xy-plane. To get to that point start from the point above the origin, where $z = 5$. Then walk x units in the x direction. Since the slope in the x direction is 2, our height increases by $2x$. Then walk y units in the y direction; since the slope in the y direction is -1, our height decreases by y units. Since our height has changed by $2x - y$ units, our z-coordinate is $5 + 2x - y$. Thus, the equation for the plane is

$$z = 5 + 2x - y.$$

For any linear function, if we know its value at a point (x_0, y_0), its slope in the x direction, and its slope in the y direction, then we can write the equation of the function. This is just like the equation of a line in the one-variable case, except that there are two slopes instead of one.

> If a plane has slope m in the x direction, slope n in the y direction, and passes through the point (x_0, y_0, z_0), then its equation is
>
> $$z = z_0 + m(x - x_0) + n(y - y_0).$$
>
> This plane is the graph of the linear function
>
> $$f(x, y) = z_0 + m(x - x_0) + n(y - y_0).$$
>
> If we write $c = z_0 - mx_0 - ny_0$, then we can write $f(x, y)$ in the equivalent form
>
> $$f(x, y) = c + mx + ny.$$

Just as in 2-dimensional space a line is determined by 2 points, so in 3-dimensional space a plane is determined by 3 points, provided they do not lie on a line.

Example 2 Find the equation of the plane that passes through the three points $(1, 0, 1)$, $(1, -1, 3)$, and $(3, 0, -1)$.

Solution The first two points have the same x-coordinate, so we use them to find the slope of the plane in the y-direction. As the y-coordinate changes from 0 to -1, the z-coordinate changes from 1 to 3, so the slope in the y-direction is $n = \Delta z/\Delta y = (3 - 1)/(-1 - 0) = -2$. The first and third points have the same y-coordinate, so we use them to find the slope in the x-direction; it is $m = \Delta z/\Delta x = (-1 - 1)/(3 - 1) = -1$. Because the plane passes through $(1, 0, 1)$, its equation is

$$z = 1 - (x - 1) - 2(y - 0) \quad \text{or} \quad z = 2 - x - 2y.$$

You should check that this equation is also satisfied by the points $(1, -1, 3)$ and $(3, 0, -1)$.

The previous example was made easier by the fact that two of the points had the same x-coordinate and two of them had the same y-coordinate. An alternative method, which works for any three points, is to substitute the x, y, and z-values of each of the three points into the equation $z = c + mx + ny$. The resulting three equations in c, m, n can then be solved simultaneously. See Problem 2 on page 41.

Linear Functions from a Numerical Point of View

To avoid flying planes with too many empty seats, airlines sell some tickets at full price and some at a discount. Table 11.10 shows an airline's revenue in dollars from tickets sold on a particular route, as a function of the number of full-price tickets sold, f, and the number of discount tickets sold, d.

TABLE 11.10 *Revenue from ticket sales (dollars)*

		Full-price tickets (f)			
		100	200	300	400
	200	39,700	63,600	87,500	111,400
	400	55,500	79,400	103,300	127,200
Discount tickets (d)	600	71,300	95,200	119,100	143,000
	800	87,100	111,000	134,900	158,800
	1000	102,900	126,800	150,700	174,600

Looking down any column, we see that the revenue jumps by $15,800 for each extra 200 discount tickets. Thus, each column is a linear function of the number of discount tickets sold. In addition, every column has the same slope, which is $15,800/200 = 79$ dollars/ticket. This is the price of a discount ticket. Similarly, each row is a linear function and all the rows have the same slope, 239, which is the price of a full-fare ticket. Thus, R is a linear function of f and d, given by:

$$R = 239f + 79d.$$

We have the following general result:

A **linear function** can be recognized from its table by the following features:
- Each row and each column is linear.
- All the rows have the same slope.
- All the columns have the same slope (although the slope of the rows and the slope of the columns are generally different).

Example 3   The table contains some values of a linear function. Fill in the blank and give a formula for the function.

$$f(2,15) = 0.5$$

$x \backslash y$	1.5	2.0
2	0.5	1.5
3	−0.5	?

Solution   In the first column the function decreases by 1 (from 0.5 to −0.5) as x goes from 2 to 3. Since the function is linear, it must decrease by the same amount in the second column. So the missing entry must be $1.5 − 1 = 0.5$. The slope of the function in the x-direction is −1. The slope in the y-direction is 2, since in each row the function increases by 1 when y increases by 0.5. From the table we get $f(2, 1.5) = 0.5$. Therefore, the formula is

$$f(x, y) = 0.5 − (x − 2) + 2(y − 1.5) = −0.5 − x + 2y.$$

What Does the Contour Diagram of a Linear Function Look Like?

Consider the function for airline revenue given in Table 11.10. Its formula is

$$R = 239f + 79d,$$

where f is the number of full-fares and d is the number of discount fares sold. Figure 11.74 gives the contour diagram for this function.

Notice that the contours are parallel straight lines. What is the practical significance of the slope? Consider the contour $R = 100,000$; that means we are looking at combinations of ticket sales that yield $100,000 in revenue. If we move down and to the right on the contour, the f-coordinate increases and the d-coordinate decreases, so we sell more full-fares and fewer discount fares. This makes sense, because to receive a fixed revenue of $100,000, we must sell more full-fares if we sell fewer discount fares. The exact trade-off depends on the slope of the contour; the diagram shows that each contour has a slope of about −3. This means that for a fixed revenue, we must sell three discount fares to replace one full-fare. This can also be seen by comparing prices. Each full fare brings in $239; to earn the same amount in discount fares we need to sell $239/79 \approx 3.03 \approx 3$ fares. Since the price ratio is independent of how many of each type of fare we sell, this slope remains constant over the whole contour map; thus, the contours are all parallel straight lines.

You should also observe that the contours are evenly spaced. This means that no matter which contour we are on, a fixed increase in one of the variables will cause the same increase in the value of the function. In terms of revenue, this says that no matter how many fares we have sold, an extra fare, whether full or discount, will bring the same revenue as it did before.

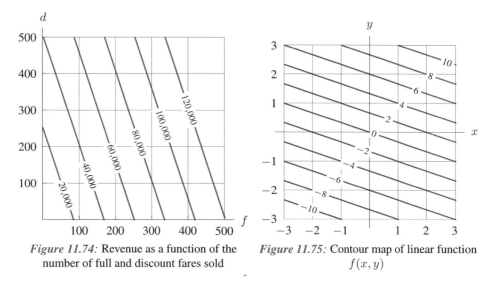

Figure 11.74: Revenue as a function of the number of full and discount fares sold

Figure 11.75: Contour map of linear function $f(x, y)$

Example 4   Find the equation of the linear function whose contour diagram is in Figure 11.75.

Solution   Suppose we start at the origin on the $z = 0$ contour. Moving 2 units in the y direction takes us to the $z = 6$ contour; so the slope in the y direction is $\Delta z / \Delta y = 6/2 = 3$. Similarly, a move of 2 in the x-direction from the origin takes us to the $z = 2$ contour, so the slope in the x direction is $\Delta z / \Delta x = 2/2 = 1$. Since $f(0, 0) = 0$, we have $f(x, y) = x + 3y$.

Problems for Section 11.5

1. Suppose that z is a linear function of x and y with slope 2 in the x direction and slope 3 in the y direction.

   (a) A change of 0.5 in x and -0.2 in y produces what change in z?
   (b) If $z = 2$ when $x = 5$ and $y = 7$, what is the value of z when $x = 4.9$ and $y = 7.2$?

2. Find the equation of the linear function $z = c + mx + ny$ whose graph contains the points $(0, 0, 0)$, $(0, 2, -1)$, and $(-3, 0, -4)$.

3. Find the linear function whose graph is the plane through the points $(4, 0, 0), (0, 3, 0)$ and $(0, 0, 2)$.

4. Find an equation for the plane containing the line in the xy-plane where $y = 1$, and the line in the xz-plane where $z = 2$.

5. Find the equation of the linear function $z = c + mx + ny$ whose graph intersects the xz-plane in the line $z = 3x + 4$ and intersects the yz-plane in the line $z = y + 4$.

6. Find the linear function $z = c + mx + ny$ whose graph intersects the xy-plane in the line $y = 3x + 4$ and contains the point $(0, 0, 5)$.

7. Is the function represented in Table 11.11 linear? Give reasons for your answer.

TABLE 11.11

$u \backslash v$	1.1	1.2	1.3	1.4
3.2	11.06	12.06	13.06	14.06
3.4	11.75	12.82	13.89	14.96
3.6	12.44	13.58	14.72	15.86
3.8	13.13	14.34	15.55	16.76
4.0	13.82	15.10	16.38	17.66

Figure 11.76

8. A discount music store sells compact discs at one price and cassette tapes at another price. Figure 11.76 shows the revenue (in dollars) of the music store as a function of the number, t, of tapes and the number, c, of compact discs that it sells. What is the price of tapes? What is the price of compact discs?

For Problems 9–10, find equations for the linear functions with the given tables.

9.

$x \backslash y$	-1	0	1	2
0	1.5	1	0.5	0
1	3.5	3	2.5	2
2	5.5	5	4.5	4
3	7.5	7	6.5	6

$f(-1,0)=$

10.

$x \backslash y$	10	20	30	40
100	3	6	9	12
200	2	5	8	11
300	1	4	7	10
400	0	3	6	9

Problems 11–12 each contain a partial table of values for a linear function. Fill in the blanks.

11.

$x \backslash y$	0.0	1.0
0.0	-1.0	1.0
2.0	3.0	5.0

12.

$x \backslash y$	-1.0	0.0	1.0
2.0	4.0		
3.0		3.0	5.0

For Problems 13–14, find equations for the linear functions with the given contour diagrams.

13.

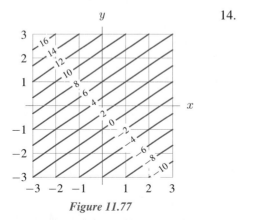

Figure 11.77

14.

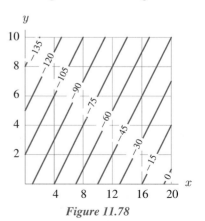

Figure 11.78

It is difficult to sketch a plane that is the graph of a linear function by hand. One method that works if the x, y, and z-intercepts of the graph are positive is to plot the intercepts and join them by a triangle as shown in Figure 11.79; this shows the part of the graph in the octant where $x \geq 0$, $y \geq 0$, $z \geq 0$. If the intercepts are not all positive, the same method works if the x, y, and z-axes are drawn from a different perspective. Use this method to sketch the graphs of the linear functions in Problems 15–17.

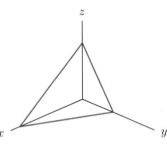

Figure 11.79: The graph of a linear function with positive x, y, and z-intercepts

15. $z = 6 - 2x - 3y$ 16. $z = 4 + x - 2y$ 17. $z = 2 - 2x + y$

18. A manufacturer makes two products out of two raw materials. Let q_1, q_2 be the quantities sold of the two products, p_1, p_2 their prices, and m_1, m_2 the quantities purchased of the two raw materials. Which of the following functions do you expect to be linear, and why? In each case, assume that all variables except the ones mentioned are held fixed.

    (a) Expenditure on raw materials as a function of m_1 and m_2.
    (b) Revenue as a function of q_1 and q_2.
    (c) Revenue as a function of p_1 and q_1.

19. A college admissions office uses the following linear equation to predict the grade point average of an incoming student:

$$z = 0.003x + 0.8y - 4,$$

where z is the predicted college GPA on a scale of 0 to 4.3, and x is the sum of the student's SAT Math and SAT Verbal on a scale of 400 to 1600, and y is the student's high school GPA on a scale of 0 to 4.3. The college admits students whose predicted GPA is at least 2.3.

    (a) Will a student with SATs of 1050 and high school GPA of 3.0 be admitted?
    (b) Will every student with SATs of 1600 be admitted?
    (c) Will every student with a high school GPA of 4.3 be admitted?
    (d) Draw a contour diagram for the predicted GPA z with $400 \leq x \leq 1600$ and $0 \leq y \leq 4.3$. Shade the points corresponding to students who will be admitted.
    (e) Which is more important, an extra 100 points on the SAT or an extra 0.5 of high school GPA?

20. Let f be the linear function $f(x, y) = c + mx + ny$, where c, m, n are constants and $n \neq 0$.

    (a) Show that all the contours of f are lines of slope $-m/n$.
    (b) Show that, for all x and y

$$f(x + n, y - m) = f(x, y)$$

    (c) Explain the relation between parts (a) and (b).

11.6 FUNCTIONS OF MORE THAN TWO VARIABLES

In applications of calculus, functions of any number of variables can arise. The density of matter in the universe is a function of three variables, since it takes three numbers to specify a point in space. If we want to study the evolution of this density over time, we need to add the fourth variable, time. We need to be able to apply calculus to functions of arbitrarily many variables.

The main problem with functions of more than two variables is that it is hard to visualize them. The graph of a function of one variable is a curve in 2-space, the graph of a function of two variables is a surface in 3-space, and so the graph of a function of three variables would be a solid in 4-space. We can't easily visualize 4-space, or any higher dimensional space, and so we won't use the graphs of functions of three or more variables.

On the other hand, the idea of cross-sectioning a function by keeping one variable fixed allows us to represent a function of three variables. It is also possible to give contour diagrams for these functions, only now the contours are surfaces in 3-space.

Representing a Function of Three Variables Using Contour Diagrams

One way to analyze a function of three variables, $f(x, y, z)$, is to look at cross-sections with one variable fixed. Suppose we keep z fixed and view f as a function of the remaining two variables, x and y. For each fixed value, $z = c$, we represent the two-variable function $f(x, y, c)$ by a contour diagram. As c varies, we get a whole collection of contour diagrams.

Example 1   A pond is 30 meters deep in the middle and 200 meters across. The pond is infested with algae. Suppose the density of algae at a point z meters below, x meters east, and y meters north of the center of the surface of the pond is approximated by the formula

$$f(x, y, z) = \frac{1}{10}\left(50 + \sqrt{x^2 + y^2}\right)(30 - z),$$

where density is measured in gm/m³. Draw contour diagrams for f at the surface of the pond and at a depth of 10 m. Describe in words how the density of algae varies with depth and distance from the center of the pond.

Solution   At the surface of the pond, $z = 0$, so the formula for the $z = 0$ cross-section is

$$f(x, y, 0) = \frac{1}{10}\left(50 + \sqrt{x^2 + y^2}\right)(30 - 0) = 150 + 3\sqrt{x^2 + y^2}.$$

Notice that we now have a function of two variables. The contours of this function are circles, since $f(x, y, 0)$ is constant when $\sqrt{x^2 + y^2}$ is constant. The contour diagram is shown in Figure 11.80.
At a depth of $z = 10$, we have

$$f(x, y, 10) = \frac{1}{10}\left(50 + \sqrt{x^2 + y^2}\right)(30 - 10) = 100 + 2\sqrt{x^2 + y^2}.$$

The contour diagram for the $z = 10$ cross-section is shown in Figure 11.81. Comparing the two contour diagrams we see that there is more algae near the surface than deeper down in the pond and that there is more algae near the shoreline than near the center of the pond.

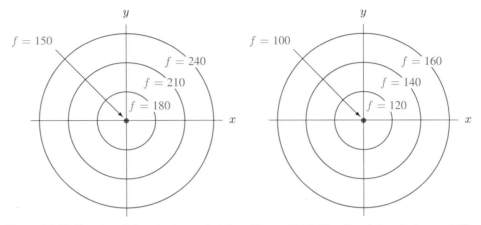

Figure 11.80: Density of algae in the pond at the surface: Contour diagram for $z = 0$

Figure 11.81: Density of algae in the pond 10 meters down: Contour diagram for $z = 10$

For weather forecasting, it is important to know the barometric pressure and temperature both at the surface of the earth and also aloft in the atmosphere above the earth. On TV, you may hear about a "surface" low pressure system or an "upper level" low pressure system (upper level lows influence the jet stream). The surface and upper level maps studied by meteorologists are contour diagrams at different altitudes of the pressure or temperature function. These maps are similar to the contour diagrams of algae density of the previous example.

Representing a Function of Three Variables Using Tables

In the previous example, we viewed the cross-sections of a three-variable function $f(x, y, z)$ as contour diagrams of a two-variable function. We can also view the cross-sections as tables of values.

Example 2   In Section 11.1, the consumption of beef as a function of the price, p, of beef and household income, I, was given in a table of values. In fact, a household's beef consumption depends on many other things as well, such as the prices of competing products. Let us view the consumption, C, of beef as a function of the price q of a competing meat, say chicken, as well as the price of beef and household income. That is, $C = f(I, p, q)$. This function can be given by a collection of tables, one for each different value of the price, q, of chicken. Tables 11.12 and 11.13 show two different cross-sections of the consumption function f with q held fixed. Explain how the two tables are related.

TABLE 11.12 *Beef consumption when price of chicken $q = 1.50$*

		p		
		3.00	3.50	4.00
	20	2.65	2.59	2.51
I	60	5.11	5.00	4.97
	100	5.80	5.77	5.60

TABLE 11.13 *Beef consumption when price of chicken $q = 2.00$*

		p		
		3.00	3.50	4.00
	20	2.75	2.75	2.71
I	60	5.21	5.12	5.11
	100	5.80	5.77	5.60

Solution   Comparing tables shows that for households with large incomes (say $I = 100$), changes in the price of chicken have little effect on the amount of beef consumed (because price is not a factor in their buying). However for small incomes (say $I = 20$), an increase in the price of chicken (from $q = 1.50$ to $q = 2.00$) causes families to spend more on beef.

Representing a Function of Three Variables Using a Family of Level Surfaces

A function of two variables, $f(x, y)$, can be represented by a *family* of level curves of the form $f(x, y) = c$ for various values of the constant, c.

> A function of three variables, $f(x, y, z)$, can be represented by a *family* of surfaces of the form $f(x, y, z) = c$, each one of which is called a *level surface*.

Example 3   Suppose the temperature, in $°C$, at a point (x, y, z) is given by $T = f(x, y, z) = x^2 + y^2 + z^2$. What do the level surfaces of the function f look like and what do they mean in terms of temperature?

Solution   The level surface corresponding to $T = 100$ is the set of all points where the temperature is $100°C$. That is, where $f(x, y, z) = 100$, so

$$x^2 + y^2 + z^2 = 100.$$

This is the equation of a sphere of radius 10, with center at the origin. Similarly, the level surface corresponding to $T = 200$ is the sphere with radius $\sqrt{200}$. The other level surfaces will be concentric spheres. The temperature is constant on each sphere. We may view the temperature distribution as a set of nested spheres, like concentric layers of an onion, each one labeled with a different temperature, starting from low temperatures in the middle and getting hotter as we go out from the center. (See Figure 11.82.) The level surfaces become more closely spaced as we move farther from the origin because the temperature increases more rapidly the farther we get from the origin.

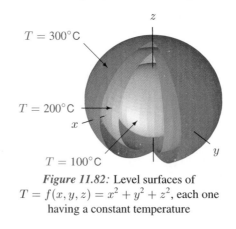

Figure 11.82: Level surfaces of
$T = f(x, y, z) = x^2 + y^2 + z^2$, each one
having a constant temperature

In general, the level surfaces of a function are nested together in some way, so it is often difficult to draw them. We generally use cut-outs to show the inner surfaces, as we did in Figure 11.82.

Example 4   What do the level surfaces of $f(x, y, z) = x^2 + y^2$ and $g(x, y, z) = z - y$ look like?

Solution   The level surface of f corresponding to the constant c is the surface consisting of all points satisfying the equation

$$x^2 + y^2 = c.$$

Since there is no z-coordinate in the equation, z can take any value whatsoever. For $c > 0$, this is a circular cylinder of radius $\sqrt{c}$ around the z-axis. The level surfaces are concentric cylinders; the narrow ones near the z-axis are the ones where f has small values, and the wider ones are where it has larger values. See Figure 11.83. The level surface of g corresponding to the constant c is the surface

$$z - y = c.$$

This time there is no x variable, so this surface is the one you get by taking each point on the straight line $z - y = c$ in the yz-plane and letting x roam back and forth. You get a plane which cuts the yz-plane diagonally; the x-axis is parallel to this plane. See Figure 11.84.

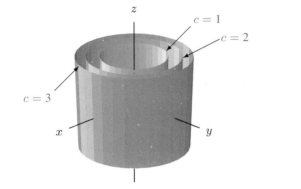

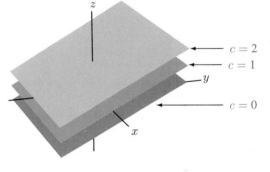

Figure 11.83: Level surfaces of $f(x, y, z) = x^2 + y^2$

Figure 11.84: Level surfaces of $g(x, y, z) = z - y$

Example 5   What do the level surfaces of $f(x, y, z) = x^2 + y^2 - z^2$ look like?

Solution   From Section 11.4, you may recall that the two-variable quadratic function $g(x, y) = x^2 - y^2$ has a saddle-shaped graph and three types of contours. The contour equation $x^2 - y^2 = c$ gives a hyperbola opening right-left when $c > 0$, a hyperbola opening up-down when $c < 0$, and a pair of intersecting lines when $c = 0$. Similarly, the three-variable quadratic function $f(x, y, z) = x^2 + y^2 - z^2$ has three types of level surfaces depending on the value of c in the equation $x^2 + y^2 - z^2 = c$. Suppose that $c > 0$, say $c = 1$. Rewrite the equation as $x^2 + y^2 = z^2 + 1$ and think of what happens as we cross-section the surface perpendicular to the z-axis by holding z fixed. The result is a circle, $x^2 + y^2 = $ constant, of radius at least 1 (since the constant $z^2 + 1 \geq 1$). The circles get larger as z gets larger. If we took the $x = 0$ cross-section instead we would get the hyperbola $y^2 - z^2 = 1$. The result is shown in Figure 11.88, with $a = b = c = 1$.

Suppose instead $c < 0$, say $c = -1$. Then the horizontal cross-sections of $x^2 + y^2 = z^2 - 1$ are again circles except that the radii shrink to 0 at $z = \pm 1$ and between $z = -1$ and $z = 1$ there are no cross-sections at all. The result is shown in Figure 11.89 with $a = b = c = 1$.

When $c = 0$, we get the equation $x^2 + y^2 = z^2$. Again the horizontal cross-sections are circles, this time with the radius shrinking down to exactly 0 when $z = 0$. The resulting surface, shown in Figure 11.90 with $a = b = c = 1$, is the cone $z = \sqrt{x^2 + y^2}$ studied in Section 11.4, together with the lower cone $z = -\sqrt{x^2 + y^2}$.

A Catalog of Surfaces

For later reference, here is a small catalog of the surfaces we have encountered.

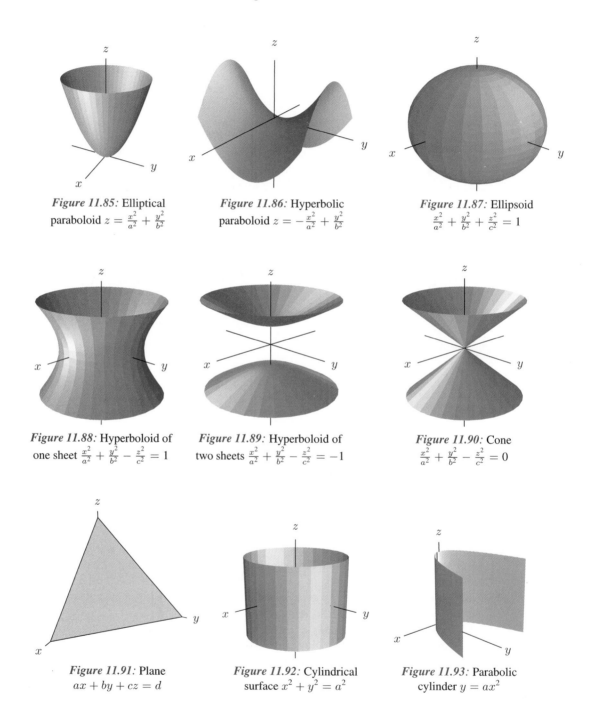

Figure 11.85: Elliptical paraboloid $z = \frac{x^2}{a^2} + \frac{y^2}{b^2}$

Figure 11.86: Hyperbolic paraboloid $z = -\frac{x^2}{a^2} + \frac{y^2}{b^2}$

Figure 11.87: Ellipsoid $\frac{x^2}{a^2} + \frac{y^2}{b^2} + \frac{z^2}{c^2} = 1$

Figure 11.88: Hyperboloid of one sheet $\frac{x^2}{a^2} + \frac{y^2}{b^2} - \frac{z^2}{c^2} = 1$

Figure 11.89: Hyperboloid of two sheets $\frac{x^2}{a^2} + \frac{y^2}{b^2} - \frac{z^2}{c^2} = -1$

Figure 11.90: Cone $\frac{x^2}{a^2} + \frac{y^2}{b^2} - \frac{z^2}{c^2} = 0$

Figure 11.91: Plane $ax + by + cz = d$

Figure 11.92: Cylindrical surface $x^2 + y^2 = a^2$

Figure 11.93: Parabolic cylinder $y = ax^2$

(These are viewed as equations in three variables x, y, and z)

How Surfaces Can Represent Functions of Two Variables and Functions of Three Variables

You may have noticed that we have used surfaces to represent functions in two different ways. First, we used a *single* surface to represent a two-variable function $z = f(x, y)$. Second, we used a *family* of level surfaces to represent a three-variable function $w = F(x, y, z)$. These level surfaces have equation $F(x, y, z) = c$.

What is the relation between these two uses of surfaces? For example, consider the equation

$$z = x^2 + y^2 + 3.$$

Define

$$G(x, y, z) = x^2 + y^2 + 3 - z$$

The points which satisfy $z = x^2 + y^2 + 3$ also satisfy $x^2 + y^2 + 3 - z = 0$. Thus the surface $z = x^2 + y^2 + 3$ is the same as the level surface

$$G(x, y, z) = x^2 + y^2 + 3 - z = 0.$$

Thus, we have the following result:

> A single surface representing a two-variable function $z = f(x, y)$ can always be thought of as one member of the family of level surfaces representing a three-variable function $G(x, y, z) = f(x, y) - z$. The graph of $z = f(x, y)$ is the level surface $G = 0$.

Conversely, a single member of a family of level surfaces can be regarded as the graph of a function of the form $z = f(x, y)$ if it is possible to solve for z. For example, if $F(x, y, z) = x^2 + y^2 + z^2$, then one member of the family of level surfaces is the sphere

$$x^2 + y^2 + z^2 = 1.$$

This equation defines z implicitly as a function of x and y. Solving it gives two functions

$$z = \sqrt{1 - x^2 - y^2} \qquad \text{and} \qquad z = -\sqrt{1 - x^2 - y^2}.$$

The graph of the first function is the top half of the sphere and the graph of the second function is the bottom half.

Problems for Section 11.6

1. Hot water is entering a rectangular swimming pool at the surface of the pool in one corner. Sketch possible contour diagrams for the temperature of the pool at the surface and the temperature one meter below the surface.

2. Figure 11.94 shows contour diagrams of temperature in degrees Celsius in a room at three different times. Describe the heat flow in the room. What could be causing this?

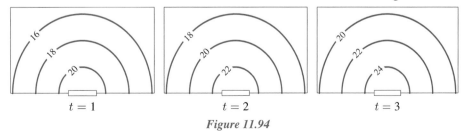

Figure 11.94

3. Match the following functions with the level surfaces in Figure 11.95.

    (a) $f(x, y, z) = y^2 + z^2$ (b) $h(x, y, z) = x^2 + z^2$.

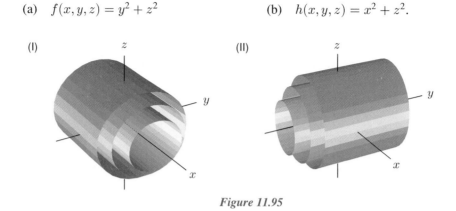

Figure 11.95

4. Describe in words the level surfaces of $f(x, y, z) = \sin(x + y + z)$.

5. Draw contour diagrams for three different cross-sections with t fixed, of the function

$$f(x, y, t) = \cos t \cos \sqrt{x^2 + y^2}, \quad 0 \le \sqrt{x^2 + y^2} \le \pi/2.$$

6. The height (in meters) of the water above the bottom of a pond at time t is given by the function $h(x, y, t) = 20 + \sin(x + y - t)$, where x and y are measured horizontally with the positive y-axis north and the positive x-axis east, and where t is in seconds. By considering contour diagrams for different t values, describe the motion of the water surface in the pond.

7. Describe in words the level surfaces of $g(x, y, z) = e^{-(x^2+y^2+z^2)}$.

In Problems 8–9, give the equation of the linear function $f(x, y, z) = ax + by + cz + d$ that has the given values for the cross-sections with $z = 1$ and $z = 4$.

8.

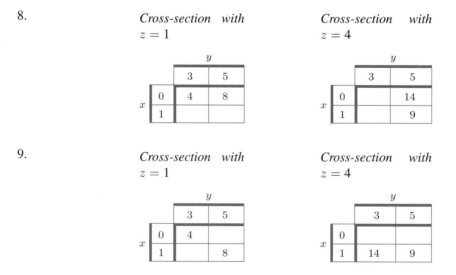

Cross-section with $z = 1$

	y	
	3	5
0	4	8
1		

Cross-section with $z = 4$

	y	
	3	5
0		14
1		9

9.

Cross-section with $z = 1$

	y	
	3	5
0	4	
1		8

Cross-section with $z = 4$

	y	
	3	5
0		
1	14	9

10. In Problems 8–9, suppose the two tables had only three values filled in. Could you determine f? Suppose the two tables had five values filled in. Could you always determine f?

11. Find the linear function $f(x, y, z) = ax + by + cz + d$ that has the contour diagrams for the cross-sections with $z = 3$ and $z = 4$ shown in Figures 11.96 and 11.97.

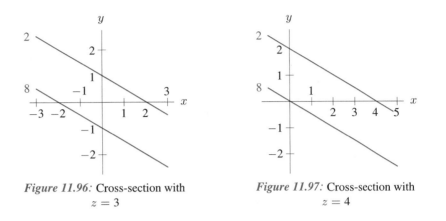

Figure 11.96: Cross-section with $z = 3$

Figure 11.97: Cross-section with $z = 4$

12. What do the level surfaces of $f(x, y, z) = x^2 - y^2 + z^2$ look like? [Hint: Use cross-sections with y constant instead of cross-sections with z constant.]

Use the catalog of surfaces to identify the surfaces in Problems 13–19.

13. $-x^2 - y^2 + z^2 = 1$ 14. $-x^2 + y^2 - z^2 = 0$ 15. $x^2 + y^2 - z = 0$

16. $x^2 + z^2 = 1$ 17. $x^2 + y^2/4 + z^2 = 1$ 18. $x + y = 1$

19. $(x - 1)^2 + y^2 + z^2 = 1$

20. Describe the surface $x^2 + y^2 = (2 + \sin z)^2$. In general, if $f(z) \geq 0$ for all z, describe the surface $x^2 + y^2 = (f(z))^2$.

11.7   LIMITS AND CONTINUITY

The sheer vertical face of Half Dome, in Yosemite National Park in California, was caused by glacial activity during the Ice Age. (See Figure 11.98.) The height of the terrain rises abruptly by nearly 1000 feet as we scale the rock from the west, whereas it is possible to make a gradual climb to the top from the east.

If we consider the function h giving the height of the terrain above sea level in terms of longitude and latitude, then h has a *discontinuity* along the path at the base of the cliff of Half Dome. Looking at the contour map of the region in Figure 11.99, we see that in most places a small change in position results in a small change in height, except near the cliff. There, no matter how small a step we take, we get a large change in height. (You can see how crowded the contours get near the cliff; some end abruptly along the discontinuity.)

This geological feature illustrates the ideas of continuity and discontinuity. Roughly speaking, a function is said to be *continuous* at a point if its values at places near the point are close to the value at the point. If this is not the case, the function is said to be *discontinuous*.

The property of continuity is one that, practically speaking, we usually assume of the functions we are studying. Informally, we expect (except under special circumstances) that values of a function do not change drastically when making small changes to the input variables. Whenever we model a

Figure 11.98: Half Dome in Yosemite National Park

Figure 11.99: A contour map of Half Dome

one-variable function by an unbroken curve, we are making this assumption. Even when functions come to us as tables of data, we usually make the assumption that the missing function values between data points are close to the measured ones.

In this section we study limits and continuity a bit more formally in the context of functions of several variables. For simplicity we study these concepts for functions of two variables, but our discussion can be adapted to functions of three or more variables.

One can show that sums, products, and compositions of continuous functions are continuous, while the quotient of two continuous functions is continuous everywhere the denominator function is nonzero. Thus, each of the functions

$$\cos(x^2 y), \qquad \ln(x^2 + y^2), \qquad \frac{e^{x+y}}{x + y}, \qquad \ln(\sin(x^2 + y^2))$$

is continuous at all points (x, y) where it is defined.

As for functions of one variable, the graph of a continuous function over an unbroken domain is unbroken—that is, the surface has no holes or rips in it.

Example 1   From Figures 11.100–11.103, which of the following functions appear to be continuous at $(0, 0)$?

(a) $f(x, y) = \begin{cases} \dfrac{x^2 y}{x^2 + y^2}, & (x, y) \neq (0, 0), \\ 0, & (x, y) = (0, 0). \end{cases}$

(b) $g(x, y) = \begin{cases} \dfrac{x^2}{x^2 + y^2}, & (x, y) \neq (0, 0), \\ 0, & (x, y) = (0, 0). \end{cases}$

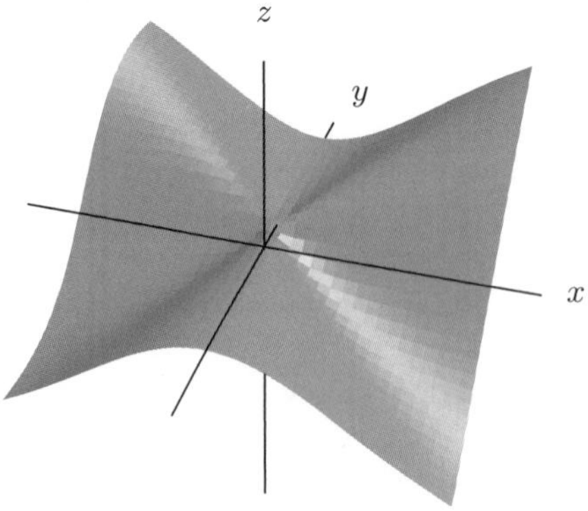

Figure 11.100: Graph of $z = x^2 y/(x^2 + y^2)$ 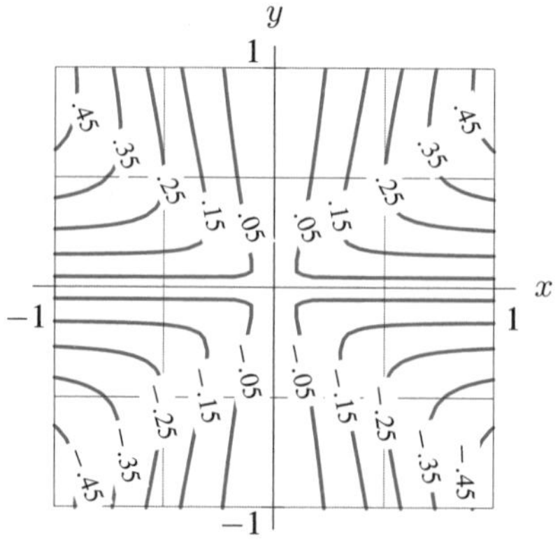   *Figure 11.101:* Contour diagram of $z = x^2 y/(x^2 + y^2)$

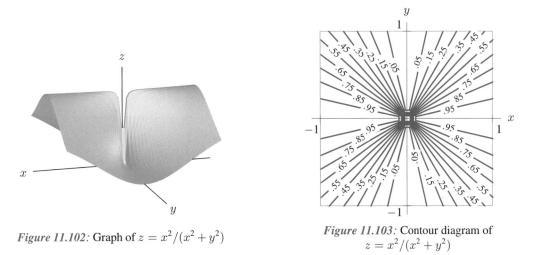

Figure 11.102: Graph of $z = x^2/(x^2 + y^2)$

Figure 11.103: Contour diagram of $z = x^2/(x^2 + y^2)$

Solution (a) The graph and contour diagram of f in Figures 11.100 and 11.101 suggest that f is close to 0 when (x, y) is close to $(0, 0)$. That is, the figures suggest that f is continuous at the point $(0, 0)$; the graph appears to have no rips or holes there.

However, the figures cannot tell us for sure whether f is continuous. To be certain we must investigate the limit analytically, as is done in Example 2(a) on page 54.

(b) The graph of g and its contours near $(0, 0)$ in Figure 11.102 and 11.103 suggest that g behaves differently from f: The contours of g seem to "crash" at the origin and the graph rises rapidly from 0 to 1 near $(0, 0)$. Small changes in (x, y) near $(0, 0)$ can yield large changes in g, so we expect that g is not continuous at the point $(0, 0)$. Again, a more precise analysis is given in Example 2(b) on page 54.

The previous example suggests that continuity *at* a point depends on a function's behavior *near* the point. To study behavior near a point more formally we need to define the limit of a function of two variables. Suppose that $f(x, y)$ is a function defined on a set in 2-space, not necessarily containing the point (a, b), but containing points (x, y) arbitrarily close to (a, b); suppose that L is a number.

> The function f has a **limit** L at the point (a, b), written
> $$\lim_{(x,y) \to (a,b)} f(x, y) = L,$$
> if the difference $|f(x, y) - L|$ is as small as we wish whenever the distance from the point (x, y) to the point (a, b) is sufficiently small, but not zero.

We define continuity for functions of two variables in the same way as for functions of one variable:

> A function f is **continuous at the point** (a, b) if
>
> $$\lim_{(x,y)\to(a,b)} f(x, y) = f(a, b).$$
>
> A function is **continuous** if it is continuous at each point of its domain.

Thus, if f is continuous at the point (a, b), then f must be defined at (a, b) and the limit, $\lim_{(x,y)\to(a,b)} f(x, y)$, must exist and be equal to the value $f(a, b)$. If a function is defined at a point (a, b) but is not continuous there, then we say that f is *discontinuous* at (a, b).

We now apply the definition of continuity to the functions in Example 1, showing that f is continuous at $(0, 0)$ and that g is discontinuous at $(0, 0)$.

Example 2   Let f and g be the functions defined everywhere on 2-space except at the origin as follows

(a) $f(x, y) = \dfrac{x^2 y}{x^2 + y^2}$ (b) $g(x, y) = \dfrac{x^2}{x^2 + y^2}$

Use the definition of the limit to show that $\lim_{(x,y)\to(0,0)} f(x, y) = 0$ and that $\lim_{(x,y)\to(0,0)} g(x, y)$ does not exist.

Solution   (a)   The graph and contour diagram of f both suggest that $\lim_{(x,y)\to(0,0)} f(x, y) = 0$. To use the definition of the limit, we must estimate $|f(x, y) - L|$ with $L = 0$:

$$|f(x, y) - L| = \left| \frac{x^2 y}{x^2 + y^2} - 0 \right| = \left| \frac{x^2}{x^2 + y^2} \right| |y| \leq |y| \leq \sqrt{x^2 + y^2},$$

Now $\sqrt{x^2 + y^2}$ is the distance from (x, y) to $(0, 0)$. Thus, to make $|f(x, y) - 0| < 0.001$, for example, we need only require (x, y) be within 0.001 of $(0, 0)$. More generally, for any positive number u, no matter how small, we are sure that $|f(x, y) - 0| < u$ whenever (x, y) is no farther than u from $(0, 0)$. This is what we mean by saying that the difference $|f(x, y) - 0|$ can be made as small as we wish by choosing the distance to be sufficiently small. Thus, we conclude that

$$\lim_{(x,y)\to(0,0)} \frac{x^2 y}{x^2 + y^2} = 0.$$

Notice that the function f has a limit at the point $(0, 0)$ even though f was not defined at $(0, 0)$. To make f continuous at $(0, 0)$ we must define its value there to be 0, as we did in Example 1.

(b)   Although the formula defining the function g looks similar to that of f, we saw in Example 1 that g's behavior near the origin is quite different. If we consider points $(x, 0)$ lying along the x-axis near $(0, 0)$, then the values $g(x, 0)$ are equal to 1, while if we consider points $(0, y)$ lying along the y-axis near $(0, 0)$, then the values $g(0, y)$ are equal to 0. Thus, within any disk (no matter how small) centered at the origin, there are points where $g = 0$ and points where $g = 1$. Therefore the limit $\lim_{(x,y)\to(0,0)} g(x, y)$ does not exist.

While the notions of limit and continuity look formally the same for one- and two-variable functions, they are somewhat more subtle in the multivariable case. The reason for this is that on the line (1-space), we can approach a point from just two directions (left or right) but in 2-space there are an infinite number of ways to approach a given point.

Problems for Section 11.7

1. Show that the function f does not have a limit at $(0,0)$ by examining the limits of f as $(x,y) \to (0,0)$ along the curve $y = kx^2$ for different values of k. The function is given by

$$f(x,y) = \frac{x^2}{x^2 + y}, \qquad x^2 + y \neq 0.$$

2. Show that the function f does not have a limit at $(0,0)$ by examining the limits of f as $(x,y) \to (0,0)$ along the line $y = x$ and along the parabola $y = x^2$. The function is given by

$$f(x,y) = \frac{x^2 y}{x^4 + y^2}, \qquad (x,y) \neq (0,0).$$

3. Consider the following function:

$$f(x,y) = \begin{cases} \dfrac{xy(x^2 - y^2)}{x^2 + y^2}, & (x,y) \neq (0,0), \\ 0, & (x,y) = (0,0). \end{cases}$$

   (a) Use a computer to draw the graph and the contour diagram of f.
   (b) Do your answers to part (a) suggest that f is continuous at $(0,0)$? Explain your answer.

4. Consider the function f, whose graph and contour diagram are in Figures 11.104 and 11.105, and which is given by

$$f(x,y) = \begin{cases} \dfrac{xy}{x^2 + y^2}, & (x,y) \neq (0,0), \\ 0, & (x,y) = (0,0). \end{cases}$$

   (a) Show that $f(0,y)$ and $f(x,0)$ are each continuous functions of one variable.
   (b) Show that rays emanating from the origin are contained in contours of f.
   (c) Is f continuous at $(0,0)$?

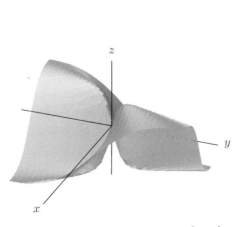

Figure 11.104: Graph of $z = xy/(x^2 + y^2)$

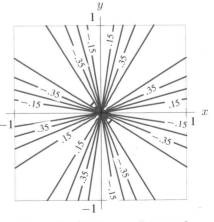

Figure 11.105: Contour diagram of $z = xy/(x^2 + y^2)$

For Problems 5–9 compute the limits of the functions $f(x, y)$ as $(x, y) \to (0, 0)$. You may assume that polynomials, exponentials, logarithmic, and trigonometric functions are continuous.

5. $f(x, y) = x^2 + y^2$

6. $f(x, y) = e^{-x-y}$

7. $f(x, y) = \dfrac{x}{x^2 + 1}$

8. $f(x, y) = \dfrac{x + y}{(\sin y) + 2}$

9. $f(x, y) = \dfrac{\sin(x^2 + y^2)}{x^2 + y^2}$ [Hint: You may assume that $\lim_{t \to 0}(\sin t)/t = 1$.]

For the functions in Problems 10–12, show that $\lim_{(x,y) \to (0,0)} f(x, y)$ does not exist.

10. $f(x, y) = \dfrac{x + y}{x - y}$, $x \neq y$

11. $f(x, y) = \dfrac{x^2 - y^2}{x^2 + y^2}$

12. $f(x, y) = \dfrac{xy}{|xy|}$, $x \neq 0$ and $y \neq 0$

13. Show that the contours of the function g defined in Example 1(b) on page 52 are rays emanating from the origin. Find the slope of the contour $g(x, y) = c$.

14. Explain why the following function is not continuous along the line $y = 0$.

$$f(x, y) = \begin{cases} 1 - x, & y \geq 0, \\ -2, & y < 0, \end{cases}$$

In Problems 15–16, determine whether there is a value for c making the function continuous everywhere. If so, find it. If not, explain why not.

15. $f(x, y) = \begin{cases} c + y, & x \leq 3, \\ 5 - y, & x > 3. \end{cases}$

16. $f(x, y) = \begin{cases} c + y, & x \leq 3, \\ 5 - x, & x > 3. \end{cases}$

REVIEW PROBLEMS FOR CHAPTER ELEVEN

1. Describe the set of points whose x coordinate is 2 and whose y coordinate is 1.

2. Find the center and radius of the sphere with equation $x^2 + 4x + y^2 - 6y + z^2 + 12z = 0$.

3. Find the equation of the plane through the points $(0, 0, 2), (0, 3, 0), (5, 0, 0)$.

4. Find a linear function whose graph is the plane that intersects the xy-plane along the line $y = 2x + 2$ and contains the point $(1, 2, 2)$.

5. Consider the function $f(r, h) = \pi r^2 h$ which gives the volume of a cylinder of radius r and height h. Sketch the cross-sections of f, first by keeping h fixed, then by keeping r fixed.

6. Consider the function $z = \cos \sqrt{x^2 + y^2}$.

   (a) Sketch the level curves of this function.

   (b) Sketch a cross-section through the surface $z = \cos \sqrt{x^2 + y^2}$ in the plane containing the x- and z-axes. Put units on your axes.

   (c) Sketch the cross-section through the surface $z = \cos \sqrt{x^2 + y^2}$ in the plane containing the z-axis and the line $y = x$ in the xy-plane.

For Problems 7–10, use a computer or calculator to sketch the graph of a function with the given shapes. Include the axes and the equation used to generate it in your sketch.

7. A cone of circular cross-section opening downward and with its vertex at the origin.

8. A bowl which opens upward and has its vertex at 5 on the z-axis.

9. A plane which has its x, y, and z intercepts all positive.

10. A parabolic cylinder opening upward from along the line $y = x$ in the xy-plane.

Decide if the statements in Problems 11–15 must be true, might be true, or could not be true. The function $z = f(x, y)$ is defined everywhere.

11. The level curves corresponding to $z = 1$ and $z = -1$ cross at the origin.

12. The level curve $z = 1$ consists of the circle $x^2 + y^2 = 2$ and the circle $x^2 + y^2 = 3$, but no other points.

13. The level curve $z = 1$ consists of two lines which intersect at the origin.

14. If $z = e^{-(x^2+y^2)}$, there is a level curve for every value of z.

15. If $z = e^{-(x^2+y^2)}$, there is a level curve through every point (x, y).

For each of the functions in Problems 16–19, make a contour plot in the region $-2 < x < 2$ and $-2 < y < 2$. In each case, what is the equation and the shape of the contour lines?

16. $z = \sin y$ 17. $z = 3x - 5y + 1$ 18. $z = 2x^2 + y^2$ 19. $z = e^{-2x^2-y^2}$

20. Suppose you are in a room 30 feet long with a heater at one end. In the morning the room is $65°$F. You turn on the heater, which quickly warms up to $85°$F. Let $H(x, t)$ be the temperature x feet from the heater, t minutes after the heater is turned on. Figure 11.106 shows the contour diagram for H. How warm is it 10 feet from the heater 5 minutes after it was turned on? 10 minutes after it was turned on?

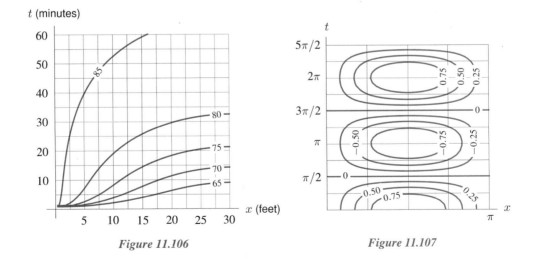

Figure 11.106   *Figure 11.107*

21. Using the contour diagram in Figure 11.106, sketch the graphs of the one-variable functions $H(x, 5)$ and $H(x, 20)$. Interpret the two graphs in practical terms, and explain the difference between them.

22. Figure 11.107 shows the contour diagram for the vibrating string function from page 8:

$$f(x, t) = \cos t \sin x, \quad 0 \le x \le \pi.$$

Using the diagram, describe in words the cross-sections of f with t fixed and the cross-sections of f with x fixed. Explain what you see in terms of the behavior of the string.

Find equations for the linear functions with the contour diagrams in Problems 23–24.

23.

24.

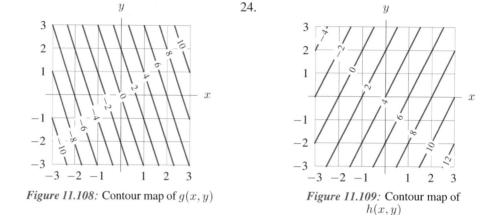

Figure 11.108: Contour map of $g(x, y)$

Figure 11.109: Contour map of $h(x, y)$

25. Figure 11.110 shows the contours of light intensity as a function of location and time in a microscopic wave-guide.

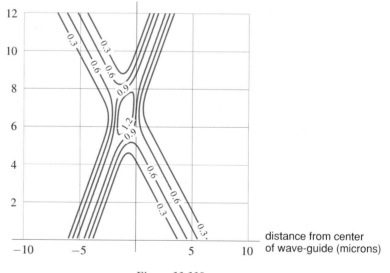

Figure 11.110

(a) Draw graphs showing intensity as a function of location at times 0, 2, 4, 6, 8, and 10 nanoseconds.

(b) If you could create an animation showing how the graph of intensity as a function of location varies as time passes, what would it look like?

(c) Draw a graph of intensity as a function of time at locations -5, 0 and 5 microns from center of wave-guide.

(d) Describe what the light beams are doing in this wave-guide.

CHAPTER TWELVE

A FUNDAMENTAL TOOL: VECTORS

In one-variable calculus we represented quantities such as velocity by numbers. However, to specify the velocity of a moving object in space, we need to say how fast it is moving and in what direction it is moving. In this chapter *vectors* are used to represent quantities that have direction as well as magnitude.

12.1 DISPLACEMENT VECTORS

Suppose you are a pilot planning a flight from Dallas to Pittsburgh. There are two things you must know: the distance to be traveled (so you have enough fuel to make it) and in what direction to go (so you don't miss Pittsburgh). Both these quantities together specify the displacement or *displacement vector* between the two cities.

> The **displacement vector** from one point to another is an arrow with its tail at the first point and its tip at the second. The **magnitude** (or length) of the displacement vector is the distance between the points, and is represented by the length of the arrow. The **direction** of the displacement vector is the direction of the arrow.

Figure 12.1 shows the displacement vectors from Dallas to Pittsburgh, from Albuquerque to Oshkosh, and from Los Angeles to Buffalo, SD. These displacement vectors have the same length and the same direction. We say that the displacement vectors between the corresponding cities are the same, even though they do not coincide. In other words

> Displacement vectors which point in the same direction and have the same magnitude are considered to be the same, even if they do not coincide.

Figure 12.1: Displacement vectors between cities

Notation and Terminology

The displacement vector is our first example of a vector. Vectors have both magnitude and direction; in comparison, a quantity specified only by a number, but no direction, is called a *scalar*.[1] For instance, the time taken by the flight from Dallas to Pittsburgh is a scalar quantity. Displacement is a vector since it requires both distance and direction to specify it.

In this book, vectors are written with an arrow over them, $\vec{v}$, to distinguish them from scalars. Other books use a bold **v** to denote a vector. We use the notation $\overrightarrow{PQ}$ to denote the displacement vector from a point P to a point Q. The magnitude, or length, of a vector $\vec{v}$ is written $\|\vec{v}\|$.

Addition and Subtraction of Displacement Vectors

Suppose NASA commands a robot on Mars to move 75 meters in one direction and then 50 meters in another direction. (See Figure 12.2.) Where does the robot end up? Suppose the displacements are represented by the vectors $\vec{v}$ and $\vec{w}$, respectively. Then the sum $\vec{v} + \vec{w}$ gives the final position.

---
[1]So named by W. R. Hamilton because they are merely numbers on the *scale* from $-\infty$ to ∞.

The **sum**, $\vec{v} + \vec{w}$, of two vectors $\vec{v}$ and $\vec{w}$ is the combined displacement resulting from first applying $\vec{v}$ and then $\vec{w}$. (See Figure 12.3.) The sum $\vec{w} + \vec{v}$ gives the same displacement.

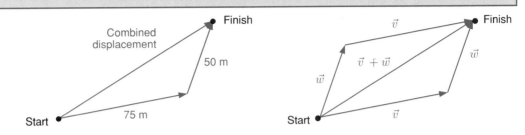

Figure 12.2: Sum of displacements of robots on Mars   *Figure 12.3:* The sum $\vec{v} + \vec{w} = \vec{w} + \vec{v}$

Suppose two different robots start from the same location. One moves along a displacement vector $\vec{v}$ and the second along a displacement vector $\vec{w}$. What is the displacement vector, $\vec{x}$, from the first robot to the second? (See Figure 12.4.) Since $\vec{v} + \vec{x} = \vec{w}$, we define $\vec{x}$ to be the difference $\vec{x} = \vec{w} - \vec{v}$. In other words, $\vec{w} - \vec{v}$ gets you from $\vec{v}$ to $\vec{w}$.

The **difference**, $\vec{w} - \vec{v}$, is the displacement vector which when added to $\vec{v}$ gives $\vec{w}$. That is, $\vec{w} = \vec{v} + (\vec{w} - \vec{v})$. (See Figure 12.4.)

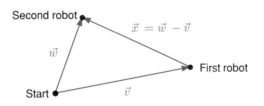

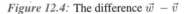

Figure 12.4: The difference $\vec{w} - \vec{v}$

If the robot ends up where it started, then its total displacement vector is the *zero vector*, $\vec{0}$. The zero vector has no direction.

The **zero vector**, $\vec{0}$, is a displacement vector with zero length.

Scalar Multiplication of Displacement Vectors

If $\vec{v}$ represents a displacement vector, the vector $2\vec{v}$ represents a displacement of twice the magnitude in the same direction as $\vec{v}$. Similarly, $-2\vec{v}$ represents a displacement of twice the magnitude in the opposite direction. (See Figure 12.5.)

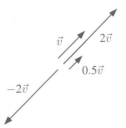

Figure 12.5: Scalar multiples of the vector $\vec{v}$

> If λ is a scalar and $\vec{v}$ is a displacement vector, the **scalar multiple of** $\vec{v}$ **by** λ, written $\lambda\vec{v}$, is the displacement vector with the following properties:
> - The displacement vector $\lambda\vec{v}$ is parallel to $\vec{v}$, pointing in the same direction if $\lambda > 0$, and in the opposite direction if $\lambda < 0$.
> - The magnitude of $\lambda\vec{v}$ is $|\lambda|$ times the magnitude of $\vec{v}$, that is, $\|\lambda\vec{v}\| = |\lambda| \|\vec{v}\|$.

Note that $|\lambda|$ represents the absolute value of the scalar λ while $\|\lambda\vec{v}\|$ represents the magnitude of the vector $\lambda\vec{v}$.

Example 1   Explain why $\vec{w} - \vec{v} = \vec{w} + (-1)\vec{v}$.

Solution   The vector $(-1)\vec{v}$ has the same magnitude as $\vec{v}$, but points in the opposite direction. Figure 12.6 shows that the combined displacement $\vec{w} + (-1)\vec{v}$ is the same as the displacement $\vec{w} - \vec{v}$.

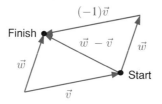

Figure 12.6: Explanation for why $\vec{w} - \vec{v} = \vec{w} + (-1)\vec{v}$

Parallel Vectors

Two vectors $\vec{v}$ and $\vec{w}$ are *parallel* if one is a scalar multiple of the other, that is, if $\vec{w} = \lambda\vec{v}$.

Components of Displacement Vectors: The Vectors $\vec{i}$, $\vec{j}$, and $\vec{k}$

Suppose that you live in a city with equally spaced streets running east-west and north-south and that you want to tell someone how to get from one place to another. You'd be likely to tell them how many blocks east-west and how many blocks north-south to go. For example, to get from P to Q in Figure 12.7, we go 4 blocks east and 1 block south. If $\vec{i}$ and $\vec{j}$ are as shown in Figure 12.7, then the displacement vector from P to Q is $4\vec{i} - \vec{j}$.

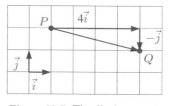

Figure 12.7: The displacement vector from P to Q is $4\vec{i} - \vec{j}$

We extend the same idea to 3-dimensions. First we choose a Cartesian system of coordinate axes. The three vectors of length 1 shown in Figure 12.8 are the vector $\vec{i}$, which points along the positive x-axis, the vector $\vec{j}$, along the positive y-axis, and the vector $\vec{k}$, along the positive z-axis.

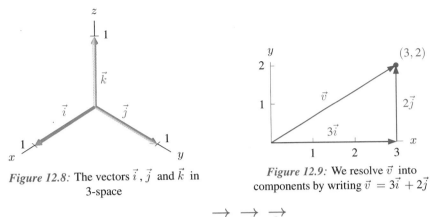

Figure 12.8: The vectors $\vec{i}$, $\vec{j}$ and $\vec{k}$ in 3-space

Figure 12.9: We resolve $\vec{v}$ into components by writing $\vec{v} = 3\vec{i} + 2\vec{j}$

Writing Displacement Vectors Using $\overrightarrow{i}$, $\overrightarrow{j}$, $\overrightarrow{k}$

Any displacement in 3-space or the plane can be expressed as a combination of displacements in the coordinate directions. For example, Figure 12.9 shows that the displacement vector $\vec{v}$ from the origin to the point $(3, 2)$ can be written as a sum of displacement vectors along the x and y-axes:

$$\vec{v} = 3\vec{i} + 2\vec{j}.$$

This is called *resolving $\vec{v}$ into components*. In general:

> We **resolve** $\vec{v}$ into components by writing $\vec{v}$ in the form
>
> $$\vec{v} = v_1\vec{i} + v_2\vec{j} + v_3\vec{k}.$$
>
> We call $v_1\vec{i}$, $v_2\vec{j}$, and $v_3\vec{k}$ the **components** of $\vec{v}$.

An Alternative Notation for Vectors

Many people write a vector in 3-dimensions as a string of three numbers, that is, as

$$\vec{v} = (v_1, v_2, v_3) \quad \text{instead of} \quad \vec{v} = v_1\vec{i} + v_2\vec{j} + v_3\vec{k}.$$

Since the first notation can be confused with a point and the second cannot, we usually use the second form.

Example 2 Resolve the displacement vector, $\vec{v}$, from the point $P_1 = (2, 4, 10)$ to the point $P_2 = (3, 7, 6)$ into components.

Solution To get from P_1 to P_2, we move 1 unit in the positive x-direction, 3 units in the positive y-direction, and 4 units in the negative z-direction. Hence $\vec{v} = \vec{i} + 3\vec{j} - 4\vec{k}$.

Example 3 Decide whether the vector $\vec{v} = 2\vec{i} + 3\vec{j} + 5\vec{k}$ is parallel to each of the following vectors:

$$\vec{w} = 4\vec{i} + 6\vec{j} + 10\vec{k}, \quad \vec{a} = -\vec{i} - 1.5\vec{j} - 2.5\vec{k}, \quad \vec{b} = 4\vec{i} + 6\vec{j} + 9\vec{k}$$

Solution Since $\vec{w} = 2\vec{v}$ and $\vec{a} = -0.5\vec{v}$, the vectors $\vec{v}$, $\vec{w}$, and $\vec{a}$ are parallel. However, $\vec{b}$ is not a multiple of $\vec{v}$ (since, for example, $4/2 \neq 9/5$), so $\vec{v}$ and $\vec{b}$ are not parallel.

In general, Figure 12.10 shows us how to express the displacement vector between two points in components:

Components of Displacement Vectors

The displacement vector from the point $P_1 = (x_1, y_1, z_1)$ to the point $P_2 = (x_2, y_2, z_2)$ is given in components by

$$\overrightarrow{P_1P_2} = (x_2 - x_1)\vec{i} + (y_2 - y_1)\vec{j} + (z_2 - z_1)\vec{k}.$$

Position Vectors: Displacement of a Point from the Origin

A displacement vector whose tail is at the origin is called a *position vector*. Thus, any point (x_0, y_0, z_0) in space has associated with it the position vector $\vec{r}_0 = x_0\vec{i} + y_0\vec{j} + z_0\vec{k}$. (See Figure 12.11.) In general, a position vector gives the displacement of a point from the origin.

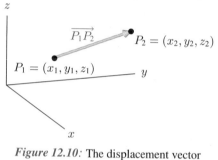

Figure 12.10: The displacement vector
$\overrightarrow{P_1P_2} = (x_2 - x_1)\vec{i} + (y_2 - y_1)\vec{j} + (z_2 - z_1)\vec{k}$

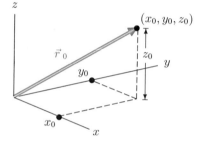

Figure 12.11: The position vector
$\vec{r}_0 = x_0\vec{i} + y_0\vec{j} + z_0\vec{k}$

The Components of the Zero Vector

The zero displacement vector has magnitude equal to zero and is written $\vec{0}$. So $\vec{0} = 0\vec{i} + 0\vec{j} + 0\vec{k}$.

The Magnitude of a Vector in Components

For a vector, $\vec{v} = v_1\vec{i} + v_2\vec{j}$, the Pythagorean theorem is used to find its magnitude, $\|\vec{v}\|$. (See Figure 12.12.)

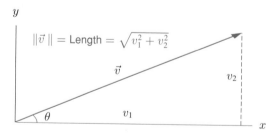

Figure 12.12: Magnitude, $\|\vec{v}\|$, of a 2-dimensional vector, $\vec{v}$

In 3-dimensions, for a vector $\vec{v} = v_1\vec{i} + v_2\vec{j} + v_3\vec{k}$, we have

---

Magnitude of $\vec{v}$ $= \|\vec{v}\| = $ Length of the arrow $= \sqrt{v_1^2 + v_2^2 + v_3^2}.$

For instance, if $\vec{v} = 3\vec{i} - 4\vec{j} + 5\vec{k}$, then $\|\vec{v}\| = \sqrt{3^2 + (-4)^2 + 5^2} = \sqrt{50}$.

Addition and Scalar Multiplication of Vectors in Components

Suppose the vectors $\vec{v}$ and $\vec{w}$ are given in components:

$$\vec{v} = v_1\vec{i} + v_2\vec{j} + v_3\vec{k} \quad \text{and} \quad \vec{w} = w_1\vec{i} + w_2\vec{j} + w_3\vec{k}.$$

Then

$$\vec{v} + \vec{w} = (v_1 + w_1)\vec{i} + (v_2 + w_2)\vec{j} + (v_3 + w_3)\vec{k},$$

and

$$\lambda\vec{v} = \lambda v_1\vec{i} + \lambda v_2\vec{j} + \lambda v_3\vec{k}.$$

Figures 12.13 and 12.14 illustrate these properties in two dimensions. Finally, $\vec{v} - \vec{w} = \vec{v} + (-1)\vec{w}$, so we can write $\vec{v} - \vec{w} = (v_1 - w_1)\vec{i} + (v_2 - w_2)\vec{j} + (v_3 - w_3)\vec{k}$.

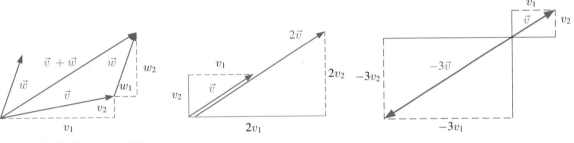

Figure 12.13: Sum $\vec{v} + \vec{w}$ in components

Figure 12.14: Scalar multiples of vectors showing $\vec{v}$, $2\vec{v}$, and $-3\vec{v}$

How to Resolve a Vector into Components

You may wonder how we find the components of a 2-dimensional vector, given its length and direction. Suppose the vector $\vec{v}$ has length v and makes an angle of θ with the x-axis, measured counterclockwise, as in Figure 12.15. If $\vec{v} = v_1\vec{i} + v_2\vec{j}$, Figure 12.15 shows that

$$v_1 = v\cos\theta \quad \text{and} \quad v_2 = v\sin\theta.$$

Thus, we resolve $\vec{v}$ into components by writing

$$\vec{v} = (v\cos\theta)\vec{i} + (v\sin\theta)\vec{j}.$$

Vectors in 3-space are resolved using direction cosines; see Problem 29 on page 95.

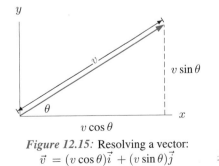

Figure 12.15: Resolving a vector:
$\vec{v} = (v\cos\theta)\vec{i} + (v\sin\theta)\vec{j}$

Example 4   Resolve $\vec{v}$ into components if $v = 2$ and $\theta = \pi/6$.

Solution   We have $\vec{v} = 2\cos(\pi/6)\vec{i} + 2\sin(\pi/6)\vec{j} = 2(\sqrt{3}/2)\,\vec{i} + 2(1/2)\vec{j} = \sqrt{3}\vec{i} + \vec{j}$.

Unit Vectors

A *unit vector* is a vector whose magnitude is 1. The vectors $\vec{i}$, $\vec{j}$, and $\vec{k}$ are unit vectors in the directions of the coordinate axes. It is often helpful to find a unit vector in the same direction as a given vector $\vec{v}$. Suppose that $\|\vec{v}\| = 10$; a unit vector in the same direction as $\vec{v}$ is $\vec{v}/10$. In general, a unit vector in the direction of any nonzero vector $\vec{v}$ is

$$\vec{u} = \frac{\vec{v}}{\|\vec{v}\|}.$$

Example 5   Find a unit vector, $\vec{u}$, in the direction of the vector $\vec{v} = \vec{i} + 3\vec{j}$.

Solution   If $\vec{v} = \vec{i} + 3\vec{j}$, then $\|\vec{v}\| = \sqrt{1^2 + 3^2} = \sqrt{10}$. Thus, a unit vector in the same direction is given by

$$\vec{u} = \frac{\vec{v}}{\sqrt{10}} = \frac{1}{\sqrt{10}}(\vec{i} + 3\vec{j}) = \frac{1}{\sqrt{10}}\vec{i} + \frac{3}{\sqrt{10}}\vec{j} \approx 0.32\vec{i} + 0.95\vec{j}.$$

Example 6   Find a unit vector at the point (x, y, z) that points radially outward away from the origin.

Solution   The vector from the origin to (x, y, z) is the position vector

$$\vec{r} = x\vec{i} + y\vec{j} + z\vec{k}.$$

Thus, if we put its tail at (x, y, z) it will point away from the origin. Its magnitude is

$$\|\vec{r}\| = \sqrt{x^2 + y^2 + z^2},$$

so a unit vector pointing in the same direction is

$$\frac{\vec{r}}{\|\vec{r}\|} = \frac{x\vec{i} + y\vec{j} + z\vec{k}}{\sqrt{x^2 + y^2 + z^2}} = \frac{x}{\sqrt{x^2 + y^2 + z^2}}\vec{i} + \frac{y}{\sqrt{x^2 + y^2 + z^2}}\vec{j} + \frac{z}{\sqrt{x^2 + y^2 + z^2}}\vec{k}.$$

Problems for Section 12.1

1.   The vectors $\vec{w}$ and $\vec{u}$ are in Figure 12.16. Match the vectors $\vec{p}, \vec{q}, \vec{r}, \vec{s}, \vec{t}$ with five of the following vectors: $\vec{u} + \vec{w}$, $\vec{u} - \vec{w}$, $\vec{w} - \vec{u}$, $2\vec{w} - \vec{u}$, $\vec{u} - 2\vec{w}$, $2\vec{w}$, $-2\vec{w}$, $2\vec{u}$, $-2\vec{u}$, $-\vec{w}$, $-\vec{u}$.

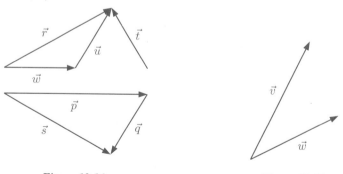

Figure 12.16   *Figure 12.17*

2. Given the displacement vectors $\vec{v}$ and $\vec{w}$ in Figure 12.17, draw the following vectors:

    (a) $\vec{v} + \vec{w}$ (b) $\vec{v} - \vec{w}$ (c) $2\vec{v}$ (d) $2\vec{v} + \vec{w}$ (e) $\vec{v} - 2\vec{w}$.

For Problems 3–8, perform the indicated operations on the following vectors:

$$\vec{a} = 2\vec{j} + \vec{k}, \quad \vec{b} = -3\vec{i} + 5\vec{j} + 4\vec{k}, \quad \vec{c} = \vec{i} + 6\vec{j},$$

$$\vec{x} = -2\vec{i} + 9\vec{j}, \quad \vec{y} = 4\vec{i} - 7\vec{j}, \quad \vec{z} = \vec{i} - 3\vec{j} - \vec{k}.$$

3. $\|\vec{z}\|$ 4. $\vec{a} + \vec{z}$ 5. $5\vec{b}$

6. $2\vec{c} + \vec{x}$ 7. $\|\vec{y}\|$ 8. $2\vec{a} + 7\vec{b} - 5\vec{z}$

For Problems 9–12, perform the indicated computation.

9. $(4\vec{i} + 2\vec{j}) - (3\vec{i} - \vec{j})$ 10. $(\vec{i} + 2\vec{j}) + (-3)(2\vec{i} + \vec{j})$

11. $-4(\vec{i} - 2\vec{j}) - 0.5(\vec{i} - \vec{k})$ 12. $2(0.45\vec{i} - 0.9\vec{j} - 0.01\vec{k}) - 0.5(1.2\vec{i} - 0.1\vec{k})$

Find the length of the vectors in Problems 13–16.

13. $\vec{v} = \vec{i} - \vec{j} + 3\vec{k}$ 14. $\vec{v} = \vec{i} - \vec{j} + 2\vec{k}$

15. $\vec{v} = 1.2\vec{i} - 3.6\vec{j} + 4.1\vec{k}$ 16. $\vec{v} = 7.2\vec{i} - 1.5\vec{j} + 2.1\vec{k}$

A cat is sitting on the ground at the point $(1, 4, 0)$ watching a squirrel at the top of a tree. The tree is one unit high and its base is at the point $(2, 4, 0)$. Find the displacement vectors in Problems 17–20.

17. From the origin to the cat.   18. From the bottom of the tree to the squirrel.

19. From the bottom of the tree to the cat.   20. From the cat to the squirrel.

21. On the graph of Figure 12.18, draw the vector $\vec{v} = 4\vec{i} + \vec{j}$ twice, once with its tail at the origin and once with its tail at the point $(3, 2)$.

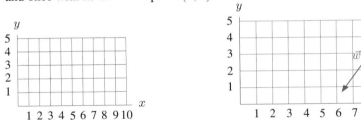

Figure 12.18   **Figure 12.19**: Scale: 1 grid length = 0.25 inches

Resolve the vectors in Problems 22–26 into components.

22. The vector shown in Figure 12.19.
23. A vector starting at the point $P = (1, 2)$ and ending at the point $Q = (4, 6)$.
24. A vector starting at the point $Q = (4, 6)$ and ending at the point $P = (1, 2)$.

25.   26.

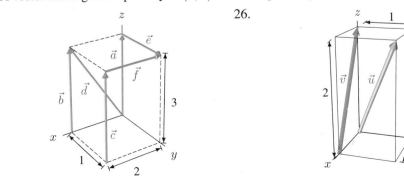

27. Find the length of the vectors $\vec{u}$ and $\vec{v}$ in Problem 26.

28. Find a vector that points in the same direction as $\vec{i} - \vec{j} + 2\vec{k}$, but has length 2.

29. (a) Find a unit vector from the point $P = (1, 2)$ and toward the point $Q = (4, 6)$.
    (b) Find a vector of length 10 pointing in the same direction.

30. Which of the following vectors are parallel?

$$\vec{u} = 2\vec{i} + 4\vec{j} - 2\vec{k}, \quad \vec{v} = \vec{i} - \vec{j} + 3\vec{k}, \quad \vec{w} = -\vec{i} - 2\vec{j} + \vec{k},$$
$$\vec{p} = \vec{i} + \vec{j} + \vec{k}, \quad \vec{q} = 4\vec{i} - 4\vec{j} + 12\vec{k}, \quad \vec{r} = \vec{i} - \vec{j} + \vec{k}.$$

31. Figure 12.20 shows a molecule with four atoms at O, A, B and C. Verify that every atom in the molecule is 2 units away from every other atom.

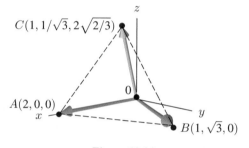

Figure 12.20

For Problems 32–33, consider the map in Figure 11.1 on page 2.

32. If you leave Topeka along the following vectors, does the temperature increase or decrease?
    (a) $\vec{u} = 3\vec{i} + 2\vec{j}$ (b) $\vec{v} = -\vec{i} - \vec{j}$ (c) $\vec{w} = -5\vec{i} - 5\vec{j}$.

33. Starting in Buffalo, sketch a vector pointing in the direction in which the temperature is increasing most rapidly. Starting in Boise, sketch a vector pointing in the direction in which the temperature is decreasing most rapidly.

34. A truck is traveling due north at 30 km/hr approaching a crossroad. On a perpendicular road a police car is traveling west toward the intersection at 40 km/hr. Suppose that both vehicles will reach the crossroad in exactly one hour. Find the vector currently representing the displacement of the truck with respect to the police car.

35. Show that the medians of a triangle intersect at a point $1/3$ of the way along each median from the side it bisects.

36. Show that the lines joining the centroid (the intersection point of the medians) of a face of the tetrahedron and the opposite vertex meet at a point $\frac{1}{4}$ of the way from each centroid to its opposite vertex.

12.2 VECTORS IN GENERAL

Besides displacement, there are many quantities that have both magnitude and direction and are added and multiplied by scalars in the same way as displacements. Any such quantity is called a *vector* and is represented by an arrow in the same manner we represent displacements. The length of the arrow is the *magnitude* of the vector, and the direction of the arrow is the direction of the vector.

Velocity Versus Speed

The speed of a moving body tells us how fast it is moving, say 80 km/hr. The speed is just a number; it is therefore a scalar. The velocity, on the other hand, tells us both how fast the body is moving and the direction of motion; it is a vector. For instance, if a car is heading northeast at 80 km/hr, then its velocity is a vector of length 80 pointing northeast.

> The **velocity vector** of a moving object is a vector whose magnitude is the speed of the object and whose direction is the direction of its motion.

Example 1   A car is traveling north at a speed of 100 km/hr, while a plane above is flying horizontally south-west at a speed of 500 km/hr. Draw the velocity vectors of the car and the plane.

Solution   Figure 12.21 shows the velocity vectors. The plane's velocity vector is five times as long as the car's, because its speed is five times as great.

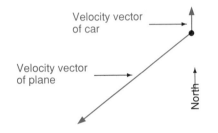

Velocity vector of car

Velocity vector of plane

North

Figure 12.21: Velocity vector of the car is 100 km/hr north and of the plane is 500 km/hr south-west

The next example illustrates that the velocity vectors for two motions add to give the velocity vector for the combined motion, just as displacements do.

Example 2   A riverboat is moving with velocity $\vec{v}$ and a speed of 8 km/hr relative to the water. In addition, the river has a current $\vec{c}$ and a speed of 1 km/hr. (See Figure 12.22.) What is the physical significance of the vector $\vec{v} + \vec{c}$?

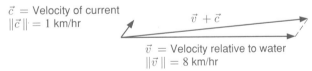

$\vec{c}$ = Velocity of current
$\|\vec{c}\| = 1$ km/hr

$\vec{v} + \vec{c}$

$\vec{v}$ = Velocity relative to water
$\|\vec{v}\| = 8$ km/hr

Figure 12.22: Boat's velocity relative to the river bed is the sum, $\vec{v} + \vec{c}$

Solution   The vector $\vec{v}$ shows how the boat is moving relative to the water, while $\vec{c}$ shows how the water is moving relative to the riverbed. During an hour, imagine that the boat first moves 8 km relative to the water, which remains still; this displacement is represented by $\vec{v}$. Then imagine the water moving 1 km while the boat remains stationary relative to the water; this displacement is represented by $\vec{c}$. The combined displacement is represented by $\vec{v} + \vec{c}$. Thus, the vector $\vec{v} + \vec{c}$ is the velocity of the boat relative to the riverbed.

Note that the effective speed of the boat is not necessarily 9 km/hr unless the boat is moving in the direction of the current. Although we add the velocity vectors, we do not necessarily add their lengths.

Scalar multiplication also makes sense for velocity vectors. For example, if $\vec{v}$ is a velocity vector, then $-2\vec{v}$ represents a velocity of twice the magnitude in the opposite direction.

Example 3   A ball is moving with velocity $\vec{v}$ when it hits a wall at a right angle and bounces straight back, with its speed reduced by 20%. Express its new velocity in terms of the old one.

Solution   The new velocity is $-0.8\vec{v}$, where the negative sign expresses the fact that the new velocity is in the direction opposite to the old.

We can represent velocity vectors in components in the same way we did on page 65.

Example 4   Represent the velocity vectors of the car and the plane in Example 1 using components. Take north to be the positive y-axis, east to be the positive x-axis, and upward to be the positive z-axis.

Solution   The car is traveling north at 100 km/hr, so the y-component of its velocity is $100\vec{j}$ and the x-component is $0\vec{i}$. Since it is traveling horizontally, the z-component is $0\vec{k}$. So we have

$$\text{Velocity of car} = 0\vec{i} + 100\vec{j} + 0\vec{k} = 100\vec{j}.$$

The plane's velocity vector also has $\vec{k}$ component equal to zero. Since it is traveling southwest, its $\vec{i}$ and $\vec{j}$ components have negative coefficients (north and east are positive). Since the plane is traveling at 500 km/hr, in one hour it is displaced $500/\sqrt{2} \approx 354$ km to the west and 354 km to the south. (See Figure 12.23.) Thus,

$$\text{Velocity of plane} = -(500\cos 45°)\vec{i} - (500\sin 45°)\vec{j} \approx -354\vec{i} - 354\vec{j}.$$

Of course, if the car were climbing a hill or if the plane were descending for a landing, then the $\vec{k}$ component would not be zero.

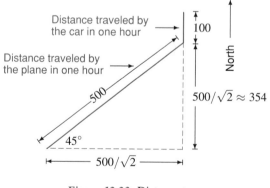

Figure 12.23: Distance traveled by the plane and car in one hour

Acceleration

Another example of a vector quantity is acceleration. Acceleration, like velocity, is specified by both a magnitude and a direction — for example, the acceleration due to gravity is 9.81 m/sec^2 vertically downward.

Force

Force is another example of a vector quantity. Suppose you push on an open door. The result depends both on how hard you push and in what direction. Thus, to specify a force we must give its magnitude (or strength) and the direction in which it is acting. For example, the gravitational force exerted on an object by the earth is a vector pointing from the object toward the center of the earth; its magnitude is the strength of the gravitational force.

Example 5   The earth travels around the sun in an ellipse. The gravitational force on the earth and the velocity of the earth are governed by the following laws:

Newton's Law of Gravitation: The magnitude of the gravitational attraction, F, between two masses m_1 and m_2 at a distance r apart is given by $F = Gm_1m_2/r^2$, where G is a constant. The force vector lies along the line between the masses.

Kepler's Second Law: The line joining a planet to the sun sweeps out equal areas in equal times.

(a)   Sketch vectors representing the gravitational force of the sun on the earth at two different positions in the earth's orbit.

(b)   Sketch the velocity vector of the earth at two points in its orbit.

Solution   (a)   Figure 12.24 shows the earth orbiting the sun. Note that the gravitational force vector always points toward the sun and is larger when the earth is closer to the sun because of the r^2 term in the denominator. (In fact, the real orbit looks much more like a circle than we have shown here.)

(b)   The velocity vector points in the direction of motion of the earth. Thus, the velocity vector is tangent to the ellipse. See Figure 12.25. Furthermore, the velocity vector is longer at points of the orbit where the planet is moving quickly, because the magnitude of the velocity vector is the speed. Kepler's Second Law enables us to determine when the earth is moving quickly and when it is moving slowly. Over a fixed period of time, say one month, the line joining the earth to the sun sweeps out a sector having a certain area. Figure 12.25 shows two sectors swept out in two different one-month time-intervals. Kepler's law says that the areas of the two sectors are the same. Thus, the earth must move farther in a month when it is close to the sun than when it is far from the sun. Therefore, the earth moves faster when it is closer to the sun and slower when it is farther away.

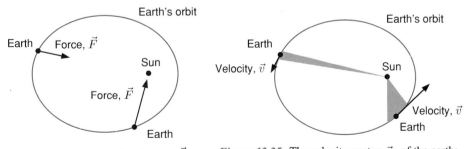

Figure 12.24: Gravitational force, $\vec{F}$, exerted by the sun on the earth: Greater magnitude closer to sun

Figure 12.25: The velocity vector, $\vec{v}$, of the earth: Greater magnitude closer to the sun

Properties of Addition and Scalar Multiplication

In general, vectors add, subtract, and are multiplied by scalars in the same way as displacement vectors. Thus, for any vectors $\vec{u}, \vec{v}$, and $\vec{w}$ and any scalars α and β, we have the following properties:

Commutativity

1. $\vec{v} + \vec{w} = \vec{w} + \vec{v}$

Associativity

2. $(\vec{u} + \vec{v}) + \vec{w} = \vec{u} + (\vec{v} + \vec{w})$

3. $\alpha(\beta\vec{v}) = (\alpha\beta)\vec{v}$

Distributivity

4. $(\alpha + \beta)\vec{v} = \alpha\vec{v} + \beta\vec{v}$

5. $\alpha(\vec{v} + \vec{w}) = \alpha\vec{v} + \alpha\vec{w}$

Identity

6. $1 \cdot \vec{v} = \vec{v}$ **8.** $\vec{v} + \vec{0} = \vec{v}$

7. $0 \cdot \vec{v} = \vec{0}$ **9.** $\vec{w} + (-1) \cdot \vec{v} = \vec{w} - \vec{v}$

Problems 16–23 at the end of this section ask for a justification of these results in terms of displacement vectors.

Using Components

Example 6   A plane, heading due east at an airspeed of 600 km/hr, experiences a wind of 50 km/hr blowing toward the northeast. Find the plane's direction and ground speed.

Solution   We choose a coordinate system with the x-axis pointing east and the y-axis pointing north. See Figure 12.26.

The airspeed tells us the speed of the plane relative to still air. Thus, the plane is moving due east with velocity $\vec{v} = 600\vec{i}$ relative to still air. In addition, the air is moving with a velocity $\vec{w}$. Writing $\vec{w}$ in components, we have

$$\vec{w} = (50\cos45°)\vec{i} + (50\sin45°)\vec{j} = 35.4\vec{i} + 35.4\vec{j}.$$

The vector $\vec{v} + \vec{w}$ represents the displacement of the plane in one hour relative to the ground. Therefore, $\vec{v} + \vec{w}$ is the velocity of the plane relative to the ground. In components, we have

$$\vec{v} + \vec{w} = 600\vec{i} + \left(35.4\vec{i} + 35.4\vec{j}\right) = 635.4\vec{i} + 35.4\vec{j}.$$

The direction of the plane's motion relative to the ground is given by the angle θ in Figure 12.26, where

$$\tan\theta = \frac{35.4}{635.4}$$

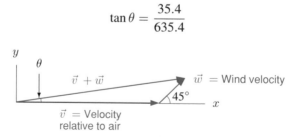

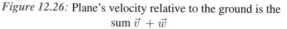

Figure 12.26: Plane's velocity relative to the ground is the sum $\vec{v} + \vec{w}$

so
$$\theta = \arctan\left(\frac{35.4}{635.4}\right) = 3.2°.$$

The ground speed is the speed of the plane relative to the ground, so

$$\text{Groundspeed} = \sqrt{635.4^2 + 35.4^2} = 636.4 \text{ km/hr}.$$

Thus, the speed of the plane relative to the ground has been increased slightly by the wind. (This is as we would expect, as the wind has a component in the direction in which the plane is traveling.) The angle θ shows how far the plane is blown off course by the wind.

Vectors in n Dimensions

Using the alternative notation $\vec{v} = (v_1, v_2, v_3)$ for a vector in 3-space, we can define a vector in n dimensions as a string of n numbers. Thus, a vector in n dimensions can be written as

$$\vec{c} = (c_1, c_2, \ldots, c_n).$$

Addition and scalar multiplication are defined by the formulas

$$\vec{v} + \vec{w} = (v_1, v_2, \ldots, v_n) + (w_1, w_2, \ldots, w_n) = (v_1 + w_1, v_2 + w_2, \ldots, v_n + w_n)$$

and

$$\lambda\vec{v} = \lambda(v_1, v_2, \ldots, v_n) = (\lambda v_1, \lambda v_2, \ldots, \lambda v_n).$$

Why Do We Want Vectors in n Dimensions?

Vectors in two and three dimensions can be used to model displacement, velocities, or forces. But what about vectors in n dimensions? There is another interpretation of 3-dimensional vectors (or 3-vectors) which is useful: they can be thought of as listing 3 different quantities — for example, the displacements parallel to the x, y, and z axes. Similarly, the n-vector

$$\vec{c} = (c_1, c_2, \ldots, c_n)$$

can be thought of as a way of keeping n different quantities organized. For example, a *population* vector $\vec{N}$ shows the number of children and adults in a population:

$$\vec{N} = (\text{Number of children, Number of adults}),$$

or, if we are interested in a more detailed breakdown of ages, we might give the number in each ten-year age bracket in the population (up to age 110) in the form

$$\vec{N} = (N_1, N_2, N_3, N_4, \ldots, N_{10}, N_{11}),$$

where N_1 is the population aged 0–9, and N_2 is the population aged 10–19, and so on.

A *consumption* vector ,

$$\vec{q} = (q_1, q_2, \ldots, q_n)$$

shows the quantities $q_1, q_2, \ldots, q_n$ consumed of each of n different goods. A *price* vector

$$\vec{p} = (p_1, p_2, \ldots, p_n)$$

contains the prices of n different items.

In 1907, Hermann Minkowski used vectors with four components when he introduced *space-time coordinates,* whereby each event is assigned a vector position $\vec{v}$ with four coordinates, three for its position in space and one for time:

$$\vec{v} = (x, y, z, t).$$

Example 7   Suppose the vector $\vec{I}$ represents the number of copies, in thousands, made by each of four copy centers in the month of December and $\vec{J}$ represents the number of copies made at the same four copy centers during the previous eleven months (the "year-to-date"). If $\vec{I} = (25, 211, 818, 642)$, and $\vec{J} = (331, 3227, 1377, 2570)$, compute $\vec{I} + \vec{J}$. What does this sum represent?

Solution   The sum is

$$\vec{I} + \vec{J} = (25 + 331, 211 + 3227, 818 + 1377, 642 + 2570) = (356, 3438, 2195, 3212).$$

Each term in $\vec{I} + \vec{J}$ represents the sum of the number of copies made in December plus those in the previous eleven months, that is, the total number of copies made during the entire year at that particular copy center.

Example 8   The price vector $\vec{p} = (p_1, p_2, p_3)$ represents the prices in dollars of three goods. Write a vector which gives the prices of the same goods in cents.

Solution   The prices in cents are $100p_1$, $100p_2$, and $100p_3$ respectively, so the new price vector is

$$(100p_1, 100p_2, 100p_3) = 100\vec{p}.$$

Problems for Section 12.2

In Problems 1–4, say whether the given quantity is a vector or a scalar.

1. The distance from Seattle to St. Louis.
2. The population of the US.
3. The magnetic field at a point on the earth's surface.
4. The temperature at a point on the earth's surface.
5. A car is traveling at a speed of 50 km/hr. Assume the positive y-axis is north and the positive x-axis is east. Resolve the car's velocity vector (in 2-space) into components if the car is traveling in each of the following directions:
    (a) East   (b) South   (c) Southeast   (d) Northwest.
6. Which is traveling faster, a car whose velocity vector is $21\vec{i} + 35\vec{j}$, or a car whose velocity vector is $40\vec{i}$, assuming that the units are the same for both directions?
7. A moving object has velocity vector $50\vec{i} + 20\vec{j}$ in meters per second. Express the velocity in kilometers per hour.
8. A car drives clockwise around the track in Figure 12.27, slowing down at the curves and speeding up along the straight portions. Sketch velocity vectors at the points P, Q, and R.
9. A racing car drives clockwise around the track shown in Figure 12.27 at a constant speed. At what point on the track does the car have the longest acceleration vector, and in roughly what direction is it pointing? (Recall that acceleration is the rate of change of velocity.)

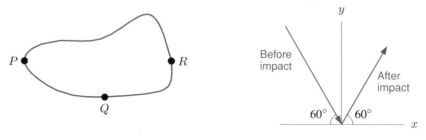

P R

Q

Figure 12.27

y

Before impact

After impact

$60°$ $60°$

x

Figure 12.28

10. A particle moving with speed v hits a barrier at an angle of $60°$ and bounces off at an angle of $60°$ in the opposite direction with speed reduced by 20 percent, as shown in Figure 12.28. Find the velocity vector of the object after impact.

11. There are five students in a class. Their scores on the midterm (out of 100) are given by the vector $\vec{v} = (73, 80, 91, 65, 84)$. Their scores on the final (out of 100) are given by $\vec{w} = (82, 79, 88, 70, 92)$. If the final counts twice as much as the midterm, find a vector giving the total scores (as a percentage) of the students.

12. Shortly after taking off, a plane is climbing northwest through still air at an airspeed of 200 km/hr, and rising at a rate of 300 m/min. Resolve into components its velocity vector in a coordinate system in which the x-axis points east, the y-axis points north, and the z-axis points up.

13. An airplane is heading northeast at an airspeed of 700 km/hr, but there is a wind blowing from the west at 60 km/hr. In what direction does the plane end up flying? What is its speed relative to the ground?

14. An airplane is flying at an airspeed of 600 km/hr in a cross-wind that is blowing from the northeast at a speed of 50 km/hr. In what direction should the plane head to end up going due east?

15. A plane is heading due east and climbing at the rate of 80 km/hr. If its airspeed is 480 km/hr and there is a wind blowing 100 km/hr to the northeast, what is the ground speed of the plane?

Use the geometric definition of addition and scalar multiplication to explain each of the properties in Problems 16–23.

16. $\vec{w} + \vec{v} = \vec{v} + \vec{w}$

17. $(\alpha + \beta)\vec{v} = \alpha\vec{v} + \beta\vec{v}$

18. $\alpha(\vec{v} + \vec{w}) = \alpha\vec{v} + \alpha\vec{w}$

19. $(\vec{u} + \vec{v}) + \vec{w} = \vec{u} + (\vec{v} + \vec{w})$

20. $\alpha(\beta\vec{v}) = (\alpha\beta)\vec{v}$

21. $\vec{v} + \vec{0} = \vec{v}$

22. $1\vec{v} = \vec{v}$

23. $\vec{v} + (-1)\vec{w} = \vec{v} - \vec{w}$

12.3 THE DOT PRODUCT

We have seen how to add vectors; can we multiply two vectors together? In the next two sections we will see two different ways of doing so: the *scalar product* (or *dot product*) which produces a scalar, and the *vector product* (or *cross product*), which produces a vector.

Definition of the Dot Product

The dot product links geometry and algebra. We already know how to calculate the length of a vector from its components; the dot product gives us a way of computing the angle between two vectors. For any two vectors $\vec{v} = v_1\vec{i} + v_2\vec{j} + v_3\vec{k}$ and $\vec{w} = w_1\vec{i} + w_2\vec{j} + w_3\vec{k}$, shown in Figure 12.29, we define a scalar as follows:

The following two definitions of the **dot product**, or **scalar product**, $\vec{v} \cdot \vec{w}$, are equivalent:
- **Geometric definition**

  $\vec{v} \cdot \vec{w} = \|\vec{v}\|\|\vec{w}\|\cos\theta$ where θ is the angle between $\vec{v}$ and $\vec{w}$ and $0 \leq \theta \leq \pi$.
- **Algebraic definition**

  $\vec{v} \cdot \vec{w} = v_1w_1 + v_2w_2 + v_3w_3$.

Notice that the dot product of two vectors is a *number*.

Why don't we give just one definition of $\vec{v} \cdot \vec{w}$? The reason is that both definitions are equally important; the geometric definition gives us a picture of what the dot product means and the algebraic definition gives us a way of calculating it.

How do we know the two definitions are equivalent — that is, they really do define the same thing? First, we observe that the two definitions give the same result in a particular example. Then we show why they are equivalent in general.

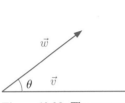

Figure 12.29: The vectors $\vec{v}$ and $\vec{w}$

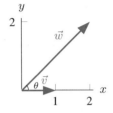

Figure 12.30: Calculating the dot product of the vectors $v = \vec{i}$ and $\vec{w} = 2\vec{i} + 2\vec{j}$ geometrically and algebraically gives the same result

---

Example 1   Suppose $\vec{v} = \vec{i}$ and $\vec{w} = 2\vec{i} + 2\vec{j}$. Compute $\vec{v} \cdot \vec{w}$ both geometrically and algebraically.

Solution   To use the geometric definition, see Figure 12.30. The angle between the vectors is $\pi/4$, or $45°$, and the lengths of the vectors are given by

$$\|\vec{v}\| = 1 \quad \text{and} \quad \|\vec{w}\| = 2\sqrt{2}.$$

Thus,

$$\vec{v} \cdot \vec{w} = \|\vec{v}\|\|\vec{w}\|\cos\theta = 1 \cdot 2\sqrt{2}\cos\left(\frac{\pi}{4}\right) = 2.$$

Using the algebraic definition, we get the same result:

$$\vec{v} \cdot \vec{w} = 1 \cdot 2 + 0 \cdot 2 = 2.$$

Why the Two Definitions of the Dot Product Give the Same Result

In the previous example, the two definitions give the same value for the dot product. To show that the geometric and algebraic definitions of the dot product always give the same result, we must show that, for any vectors $\vec{v} = v_1\vec{i} + v_2\vec{j} + v_3\vec{k}$ and $\vec{w} = w_1\vec{i} + w_2\vec{j} + w_3\vec{k}$ with an angle θ between them:

$$\|\vec{v}\|\|\vec{w}\|\cos\theta = v_1 w_1 + v_2 w_2 + v_3 w_3.$$

One method follows; a method which does not use trigonometry is given in Problem 33 on page 85.

Using the Law of Cosines. Suppose that $0 < \theta < \pi$, so that the vectors $\vec{v}$ and $\vec{w}$ form a triangle. (See Figure 12.31.) By the Law of Cosines, we have

$$\|\vec{v} - \vec{w}\|^2 = \|\vec{v}\|^2 + \|\vec{w}\|^2 - 2\|\vec{v}\|\|\vec{w}\|\cos\theta.$$

This result is also true for $\theta = 0$ and $\theta = \pi$. We calculate the lengths using components:

$$\|\vec{v}\|^2 = v_1^2 + v_2^2 + v_3^2$$
$$\|\vec{w}\|^2 = w_1^2 + w_2^2 + w_3^2$$
$$\|\vec{v} - \vec{w}\|^2 = (v_1 - w_1)^2 + (v_2 - w_2)^2 + (v_3 - w_3)^2$$
$$= v_1^2 - 2v_1 w_1 + w_1^2 + v_2^2 - 2v_2 w_2 + w_2^2 + v_3^2 - 2v_3 w_3 + w_3^2.$$

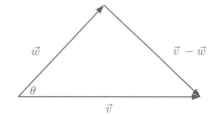

Figure 12.31: Triangle used in the justification of
$$\|\vec{v}\,\|\|\vec{w}\,\|\cos\theta = v_1 w_1 + v_2 w_2 + v_3 w_3$$

Substituting into the Law of Cosines and canceling, we see that

$$-2v_1 w_1 - 2v_2 w_2 - 2v_3 w_3 = -2\|\vec{v}\,\|\|\vec{w}\,\|\cos\theta.$$

Therefore we have the result we wanted, namely that:

$$v_1 w_1 + v_2 w_2 + v_3 w_3 = \|\vec{v}\,\|\|\vec{w}\,\|\cos\theta.$$

Properties of the Dot Product

The following properties of the dot product can be justified using the algebraic definition; see Problem 26 on page 84. For a geometric interpretation of Property 3, see Problem 30.

> **Properties of the Dot Product**. For any vectors $\vec{u}$, $\vec{v}$, and $\vec{w}$ and any scalar λ,
>
> 1. $\vec{v} \cdot \vec{w} = \vec{w} \cdot \vec{v}$
> 2. $\vec{v} \cdot (\lambda\vec{w}) = \lambda(\vec{v} \cdot \vec{w}) = (\lambda\vec{v}) \cdot \vec{w}$
> 3. $(\vec{v} + \vec{w}) \cdot \vec{u} = \vec{v} \cdot \vec{u} + \vec{w} \cdot \vec{u}$

Perpendicularity, Magnitude, and Dot Products

Two vectors are perpendicular if the angle between them is $\pi/2$ or $90°$. Since $\cos(\pi/2) = 0$, if $\vec{v}$ and $\vec{w}$ are perpendicular, then $\vec{v} \cdot \vec{w} = 0$. Conversely, provided that $\vec{v} \cdot \vec{w} = 0$, then $\cos\theta = 0$, so $\theta = \pi/2$ and the vectors are perpendicular. Thus, we have the following result:

> Two nonzero vectors $\vec{v}$ and $\vec{w}$ are **perpendicular**, or **orthogonal**, if and only if
>
> $$\vec{v} \cdot \vec{w} = 0.$$

For example: $\vec{i} \cdot \vec{j} = 0, \vec{j} \cdot \vec{k} = 0, \vec{i} \cdot \vec{k} = 0$.

If we take the dot product of a vector with itself, then $\theta = 0$ and $\cos\theta = 1$. For any vector $\vec{v}$:

> Magnitude and dot product are related as follows:
>
> $$\vec{v} \cdot \vec{v} = \|\vec{v}\,\|^2.$$

For example: $\vec{i} \cdot \vec{i} = 1, \vec{j} \cdot \vec{j} = 1, \vec{k} \cdot \vec{k} = 1$.

Using the Dot Product

Depending on the situation, one definition of the dot product may be more convenient to use than the other. In Example 2 which follows, the geometric definition is the only one which can be used because we are not given components. In Example 3, the algebraic definition is used.

Example 2   Suppose the vector $\vec{b}$ is fixed and has length 2; the vector $\vec{a}$ is free to rotate and has length 3. What are the maximum and minimum values of the dot product $\vec{a} \cdot \vec{b}$ as the vector $\vec{a}$ rotates through all possible positions? What positions of $\vec{a}$ and $\vec{b}$ lead to these values?

Solution   The geometric definition gives $\vec{a} \cdot \vec{b} = \|\vec{a}\|\|\vec{b}\|\cos\theta = 3 \cdot 2\cos\theta = 6\cos\theta$. Thus, the maximum value of $\vec{a} \cdot \vec{b}$ is 6, and it occurs when $\cos\theta = 1$ so $\theta = 0$, that is, when $\vec{a}$ and $\vec{b}$ point in the same direction. The minimum value of $\vec{a} \cdot \vec{b}$ is -6, and it occurs when $\cos\theta = -1$ so $\theta = \pi$, that is, when $\vec{a}$ and $\vec{b}$ point in opposite directions. (See Figure 12.32.)

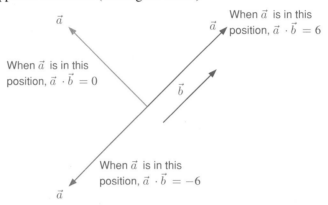

Figure 12.32: Maximum and minimum values of $\vec{a} \cdot \vec{b}$ obtained from a fixed vector $\vec{b}$ of length 2 and rotating vector $\vec{a}$ of length 3

Example 3   Which pairs from the following list of 3-dimensional vectors are perpendicular to one another?
$$\vec{u} = \vec{i} + \sqrt{3}\,\vec{k}, \quad \vec{v} = \vec{i} + \sqrt{3}\,\vec{j}, \quad \vec{w} = \sqrt{3}\,\vec{i} + \vec{j} - \vec{k}.$$

Solution   The geometric definition tells us that two vectors are perpendicular if and only if their dot product is zero. Since the vectors are given in components, we calculate dot products using the algebraic definition:
$$\vec{v} \cdot \vec{u} = (\vec{i} + \sqrt{3}\,\vec{j} + 0\vec{k}) \cdot (\vec{i} + 0\vec{j} + \sqrt{3}\,\vec{k}) = 1 \cdot 1 + \sqrt{3} \cdot 0 + 0 \cdot \sqrt{3} = 1,$$
$$\vec{v} \cdot \vec{w} = (\vec{i} + \sqrt{3}\,\vec{j} + 0\vec{k}) \cdot (\sqrt{3}\,\vec{i} + \vec{j} - \vec{k}) = 1 \cdot \sqrt{3} + \sqrt{3} \cdot 1 + 0(-1) = 2\sqrt{3},$$
$$\vec{w} \cdot \vec{u} = (\sqrt{3}\,\vec{i} + \vec{j} - \vec{k}) \cdot (\vec{i} + 0\vec{j} + \sqrt{3}\,\vec{k}) = \sqrt{3} \cdot 1 + 1 \cdot 0 + (-1) \cdot \sqrt{3} = 0.$$

So the only two vectors which are perpendicular are $\vec{w}$ and $\vec{u}$.

Normal Vectors and the Equation of a Plane

In Section 11.5 we wrote the equation of a plane given its x-slope, y-slope and z-intercept. Now we write the equation of a plane using a vector and a point lying on the plane. A *normal vector* to a plane is a vector that is perpendicular to the plane, that is, it is perpendicular to every displacement vector between any two points in the plane. Let $\vec{n} = a\vec{i} + b\vec{j} + c\vec{k}$ be a normal vector to the plane, let $P_0 = (x_0, y_0, z_0)$ be a fixed point in the plane, and let $P = (x, y, z)$ be any other point in the plane. Then $\overrightarrow{P_0P} = (x - x_0)\vec{i} + (y - y_0)\vec{j} + (z - z_0)\vec{k}$ is a vector whose head and tail both lie in the plane. (See Figure 12.33.) Thus, the vectors $\vec{n}$ and $\overrightarrow{P_0P}$ are perpendicular, so $\vec{n} \cdot \overrightarrow{P_0P} = 0$. The algebraic definition of the dot product gives $\vec{n} \cdot \overrightarrow{P_0P} = a(x - x_0) + b(y - y_0) + c(z - z_0)$, so we obtain the following result:

The **equation of the plane** with normal vector $\vec{n} = a\vec{i} + b\vec{j} + c\vec{k}$ and containing the point $P_0 = (x_0, y_0, z_0)$ is

$$a(x - x_0) + b(y - y_0) + c(z - z_0) = 0.$$

Letting $d = ax_0 + by_0 + cz_0$ (a constant), we can write the equation of the plane in the form

$$ax + by + cz = d.$$

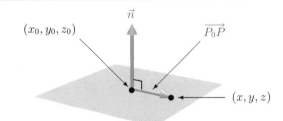

Figure 12.33: Plane with normal $\vec{n}$ and containing a fixed point (x_0, y_0, z_0)

Example 4   Find the equation of the plane perpendicular to $-\vec{i} + 3\vec{j} + 2\vec{k}$ and passing through the point $(1, 0, 4)$.

Solution   The equation of the plane is

$$-(x - 1) + 3(y - 0) + 2(z - 4) = 0,$$

which simplifies to

$$-x + 3y + 2z = 7.$$

Example 5   Find a normal vector to the plane with equation   (a) $x - y + 2z = 5$ (b) $z = 0.5x + 1.2y.$

Solution   (a)   Since the coefficients of $\vec{i}$, $\vec{j}$, and $\vec{k}$ in a normal vector are the coefficients of x, y, z in the equation of the plane, a normal vector is $\vec{n} = \vec{i} - \vec{j} + 2\vec{k}$.

(b)   Before we can find a normal vector, we rewrite the equation of the plane in the form

$$0.5x + 1.2y - z = 0.$$

Thus, a normal vector is $\vec{n} = 0.5\vec{i} + 1.2\vec{j} - \vec{k}$.

The Dot Product in n Dimensions

The algebraic definition of the dot product can be extended to vectors in higher dimensions.

If $\vec{u} = (u_1, \ldots, u_n)$ and $\vec{v} = (v_1, \ldots, v_n)$ then the dot product of $\vec{u}$ and $\vec{v}$ is the **scalar**

$$\vec{u} \cdot \vec{v} = u_1 v_1 + \ldots + u_n v_n.$$

Example 6   A video store sells videos, tapes, CD's, and computer games. We define the quantity vector $\vec{q} = (q_1, q_2, q_3, q_4)$, where q_1, q_2, q_3, q_4 denote the quantities sold of each of the items, and the price vector $\vec{p} = (p_1, p_2, p_3, p_4)$, where p_1, p_2, p_3, p_4 denote the price per unit of each item. What does the dot product $\vec{p} \cdot \vec{q}$ represent?

Solution   The dot product is $\vec{p} \cdot \vec{q} = p_1 q_1 + p_2 q_2 + p_3 q_3 + p_4 q_4$. The quantity $p_1 q_1$ represents the revenue received by the store for the videos, $p_2 q_2$ represents the revenue for the tapes, and so on. The dot product represents the total revenue received by the store for the sale of these four items.

Resolving a Vector into Components: Projections

In Section 12.1, we resolved a vector into components parallel to the axes. Now we see how to resolve a vector, $\vec{v}$, into components, called $\vec{v}_{\text{parallel}}$ and $\vec{v}_{\text{perp}}$, which are parallel and perpendicular, respectively, to a given nonzero vector, $\vec{u}$. (See Figure 12.34.)

The projection of $\vec{v}$ on $\vec{u}$, written $\vec{v}_{\text{parallel}}$, measures (in some sense) how much the vector $\vec{v}$ is aligned with the vector $\vec{u}$. The length of $\vec{v}_{\text{parallel}}$ is the length of the shadow cast by $\vec{v}$ on a line in the direction of $\vec{u}$.

To compute $\vec{v}_{\text{parallel}}$, we assume $\vec{u}$ is a unit vector. (If not, create one by dividing by its length.) Then Figure 12.34(a) shows that, if $0 \leq \theta \leq \pi/2$:

$$\|\vec{v}_{\text{parallel}}\| = \|\vec{v}\| \cos \theta = \vec{v} \cdot \vec{u} \qquad (\text{since } \|\vec{u}\| = 1).$$

Now $\vec{v}_{\text{parallel}}$ is a scalar multiple of $\vec{u}$, and since $\vec{u}$ is a unit vector,

$$\vec{v}_{\text{parallel}} = (\|\vec{v}\| \cos \theta)\vec{u} = (\vec{v} \cdot \vec{u})\vec{u}.$$

A similar argument shows that if $\pi/2 < \theta \leq \pi$, as in Figure 12.34(b), this formula for $\vec{v}_{\text{parallel}}$ still holds. The vector $\vec{v}_{\text{perp}}$ is specified by

$$\vec{v}_{\text{perp}} = \vec{v} - \vec{v}_{\text{parallel}}.$$

Thus, we have the following results:

Projection of $\vec{v}$ on the Line in the Direction of the Unit Vector $\vec{u}$

If $\vec{v}_{\text{parallel}}$ and $\vec{v}_{\text{perp}}$ are components of $\vec{v}$ which are parallel and perpendicular, respectively, to $\vec{u}$, then

$$\text{Projection of } \vec{v} \text{ on to } \vec{u} = \vec{v}_{\text{parallel}} = (\vec{v} \cdot \vec{u})\vec{u} \qquad \text{provided } \|\vec{u}\| = 1$$

and $\vec{v} = \vec{v}_{\text{parallel}} + \vec{v}_{\text{perp}}$ so $\vec{v}_{\text{perp}} = \vec{v} - \vec{v}_{\text{parallel}}.$

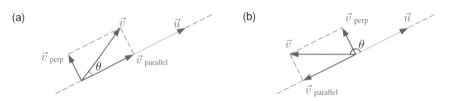

Figure 12.34: Resolving $\vec{v}$ into components parallel and perpendicular to $\vec{u}$
(a) $0 < \theta < \pi/2$ (b) $\pi/2 < \theta < \pi$

Example 7   Figure 12.35 shows the force the wind exerts on the sail of a sailboat. Find the component of the force in the direction in which the sailboat is traveling.

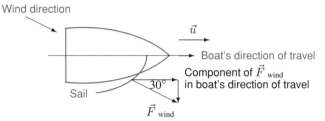

Figure 12.35: Wind moving a sailboat

Solution   Let $\vec{u}$ be a unit vector in the direction of travel. The force of the wind on the sail makes an angle of $30°$ with $\vec{u}$. Thus, the component of this force in the direction of $\vec{u}$ is

$$\vec{F}_{\text{parallel}} = (\vec{F} \cdot \vec{u})\vec{u} = \|\vec{F}\|(\cos 30°)\vec{u} = 0.87\|\vec{F}\|\vec{u}.$$

Thus, the boat is being pushed forward with about 87% of the total force due to the wind. (In fact, the interaction of wind and sail is much more complex than this model suggests.)

A Physical Interpretation of the Dot Product: Work

In physics, the word "work" has a slightly different meaning from its everyday meaning. In physics, when a force of magnitude F acts on an object through a distance d, we say the *work*, W, done by the force is

$$W = Fd,$$

provided the force and the displacement are in the same direction. For example, if a 1 kg body falls 10 meters under the force of gravity, which is 9.8 newtons, then the work done by gravity is

$$W = (9.8 \text{ newtons}) \cdot (10 \text{ meters}) = 98 \text{ joules}.$$

What if the force and the displacement are not in the same direction? Suppose a force $\vec{F}$ acts on an object as it moves along a displacement vector $\vec{d}$. Let θ be the angle between $\vec{F}$ and $\vec{d}$. First, we assume $0 \le \theta \le \pi/2$. Figure 12.36 shows how we can resolve $\vec{F}$ into components that are parallel and perpendicular to $\vec{d}$:

$$\vec{F} = \vec{F}_{\text{parallel}} + \vec{F}_{\text{perp}},$$

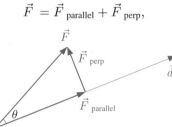

Figure 12.36: Resolving the force $\vec{F}$ into two forces, one parallel to $\vec{d}$, one perpendicular to $\vec{d}$

Then the work done by $\vec{F}$ is defined to be

$$W = \|\vec{F}_{\text{parallel}}\| \, \|\vec{d}\|.$$

We see from Figure 12.36 that $\vec{F}_{\text{parallel}}$ has magnitude $\|\vec{F}\|\cos\theta$. So the work is given by the dot product:

$$W = (\|\vec{F}\| \cos \theta)\|\vec{d}\| = \|\vec{F}\|\|\vec{d}\| \cos \theta = \vec{F} \cdot \vec{d}.$$

The formula $W = \vec{F} \cdot \vec{d}$ holds when $\pi/2 < \theta \le \pi$ also. In that case, the work done by the force is negative and the object is moving against the force. Thus, we have the following definition:

> The **work**, W, done by a force $\vec{F}$ acting on an object through a displacement $\vec{d}$ is given by
>
> $$W = \vec{F} \cdot \vec{d}.$$

Notice that if the vectors $\vec{F}$ and $\vec{d}$ are parallel and in the same direction, with magnitudes F and d, then $\cos \theta = \cos 0 = 1$, so $W = \|\vec{F}\|\|\vec{d}\| = Fd$, which is the original definition. When the vectors are perpendicular, $\cos \theta = \cos \frac{\pi}{2} = 0$, so $W = 0$ and no work is done in the technical definition of the word. For example, if you carry a heavy box across the room at the same horizontal height, no work is done by gravity because the force of gravity is vertical but the motion is horizontal.

Problems for Section 12.3

For Problems 1–6, perform the following operations on the given 3-dimensional vectors.

$$\vec{a} = 2\vec{j} + \vec{k} \qquad \vec{b} = -3\vec{i} + 5\vec{j} + 4\vec{k} \qquad \vec{c} = \vec{i} + 6\vec{j} \qquad \vec{y} = 4\vec{i} - 7\vec{j} \qquad \vec{z} = \vec{i} - 3\vec{j} - \vec{k}$$

1. $\vec{c} \cdot \vec{y}$

2. $\vec{a} \cdot \vec{z}$

3. $\vec{a} \cdot \vec{b}$

4. $(\vec{a} \cdot \vec{b})\vec{a}$

5. $(\vec{a} \cdot \vec{y})(\vec{c} \cdot \vec{z})$

6. $((\vec{c} \cdot \vec{c})\vec{a}) \cdot \vec{a}$

7. Compute the angle between the vectors $\vec{i} + \vec{j} + \vec{k}$ and $\vec{i} - \vec{j} - \vec{k}$.

8. Which pairs of the vectors $\sqrt{3}\vec{i} + \vec{j}$, $3\vec{i} + \sqrt{3}\vec{j}$, $\vec{i} - \sqrt{3}\vec{j}$ are parallel and which are perpendicular?

9. For what values of t are $\vec{u} = t\vec{i} - \vec{j} + \vec{k}$ and $\vec{v} = t\vec{i} + t\vec{j} - 2\vec{k}$ perpendicular? Are there values of t for which $\vec{u}$ and $\vec{v}$ are parallel?

In Problems 10–12, find a normal vector to the given plane.

10. $2x + y - z = 5$

11. $z = 3x + 4y - 7$

12. $2(x - z) = 3(x + y)$

In Problems 13–17, find an equation of a plane that satisfies the given conditions.

13. Perpendicular to the vector $-\vec{i} + 2\vec{j} + \vec{k}$ and passing through the point $(1, 0, 2)$.

14. Perpendicular to the vector $5\vec{i} + \vec{j} - 2\vec{k}$ and passing through the point $(0, 1, -1)$.

15. Perpendicular to the vector $2\vec{i} - 3\vec{j} + 7\vec{k}$ and passing through the point $(1, -1, 2)$.

16. Parallel to the plane $2x + 4y - 3z = 1$ and through the point $(1, 0, -1)$.

17. Through the point $(-2, 3, 2)$ and parallel to $3x + y + z = 4$.

18. Let S be the triangle with vertices $A = (2, 2, 2)$, $B = (4, 2, 1)$, and $C = (2, 3, 1)$.
    (a) Find the length of the shortest side of S.
    (b) Find the cosine of the angle BAC at vertex A.

19. Write $\vec{a} = 3\vec{i} + 2\vec{j} - 6\vec{k}$ as the sum of two vectors, one parallel to $\vec{d} = 2\vec{i} - 4\vec{j} + \vec{k}$ and the other perpendicular to $\vec{d}$.

20. Find the points where the plane $z = 5x - 4y + 3$ intersects each of the coordinate axes. Then find the lengths of the sides and the angles of the triangle formed by these points.

21. Find the angle between the planes $5(x-1)+3(y+2)+2z = 0$ and $x+3(y-1)+2(z+4) = 0$.

22. A basketball gymnasium is 25 meters high, 80 meters wide and 200 meters long. For a half time stunt, the cheerleaders want to run two strings, one from each of the two corners above one basket to the diagonally opposite corners of the gym floor. What is the cosine of the angle made by the strings as they cross?

23. A consumption vector of three goods is defined by $\vec{x} = (x_1, x_2, x_3)$, where x_1, x_2 and x_3 are the quantities consumed of the three goods. Consider a budget constraint represented by the equation $\vec{p} \cdot \vec{x} = k$, where $\vec{p}$ is the price vector of the three goods and k is a constant. Show that the difference between two consumption vectors corresponding to points satisfying the same budget constraint is perpendicular to the price vector $\vec{p}$.

24. A 100-meter dash is run on a track in the direction of the vector $\vec{v} = 2\vec{i} + 6\vec{j}$. The wind velocity $\vec{w}$ is $5\vec{i} + \vec{j}$ km/h. The rules say that a legal wind speed measured in the direction of the dash must not exceed 5 km/h. Will the race results be disqualified due to an illegal wind? Justify your answer.

25. Recall that in 2 or 3 dimensions, if θ is the angle between $\vec{v}$ and $\vec{w}$, the dot product is given by

$$\vec{v} \cdot \vec{w} = \|\vec{v}\|\|\vec{w}\| \cos \theta.$$

We use this relationship to define the angle between two vectors in n-dimensions. If $\vec{v}$, $\vec{w}$ are n-vectors, then the dot product, $\vec{v} \cdot \vec{w} = v_1 w_1 + v_2 w_2 + \cdots + v_n w_n$, is used to define[2] the angle θ by

$$\cos \theta = \frac{\vec{v} \cdot \vec{w}}{\|\vec{v}\|\|\vec{w}\|} \qquad \text{provided } \|\vec{v}\|, \|\vec{w}\| \neq 0.$$

We now use this idea of angle to measure how close two populations are to one another genetically. Table 12.1 shows the relative frequencies of four alleles (variants of a gene) in four populations.

TABLE 12.1

Allele	Eskimo	Bantu	English	Korean
A_1	0.29	0.10	0.20	0.22
A_2	0.00	0.08	0.06	0.00
B	0.03	0.12	0.06	0.20
O	0.67	0.69	0.66	0.57

Let $\vec{a}_1$ be the 4-vector showing the relative frequencies in the Eskimo population;

$\vec{a}_2$ be the 4-vector showing the relative frequencies in the Bantu population;

$\vec{a}_3$ be the 4-vector showing the relative frequencies in the English population;

$\vec{a}_4$ be the 4-vector showing the relative frequencies in the Korean population.

---

[2]The result of Problem 34 shows that the quantity on the right-hand side of this equation is between -1 and 1, so this definition makes sense.

The genetic distance between two populations is defined as the angle between the corresponding vectors. Using this definition, is the English population closer genetically to the Bantus or to the Koreans? Explain.[3]

26.   Show why each of the properties of the dot product in the box on page 77 follows from the algebraic definition of the dot product:

$$\vec{v} \cdot \vec{w} = v_1 w_1 + v_2 w_2 + v_3 w_3$$

27.   What does Property 2 of the dot product in the box on page 77 say geometrically?

28.   Show that the vectors $(\vec{b} \cdot \vec{c})\vec{a} - (\vec{a} \cdot \vec{c})\vec{b}$ and $\vec{c}$ are perpendicular.

29.   Show that if $\vec{u}$ and $\vec{v}$ are two vectors such that

$$\vec{u} \cdot \vec{w} = \vec{v} \cdot \vec{w}$$

for every vector $\vec{w}$, then

$$\vec{u} = \vec{v}.$$

30.   Figure 12.37 shows that, given three vectors $\vec{u}$, $\vec{v}$, and $\vec{w}$, the sum of the components of $\vec{v}$ and $\vec{w}$ in the direction of $\vec{u}$ is the component of $\vec{v} + \vec{w}$ in the direction of $\vec{u}$. (Although the figure is drawn in two dimensions, this result is also true in three dimensions.) Use this figure to explain why the geometric definition of the dot product satisfies $(\vec{v} + \vec{w}) \cdot \vec{u} = \vec{v} \cdot \vec{u} + \vec{w} \cdot \vec{u}$.

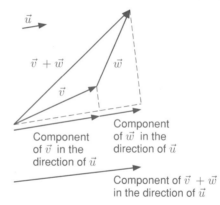

Figure 12.37: The component of $\vec{v} + \vec{w}$ in the direction of $\vec{u}$ is the sum of the components of $\vec{v}$ and $\vec{w}$ in that direction

31.   (a)   Using the geometric definition of the dot product, show that

$$\vec{u} \cdot (-\vec{v}) = -(\vec{u} \cdot \vec{v}).$$

   [Hint: What happens to the angle when you multiply $\vec{v}$ by -1?]

   (b)   Using the geometric definition of the dot product, show that for any negative scalar λ

$$\vec{u} \cdot (\lambda\vec{v}) = \lambda(\vec{u} \cdot \vec{v})$$
$$(\lambda\vec{u}) \cdot \vec{v} = \lambda(\vec{u} \cdot \vec{v}).$$

32.   The Law of Cosines for a triangle with side lengths a, b, and c, and with angle C opposite side c, says

$$c^2 = a^2 + b^2 - 2ab\cos C.$$

On page 76, we used the Law of Cosines to show that the two definitions of the dot product

[3] Adapted from Cavalli-Sforza and Edwards, "Models and Estimation Procedures," Am J. Hum. Genet., Vol. 19 (1967), pp. 223-57.

are equivalent. In this problem, you will use the geometric definition of the dot product and its properties in the box on page 77 to prove the Law of Cosines. [Hint: Let $\vec{u}$ and $\vec{v}$ be the displacement vectors from C to the other two vertices, and express c^2 in terms of $\vec{u}$ and $\vec{v}$.]

33. Use the following steps and the results of Problems 30–31 to show (without trigonometry) that the geometric and algebraic definitions of the dot product are equivalent.

    Follow these steps. Let $\vec{u} = u_1\vec{i} + u_2\vec{j} + u_3\vec{k}$ and $\vec{v} = v_1\vec{i} + v_2\vec{j} + v_3\vec{k}$ be any vectors. Write $(\vec{u} \cdot \vec{v})_{\text{geom}}$ for the result of the dot product computed geometrically. Substitute $\vec{u} = u_1\vec{i} + u_2\vec{j} + u_3\vec{k}$ and use Problems 30–31 to expand $(\vec{u} \cdot \vec{v})_{\text{geom}}$. Next substitute for $\vec{v}$ and expand. Finally, calculate geometrically the dot products $\vec{i} \cdot \vec{i}$, $\vec{i} \cdot \vec{j}$, etc.

34. Suppose that $\vec{v}$ and $\vec{w}$ are any two vectors. Consider the following function of t:

$$q(t) = (\vec{v} + t\vec{w}) \cdot (\vec{v} + t\vec{w})$$

    (a) Explain why $q(t) \geq 0$ for all real t.
    (b) Expand $q(t)$ as a quadratic polynomial in t using the properties on page 77.
    (c) Using the discriminant of the quadratic, show that,

$$|\vec{v} \cdot \vec{w}| \leq \|\vec{v}\| \|\vec{w}\|.$$

12.4 THE CROSS PRODUCT

In the previous section we combined two vectors to get a number, the dot product. In this section we see another way of combining two vectors, this time to get a vector, the *cross product*. Any two vectors in 3-space form a parallelogram. We define the cross product using this parallelogram.

The Area of a Parallelogram

Consider the parallelogram formed by the vectors $\vec{v}$ and $\vec{w}$ with an angle of θ between them. Then Figure 12.38 shows

$$\text{Area of parallelogram} = \text{Base} \cdot \text{Height} = \|\vec{v}\| \|\vec{w}\| \sin\theta.$$

How would we compute the area of the parallelogram if we were given $\vec{v}$ and $\vec{w}$ in components, $\vec{v} = v_1\vec{i} + v_2\vec{j} + v_3\vec{k}$ and $\vec{w} = w_1\vec{i} + w_2\vec{j} + w_3\vec{k}$? Problem 27 on page 92 shows that if $\vec{v}$ and $\vec{w}$ are in the xy-plane, so $v_3 = w_3 = 0$, then

$$\text{Area of parallelogram} = |v_1 w_2 - v_2 w_1|.$$

What if $\vec{v}$ and $\vec{w}$ do not lie in the xy-plane? The cross product will enable us to compute the area of the parallelogram formed by any two vectors.

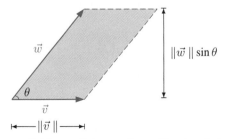

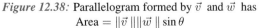

Figure 12.38: Parallelogram formed by $\vec{v}$ and $\vec{w}$ has
Area $= \|\vec{v}\| \|\vec{w}\| \sin\theta$

Definition of the Cross Product

We define the cross product of the vectors $\vec{v}$ and $\vec{w}$, written $\vec{v} \times \vec{w}$, to be a vector perpendicular to both $\vec{v}$ and $\vec{w}$. The magnitude of this vector is the area of the parallelogram formed by the two vectors. The direction of $\vec{v} \times \vec{w}$ is given by the normal vector, $\vec{n}$, to the plane defined by $\vec{v}$ and $\vec{w}$. If we require that $\vec{n}$ be a unit vector, there are two choices for $\vec{n}$, pointing out of the plane in opposite directions. We pick one by the following rule (see Figure 12.39):

> **The right-hand rule:** Place $\vec{v}$ and $\vec{w}$ so that their tails coincide and curl the fingers of your right hand through the smaller of the two angles from $\vec{v}$ to $\vec{w}$; your thumb points in the direction of the normal vector, $\vec{n}$.

Like the dot product, there are two equivalent definitions of the cross product:

> The following two definitions of the **cross product** or **vector product** $\vec{v} \times \vec{w}$ are equivalent:
> - **Geometric definition**
>   If $\vec{v}$ and $\vec{w}$ are not parallel, then
>
> $$\vec{v} \times \vec{w} = \left(\begin{array}{c}\text{Area of parallelogram} \\ \text{with edges } \vec{v} \text{ and } \vec{w}\end{array}\right) \vec{n} = (\|\vec{v}\|\|\vec{w}\| \sin \theta)\vec{n} \,,$$
>
>   where $0 \leq \theta \leq \pi$ is the angle between $\vec{v}$ and $\vec{w}$ and $\vec{n}$ is the unit vector perpendicular to $\vec{v}$ and $\vec{w}$ pointing in the direction given by the right-hand rule. If $\vec{v}$ and $\vec{w}$ are parallel, then $\vec{v} \times \vec{w} = \vec{0}$.
> - **Algebraic definition**
>
> $$\vec{v} \times \vec{w} = (v_2 w_3 - v_3 w_2)\vec{i} + (v_3 w_1 - v_1 w_3)\vec{j} + (v_1 w_2 - v_2 w_1)\vec{k}$$
>
>   where $\vec{v} = v_1\vec{i} + v_2\vec{j} + v_3\vec{k}$ and $\vec{w} = w_1\vec{i} + w_2\vec{j} + w_3\vec{k}$.

Notice that the magnitude of the $\vec{k}$ component is the area of a 2-dimensional parallelogram and the other components have a similar form. Problems 25 and 26 at the end of this section show that the geometric and algebraic definitions of the cross product give the same result.

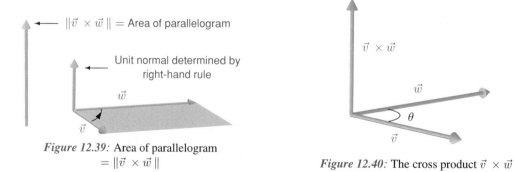

Figure 12.39: Area of parallelogram $= \|\vec{v} \times \vec{w}\|$

Figure 12.40: The cross product $\vec{v} \times \vec{w}$

Unlike the dot product, the cross product is only defined for three-dimensional vectors. The geometric definition shows us that the cross product is *rotation invariant*. Imagine the two vectors $\vec{v}$ and $\vec{w}$ as two metal rods welded together. Attach a third rod whose direction and length correspond to $\vec{v} \times \vec{w}$. (See Figure 12.40.) Then, no matter how we turn this set of rods, the third will still be the cross product of the first two.

The algebraic definition is more easily remembered by writing it as a 3×3 determinant. (See Appendix C.)

$$\vec{v} \times \vec{w} = \begin{vmatrix} \vec{i} & \vec{j} & \vec{k} \\ v_1 & v_2 & v_3 \\ w_1 & w_2 & w_3 \end{vmatrix} = (v_2 w_3 - v_3 w_2)\vec{i} + (v_3 w_1 - v_1 w_3)\vec{j} + (v_1 w_2 - v_2 w_1)\vec{k}.$$

Example 1 Find $\vec{i} \times \vec{j}$ and $\vec{j} \times \vec{i}$.

Solution The vectors $\vec{i}$ and $\vec{j}$ both have magnitude 1 and the angle between them is $\pi/2$. By the right-hand rule, the vector $\vec{i} \times \vec{j}$ is in the direction of $\vec{k}$, so $\vec{n} = \vec{k}$ and we have

$$\vec{i} \times \vec{j} = \left(\|\vec{i}\| \|\vec{j}\| \sin \frac{\pi}{2} \right) \vec{k} = \vec{k}.$$

Similarly, the right-hand rule says that the direction of $\vec{j} \times \vec{i}$ is $-\vec{k}$, so

$$\vec{j} \times \vec{i} = (\|\vec{j}\| \|\vec{i}\| \sin \frac{\pi}{2})(-\vec{k}) = -\vec{k}.$$

Similar calculations show that $\vec{j} \times \vec{k} = \vec{i}$ and $\vec{k} \times \vec{i} = \vec{j}$.

Example 2 For any vector $\vec{v}$, find $\vec{v} \times \vec{v}$.

Solution Since $\vec{v}$ is parallel to itself, $\vec{v} \times \vec{v} = \vec{0}$.

Example 3 Find the cross product of $\vec{v} = 2\vec{i} + \vec{j} - 2\vec{k}$ and $\vec{w} = 3\vec{i} + \vec{k}$ and check that the cross product is perpendicular to both $\vec{v}$ and $\vec{w}$.

Solution Writing $\vec{v} \times \vec{w}$ as a determinant and expanding it into three two-by-two determinants, we have

$$\vec{v} \times \vec{w} = \begin{vmatrix} \vec{i} & \vec{j} & \vec{k} \\ 2 & 1 & -2 \\ 3 & 0 & 1 \end{vmatrix} = \vec{i} \begin{vmatrix} 1 & -2 \\ 0 & 1 \end{vmatrix} - \vec{j} \begin{vmatrix} 2 & -2 \\ 3 & 1 \end{vmatrix} + \vec{k} \begin{vmatrix} 2 & 1 \\ 3 & 0 \end{vmatrix}$$

$$= \vec{i} \left(1(1) - 0(-2) \right) - \vec{j} \left(2(1) - 3(-2) \right) + \vec{k} \left(2(0) - 3(1) \right)$$

$$= \vec{i} - 8\vec{j} - 3\vec{k}.$$

To check that $\vec{v} \times \vec{w}$ is perpendicular to $\vec{v}$, we compute the dot product:

$$\vec{v} \cdot (\vec{v} \times \vec{w}) = (2\vec{i} + \vec{j} - 2\vec{k}) \cdot (\vec{i} - 8\vec{j} - 3\vec{k}) = 2 - 8 + 6 = 0.$$

Similarly,

$$\vec{w} \cdot (\vec{v} \times \vec{w}) = (3\vec{i} + 0\vec{j} + \vec{k}) \cdot (\vec{i} - 8\vec{j} - 3\vec{k}) = 3 + 0 - 3 = 0.$$

Thus, $\vec{v} \times \vec{w}$ is perpendicular to both $\vec{v}$ and $\vec{w}$.

Properties of the Cross Product

The right-hand rule tells us that $\vec{v} \times \vec{w}$ and $\vec{w} \times \vec{v}$ point in opposite directions. The magnitudes of $\vec{v} \times \vec{w}$ and $\vec{w} \times \vec{v}$ are the same, so $\vec{w} \times \vec{v} = -(\vec{v} \times \vec{w})$. (See Figure 12.41.)

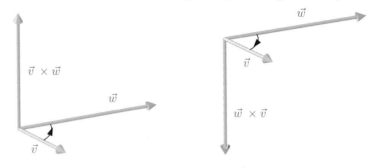

Figure 12.41: Diagram showing $\vec{v} \times \vec{w} = -(\vec{w} \times \vec{v})$

This explains the first of the following properties. The other two are derived in Problems 17, 21, and 25 at the end of this section.

Properties of the Cross Product

For vectors $\vec{u}, \vec{v}, \vec{w}$ and scalar λ
1. $\vec{w} \times \vec{v} = -(\vec{v} \times \vec{w})$
2. $(\lambda\vec{v}) \times \vec{w} = \lambda(\vec{v} \times \vec{w}) = \vec{v} \times (\lambda\vec{w})$
3. $\vec{u} \times (\vec{v} + \vec{w}) = \vec{u} \times \vec{v} + \vec{u} \times \vec{w}.$

The Equivalence of the Two Definitions of the Cross Product

Problems 22 and 25 on page 92 use geometric arguments to show that the cross product distributes over addition. Problem 26 then shows how the formula in the algebraic definition of the cross product can be derived from the geometric definition.

The Equation of a Plane Through Three Points

The equation of a plane is determined by a point $P_0 = (x_0, y_0, z_0)$ on the plane, and a normal vector, $\vec{n} = a\vec{i} + b\vec{j} + c\vec{k}$:

$$a(x - x_0) + b(y - y_0) + c(z - z_0) = 0.$$

However, a plane can also be determined by three points on it (provided they do not lie on a line). In that case we can find an equation of the plane by first determining two vectors in the plane and then finding a normal vector using the cross product, as in the following example.

Example 4   Find an equation of the plane containing the points $P = (1, 3, 0)$, $Q = (3, 4, -3)$, and $R = (3, 6, 2)$.

Solution   Since the points P and Q are in the plane, the displacement vector between them, $\overrightarrow{PQ}$, is in the plane, where

$$\overrightarrow{PQ} = (3 - 1)\vec{i} + (4 - 3)\vec{j} + (-3 - 0)\vec{k} = 2\vec{i} + \vec{j} - 3\vec{k}.$$

The displacement vector $\overrightarrow{PR}$ is also in the plane, where

$$\overrightarrow{PR} = (3 - 1)\vec{i} + (6 - 3)\vec{j} + (2 - 0)\vec{k} = 2\vec{i} + 3\vec{j} + 2\vec{k}.$$

Thus, a normal vector, $\vec{n}$, to the plane is given by

$$\vec{n} = \overrightarrow{PQ} \times \overrightarrow{PR} = \begin{vmatrix} \vec{i} & \vec{j} & \vec{k} \\ 2 & 1 & -3 \\ 2 & 3 & 2 \end{vmatrix} = 11\vec{i} - 10\vec{j} + 4\vec{k}.$$

Since the point $(1, 3, 0)$ is on the plane, the equation of the plane is

$$11(x - 1) - 10(y - 3) + 4(z - 0) = 0,$$

which simplifies to

$$11x - 10y + 4z = -19.$$

You should check that P, Q, and R satisfy the equation of the plane.

Areas and Volumes Using the Cross Product and Determinants

We can use the cross product to calculate the area of the parallelogram with sides $\vec{v}$ and $\vec{w}$. We say that $\vec{v} \times \vec{w}$ is the *area vector* of the parallelogram. The geometric definition of the cross product tells us that $\vec{v} \times \vec{w}$ is normal to the parallelogram and gives us the following result:

> **Area of a parallelogram** with edges $\vec{v} = v_1\vec{i} + v_2\vec{j} + v_3\vec{k}$ and $\vec{w} = w_1\vec{i} + w_2\vec{j} + w_3\vec{k}$ is given by
>
> $$\text{Area} = \|\vec{v} \times \vec{w}\|, \qquad \text{where} \quad \vec{v} \times \vec{w} = \begin{vmatrix} \vec{i} & \vec{j} & \vec{k} \\ v_1 & v_2 & v_3 \\ w_1 & w_2 & w_3 \end{vmatrix}.$$

Example 5   Find the area of the parallelogram with edges $\vec{v} = 2\vec{i} + \vec{j} - 3\vec{k}$ and $\vec{w} = \vec{i} + 3\vec{j} + 2\vec{k}$.

Solution   We calculate the cross product:

$$\vec{v} \times \vec{w} = \begin{vmatrix} \vec{i} & \vec{j} & \vec{k} \\ 2 & 1 & -3 \\ 1 & 3 & 2 \end{vmatrix} = (2 + 9)\vec{i} - (4 + 3)\vec{j} + (6 - 1)\vec{k} = 11\vec{i} - 7\vec{j} + 5\vec{k}.$$

The area of the parallelogram with edges $\vec{v}$ and $\vec{w}$ is the magnitude of the vector $\vec{v} \times \vec{w}$:

$$\text{Area} = \|\vec{v} \times \vec{w}\| = \sqrt{11^2 + (-7)^2 + 5^2} = \sqrt{195}.$$

Volume of a Parallelepiped

Consider the parallelepiped with sides formed by $\vec{a}$, $\vec{b}$, and $\vec{c}$. (See Figure 12.42.) Since the base is formed by the vectors $\vec{b}$ and $\vec{c}$, we have

$$\text{Area of base of parallelepiped} = \|\vec{b} \times \vec{c}\|.$$

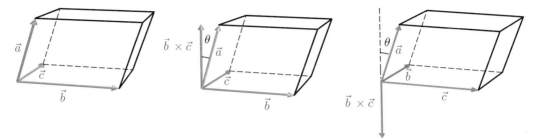

Figure 12.42: Volume of a Parallelepiped

Figure 12.43: The vectors $\vec{a}$, $\vec{b}$, $\vec{c}$ are called a right-handed set

Figure 12.44: The vectors $\vec{a}$, $\vec{b}$, $\vec{c}$ are called a left-handed set

The vectors $\vec{a}$, $\vec{b}$, and $\vec{c}$ can be arranged either as in Figure 12.43 or as in Figure 12.44. In either case,

$$\text{Height of parallelepiped} = \|\vec{a}\| \cos\theta,$$

where θ is the angle between $\vec{a}$ and $\vec{b} \times \vec{c}$. In Figure 12.43 the angle θ is less than $\pi/2$, so the product, $(\vec{b} \times \vec{c}) \cdot \vec{a}$, called the *triple product*, is positive. Thus, in this case

$$\text{Volume of parallelepiped} = \text{Base} \cdot \text{Height} = \|\vec{b} \times \vec{c}\| \cdot \|\vec{a}\| \cos\theta = (\vec{b} \times \vec{c}) \cdot \vec{a}.$$

In Figure 12.44, the angle, $\pi - \theta$, between $\vec{a}$ and $\vec{b} \times \vec{c}$ is more than $\pi/2$, so the product $(\vec{b} \times \vec{c}) \cdot \vec{a}$ is negative. Thus, in this case we have

$$\text{Volume} = \text{Base} \cdot \text{Height} = \|\vec{b} \times \vec{c}\| \cdot \|\vec{a}\| \cos\theta = -\|\vec{b} \times \vec{c}\| \cdot \|\vec{a}\| \cos(\pi - \theta)$$
$$= -(\vec{b} \times \vec{c}) \cdot \vec{a} = \left| (\vec{b} \times \vec{c}) \cdot \vec{a} \right|.$$

Therefore, in both cases the volume is given by $\left| (\vec{b} \times \vec{c}) \cdot \vec{a} \right|$. Using determinants, we can write

---

Volume of a parallelepiped with edges $\vec{a}$, $\vec{b}$, $\vec{c}$ is given by

$$\text{Volume} = \left| (\vec{b} \times \vec{c}) \cdot \vec{a} \right| = \text{Absolute value of the determinant} \begin{vmatrix} a_1 & a_2 & a_3 \\ b_1 & b_2 & b_3 \\ c_1 & c_2 & c_3 \end{vmatrix}.$$

Problems for Section 12.4

1. Find $\vec{k} \times \vec{j}$.

2. Does $\vec{i} \times \vec{i} = \vec{i} \cdot \vec{i}$? Explain your answer.

In Problems 3–6, find $\vec{a} \times \vec{b}$.

3. $\vec{a} = \vec{i} + \vec{k}$ and $\vec{b} = \vec{i} + \vec{j}$.
4. $\vec{a} = -\vec{i}$ and $\vec{b} = \vec{j} + \vec{k}$.

5. $\vec{a} = \vec{i} + \vec{j} + \vec{k}$ and $\vec{b} = \vec{i} + \vec{j} + -\vec{k}$.
6. $\vec{a} = 2\vec{i} - 3\vec{j} + \vec{k}$ and $\vec{b} = \vec{i} + 2\vec{j} - \vec{k}$.

7. Given $\vec{a} = 3\vec{i} + \vec{j} - \vec{k}$ and $\vec{b} = \vec{i} - 4\vec{j} + 2\vec{k}$, find $\vec{a} \times \vec{b}$ and check that $\vec{a} \times \vec{b}$ is perpendicular to both $\vec{a}$ and $\vec{b}$.

8. If $\vec{v} \times \vec{w} = 2\vec{i} - 3\vec{j} + 5\vec{k}$, and $\vec{v} \cdot \vec{w} = 3$, find $\tan\theta$ where θ is the angle between $\vec{v}$ and $\vec{w}$.

9. Suppose $\vec{a}$ is a fixed vector of length 3 in the direction of the positive x-axis and the vector $\vec{b}$ of length 2 is free to rotate in the xy-plane. What are the maximum and minimum values of the magnitude of $\vec{a} \times \vec{b}$? In what direction is $\vec{a} \times \vec{b}$ as $\vec{b}$ rotates?

10. You are using a jet pilot dogfight simulator. Your monitor tells you that two missiles have honed in on your plane along the directions $3\vec{i} + 5\vec{j} + 2\vec{k}$ and $\vec{i} - 3\vec{j} - 2\vec{k}$. In what direction should you turn to have the maximum chance of avoiding both missiles?

Find an equation for the plane through the points in Problems 11–12.

11. $(1, 0, 0), (0, 1, 0), (0, 0, 1)$.

12. $(3, 4, 2), (-2, 1, 0), (0, 2, 1)$.

13. Given the points $P = (0, 1, 0)$, $Q = (-1, 1, 2)$, $R = (2, 1, -1)$, find

    (a) The area of the triangle PQR.
    (b) The equation for a plane that contains P, Q, and R.

14. Find a vector parallel to the intersection of the planes $2x - 3y + 5z = 2$ and $4x + y - 3z = 7$.

15. Find the equation of the plane through the origin which is perpendicular to the line of intersection of the planes in Problem 14.

16. Find the equation of the plane through the point $(4, 5, 6)$ which is perpendicular to the line of intersection of the planes in Problem 14.

17. Use the algebraic definition for the cross product to check that

$$\vec{a} \times (\vec{b} + \vec{c}) = (\vec{a} \times \vec{b}) + (\vec{a} \times \vec{c}).$$

18. In this problem, we arrive at the algebraic definition for the cross product by a different route. Let $\vec{a} = a_1\vec{i} + a_2\vec{j} + a_3\vec{k}$ and $\vec{b} = b_1\vec{i} + b_2\vec{j} + b_3\vec{k}$. We seek a vector $\vec{v} = x\vec{i} + y\vec{j} + z\vec{k}$ which is perpendicular to both $\vec{a}$ and $\vec{b}$. Use this requirement to construct two equations for x, y, and z. Eliminate x and solve for y in terms of z. Then eliminate y and solve for x in terms of z. Since z can be any value whatsoever (the direction of $\vec{v}$ is unaffected), select the value for z which eliminates the denominator in the equation you obtained. How does the resulting expression for $\vec{v}$ compare to the formula we derived on page 87?

19. Suppose $\vec{a}$ and $\vec{b}$ are vectors in the xy-plane, such that $\vec{a} = a_1\vec{i} + a_2\vec{j}$ and $\vec{b} = b_1\vec{i} + b_2\vec{j}$ with $0 < a_2 < a_1$ and $0 < b_1 < b_2$.

    (a) Sketch $\vec{a}$ and $\vec{b}$ and the vector $\vec{c} = -a_2\vec{i} + a_1\vec{j}$. Shade the parallelogram formed by $\vec{a}$ and $\vec{b}$.
    (b) What is the relation between $\vec{a}$ and $\vec{c}$? [Hint: Find $\vec{c} \cdot \vec{a}$ and $\vec{c} \cdot \vec{c}$.]
    (c) Find $\vec{c} \cdot \vec{b}$.
    (d) Explain why $\vec{c} \cdot \vec{b}$ gives the area of the parallelogram formed by $\vec{a}$ and $\vec{b}$.
    (e) Verify that in this case $\vec{a} \times \vec{b} = (a_1 b_2 - a_2 b_1)\vec{k}$.

20. If $\vec{a} + \vec{b} + \vec{c} = \vec{0}$, show that

$$\vec{a} \times \vec{b} = \vec{b} \times \vec{c} = \vec{c} \times \vec{a}.$$

Geometrically, what does this imply about $\vec{a}$, $\vec{b}$, and $\vec{c}$?

21. If $\vec{v}$ and $\vec{w}$ are nonzero vectors, use the geometric definition of the cross product to explain why

$$(\lambda\vec{v}) \times \vec{w} = \lambda(\vec{v} \times \vec{w}) = \vec{v} \times (\lambda\vec{w}).$$

Consider the cases $\lambda > 0$, and $\lambda = 0$, and $\lambda < 0$ separately.

22. Use a parallelepiped to show that $\vec{a} \cdot (\vec{b} \times \vec{c}) = (\vec{a} \times \vec{b}) \cdot \vec{c}$ for any vectors $\vec{a}$, $\vec{b}$, and $\vec{c}$.

23. Show that $\|\vec{a} \times \vec{b}\|^2 = \|\vec{a}\|^2 \|\vec{b}\|^2 - (\vec{a} \cdot \vec{b})^2$.

24. For vectors $\vec{a}$ and $\vec{b}$, let $\vec{c} = \vec{a} \times (\vec{b} \times \vec{a})$.

    (a) Show that $\vec{c}$ lies in the plane containing $\vec{a}$ and $\vec{b}$.
    (b) Show that $\vec{a} \cdot \vec{c} = 0$ and $\vec{b} \cdot \vec{c} = \|\vec{a}\|^2 \|\vec{b}\|^2 - (\vec{a} \cdot \vec{b})^2$. [Hint: Use Problems 22 and 23.]
    (c) Show that

$$\vec{a} \times (\vec{b} \times \vec{a}) = \|\vec{a}\|^2 \vec{b} - (\vec{a} \cdot \vec{b})\vec{a}.$$

25. Use the result of Problem 22 to show that the cross product distributes over addition. First, use distributivity for the dot product to show that for any vector $\vec{d}$,

$$[(\vec{a} + \vec{b}) \times \vec{c}] \cdot \vec{d} = [(\vec{a} \times \vec{c}) + (\vec{b} \times \vec{c})] \cdot \vec{d}.$$

Next, show that for any vector $\vec{d}$,

$$[((\vec{a} + \vec{b}) \times \vec{c}) - (\vec{a} \times \vec{c}) - (\vec{b} \times \vec{c})] \cdot \vec{d} = 0.$$

Finally, explain why you can conclude that

$$(\vec{a} + \vec{b}) \times \vec{c} = (\vec{a} \times \vec{c}) + (\vec{b} \times \vec{c}).$$

26. Use the fact that $\vec{i} \times \vec{i} = \vec{0}, \vec{i} \times \vec{j} = \vec{k}, \vec{i} \times \vec{k} = -\vec{j}$, and so on, together with the properties of the cross product on page 88 to derive the algebraic definition for the cross product.

27. Let $\vec{a} = a_1 \vec{i} + a_2 \vec{j}$ and $\vec{b} = b_1 \vec{i} + b_2 \vec{j}$ be two nonparallel vectors in 2-space, as in Figure 12.45.

    (a) Use the identity $\sin(\beta - \alpha) = (\sin \beta \cos \alpha - \cos \beta \sin \alpha)$ to verify the formula for the area of the parallelogram formed by $\vec{a}$ and $\vec{b}$:

$$\text{Area of parallelogram} = |a_1 b_2 - a_2 b_1|.$$

    (b) Show that $a_1 b_2 - a_2 b_1$ is positive when the rotation from $\vec{a}$ to $\vec{b}$ is counterclockwise, and negative when it is clockwise.

28. Consider the tetrahedron determined by three vectors $\vec{a}, \vec{b}, \vec{c}$ in Figure 12.46. The *area vector* of a face is a vector perpendicular to the face, pointing outward, whose magnitude is the area of the face. Show that the sum of the four outward pointing area vectors of the faces equals the zero vector.

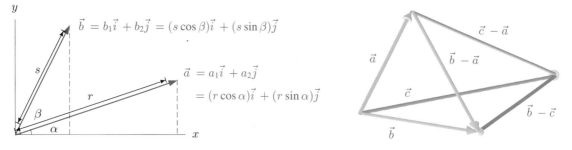

Figure 12.45

Figure 12.46

REVIEW PROBLEMS FOR CHAPTER TWELVE

1. Given the displacement vectors $\vec{u}$ and $\vec{v}$ in Figure 12.47, draw the following vectors:
   (a) $(\vec{u} + \vec{v}) + \vec{u}$ (b) $\vec{v} + (\vec{v} + \vec{u})$ (c) $(\vec{u} + \vec{u}) + \vec{u}$.

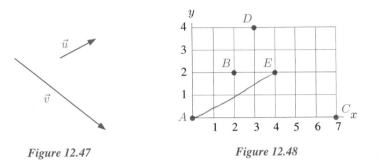

Figure 12.47   Figure 12.48

2. Figure 12.48 shows five points A, B, C, D, and E.
   (a) Read off the coordinates of the five points and thus resolve into components the following
        two vectors: $\vec{u} = (2.5)\overrightarrow{AB} + (-0.8)\overrightarrow{CD}$, $\vec{v} = (2.5)\overrightarrow{BA} - (-0.8)\overrightarrow{CD}$
   (b) What is the relation between $\vec{u}$ and $\vec{v}$? Why was this to be expected?

3. Find the components of a vector $\vec{p}$ which has the same direction as $\overrightarrow{EA}$ in Figure 12.48 and
   whose length equals two units.

4. For each of the four statements below, answer the following questions: Does the statement
   make sense? If yes, is it true for all possible choices of $\vec{a}$ and $\vec{b}$? If no, why not? Use complete
   sentences for your answers.
   (a) $\vec{a} + \vec{b} = \vec{b} + \vec{a}$ (b) $\vec{a} + \|\vec{b}\| = \|\vec{a} + \vec{b}\|$ (c) $\|\vec{b} + \vec{a}\| = \|\vec{a} + \vec{b}\|$
   (d) $\|\vec{a} + \vec{b}\| = \|\vec{a}\| + \|\vec{b}\|$.

5. Two adjacent sides of a regular hexagon are given as the vectors $\vec{u}$ and $\vec{v}$ in Figure 12.49.
   Label the remaining sides in terms of $\vec{u}$ and $\vec{v}$.

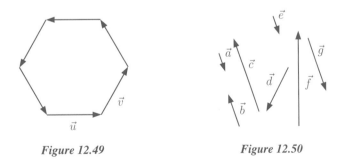

Figure 12.49   Figure 12.50

6. Figure 12.50 shows seven vectors $\vec{a}$, $\vec{b}$, $\vec{c}$, $\vec{d}$, $\vec{e}$, $\vec{f}$ and $\vec{g}$.
   (a) Are any of these two vectors equal? Write down all equal pairs.
   (b) Can you find a scalar x so that $\vec{a} = x\vec{g}$? If yes, find such an x; if no, explain why not.
   (c) Same question as in part (b), but for the equation $\vec{b} = x\vec{d}$.
   (d) Can you solve the equation $\vec{f} = u\vec{c} + v\vec{d}$ for the scalars u and v? If yes, find them; if
        no, explain why not.

7. For what values of t are the following pairs of vectors parallel?
   (a) $2\vec{i} + (t^2 + \frac{2}{3}t + 1)\vec{j} + t\vec{k}$, $\qquad 6\vec{i} + 8\vec{j} + 3\vec{k}$
   (b) $t\vec{i} + \vec{j} + (t - 1)\vec{k}$, $\qquad 2\vec{i} - 4\vec{j} + \vec{k}$
   (c) $2t\vec{i} + t\vec{j} + t\vec{k}$, $\qquad 6\vec{i} + 3\vec{j} + 3\vec{k}$.

Use the geometric definition to compute the cross products in Problems 8–9.

8. $2\vec{i} \times (\vec{i} + \vec{j})$

9. $(\vec{i} + \vec{j}) \times (\vec{i} - \vec{j})$

Compute the cross products for Problems 10–11, using the algebraic definition.

10. $\left((\vec{i} + \vec{j}) \times \vec{i}\right) \times \vec{j}$

11. $(\vec{i} + \vec{j}) \times (\vec{i} \times \vec{j})$

12. True or false? $\vec{a} \times \vec{b} = -(\vec{b} \times \vec{a})$ for all $\vec{a}$ and $\vec{b}$. Explain your answer.

13. Find the area of the triangle with vertices $P = (-2, 2, 0)$, $Q = (1, 3, -1)$, and $R = (-4, 2, 1)$ using the cross product.

14. Find the equation of the plane through the origin which is parallel to $z = 4x - 3y + 8$.

15. Find a vector normal to the plane $4(x - 1) + 6(z + 3) = 12$.

16. Find the equation of the plane through the points $(0, 0, 2), (0, 3, 0), (5, 0, 0)$.

17. Consider the plane $5x - y + 7z = 21$.
    (a) Find a point on the x-axis on this plane.
    (b) Find two other points on the plane.
    (c) Find a vector perpendicular to the plane.
    (d) Find a vector parallel to the plane.

18. Given the points $P = (1, 2, 3)$, $Q = (3, 5, 7)$, and $R = (2, 5, 3)$, find:
    (a) A unit vector perpendicular to a plane containing P, Q, R.
    (b) The angle between PQ and PR.
    (c) The area of the triangle PQR.
    (d) The distance from R to the line through P and Q.

19. Find all vectors $\vec{v}$ in the plane such that $\|\vec{v}\| = 1$ and $\|\vec{v} + \vec{i}\| = 1$.

20. Find all vectors $\vec{w}$ in 3-space such that $\|\vec{w}\| = 1$ and $\|\vec{w} + \vec{i}\| = 1$. Describe this set geometrically.

21. Is the collection of populations of each of the 50 states a vector or scalar quantity?

22. The price vector of beans, rice, and tofu is $(0.30, 0.20, 0.50)$ in dollars per pound. Express it in dollars per ounce.

23. An object is attached by an inelastic string to a fixed point and rotates 30 times per minute in a horizontal plane. Show that the speed of the object is constant but the velocity is not. What does this imply about the acceleration?

24. An object is moving counterclockwise at a constant speed around the circle $x^2 + y^2 = 1$, where x and y are measured in meters. It completes one revolution every minute.
    (a) What is its speed?
    (b) What is its velocity vector 30 seconds after it passes the point $(1, 0)$? Does your answer change if the object is moving clockwise? Explain.

25. In the game of laser tag, you shoot a harmless laser gun and try to hit a target worn at the waist by other players. Suppose you are standing at the origin of a three dimensional coordinate system and that the xy-plane is the floor. Suppose that waist-high is 3 feet above floor level and that eye level is 5 feet above the floor. Three of your friends are your opponents. One is standing so that his target is 30 feet along the x-axis, the other lying down so that his target is at the point $x = 20$, $y = 15$, and the third lying in ambush so that his target is at a point 8 feet above the point $x = 12$, $y = 30$.

    (a) If you aim with your gun at eye level, find the vector from your gun to each of the three targets.

    (b) If you shoot from waist height, with your gun one foot to the right of the center of your body as you face along the x-axis, find the vector from your gun to each of the three targets.

26. An airport is at the point $(200, 10, 0)$ and an approaching plane is at the point $(550, 60, 4)$. Assume that the xy-plane is horizontal, with the x-axis pointing eastward and the y-axis pointing northward. Also assume that the z-axis is upward and that all distances are measured in kilometers. The plane flies due west at a constant altitude at a speed of 500 km/h for half an hour. It then descends at 200 km/h, heading straight for the airport.

    (a) Find the velocity vector of the plane while it is flying at constant altitude.

    (b) Find the coordinates of the point at which the plane starts to descend.

    (c) Find a vector representing the velocity of the plane when it is descending.

27. A large ship is being towed by two tugs. The larger tug exerts a force which is 25% greater than the smaller tug and at an angle of 30 degrees north of east. Which direction must the smaller tug pull to ensure that the ship travels due east?

28. A man wishes to row the shortest possible distance from north to south across a river which is flowing at 4 km/hr from the east. He can row at 5 km/hr.

    (a) In which direction should he steer?

    (b) If there is a wind of 10 km/hr from the southwest, in which direction should he steer to try and go directly across the river? What happens?

29. (a) A vector $\vec{v}$ of magnitude v makes an angle α with the positive x-axis, angle β with the positive y-axis, and angle γ with the positive z-axis. Show that
    $$\vec{v} = v \cos \alpha \vec{i} + v \cos \beta \vec{j} + v \cos \gamma \vec{k}.$$

    (b) $\cos \alpha$, $\cos \beta$, and $\cos \gamma$ are called *direction cosines*. Show that
    $$\cos^2 \alpha + \cos^2 \beta + \cos^2 \gamma = 1.$$

30. Find the vector $\vec{v}$ with all of the following properties:

    (i) Magnitude 10   (ii) Angle of $45°$ with positive x-axis
    (iii) Angle of $75°$ with positive y-axis   (iv) Positive $\vec{k}$-component.

31. Using vectors, show that the perpendicular bisectors of a triangle intersect at a point.

32. Find the distance from the point $P = (2, -1, 3)$ to the plane $2x + 4y - z = -1$.

33. Find an equation of the plane passing through the three points $(1, 1, 1)$, $(1, 4, 5)$, $(-3, -2, 0)$. Find the distance from the origin to the plane.

34. Two lines in space are skew if they are not parallel and do not intersect. Determine the minimum distance between two such lines.

CHAPTER THIRTEEN

DIFFERENTIATING FUNCTIONS OF MANY VARIABLES

For a function of one variable, $y = f(x)$, the derivative $dy/dx = f'(x)$ gives the rate of change of y with respect to x. For a function of two variables, $z = f(x, y)$, there is no such thing as *the* rate of change, since x and y can each vary while the other is held fixed or both can vary at once. However, we can consider the rate of change with respect to each one of the independent variables. This chapter introduces these *partial derivatives* and several ways they can be used to get a complete picture of the way the function varies.

13.1   THE PARTIAL DERIVATIVE

The derivative of a one-variable function measures its rate of change. In this section we see how a two variable function has two rates of change: one as x changes (with y held constant) and one as y changes (with x held constant).

Rate of Change of Temperature in a Metal Rod: a One-Variable Problem

Imagine an unevenly heated metal rod lying along the x-axis, with its left end at the origin and x measured in meters. (See Figure 13.1.) Let $u(x)$ be the temperature (in °C) of the rod at the point x. Table 13.1 gives values of $u(x)$. We see that the temperature increases as we move along the rod, reaching its maximum at $x = 4$, after which it starts to decrease.

Figure 13.1: Unevenly heated metal rod

TABLE 13.1   *Temperature $u(x)$ of the rod*

x (m)	0	1	2	3	4	5
$u(x)$ (°C)	125	128	135	160	175	160

Example 1   Estimate the derivative $u'(2)$ using Table 13.1 and explain what the answer means in terms of temperature.

Solution   The derivative $u'(2)$ is defined as a limit of difference quotients:

$$u'(2) = \lim_{h \to 0} \frac{u(2+h) - u(2)}{h}.$$

Choosing $h = 1$ so that we can use the data in Table 13.1, we get

$$u'(2) \approx \frac{u(2+1) - u(2)}{1} = \frac{160 - 135}{1} = 25.$$

This means that the temperature increases at a rate of approximately 25°C per meter as we go from left to right, past $x = 2$.

Rate of Change of Temperature in a Metal Plate

Imagine an unevenly heated thin rectangular metal plate lying in the xy-plane with its lower left corner at the origin and x and y measured in meters. The temperature (in °C) at the point (x, y) is $T(x, y)$. See Figure 13.2 and Table 13.2. How does T vary near the point $(2, 1)$? We consider the horizontal line $y = 1$ containing the point $(2, 1)$. The temperature along this line is the cross-section, $T(x, 1)$, of the function $T(x, y)$ with $y = 1$. Suppose we write $u(x) = T(x, 1)$.

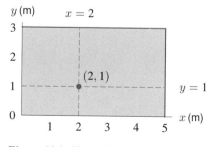

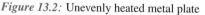

Figure 13.2: Unevenly heated metal plate

TABLE 13.2   *Temperature (°C) of a metal plate*

y (m)						
3	85	90	**110**	135	155	180
2	100	110	**120**	145	190	170
1	**125**	**128**	**135**	**160**	**175**	**160**
0	120	135	**155**	160	160	150
	0	1	2	3	4	5

x (m)

What is the meaning of the derivative $u'(2)$? It is the rate of change of temperature T *in the x-direction* at the point $(2, 1)$, keeping y fixed. Denote this rate of change by $T_x(2, 1)$, so that

$$T_x(2, 1) = u'(2) = \lim_{h \to 0} \frac{u(2 + h) - u(2)}{h} = \lim_{h \to 0} \frac{T(2 + h, 1) - T(2, 1)}{h}.$$

We call $T_x(2, 1)$ the *partial derivative of T with respect to x at the point* $(2, 1)$. Taking $h = 1$, we can read values of T from the row with $y = 1$ in Table 13.2, giving

$$T_x(2, 1) \approx \frac{T(3, 1) - T(2, 1)}{1} = \frac{160 - 135}{1} = 25°\text{C/m}.$$

The fact that $T_x(2, 1)$ is positive means that the temperature of the plate is increasing as we move past the point $(2, 1)$ in the direction of increasing x (that is, horizontally from left to right in Figure 13.2).

Example 2   Estimate the rate of change of T in the y-direction at the point $(2, 1)$.

Solution   The temperature along the line $x = 2$ is the cross-section of T with $x = 2$, that is, the function $v(y) = T(2, y)$. If we denote the rate of change of T in the y-direction at $(2, 1)$ by $T_y(2, 1)$, then

$$T_y(2, 1) = v'(1) = \lim_{h \to 0} \frac{v(1 + h) - v(1)}{h} = \lim_{h \to 0} \frac{T(2, 1 + h) - T(2, 1)}{h}.$$

We call $T_y(2, 1)$ the *partial derivative of T with respect to y at the point* $(2, 1)$. Taking $h = 1$ so that we can use the column with $x = 2$ in Table 13.2, we get

$$T_y(2, 1) \approx \frac{T(2, 1 + 1) - T(2, 1)}{1} = \frac{120 - 135}{1} = -15°\text{C/m}.$$

The fact that $T_y(2, 1)$ is negative means that the temperature decreases as y increases.

Definition of the Partial Derivative

We study the influence of x and y separately on the value of the function $f(x, y)$ by holding one fixed and letting the other vary. This leads to the following definitions.

Partial Derivatives of f With Respect to x and y

For all points at which the limits exist, we define the **partial derivatives at the point $(\mathbf{a}, \mathbf{b})$** by

$$f_x(a, b) = \begin{array}{c} \text{Rate of change of } f \text{ with respect to } x \\ \text{at the point } (a, b) \end{array} = \lim_{h \to 0} \frac{f(a + h, b) - f(a, b)}{h},$$

$$f_y(a, b) = \begin{array}{c} \text{Rate of change of } f \text{ with respect to } y \\ \text{at the point } (a, b) \end{array} = \lim_{h \to 0} \frac{f(a, b + h) - f(a, b)}{h}.$$

If we let a and b vary, we have the **partial derivative functions** $f_x(x, y)$ and $f_y(x, y)$.

Just as with ordinary derivatives, there is an alternative notation:

Alternative Notation for Partial Derivatives

If $z = f(x, y)$, we can write

$$f_x(x, y) = \frac{\partial z}{\partial x} \quad \text{and} \quad f_y(x, y) = \frac{\partial z}{\partial y},$$

$$f_x(a, b) = \frac{\partial z}{\partial x}\bigg|_{(a,b)} \quad \text{and} \quad f_y(a, b) = \frac{\partial z}{\partial y}\bigg|_{(a,b)}.$$

We use the symbol ∂ to distinguish partial derivatives from ordinary derivatives. In cases where the independent variables have names different from x and y, we adjust the notation accordingly. For example, the partial derivatives of $f(u, v)$ are denoted by f_u and f_v.

Visualizing Partial Derivatives on a Graph

The ordinary derivative of a one-variable function is the slope of its graph. How do we visualize the partial derivative $f_x(a, b)$? The graph of the one-variable function $f(x, b)$ is the curve where the vertical plane $y = b$ cuts the graph of $f(x, y)$. (See Figure 13.3.) Thus, $f_x(a, b)$ is the slope of the tangent line to this curve at $x = a$.

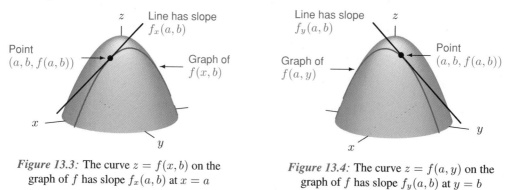

Figure 13.3: The curve $z = f(x, b)$ on the graph of f has slope $f_x(a, b)$ at $x = a$

Figure 13.4: The curve $z = f(a, y)$ on the graph of f has slope $f_y(a, b)$ at $y = b$

Similarly, the graph of the function $f(a, y)$ is the curve where the vertical plane $x = a$ cuts the graph of f, and the partial derivative $f_y(a, b)$ is the slope of this curve at $y = b$. (See Figure 13.4.)

Example 3   At each point labeled on the graph of the surface $z = f(x, y)$ in Figure 13.5, say whether each partial derivative is positive or negative.

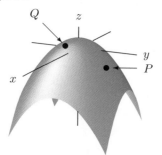

Figure 13.5: Decide the signs of f_x and f_y at P and Q

Solution   The positive x-axis points out of the page. Imagine heading off in this direction from the point marked P; we descend steeply. So the partial derivative with respect to x is negative at P, with quite a large absolute value. The same is true for the partial derivative with respect to y at P, since there is also a steep descent in the positive y-direction.

At the point marked Q, heading in the positive x-direction results in a gentle descent, whereas heading in the positive y-direction results in a gentle ascent. Thus, the partial derivative f_x at Q is negative but small (that is, near zero), and the partial derivative f_y is positive but small.

Estimating Partial Derivatives From a Contour Diagram

The graph of a function $f(x, y)$ often makes clear the sign of the partial derivatives. However, numerical estimates of these derivatives are more easily made from a contour diagram than a surface graph. If we move parallel to one of the axes on a contour diagram, the partial derivative is the rate of change of the value of the function on the contours. For example, if the values on the contours are increasing as we move in the positive direction, then the partial derivative must be positive.

Example 4   Figure 13.6 shows the contour diagram for the temperature $H(x, t)$ (in °C) in a room as a function of distance x (in meters) from a heater and time t (in minutes) after the heater has been turned on. What are the signs of $H_x(10, 20)$ and $H_t(10, 20)$? Estimate these partial derivatives and explain the answers in practical terms.

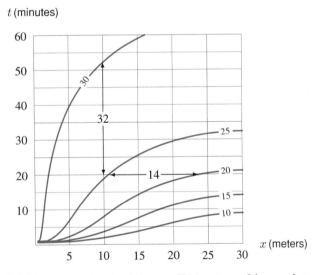

Figure 13.6: Temperature in a heated room: Heater at $x = 0$ is turned on at $t = 0$

Solution   The point $(10, 20)$ is nearly on the $H = 25$ contour. As x increases, we move towards the $H = 20$ contour, so H is decreasing and $H_x(10, 20)$ is negative. This makes sense because the $H = 30$ contour is to the left: As we move further from the heater, the temperature drops. On the other hand, as t increases, we move towards the $H = 30$ contour, so H is increasing; as t decreases H decreases. Thus, $H_t(10, 20)$ is positive. This says that as time passes, the room warms up.

To estimate the partial derivatives, use a difference quotient. Looking at the contour diagram, we see there is a point on the $H = 20$ contour about 14 units to the right of the point $(10, 20)$. Hence, H decreases by 5 when x increases by 14, so we find

$$\text{Rate of change of } H \text{ with respect to } x = H_x(10, 20) \approx \frac{-5}{14} \approx -0.36°\text{C/meter.}$$

This means that near the point 10 m from the heater, after 20 minutes the temperature drops about 0.36, or one third, of a degree, for each meter we move away from the heater.

To estimate $H_t(10, 20)$, we notice that the $H = 30$ contour is about 32 units directly above the point $(10, 20)$. So H increases by 5 when t increases by 32. Hence,

$$\text{Rate of change of } H \text{ with respect to } t = H_t(10, 20) \approx \frac{5}{32} = 0.16°\text{C/meter.}$$

This means that after 20 minutes the temperature is going up about 0.16, or 1/6, of a degree each minute at the point 10 m from the heater.

Using Units to Interpret Partial Derivatives

The meaning of a partial derivative can often be explained using units.

Example 5   Suppose that your weight w in pounds is a function $f(c, n)$ of the number c of calories you consume daily and the number n of minutes you exercise daily. Using the units for w, c and n, interpret in everyday terms the statements

$$\frac{\partial w}{\partial c}(2000, 15) = 0.02 \quad \text{and} \quad \frac{\partial w}{\partial n}(2000, 15) = -0.025.$$

Solution   The units of $\partial w/\partial c$ are pounds per calorie. The statement

$$\frac{\partial w}{\partial c}(2000, 15) = 0.02$$

means that if you are presently consuming 2000 calories daily and exercising 15 minutes daily, you will weigh 0.02 pounds more for each extra calorie you consume daily, or about 2 pounds for each extra 100 calories per day. The units of $\partial w/\partial n$ are pounds per minute. The statement

$$\frac{\partial w}{\partial n}(2000, 15) = -0.025$$

means that for the same calorie consumption and number of minutes of exercise, you will weigh 0.025 pounds less for each extra minute you exercise daily, or about 1 pound less for each extra 40 minutes per day. So if you eat an extra 100 calories each day and exercise about 80 minutes more each day, your weight should remain roughly steady.

Problems for Section 13.1

1. Using difference quotients, estimate $f_x(3, 2)$ and $f_y(3, 2)$ for the function given by

$$f(x, y) = \frac{x^2}{y + 1}.$$

[Recall: A difference quotient is an expression of the form $(f(a + h, b) - f(a, b))/h$.]

2. Use difference quotients with $\Delta x = 0.1$ and $\Delta y = 0.1$ to estimate $f_x(1, 3)$ and $f_y(1, 3)$ where

$$f(x, y) = e^{-x} \sin y.$$

Then give better estimates by using $\Delta x = 0.01$ and $\Delta y = 0.01$.

3. The monthly mortgage payment in dollars, P, for a house is a function of three variables

$$P = f(A, r, N),$$

where A is the amount borrowed in dollars, r is the interest rate, and N is the number of years before the mortgage is paid off.

(a) $f(92000, 14, 30) = 1090.08$. What does this tell you, in financial terms?

(b) $\dfrac{\partial P}{\partial r}(92000, 14, 30) = 72.82$. What is the financial significance of the number 72.82?

(c) Would you expect $\partial P/\partial A$ to be positive or negative? Why?

(d) Would you expect $\partial P/\partial N$ to be positive or negative? Why?

4. Suppose you borrow $\$A$ at an interest rate of $r\%$ (per month) and pay it off over t months by making monthly payments of $\$P$, as determined by the function $P = g(A, r, t)$. In financial terms, what do the following statements tell you?

(a) $g(8000, 1, 24) = 376.59$

(b) $\dfrac{\partial g}{\partial A}(8000, 1, 24) = 0.047$

(c) $\dfrac{\partial g}{\partial r}(8000, 1, 24) = 44.83$

5. Suppose that x is the average price of a new car and that y is the average price of a gallon of gasoline. Then q_1, the number of new cars bought in a year, depends on both x and y, so $q_1 = f(x, y)$. Similarly, if q_2 is the quantity of gas bought in a year, then $q_2 = g(x, y)$.

(a) What do you expect the signs of $\partial q_1/\partial x$ and $\partial q_2/\partial y$ to be? Explain.

(b) What do you expect the signs of $\partial q_1/\partial y$ and $\partial q_2/\partial x$ to be? Explain.

6. A drug is injected into a patient's blood vessel. The function $c = f(x, t)$ represents the concentration of the drug at a distance x in the direction of the blood flow measured from the point of injection and at time t since the injection. What are the units of the following partial derivatives? What are their practical interpretations? What do you expect their signs to be?

(a) $\partial c/\partial x$ (b) $\partial c/\partial t$

7. Suppose P is your monthly car payment in dollars and $P = f(P_0, t, r)$, where $\$P_0$ is the amount you borrowed, t is the number of months it takes to pay off the loan, and $r\%$ is the interest rate. What are the units, the financial meanings, and the signs of $\partial P/\partial t$ and $\partial P/\partial r$?

8. The surface $z = f(x, y)$ is shown in Figure 13.7. The points A and B are in the xy-plane.

(a) What is the sign of $f_x(A)$?   (b) What is the sign of $f_y(A)$?

(c) Suppose P is a point in the xy-plane which moves along a straight line from A to B. How does the sign of $f_x(P)$ change? How does the sign of $f_y(P)$ change?

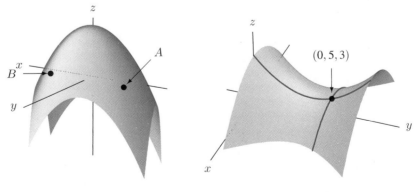

Figure 13.7

Figure 13.8

9. Consider the saddle-shaped surface $z = f(x, y)$ graphed in Figure 13.8.

(a) What is the sign of $f_x(0, 5)$?   (b) What is the sign of $f_y(0, 5)$?

For Problems 10–12, refer to Table 11.3 on page 8 giving the temperature adjusted for wind-chill, C, in °F, as a function $f(w, T)$ of the wind speed, w, in mph, and the temperature, T, in °F. The temperature adjusted for wind-chill tells you how cold it feels, as a result of the combination of wind and temperature.

10. Estimate $f_w(10, 25)$. What does your answer mean in practical terms?

11. Estimate $f_T(5, 20)$. What does your answer mean in practical terms?

12. From Table 11.3 you can see that when the temperature is 20°F, the temperature adjusted for wind-chill drops by an average of about 2.6°F with every 1 mph increase in wind speed from 5 mph to 10 mph. Which partial derivative is this telling you about?

13. Figure 13.9 shows a contour diagram for the monthly payment P as a function of the interest rate, $r\%$, and the amount, L, of a 5-year loan. Estimate $\partial P / \partial r$ and $\partial P / \partial L$ at the following points. In each case, give the units and the everyday meaning of your answer.

    (a) $r = 8, L = 4000$ (b) $r = 8, L = 6000$ (c) $r = 13, L = 7000$

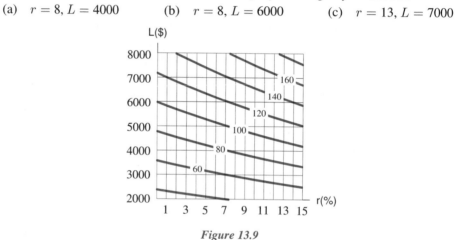

Figure 13.9

14. Figure 13.10 gives a contour diagram for the number n of foxes per square kilometer in southwestern England. Estimate $\partial n / \partial x$ and $\partial n / \partial y$ at the points marked A, B, and C, where x is kilometers east-west and y is kilometers north-south.

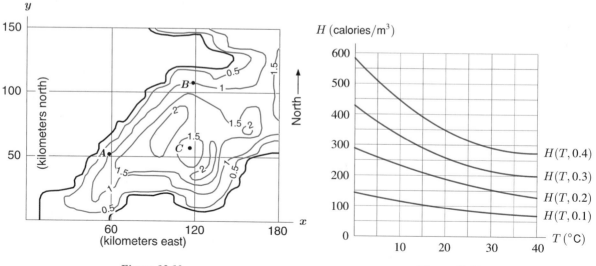

Figure 13.10   **Figure 13.11**

An airport can be cleared of fog by heating the air. The amount of heat required to do the job depends on the air temperature and the wetness of the fog. Problems 15–17 involve Figure 13.11 which shows the heat $H(T, w)$ required (in calories per cubic meter of fog) as a function of the temperature T (in degrees Celsius) and the water content w (in grams per cubic meter of fog). Note that Figure 13.11 is not a contour diagram, but shows cross-sections of H with w fixed at 0.1, 0.2, 0.3, and 0.4.

15. Use Figure 13.11 to find an approximate value for $H_T(10, 0.1)$. Interpret the partial derivative in practical terms.

16. Make a table of values for $H(T, w)$ from Figure 13.11, and use it to estimate $H_T(T, w)$ for $T = 10$, 20, and 30 and $w = 0.1, 0.2$, and 0.3.

17. Repeat Problem 16 for $H_w(T, w)$ at $T = 10$, 20, and 30 and $w = 0.1, 0.2$, and 0.3. What is the practical meaning of these partial derivatives?

18. The cardiac output, represented by c, is the volume of blood flowing through a person's heart, per unit time. The systemic vascular resistance (SVR), represented by s, is the resistance to blood flowing through veins and arteries. Let p be a person's blood pressure. Then p is a function of c and s, so $p = f(c, s)$.

    (a) What does $\partial p / \partial c$ represent?

Suppose now that $p = kcs$, where k is a constant.

    (b) Sketch the level curves of p. What do they represent? Label your axes.

    (c) For a person with a weak heart, it is desirable to have the heart pumping against less resistance, while maintaining the same blood pressure. Such a person may be given the drug Nitroglycerine to decrease the SVR and the drug Dopamine to increase the cardiac output. Represent this on a graph showing level curves. Put a point A on the graph representing the person's state before drugs are given and a point B for after.

    (d) Right after a heart attack, a patient's cardiac output drops, thereby causing the blood pressure to drop. A common mistake made by medical residents is to get the patient's blood pressure back to normal by using drugs to increase the SVR, rather than by increasing the cardiac output. On a graph of the level curves of p, put a point D representing the patient before the heart attack, a point E representing the patient right after the heart attack, and a third point F representing the patient after the resident has given the drugs to increase the SVR.

13.2 COMPUTING PARTIAL DERIVATIVES ALGEBRAICALLY

Since the partial derivative $f_x(x, y)$ is the ordinary derivative of the function $f(x, y)$ with y held constant and $f_y(x, y)$ is the ordinary derivative of $f(x, y)$ with x held constant, we can use all the differentiation formulas from one-variable calculus to find partial derivatives.

Example 1   Let $f(x, y) = \dfrac{x^2}{y + 1}$. Find $f_x(3, 2)$ algebraically.

Solution   We use the fact that $f_x(3, 2)$ equals the derivative of $f(x, 2)$ at $x = 3$. Since

$$f(x, 2) = \frac{x^2}{2 + 1} = \frac{x^2}{3},$$

differentiating with respect to x, we have

$$f_x(x, 2) = \frac{\partial}{\partial x}\left(\frac{x^2}{3}\right) = \frac{2x}{3}, \qquad \text{and so} \qquad f_x(3, 2) = 2.$$

Example 2   Compute the partial derivatives with respect to x and with respect to y for the following functions.

(a) $f(x, y) = y^2 e^{3x}$ (b) $z = (3xy + 2x)^5$ (c) $g(x, y) = e^{x+3y} \sin(xy)$

Solution   (a)   This is the product of a function of x (namely e^{3x}) and a function of y (namely y^2). When we differentiate with respect to x, we think of the function of y as a constant, and vice versa. Thus,

when taking at
partial pull out
x y-comp. and vice
versa

$$f_x(x, y) = y^2 \frac{\partial}{\partial x}\left(e^{3x}\right) = 3y^2 e^{3x},$$

$$f_y(x, y) = e^{3x} \frac{\partial}{\partial y}(y^2) = 2ye^{3x}.$$

(b)   Here we use the chain rule:

$$\frac{\partial z}{\partial x} = 5(3xy + 2x)^4 \frac{\partial}{\partial x}(3xy + 2x) = \boxed{5(3xy + 2x)^4(3y + 2),}$$

$$\frac{\partial z}{\partial y} = 5(3xy + 2x)^4 \frac{\partial}{\partial y}(3xy + 2x) = 5(3xy + 2x)^4 3x = \boxed{15x(3xy + 2x)^4.}$$

(c)   Since each function in the product is a function of both x and y, we need to use the product rule for each partial derivative:

$$g_x(x, y) = \left(\frac{\partial}{\partial x}(e^{x+3y})\right) \sin(xy) + e^{x+3y} \frac{\partial}{\partial x}(\sin(xy)) = \boxed{e^{x+3y} \sin(xy) + e^{x+3y} y \cos(xy),}$$

$$g_y(x, y) = \left(\frac{\partial}{\partial y}(e^{x+3y})\right) \sin(xy) + e^{x+3y} \frac{\partial}{\partial y}(\sin(xy)) = \boxed{3e^{x+3y} \sin(xy) + e^{x+3y} x \cos(xy).}$$

For functions of three or more variables, we find partial derivatives by the same method: Differentiate with respect to one variable, regarding the other variables as constants. For a function $f(x, y, z)$, the partial derivative $f_x(a, b, c)$ gives the rate of change of f with respect to x along the line $y = b, z = c$.

Example 3   Find all the partial derivatives of $f(x, y, z) = \dfrac{x^2 y^3}{z}$.

$(x^2 y^3) z^{-1} = -(x^2 y^3) z^{-2}$

$= -\dfrac{x^2 y^3}{z^2}$

Solution   To find $f_x(x, y, z)$, for example, we consider y and z as fixed, giving

$$f_x(x, y, z) = \frac{2xy^3}{z}, \quad \text{and} \quad f_y(x, y, z) = \frac{3x^2 y^2}{z}, \quad \text{and} \quad f_z(x, y, z) = -\frac{x^2 y^3}{z^2}.$$

Interpretation of Partial Derivatives

Example 4   A vibrating guitar string, originally at rest along the x-axis, is shown in Figure 13.12. Let x be the distance in meters from the left end of the string. At time t seconds the point x has been displaced $y = f(x, t)$ meters vertically from its rest position, where

$$y = f(x, t) = 0.003 \sin(\pi x) \sin(2765t).$$

Evaluate $f_x(0.3, 1)$ and $f_t(0.3, 1)$ and explain what each means in practical terms.

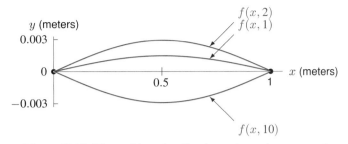

Figure 13.12: The position of a vibrating guitar string at several different times: Graph of $f(x, t)$ for $t = 1, 2, 10$.

Solution Differentiating $f(x, t) = 0.003 \sin(\pi x) \sin(2765t)$ with respect to x, we have

$$f_x(x, t) = 0.003\pi \cos(\pi x) \sin(2765t).$$

In particular, substituting $x = 0.3$ and $t = 1$ gives

$$f_x(0.3, 1) = 0.003\pi \cos(\pi(0.3)) \sin(2765) \approx 0.002.$$

To see what $f_x(0.3, 1)$ means, think about the function $f(x, 1)$. The graph of $f(x, 1)$ in Figure 13.13 is a snapshot of the string at the time $t = 1$. Thus, the derivative $f_x(0.3, 1)$ is the slope of the string at the point $x = 0.3$ at the instant when $t = 1$. Similarly, taking the derivative of $f(x, t) = 0.003 \sin(\pi x) \sin(2765t)$ with respect to t, we get

$$f_t(x, t) = (0.003)(2765) \sin(\pi x) \cos(2765t) = 8.3 \sin(\pi x) \cos(2765t).$$

Since $f(x, t)$ is in meters and t is in seconds, the derivative $f_t(0.3, 1)$ is in m/sec. Thus, substituting $x = 0.3$ and $t = 1$,

$$f_t(0.3, 1) = 8.3 \sin(\pi(0.3)) \cos(2765(1)) \approx 6 \text{ m/sec}.$$

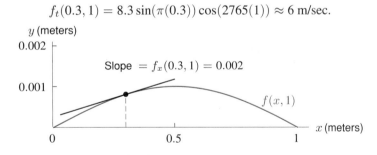

Figure 13.13: Graph of $f(x, 1)$: Snapshot of the shape of the string at $t = 1$ sec

To see what $f_t(0.3, 1)$ means, think about the function $f(0.3, t)$. The graph of $f(0.3, t)$ is a position versus time graph that tracks the up and down movement of the point on the string where $x = 0.3$. (See Figure 13.14.) The derivative $f_t(0.3, 1) = 6$ m/sec is the velocity of that point on the string at time $t = 1$. The fact that $f_t(0.3, 1)$ is positive indicates that the point is moving upward when $t = 1$.

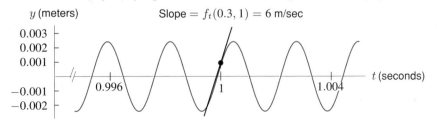
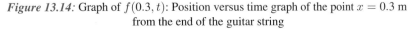

Figure 13.14: Graph of $f(0.3, t)$: Position versus time graph of the point $x = 0.3$ m from the end of the guitar string

Problems for Section 13.2

1. Let $f(x, y) = \dfrac{x^2}{y + 1}$. Find $f_y(3, 2)$ algebraically.

2. Let $f(u, v) = u(u^2 + v^2)^{3/2}$.
   (a) Use a difference quotient to approximate $f_u(1, 3)$ with $h = 0.001$.
   (b) Now evaluate $f_u(1, 3)$ exactly. Was the approximation in part (a) reasonable?

Find the indicated partial derivatives for Problems 3–34. Assume the variables are restricted to a domain on which the function is defined.

3. z_x if $z = x^2 y + 2x^5 y$

4. z_x if $z = \sin(5x^3 y - 3xy^2)$

5. g_x if $g(x, y) = \ln(y e^{xy})$

6. F_m if $F = mg$

7. $\dfrac{\partial}{\partial x}(a\sqrt{x})$

8. $\dfrac{\partial}{\partial x}(xe^{\sqrt{xy}})$

9. $\dfrac{\partial}{\partial y}(3x^5 y^7 - 32x^4 y^3 + 5xy)$

10. z_y if $z = \dfrac{3x^2 y^7 - y^2}{15xy - 8}$

11. $\dfrac{\partial A}{\partial h}$ if $A = \frac{1}{2}(a + b)h$

12. $\dfrac{\partial}{\partial m}\left(\frac{1}{2}mv^2\right)$

13. $\dfrac{\partial}{\partial B}\left(\dfrac{1}{u_0}B^2\right)$ $\quad \frac{2}{u_0}B$

14. $\dfrac{\partial}{\partial r}\left(\dfrac{2\pi r}{v}\right)$

15. F_v if $F = \dfrac{mv^2}{r}$ $\quad F_v = \frac{2mv}{r}$

16. $\dfrac{\partial}{\partial v_0}(v_0 + at)$

17. $\dfrac{\partial F}{\partial m_2}$ if $F = \dfrac{Gm_1 m_2}{r^2}$ $\quad \frac{Gm_1}{r^2}$

18. a_v if $a = \dfrac{v^2}{r}$

19. $\dfrac{\partial}{\partial T}\left(\dfrac{2\pi r}{T}\right)$

20. $\dfrac{\partial}{\partial t}\left(v_0 t + \frac{1}{2}at^2\right)$

21. u_E if $u = \dfrac{1}{2}\epsilon_0 E^2 + \dfrac{1}{2\mu_0}B^2$

22. $\dfrac{\partial f_0}{\partial L}$ if $f_0 = \dfrac{1}{2\pi\sqrt{LC}}$

23. $\dfrac{\partial y}{\partial t}$ if $y = \sin(ct - 5x)$

24. $\dfrac{\partial}{\partial M}\left(\dfrac{2\pi r^{3/2}}{\sqrt{GM}}\right)$

25. z_x if $z = \dfrac{1}{2x^2 ay} + \dfrac{3x^5 abc}{y}$

26. $\dfrac{\partial \alpha}{\partial \beta}$ if $\alpha = \dfrac{e^{x\beta - 3}}{2y\beta + 5}$

27. $\dfrac{\partial}{\partial \lambda}\left(\dfrac{x^2 y\lambda - 3\lambda^5}{\sqrt{\lambda^2 - 3\lambda + 5}}\right)$

28. $\dfrac{\partial m}{\partial v}$ if $m = \dfrac{m_0}{\sqrt{1 - v^2/c^2}}$

29. $\dfrac{\partial}{\partial w}(\sqrt{2\pi xyw - 13x^7 y^3 v})$

30. $\dfrac{\partial}{\partial w}\left(\dfrac{x^2 yw - xy^3 w^7}{w - 1}\right)^{-7/2}$

31. z_x and z_y for $z = x^7 + 2^y + x^y$

32. $z_x(2, 3)$ if $z = (\cos x) + y$

33. $\left.\dfrac{\partial z}{\partial y}\right|_{(1, 0.5)}$ if $z = e^{x+2y}\sin y$

34. $\left.\dfrac{\partial f}{\partial x}\right|_{(\pi/3, 1)}$ if $f(x, y) = x\ln(y\cos x)$

35. Consider the function $f(x, y) = x^2 + y^2$.

    (a) Estimate $f_x(2, 1)$ and $f_y(2, 1)$ using the contour diagram for f in Figure 13.15.

    (b) Estimate $f_x(2, 1)$ and $f_y(2, 1)$ from a table of values for f with $x = 1.9, 2, 2.1$ and $y = 0.9, 1, 1.1$.

    (c) Compare your estimates in parts (a) and (b) with the exact values of $f_x(2, 1)$ and $f_y(2, 1)$ found algebraically.

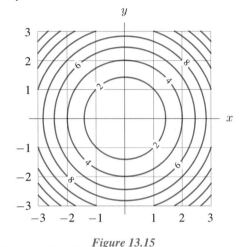

Figure 13.15

36. Show that the Cobb-Douglas function

$$Q = bK^\alpha L^{1-\alpha} \quad \text{where} \quad 0 < \alpha < 1$$

    satisfies the equation

$$K\frac{\partial Q}{\partial K} + L\frac{\partial Q}{\partial L} = Q.$$

37. Money in a bank account earns interest at a continuous rate, r. The amount of money, $\$B$, in the account depends on the amount deposited, $\$P$, and the time, t, it has been in the bank according to the formula

$$B = Pe^{rt}.$$

    Find $\partial B/\partial t$ and $\partial B/\partial P$ and interpret each in financial terms.

38. A one-meter long bar is heated unevenly, with temperature in $^\circ$C at a distance x meters from one end at time t given by

$$H(x, t) = 100e^{-0.1t}\sin(\pi x) \qquad 0 \le x \le 1.$$

    (a) Sketch a graph of H against x for $t = 0$ and $t = 1$.

    (b) Calculate $H_x(0.2, t)$ and $H_x(0.8, t)$. What is the practical interpretation (in terms of temperature) of these two partial derivatives? Explain why each one has the sign it does.

    (c) Calculate $H_t(x, t)$. What is its sign? What is its interpretation in terms of temperature?

39. Let $h(x, t) = 5 + \cos(0.5x - t)$ be the function describing the stadium wave on page 4. The value of $h(x, t)$ gives the height of the head of the spectator in seat x at time t seconds. Evaluate $h_x(2, 5)$ and $h_t(2, 5)$ and interpret each in terms of the wave.

40. Is there a function f which has the following partial derivatives? If so what is it? Are there any others?

$$f_x(x, y) = 4x^3y^2 - 3y^4,$$
$$f_y(x, y) = 2x^4y - 12xy^3.$$

13.3 LOCAL LINEARITY AND THE DIFFERENTIAL

In Sections 13.1 and 13.2 we studied a function of two variables by allowing one variable at a time to change. We now let both variables change at once to develop a linear approximation for functions of two variables.

Zooming In to See Local Linearity

For a function of one variable, local linearity means that as we zoom in on the graph, it looks like a straight line. (See Appendix A.) As we zoom in on the graph of a two-variable function, the graph usually looks like a plane, which is the graph of a linear function of two variables. (See Figure 13.16.)

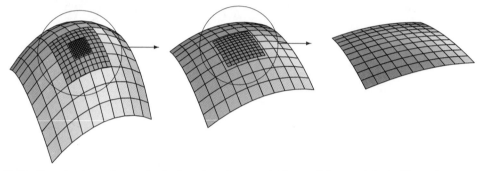

Figure 13.16: Zooming in on the graph of a function of two variables until the graph looks like a plane

Similarly, Figure 13.17 shows three successive views of the contours near a point. As we zoom in, the contours look more like equally spaced parallel lines, which are the contours of a linear function. (As we zoom in, we have to add more contours.)

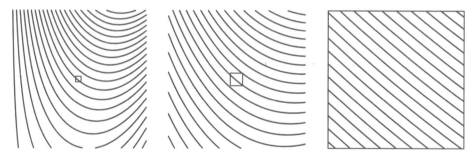

Figure 13.17: Zooming in on a contour diagram until the lines look parallel and equally spaced

This effect can also be seen numerically by zooming in with tables of values. Table 13.3 shows three tables of values for $f(x, y) = x^2 + y^3$ near $x = 2$, $y = 1$, each one a closer view than the previous one. Notice how each table looks more like the table of a linear function.

TABLE 13.3   *Zooming in on values of $f(x, y) = x^2 + y^3$ near $(2, 1)$ until the table looks linear*

		y		
		0	1	2
	1	1	2	9
x	2	4	5	12
	3	9	10	17

		y		
		0.9	1.0	1.1
	1.9	4.34	4.61	4.94
x	2.0	4.73	5.00	5.33
	2.1	5.14	5.41	5.74

		y		
		0.99	1.00	1.01
	1.99	4.93	4.96	4.99
x	2.00	4.97	5.00	5.03
	2.01	5.01	5.04	5.07

Zooming in Algebraically: Differentiability

Seeing a plane when we zoom in at a point tells us (provided the plane is not vertical) that $f(x, y)$ is closely approximated near that point by a linear function, $L(x, y)$:

$$f(x, y) \approx L(x, y).$$

The graph of the function $z = L(x, y)$ is the tangent plane at that point. Provided the approximation is sufficiently good, we say that $f(x, y)$ is *differentiable* at the point. Section 13.10 explains how to tell if the approximation is sufficiently good. The functions we encounter will be differentiable at most points in their domain.

The Tangent Plane

The plane that we see when we zoom in on a surface is called the *tangent plane* to the surface at the point. Figure 13.18 shows the graph of a function with a tangent plane.

What is the equation of the tangent plane? At the point (a, b), the x-slope of the graph of f is the partial derivative $f_x(a, b)$ and the y-slope is $f_y(a, b)$. Thus, using the equation for a plane on page 38 of Chapter 11, we have the following result:

Tangent Plane to the Surface $z = f(x, y)$ at the Point (a, b)

Assuming f is differentiable at (a, b), the equation of the tangent plane is

$$z = f(a, b) + f_x(a, b)(x - a) + f_y(a, b)(y - b).$$

Here we are thinking of a and b as fixed, so $f(a, b)$, and $f_x(a, b)$, and $f_y(a, b)$ are constants. Thus, the right side of the equation is a linear function of x and y.

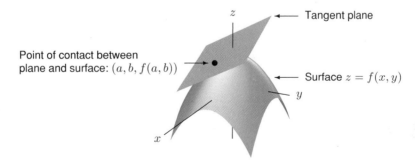

Figure 13.18: The tangent plane to the surface $z = f(x, y)$ at the point (a, b)

Example 1   Find the equation for the tangent plane to the surface $z = x^2 + y^2$ at the point $(3, 4)$. ← (a, b)

Solution   We have $f_x(x, y) = 2x$, so $f_x(3, 4) = 6$, and $f_y(x, y) = 2y$, so $f_y(3, 4) = 8$. Also, $f(3, 4) = 3^2 + 4^2 = 25$. Thus, the equation for the tangent plane at $(3, 4)$ is

$$z = 25 + 6(x - 3) + 8(y - 4) = -25 + 6x + 8y.$$

$f(a,b) \qquad f_x(a,b)(x-a) \qquad f_y(a,b)(y-b)$

Local Linearization

Since the tangent plane lies close to the surface near the point at which they meet, z-values on the tangent plane are close to values of $f(x, y)$ for points near (a, b). Thus, replacing z by $f(x, y)$ in the equation of the tangent plane, we get the following approximation:

Tangent Plane Approximation to $f(x, y)$ for (x, y) Near the Point (a, b)

Provided f is differentiable at (a, b), we can approximate $f(x, y)$:

$$f(x, y) \approx f(a, b) + f_x(a, b)(x - a) + f_y(a, b)(y - b).$$

We are thinking of a and b as fixed, so the expression on the right side is linear in x and y. The right side of this approximation is called the **local linearization** of f near $x = a, y = b$.

Figure 13.19 shows the tangent plane approximation graphically.

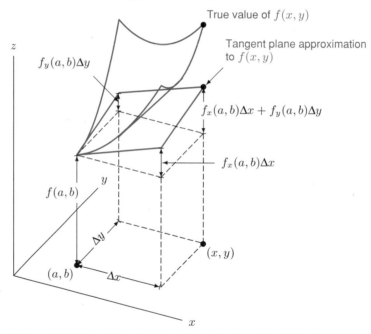

Figure 13.19: Local linearization: Approximating $f(x, y)$ by the z-value from the tangent plane

---

Example 2   Find the local linearization of $f(x, y) = x^2 + y^2$ at the point $(3, 4)$. Estimate $f(2.9, 4.2)$ and $f(2, 2)$ using the linearization and compare your answers to the true values.

Solution   Let $z = f(x, y) = x^2 + y^2$. In Example 1 on page 111, we found the equation of the tangent plane at $(3, 4)$ to be

$$z = 25 + 6(x - 3) + 8(y - 4).$$

Therefore, for (x, y) near $(3, 4)$, we have the local linearization

$$f(x, y) \approx 25 + 6(x - 3) + 8(y - 4).$$

Substituting $x = 2.9, y = 4.2$ gives

$$f(2.9, 4.2) \approx 25 + 6(-0.1) + 8(0.2) = 26.$$

This compares favorably with the true value $f(2.9, 4.2) = (2.9)^2 + (4.2)^2 = 26.05$.

However, the local linearization does not give a good approximation at points far away from $(3, 4)$. For example, if $x = 2, y = 2$, the local linearization gives

$$f(2, 2) \approx 25 + 6(-1) + 8(-2) = 3,$$

whereas the true value of the function is $f(2, 2) = 2^2 + 2^2 = 8$.

Example 3 Designing safe boilers depends on knowing how steam behaves under changes in temperature and pressure. Steam tables, such as Table 13.4, are published giving values of the function $V = f(T, P)$ where V is the volume (in ft^3) of one pound of steam at a temperature T (in °F) and pressure P (in lb/in^2).

(a) Give a linear function approximating $V = f(T, P)$ for T near 500°F and P near 24 lb/in^2.

(b) Estimate the volume of a pound of steam at a temperature of 505°F and a pressure of 24.3 lb/in^2.

TABLE 13.4 *Volume (in cubic feet) of one pound of steam at various temperatures and pressures*

		Pressure P (lb/in^2)			
		20	22	24	26
Temperature T (°F)	480	27.85	25.31	23.19	21.39
	500	28.46	25.86	23.69	21.86
	520	29.06	26.41	24.20	22.33
	540	29.66	26.95	24.70	22.79

Solution (a) We want the local linearization around the point $T = 500$, $P = 24$, which is

$$f(T, P) \approx f(500, 24) + f_T(500, 24)(T - 500) + f_P(500, 24)(P - 24).$$

We read the value $f(500, 24) = 23.69$ from the table.

Next we approximate $f_T(500, 24)$ by a difference quotient. From the $P = 24$ column, we compute the average rate of change between $T = 500$ and $T = 520$:

$$f_T(500, 24) \approx \frac{f(520, 24) - f(500, 24)}{520 - 500} = \frac{24.20 - 23.69}{20} = 0.0255.$$

Note that $f_T(500, 24)$ is positive, because steam expands when heated.

Next we approximate $f_P(500, 24)$ by looking at the $T = 500$ row and computing the average rate of change between $P = 24$ and $P = 26$:

$$f_P(500, 24) \approx \frac{f(500, 26) - f(500, 24)}{26 - 24} = \frac{21.86 - 23.69}{2} = -0.915.$$

Note that $f_P(500, 24)$ is negative, because increasing the pressure on steam decreases its volume. Using these approximations for the partial derivatives, we obtain the local linearization:

$$V = f(T, P) \approx 23.69 + 0.0255(T - 500) - 0.915(P - 24) \text{ ft}^3 \quad \begin{array}{l} \text{for } T \text{ near } 500 \text{ °F} \\ \text{and } P \text{ near } 24 \text{ lb/in}^2. \end{array}$$

(b) We are interested in the volume at $T = 505$°F and $P = 24.3$ lb/in^2. Since these values are close to $T = 500$°F and $P = 24$ lb/in^2, we use the linear relation obtained in part (a).

$$V \approx 23.69 + 0.0255(505 - 500) - 0.915(24.3 - 24) = 23.54 \text{ ft}^3.$$

Local Linearity With Three or More Variables

Local linear approximations for functions of three or more variables follow the same pattern as for functions of two variables. The local linearization of $f(x, y, z)$ at (a, b, c) is given by

$$f(x, y, z) \approx f(a, b, c) + f_x(a, b, c)(x - a) + f_y(a, b, c)(y - b) + f_z(a, b, c)(z - c).$$

The Differential

We are often interested in the change in the value of the function as we move from the point (a, b) to a nearby point (x, y). Then we use the notation

$$\Delta f = f(x, y) - f(a, b) \quad \text{and} \quad \Delta x = x - a \quad \text{and} \quad \Delta y = y - b$$

to rewrite the tangent plane approximation

$$f(x, y) \approx f(a, b) + f_x(a, b)(x - a) + f_y(a, b)(y - b)$$

in the form

$$\Delta f \approx f_x(a, b)\Delta x + f_y(a, b)\Delta y.$$

For fixed a and b, the right side of this is a linear function of Δx and Δy that can be used to estimate Δf. We call this linear function the *differential*. To define the differential in general, we introduce new variables dx and dy to represent changes in x and y.

The Differential of a Function $z = f(x, y)$

The **differential**, df (or dz), at a point (a, b) is the linear function of dx and dy given by the formula

$$df = f_x(a, b) \, dx + f_y(a, b) \, dy.$$

The differential at a general point is often written $df = f_x \, dx + f_y \, dy$.

Note that the differential, df, is a function of four variables a, b, and dx, dy.

Example 4   Compute the differentials of the following functions.

(a) $f(x, y) = x^2 e^{5y}$ (b) $z = x \sin(xy)$ (c) $f(x, y) = x \cos(2x)$

Solution   (a)   Since $f_x(x, y) = 2xe^{5y}$ and $f_y(x, y) = 5x^2 e^{5y}$, we have

$$df = 2xe^{5y} \, dx + 5x^2 e^{5y} \, dy.$$

(b)   Since $\partial z / \partial x = \sin(xy) + xy \cos(xy)$ and $\partial z / \partial y = x^2 \cos(xy)$, we have

$$dz = (\sin(xy) + xy \cos(xy)) \, dx + x^2 \cos(xy) \, dy.$$

(c)   Since $f_x(x, y) = \cos(2x) - 2x \sin(2x)$ and $f_y(x, y) = 0$, we have

$$df = (\cos(2x) - 2x \sin(2x)) \, dx + 0 \, dy = (\cos(2x) - 2x \sin(2x)) \, dx.$$

Example 5   The density ρ (in g/cm^3) of carbon dioxide gas CO_2 depends upon its temperature T (in °C) and pressure P (in atmospheres). The ideal gas model for CO_2 gives what is called the state equation

$$\rho = \frac{0.5363P}{T + 273.15}.$$

Compute the differential $d\rho$. Explain the signs of the coefficients of dT and dP.

Solution   The differential for $\rho = f(T, P)$ is

$$d\rho = f_T(T, P)\, dT + f_P(T, P)dP = \frac{-0.5363P}{(T + 273.15)^2}\, dT + \frac{0.5363}{T + 273.15}\, dP.$$

The coefficient of dT is negative because increasing the temperature expands the gas (if the pressure is kept constant) and therefore decreases its density. The coefficient of dP is positive because increasing the pressure compresses the gas (if the temperature is kept constant) and therefore increases its density.

Where Does the Notation for the Differential Come From?

We write the differential as a linear function of the new variables dx and dy. You may wonder why we chose these names for our variables. The reason is historical: The people who invented calculus thought of dx and dy as "infinitesimal" changes in x and y. The equation

$$df = f_x dx + f_y dy$$

was regarded as an infinitesimal version of the local linear approximation

$$\Delta f \approx f_x \Delta x + f_y \Delta y.$$

In spite of the problems with defining exactly what "infinitesimal" means, some mathematicians, scientists, and engineers think of the differential in terms of infinitesimals.

Figure 13.20 illustrates a way of thinking about differentials that combines the definition with this informal point of view. It shows the graph of f along with a view of the graph around the point $(a, b, f(a, b))$ under a microscope. Since f is locally linear at the point, the magnified view looks like the tangent plane. Under the microscope, we use a magnified coordinate system with its origin at the point $(a, b, f(a, b))$ and with coordinates dx, dy, and dz along the three axes. The graph of the differential df is the tangent plane, which has equation $df = f_x(a, b)\, dx + f_y(a, b)\, dy$ in the magnified coordinates.

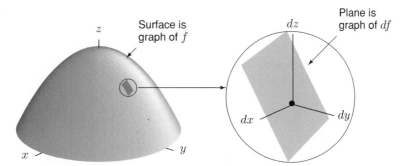

Figure 13.20: The graph of f, with a view through a microscope showing the tangent plane in the magnified coordinate system

Problems for Section 13.3

$f_y = 1 + 2y$ 　　 $z_y = e^y$ 　　 $9 + 8(x-1) + (1(y-0))$ 　 $3x - 3 + y$ 　 $6 + 3x + y$ 　 #1
$f_x = 1 + 2x$ 　　　　　　　　　　 $9 + 3x - 3 + y$

For the functions in Problems 1–3, find the equation of the tangent plane at the given point.

1. $z = e^y + x + x^2 + 6$ at the point $(1, 0, 9)$.　　2. $z = ye^{x/y}$ at the point $(1, 1, e)$.

3. $z = \frac{1}{2}(x^2 + 4y^2)$ at the point $(2, 1, 4)$.

$z = 9 + (x - 1)$ 　　　　 $z = 4 + 2(x-2) + 4(y-1)$ 　　　 $z = -4 + 2x + 4y$
　　　　　　　　　　　　 $= 4 + 2x - 4 + 4y - 4$

4. A student was asked to find the equation of the tangent plane to the surface $z = x^3 - y^2$ at the point $(x, y) = (2, 3)$. The student's answer was

$$z = 3x^2(x - 2) - 2y(y - 3) - 1.$$

   (a) At a glance, how do you know this is wrong?
   (b) What mistake did the student make?
   (c) Answer the question correctly.

5. Find the local linearization of the function $f(x, y) = x^2 y$ at the point $(3, 1)$.

6. (a) Verify the local linearity of $f(x, y) = e^{-x} \sin y$ near $x = 1$, $y = 2$ by making a table of values of f for $x = 0.9$, 1.0, 1.1 and $y = 1.9$, 2.0, 2.1. Express values of f with 4 digits after the decimal point. Then make a table of values for $x = 0.99$, 1.00, 1.01 and $y = 1.99$, 2.00. 2.01, again showing 4 digits after the decimal point. Do both tables look nearly linear? Does the second table look more linear than the first?
   (b) Give the local linearization of $f(x, y) = e^{-x} \sin y$ at $(1, 2)$, first using your tables, and second using the fact that $f_x(x, y) = -e^{-x} \sin y$ and $f_y(x, y) = e^{-x} \cos y$.

7. Give the local linearization for the monthly car-loan payment function at each of the points investigated in Problem 13 on page 104.

8. In Example 3 on page 113 we found a linear approximation for $V = f(T, p)$ near $(500, 24)$. Now find a linear approximation near $(480, 20)$.

9. In Example 3 on page 113 we found a linear approximation for $V = f(T, p)$ near $(500, 24)$.
   (a) Test the accuracy of this approximation by comparing its predicted value with the four neighboring values in the table. What do you notice? Which predicted values are accurate? Which are not? Explain your answer.
   (b) Suggest a linear approximation for $f(T, p)$ near $(500, 24)$ that does not have the property you noticed in part (a). [Hint: Estimate the partial derivatives in a different way.]

10. Figure 13.21 shows a transistor whose state at any moment is determined by the three currents i_b, i_c, and i_e and the two voltages v_b and v_c. These five quantities can be determined from measurements of i_b and v_c alone, because there are functions f and g (called the *characteristics* of the transistor) such that $i_c = f(i_b, v_c)$ and $v_b = g(i_b, v_c)$ and $i_e = -i_b - i_c$. The units are microamps (μA) for i_b, volts (V) for v_c, and milliamps (mA) for i_c.

   Use Figure 13.22 to find a linear approximation for f that is valid when the transistor is operating with i_b near $-300 \; \mu A$ and v_c near $-8V$.

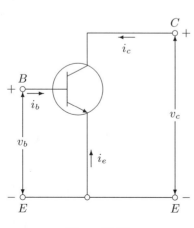

Figure 13.21

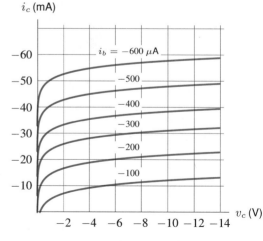

Figure 13.22: Graphs of f as a function of v_c with i_b fixed

Find the differentials of the functions in Problems 11–14.

11. $f(x, y) = \sin(xy)$

12. $z = e^{-x} \cos y$

13. $g(u, v) = u^2 + uv$

14. $h(x, t) = e^{-3t} \sin(x + 5t)$

Find differentials of the functions in Problems 15–18 at the given point.

15. $f(x, y) = xe^{-y}$ at $(1, 0)$

16. $g(x, t) = x^2 \sin(2t)$ at $(2, \pi/4)$

17. $P(L, K) = 1.01L^{0.25} K^{0.75}$ at $(100, 1)$

18. $F(m, r) = Gm/r^2$ at $(100, 10)$

19. Find the differential of $f(x, y) = \sqrt{x^2 + y^3}$ at the point $(1, 2)$. Use it to estimate $f(1.04, 1.98)$.

20. An unevenly heated plate has temperature $T(x, y)$ in °C at the point (x, y). If $T(2, 1) = 135$, and $T_x(2, 1) = 16$, and $T_y(2, 1) = -15$, estimate the temperature at the point $(2.04, 0.97)$.

21. One mole of ammonia gas is contained in a vessel which is capable of changing its volume (a compartment sealed by a piston, for example). The total energy U (in joules) of the ammonia, is a function of the volume V (in m³) of the container, and the temperature T (in K) of the gas. The differential dU is given by

$$dU = 840 \, dV + 27.32 \, dT.$$

(a) How does the energy change if the volume is held constant and the temperature is increased slightly?

(b) How does the energy change if the temperature is held constant and the volume is increased slightly?

(c) Find the approximate change in energy if the gas is compressed by 100 cm³ and heated by 2 K.

22. The coefficient, β, of thermal expansion of a liquid relates the change in the volume V (in m³) of a fixed quantity of a liquid to an increase in its temperature T (in °C):

$$dV = \beta V \, dT.$$

(a) Let ρ be the density (in kg/m³) of water as a function of temperature. Write an expression for $d\rho$ in terms of ρ and dT.

(b) The graph in Figure 13.23 shows density of water as a function of temperature. Use it to estimate β when $T = 20°C$ and when $T = 80°C$.

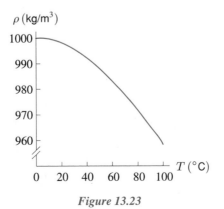

Figure 13.23

23. A pendulum of period T and length l was used to determine g from the formulas

$$g = \frac{4\pi^2 l}{T^2} \quad \text{and} \quad l = s + \frac{k^2}{s}, \quad k < s.$$

If the measurements of k and s are accurate to within 1%, find the maximum percentage error in l. If the measurement of T is accurate to 0.5%, find the maximum percentage error in the computed value of g.

24. The period, T, of oscillation in seconds of a pendulum clock is given by $T = 2\pi\sqrt{l/g}$, where g is the acceleration due to gravity. The length of the pendulum, l, depends on the temperature, t, according to the formula $l = l_0(1+\alpha t)$ where l_0 is the length of the pendulum at temperature t_0 and α is a constant which characterizes the clock. The clock is set to the correct period at the temperature t_0. How many seconds a day does the clock lose or gain when the temperature is $t_0 + \Delta t$? Show that this loss or gain is independent of l_0.

25. (a) Write a formula for the number π using only the perimeter L and the area A of a circle.
    (b) Suppose that L and A are determined experimentally. Show that if the relative, or percent, errors in the measured values of L and A are λ and μ, respectively, then the resulting relative, or percent, error in π is $2\lambda - \mu$.

13.4 GRADIENTS AND DIRECTIONAL DERIVATIVES IN THE PLANE

The Rate of Change in an Arbitrary Direction: The Directional Derivative

The partial derivatives of a function f tell us the rate of change of f in the directions parallel to the coordinate axes. In this section we see how to compute the rate of change of f in an arbitrary direction.

Example 1   Figure 13.24 shows the temperature, in $°C$, at the point (x, y). Estimate the average rate of change of temperature as we walk from point A to point B.

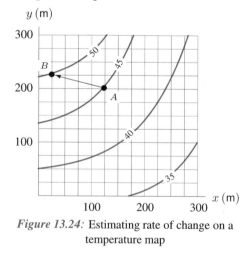

Figure 13.24: Estimating rate of change on a temperature map

Solution   At the point A we are on the $H = 45°C$ contour. At B we are on the $H = 50°C$ contour. The displacement vector from A to B has x component approximately $-100\vec{i}$ and y component approximately $25\vec{j}$, so its length is $\sqrt{(-100)^2 + 25^2} \approx 103$. Thus the temperature rises by $5°C$ as we move 103 meters, so the average rate of change of the temperature in that direction is about $5/103 \approx 0.05°C/m$.

Suppose we want to compute the rate of change of a function $f(x, y)$ at the point $P = (a, b)$ in the direction of the unit vector $\vec{u} = u_1\vec{i} + u_2\vec{j}$. For $h > 0$, consider the point $Q = (a + hu_1, b + hu_2)$ whose displacement from P is $h\vec{u}$. (See Figure 13.25.) Since $\|\vec{u}\| = 1$, the distance from P to Q is h. Thus,

$$\begin{array}{l} \text{Average rate of change} \\ \text{in } f \text{ from } P \text{ to } Q \end{array} = \frac{\text{Change in } f}{\text{Distance from } P \text{ to } Q} = \frac{f(a + hu_1, b + hu_2) - f(a, b)}{h}.$$

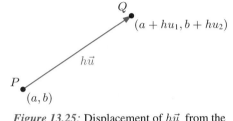

Figure 13.25: Displacement of $h\vec{u}$ from the point (a, b)

Taking the limit as $h \to 0$ gives the instantaneous rate of change and the following definition:

Directional Derivative of f at (a, b) in the Direction of a Unit Vector $\vec{u}$

If $\vec{u} = u_1\vec{i} + u_2\vec{j}$ is a unit vector, we define the directional derivative, $f_{\vec{u}}$, by

$$f_{\vec{u}}(a, b) = \begin{array}{l} \text{Rate of change} \\ \text{of } f \text{ in direction} \\ \text{of } \vec{u} \text{ at } (a, b) \end{array} = \lim_{h \to 0} \frac{f(a + hu_1, b + hu_2) - f(a, b)}{h},$$

provided the limit exists.

Notice that if $\vec{u} = \vec{i}$, so $u_1 = 1, u_2 = 0$, then the directional derivative is f_x, since

$$f_{\vec{i}}(a, b) = \lim_{h \to 0} \frac{f(a + h, b) - f(a, b)}{h} = f_x(a, b).$$

Similarly, if $\vec{u} = \vec{j}$ then the directional derivative $f_{\vec{j}} = f_y$.

What If We Do Not Have a Unit Vector?

We defined $f_{\vec{u}}$ for $\vec{u}$ a unit vector. If $\vec{v}$ is not a unit vector, $\vec{v} \neq \vec{0}$, we construct a unit vector $\vec{u} = \vec{v}/\|\vec{v}\|$ in the same direction as $\vec{v}$ and define the rate of change of f in the direction of $\vec{v}$ as $f_{\vec{u}}$.

Example 2 For each of the functions f, g, and h in Figure 13.26, decide whether the directional derivative at the indicated point is positive, negative, or zero, in the direction of the vector $\vec{v} = \vec{i} + 2\vec{j}$, and in the direction of the vector $\vec{w} = 2\vec{i} + \vec{j}$.

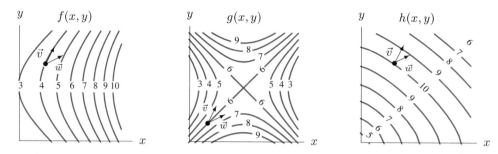

Figure 13.26: Contour diagrams of three functions with direction vectors $\vec{v} = \vec{i} + 2\vec{j}$ and $\vec{w} = 2\vec{i} + \vec{j}$ marked on each

Solution   On the contour diagram for f, the vector $\vec{v} = \vec{i} + 2\vec{j}$ appears to be tangent to the contour. Thus, in this direction, the value of the function is not changing, so the directional derivative in the direction of $\vec{v}$ is zero. The vector $\vec{w} = 2\vec{i} + \vec{j}$ points from the contour marked 4 toward the contour marked 5. Thus, the values of the function are increasing and the directional derivative in the direction of $\vec{w}$ is positive.

On the contour diagram for g, the vector $\vec{v} = \vec{i} + 2\vec{j}$ points from the contour marked 6 toward the contour marked 5, so the function is decreasing in that direction. Thus, the rate of change is negative. On the other hand, the vector $\vec{w} = 2\vec{i} + \vec{j}$ points from the contour marked 6 toward the contour marked 7, and hence the directional derivative in the direction of $\vec{w}$ is positive.

Finally, on the contour diagram for h, both vectors point from the $h = 10$ contour to the $h = 9$ contour, so both directional derivatives are negative.

Example 3   Calculate the directional derivative of $f(x, y) = x^2 + y^2$ at $(1, 0)$ in the direction of the vector $\vec{i} + \vec{j}$.

Solution   First we have to find the unit vector in the same direction as the vector $\vec{i} + \vec{j}$. Since this vector has magnitude $\sqrt{2}$, the unit vector is

$$\vec{u} = \frac{1}{\sqrt{2}}(\vec{i} + \vec{j}) = \frac{1}{\sqrt{2}}\vec{i} + \frac{1}{\sqrt{2}}\vec{j}.$$

Thus,

$$f_{\vec{u}}(1, 0) = \lim_{h \to 0} \frac{f(1 + h/\sqrt{2}, h/\sqrt{2}) - f(1, 0)}{h} = \lim_{h \to 0} \frac{(1 + h/\sqrt{2})^2 + (h/\sqrt{2})^2 - 1}{h}$$

$$= \lim_{h \to 0} \frac{\sqrt{2}h + h^2}{h} = \lim_{h \to 0} (\sqrt{2} + h) = \sqrt{2}.$$

Computing Directional Derivatives From Partial Derivatives

If f is differentiable, we will now see how to use local linearity to find a formula for the directional derivative which does not involve a limit. If $\vec{u}$ is a unit vector, the definition of $f_{\vec{u}}$ says

$$f_{\vec{u}}(a, b) = \lim_{h \to 0} \frac{f(a + hu_1, b + hu_2) - f(a, b)}{h} = \lim_{h \to 0} \frac{\Delta f}{h},$$

where $\Delta f = f(a + hu_1, b + hu_2) - f(a, b)$ is the change in f. We write Δx for the change in x, so $\Delta x = (a + hu_1) - a = hu_1$; similarly $\Delta y = hu_2$. Using local linearity, we have

$$\Delta f \approx f_x(a, b)\Delta x + f_y(a, b)\Delta y = f_x(a, b)hu_1 + f_y(a, b)hu_2.$$

Thus, dividing by h gives

$$\frac{\Delta f}{h} \approx \frac{f_x(a,b)hu_1 + f_y(a,b)hu_2}{h} = f_x(a,b)u_1 + f_y(a,b)u_2.$$

This approximation becomes exact as $h \to 0$, so we have the following formula:

$$f_{\vec{u}}(a,b) = f_x(a,b)u_1 + f_y(a,b)u_2.$$

Example 4   Use the preceding formula to compute the directional derivative in Example 3. Check that we get the same answer as before.

Solution   We calculate $f_{\vec{u}}(1,0)$, where $f(x,y) = x^2 + y^2$ and $\vec{u} = \frac{1}{\sqrt{2}}\vec{i} + \frac{1}{\sqrt{2}}\vec{j}$.

The partial derivatives are $f_x(x,y) = 2x$ and $f_y(x,y) = 2y$. So, as before

$$f_{\vec{u}}(1,0) = f_x(1,0)u_1 + f_y(1,0)u_2 = (2)\left(\frac{1}{\sqrt{2}}\right) + (0)\left(\frac{1}{\sqrt{2}}\right) = \sqrt{2}.$$

The Gradient Vector

Notice that the expression for $f_{\vec{u}}(a,b)$ can be written as a dot product of $\vec{u}$ and a new vector:

$$f_{\vec{u}}(a,b) = f_x(a,b)u_1 + f_y(a,b)u_2 = (f_x(a,b)\vec{i} + f_y(a,b)\vec{j}) \cdot (u_1\vec{i} + u_2\vec{j}).$$

The new vector, $f_x(a,b)\vec{i} + f_y(a,b)\vec{j}$, turns out to be important. Thus, we make the following definition:

> **The Gradient Vector** of a differentiable function at the point (a,b) is
>
> $$\text{grad } f(a,b) = f_x(a,b)\vec{i} + f_y(a,b)\vec{j}$$

The formula for the directional derivative can be written in terms of the gradient as follows:

> **The Directional Derivative and the Gradient**
>
> If f is differentiable at (a,b), then
>
> $$f_{\vec{u}}(a,b) = f_x(a,b)u_1 + f_y(a,b)u_2 = \text{grad } f(a,b) \cdot \vec{u},$$
>
> where $\vec{u} = u_1\vec{i} + u_2\vec{j}$ is a unit vector.

Example 5   Find the gradient vector of $f(x,y) = x + e^y$ at the point $(1,1)$.

Solution   Using the definition we have

$$\text{grad } f = f_x\vec{i} + f_y\vec{j} = \vec{i} + e^y\vec{j},$$

so at the point $(1,1)$

$$\text{grad } f(1,1) = \vec{i} + e\vec{j}.$$

Alternative Notation for the Gradient

You can think of $\dfrac{\partial f}{\partial x}\vec{i} + \dfrac{\partial f}{\partial y}\vec{j}$ as the result of applying the vector operator (pronounced "del")

$$\nabla = \frac{\partial}{\partial x}\vec{i} + \frac{\partial}{\partial y}\vec{j}$$

to the function f. Thus, we get the alternative notation

$$\text{grad } f = \nabla f.$$

What Does the Gradient Tell Us?

The fact that $f_{\vec{u}} = \text{grad } f \cdot \vec{u}$ enables us to see what the gradient vector represents. Suppose θ is the angle between the vectors $\text{grad} f$ and $\vec{u}$. At the point (a, b), we have

$$f_{\vec{u}} = \text{grad } f \cdot \vec{u} = \| \text{grad } f \| \, \|\vec{u}\| \cos\theta = \| \text{grad } f \| \cos\theta.$$

Imagine that $\text{grad } f$ is fixed and that $\vec{u}$ can rotate. (See Figure 13.27.) The maximum value of $f_{\vec{u}}$ occurs when $\cos\theta = 1$, so $\theta = 0$ and $\vec{u}$ is pointing in the direction of $\text{grad } f$. Then

$$\text{Maximum } f_{\vec{u}} = \| \text{grad } f \| \cos 0 = \| \text{grad } f \|.$$

The minimum value of $f_{\vec{u}}$ occurs when $\cos\theta = -1$, so $\theta = \pi$ and $\vec{u}$ is pointing in the direction opposite to $\text{grad } f$. Then

$$\text{Minimum } f_{\vec{u}} = \| \text{grad } f \| \cos \pi = -\| \text{grad } f \|.$$

When $\theta = \pi/2$ or $3\pi/2$, so $\cos\theta = 0$, the directional derivative is zero.

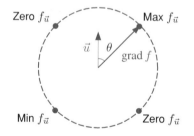

Figure 13.27: Values of the directional derivative at different angles to the gradient

Properties of The Gradient Vector

We have seen that the gradient vector points in the direction of the greatest rate of change at a point and the magnitude of the gradient vector is that rate of change.

Figure 13.28 shows that the gradient vector at any point is perpendicular to the contour through that point. Assuming f is differentiable at the point (a, b), local linearity tells us that the contours of f around the point (a, b) appear straight, parallel, and equally spaced. The greatest rate of change is obtained by moving in the direction that takes us to the next contour in the shortest possible distance, which is the direction perpendicular to the contour. Thus, we have the following:

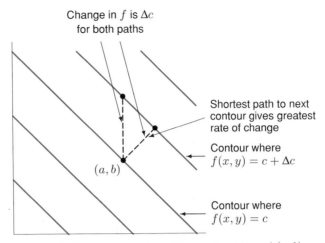

Figure 13.28: Close-up view of the contours around (a, b), showing the gradient is perpendicular to the contours

---

Geometric Properties of the Gradient Vector

If f is a differentiable function at the point (a, b) and grad $f(a, b) \neq \vec{0}$, then:

- The direction of grad $f(a, b)$ is
    - Perpendicular to the contour of f through (a, b)
    - In the direction of increasing f
- The magnitude of the gradient vector, $\| \text{grad } f \|$, is
    - The maximum rate of change of f at that point
    - Large when the contours are close together and small when they are far apart.

Examples of Directional Derivatives and Gradient Vectors

Example 6 Explain why the gradient vectors at points A and C in Figure 13.29 have the direction and the relative magnitudes they do.

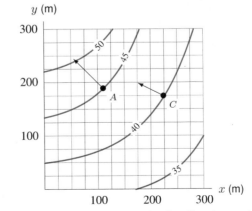

Figure 13.29: A temperature map showing directions and relative magnitudes of two gradient vectors

Solution   The gradient vector points in the direction of greatest increase of the function. This means that in Figure 13.29, the gradient points directly towards warmer temperatures. The magnitude of the gradient vector measures the rate of change. The gradient vector at A is longer than the gradient vector at C because the contours are closer together at A, so the rate of change is larger.

Example 2 on page 119 shows how the contour diagram can tell us the sign of the directional derivative. In the next example we compute the directional derivative in three directions, two close to that of the gradient vector and one that is not.

Example 7   Use the gradient to find the directional derivative of $f(x, y) = x + e^y$ at the point $(1, 1)$ in the direction of the vectors $\vec{i} - \vec{j}, \vec{i} + 2\vec{j}, \vec{i} + 3\vec{j}$.

Solution   In Example 5 we found

$$\text{grad } f(1, 1) = \vec{i} + e\vec{j}.$$

A unit vector in the direction of $\vec{i} - \vec{j}$ is $\vec{s} = (\vec{i} - \vec{j})/\sqrt{2}$, so

$$f_{\vec{s}}(1, 1) = \text{grad } f(1, 1) \cdot \vec{s} = (\vec{i} + e\vec{j}) \cdot \left(\frac{\vec{i} - \vec{j}}{\sqrt{2}}\right) = \frac{1 - e}{\sqrt{2}} \approx -1.215.$$

A unit vector in the direction of $\vec{i} + 2\vec{j}$ is $\vec{v} = (\vec{i} + 2\vec{j})/\sqrt{5}$, so

$$f_{\vec{v}}(1, 1) = \text{grad } f(1, 1) \cdot \vec{v} = (\vec{i} + e\vec{j}) \cdot \left(\frac{\vec{i} + 2\vec{j}}{\sqrt{5}}\right) = \frac{1 + 2e}{\sqrt{5}} \approx 2.879.$$

A unit vector in the direction of $\vec{i} + 3\vec{j}$ is $\vec{w} = (\vec{i} + 3\vec{j})/\sqrt{10}$, so

$$f_{\vec{w}}(1, 1) = \text{grad } f(1, 1) \cdot \vec{w} = (\vec{i} + e\vec{j}) \cdot \left(\frac{\vec{i} + 3\vec{j}}{\sqrt{10}}\right) = \frac{1 + 3e}{\sqrt{10}} \approx 2.895.$$

Now look back at the answers and compare with the value of $\| \text{grad } f\| = \sqrt{1 + e^2} \approx 2.896$. One answer is not close to this value; the other two, $f_{\vec{v}} = 2.879$ and $f_{\vec{w}} = 2.895$, are close but slightly smaller than $\| \text{grad } f\|$. Since $\| \text{grad } f\|$ is the maximum rate of change of f at the point, we would expect for *any* unit vector $\vec{u}$:

$$f_{\vec{u}}(1, 1) \leq \| \text{grad } f\|.$$

with equality when $\vec{u}$ is in the direction of grad f. Since $e \approx 2.718$, the vectors $\vec{i} + 2\vec{j}$ and $\vec{i} + 3\vec{j}$ both point roughly, but not exactly, in the direction of the gradient vector grad $f(1, 1) = \vec{i} + e\vec{j}$. Thus, the values of $f_{\vec{v}}$ and $f_{\vec{w}}$ are both close to the value of $\| \text{grad } f\|$. The direction of the vector $\vec{i} - \vec{j}$ is not close to the direction of grad f and the value of $f_{\vec{s}}$ is not close to the value of $\| \text{grad } f\|$.

Problems for Section 13.4

1. Suppose $f(x, y) = x + \ln y$. Using difference quotients as in Example 1 on page 118, estimate
   (a) The rate of change of f as you leave the point $(1, 4)$ going towards the point $(3, 5)$.
   (b) The rate of change of f as you arrive at the point $(3, 5)$.

2. Using the limit of a difference quotient, compute the rate of change of $f(x, y) = 2x^2 + y^2$ at the point $(2, 1)$ as you move in the direction of the vector $\vec{u} = (\vec{i} + \vec{j})/\sqrt{2}$.

For Problems 3–8 use Figure 13.30, showing level curves of $f(x, y)$, to estimate the directional derivatives.

3. $f_{\vec{i}}(3, 1)$

4. $f_{\vec{j}}(3, 1)$

5. $f_{\vec{u}}(3, 1)$ where $\vec{u} = (\vec{i} - \vec{j})/\sqrt{2}$

6. $f_{\vec{u}}(3, 1)$ where $\vec{u} = (-\vec{i} + \vec{j})/\sqrt{2}$

7. For what part of the rectangular region shown in Figure 13.30 is $f_{\vec{i}}$ positive?

8. For what part of the rectangular region shown in Figure 13.30 is $f_{\vec{j}}$ negative?

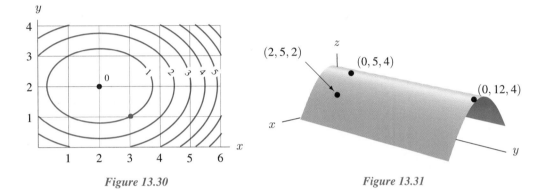

Figure 13.30   Figure 13.31

9. The surface $z = g(x, y)$ is shown in Figure 13.31. What is the sign of each of the following directional derivatives?
    (a) $g_{\vec{u}}(2, 5)$ where $\vec{u} = (\vec{i} - \vec{j})/\sqrt{2}$.
    (b) $g_{\vec{u}}(2, 5)$ where $\vec{u} = (\vec{i} + \vec{j})/\sqrt{2}$.

10. If $f(x, y) = x^2 + \ln y$, find (a) grad f (b) grad f at $(4, 1)$.

In Problems 11–22 find the gradient of the given function z. Assume the variables are restricted to a domain on which the function is defined.

11. $z = \sin(x/y)$

12. $z = xe^y$

13. $z = (x + y)e^y$

14. $z = \tan^{-1}(x/y)$

15. $z = \sin(x^2 + y^2)$

16. $z = xe^y/(x + y)$

17. $f(m, n) = m^2 + n^2$

18. $f(x, y) = \frac{3}{2}x^5 - \frac{4}{7}y^6$

19. $f(s, t) = \frac{1}{\sqrt{s}}(t^2 - 2t + 4)$

20. $f(\alpha, \beta) = \frac{2\alpha + 3\beta}{2\alpha - 3\beta}$

21. $f(\alpha, \beta) = \sqrt{5\alpha^2 + \beta}$

22. $f(x, y) = \sin(xy) + \cos(xy)$

In Problems 23–26 compute the gradient at the specified point.

23. $f(m, n) = 5m^2 + 3n^4$, at $(5, 2)$

24. $f(x, y) = x^2y + 7xy^3$, at $(1, 2)$

25. $f(x, y) = \sqrt{\tan x + y}$, at $(0, 1)$

26. $f(x, y) = \sin(x^2) + \cos y$, at $(\frac{\sqrt{\pi}}{2}, 0)$

27. Let $f(x, y) = (x + y)/(1 + x^2)$. Find the directional derivative at $P = (1, -2)$ in the direction of the vectors
    (a) $\vec{v} = 3\vec{i} - 2\vec{j}$,   (b) $\vec{v} = -\vec{i} + 4\vec{j}$.
    (c) What is the direction of greatest increase at P ?

28. A student was asked to find the directional derivative of $f(x, y) = x^2 e^y$ at the point $(1, 0)$ in the direction of $\vec{v} = 4\vec{i} + 3\vec{j}$. The student's answer was

$$f_{\vec{u}}(1, 0) = \nabla f(1, 0) \cdot \vec{u} = \frac{8}{5}\vec{i} + \frac{3}{5}\vec{j}.$$

(a) At a glance, how do you know this is wrong?
(b) What is the correct answer?

29. Find the directional derivative of $f(x, y) = e^x \tan(y) + 2x^2 y$ at the point $(0, \pi/4)$ in the following directions (a) $\vec{i} - \vec{j}$ (b) $\vec{i} + \sqrt{3}\vec{j}$.

30. Find the directional derivative of $z = x^2 y$ at the point $(1, 2)$ in the direction making an angle of $5\pi/4$ with the x-axis. In which direction is the directional derivative the largest?

31. Find the rate of change of $f(x, y) = x^2 + y^2$ at the point $(1, 2)$ in the direction of the vector $\vec{u} = 0.6\vec{i} + 0.8\vec{j}$.

32. The directional derivative of $z = f(x, y)$ at the point $(2, 1)$ in the direction toward the point $(1, 3)$ is $-2/\sqrt{5}$, and the directional derivative in the direction toward the point $(5, 5)$ is 1. Compute $\partial z/\partial x$ and $\partial z/\partial y$ at the point $(2, 1)$.

33. Consider the function $f(x, y)$. If you start at the point $(4, 5)$ and move towards the point $(5, 6)$, the directional derivative is 2. Starting at the point $(4, 5)$ and moving towards the point $(6, 6)$ gives a directional derivative of 3. Find ∇f at the point $(4, 5)$.

34. The temperature at any point in the plane is given by the function

$$T(x, y) = \frac{100}{x^2 + y^2 + 1}.$$

(a) What shape are the level curves of T?
(b) Where on the plane is it hottest? What is the temperature at that point?
(c) Find the direction of the greatest increase in temperature at the point $(3, 2)$. What is the magnitude of that greatest increase?
(d) Find the direction of the greatest decrease in temperature at the point $(3, 2)$.
(e) Find a direction at the point $(3, 2)$ in which the temperature does not increase or decrease.

35. A differentiable function $f(x, y)$ has the property that $f_x(4, 1) = 2$ and $f_y(4, 1) = -1$. Find the equation of the tangent line to the level curve of f through the point $(4, 1)$.

36. Figure 13.32 represents the level curves $f(x, y) = c$; the values of f on each curve are marked. In each of the following parts, decide whether the given quantity is positive, negative or zero. Explain your answer.

(a) The value of $\nabla f \cdot \vec{i}$ at P.   (b) The value of $\nabla f \cdot \vec{j}$ at P.
(c) $\partial f/\partial x$ at Q.   (d) $\partial f/\partial y$ at Q.

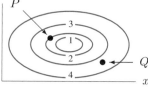

Figure 13.32

37. In Figure 13.32, which is larger: $\|\nabla f\|$ at P or $\|\nabla f\|$ at Q? Explain how you know.

38. The sketch in Figure 13.33 shows the level curves of a function $z = f(x, y)$. At the points $(1, 1)$ and $(1, 4)$ on the sketch, draw a vector representing grad f. Explain how you decided the approximate direction and length of each vector.

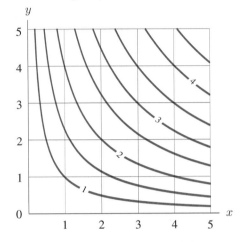

Figure 13.33: Contours of f

39. Figure 13.35 is a graph of the directional derivative, $f_{\vec{u}}$, at the point (a, b) versus θ, the angle shown in Figure 13.34.

(a) Which points on the graph in Figure 13.35 correspond to the greatest rate of increase of f? The greatest rate of decrease?

(b) Mark points on the circle in Figure 13.34 corresponding to the points P, Q, R, S.

(c) What is the amplitude of the function graphed in Figure 13.35? What is its formula?

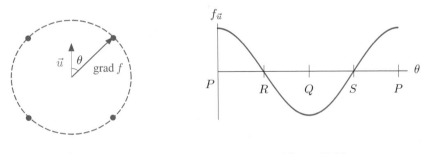

Figure 13.34   *Figure 13.35*

40. In this problem we see another way of explaining the formula $f_{\vec{u}}(a, b) = \text{grad } f(a, b) \cdot \vec{u}$. Imagine zooming in on a function $f(x, y)$ at a point (a, b). By local linearity, the contours around (a, b) look like the contours of a linear function. See Figure 13.36. Suppose you want to find the directional derivative $f_{\vec{u}}(a, b)$ in the direction of a unit vector $\vec{u}$. If you move from P to Q, a small distance h in the direction of $\vec{u}$, then the directional derivative is approximated by the difference quotient

$$\frac{\text{Change in } f \text{ between } P \text{ and } Q}{h}.$$

(a) Use the gradient to show that

$$\text{Change in } f \approx \| \text{grad } f \| (h \cos \theta).$$

(b)   Use part (a) to justify the formula $f_{\vec{u}}(a, b) = \text{grad } f(a, b) \cdot \vec{u}$.

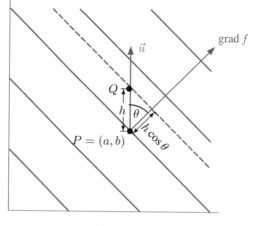

Figure 13.36

13.5 GRADIENTS AND DIRECTIONAL DERIVATIVES IN SPACE

Directional Derivatives of Functions of Three Variables

We calculate directional derivatives of a function of three variables in the same way as for a function of two variables. If the function f is differentiable at the point (a, b, c) and if $\vec{u} = u_1\vec{i} + u_2\vec{j} + u_3\vec{k}$ is a unit vector, then the rate of change of $f(x, y, z)$ in the direction of $\vec{u}$ at the point (a, b, c) is

$$f_{\vec{u}}(a, b, c) = f_x(a, b, c)u_1 + f_y(a, b, c)u_2 + f_z(a, b, c)u_3.$$

This can be justified using local linearity in the same way as for functions of two variables.

Example 1   Find the directional derivative of $f(x, y, z) = xy + z$ at the point $(-1, 0, 1)$ in the direction of the vector $\vec{v} = 2\vec{i} + \vec{k}$.

Solution   The magnitude of $\vec{v}$ is $\|\vec{v}\| = \sqrt{2^2 + 1} = \sqrt{5}$, so a unit vector in the same direction as $\vec{v}$ is

$$\vec{u} = \frac{\vec{v}}{\|\vec{v}\|} = \frac{2}{\sqrt{5}}\vec{i} + 0\vec{j} + \frac{1}{\sqrt{5}}\vec{k}.$$

The partial derivatives of f are

$$f_x(x, y, z) = y \quad \text{and} \quad f_y(x, y, z) = x \quad \text{and} \quad f_z(x, y, z) = 1.$$

Thus,

$$f_{\vec{u}}(-1, 0, 1) = f_x(-1, 0, 1)u_1 + f_y(-1, 0, 1)u_2 + f_z(-1, 0, 1)u_3$$

$$= (0)\left(\frac{2}{\sqrt{5}}\right) + (-1)(0) + (1)\left(\frac{1}{\sqrt{5}}\right) = \frac{1}{\sqrt{5}}.$$

The Gradient Vector of a Function of Three Variables

The gradient of a function of three variables is defined in the same way as for two variables:

$$\text{grad } f(a, b, c) = f_x(a, b, c)\vec{i} + f_y(a, b, c)\vec{j} + f_z(a, b, c)\vec{k}.$$

Geometrically, the gradient is the vector pointing in the direction of greatest increase of f, whose magnitude is the rate of increase in that direction.

Just as the gradient vector of a function of two variables is perpendicular to the level curves, so the gradient of a function of three variables is perpendicular to the level surfaces. The reason is the same: The function has the same value all over a level surface, so to change the value as quickly as possible we need to move directly away from the level surface, that is, perpendicular to the surface.

Example 2   Let $f(x, y, z) = x^2 + y^2$ and $g(x, y, z) = -x^2 - y^2 - z^2$. What can we say about the direction of the following vectors?

(a) grad $f(0, 1, 1)$ (b) grad $f(1, 0, 1)$ (c) grad $g(0, 1, 1)$ (d) grad $g(1, 0, 1)$.

Solution   The cylinder $x^2 + y^2 = 1$ in Figure 13.37 is a level surface of f and contains both the points $(0, 1, 1)$ and $(1, 0, 1)$. Since the value of f does not change at all in the z-direction, all the gradient vectors are horizontal. They are perpendicular to the cylinder and point outward because the value of f increases as we move out.

Similarly, the points $(0, 1, 1)$ and $(1, 0, 1)$ also lie on the same level surface of g, namely $g(x, y, z) = -x^2 - y^2 - z^2 = -2$, which is the sphere $x^2 + y^2 + z^2 = 2$. Part of this level surface is shown in Figure 13.38. This time the gradient vectors point inward, since the negative signs mean that the function increases (from large negative values to small negative values) as we move inward.

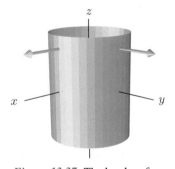

Figure 13.37: The level surface
$f(x, y, z) = x^2 + y^2 = 1$ with two gradient vectors

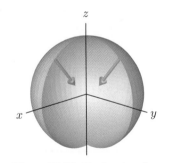

Figure 13.38: The level surface
$g(x, y, z) = -x^2 - y^2 - z^2 = -2$ with two gradient vectors

---

Example 3   Consider the functions $f(x, y) = 4 - x^2 - 2y^2$ and $g(x, y) = 4 - x^2$. Calculate a vector perpendicular to each of the following:

(a)   The level curve of f at the point $(1, 1)$ (b)   The surface $z = f(x, y)$ at the point $(1, 1, 1)$
(c)   The level curve of g at the point $(1, 1)$ (d)   The surface $z = g(x, y)$ at the point $(1, 1, 3)$

Solution   (a)   The vector we want is a 2-vector in the plane. Since grad $f = -2x\vec{i} - 4y\vec{j}$, we have

$$\text{grad } f(1, 1) = -2\vec{i} - 4\vec{j}.$$

Any nonzero multiple of this vector is perpendicular to the level curve at the point $(1, 1)$.

(b)   In this case we want a 3-vector in space. To find it we rewrite $z = 4 - x^2 - 2y^2$ as the level surface of the function F, where

$$F(x, y, z) = 4 - x^2 - 2y^2 - z = 0.$$

Then

$$\text{grad } F = -2x\vec{i} - 4y\vec{j} - \vec{k},$$

so

$$\text{grad } F(1, 1, 1) = -2\vec{i} - 4\vec{j} - \vec{k},$$

and grad $F(1, 1, 1)$ is perpendicular to the surface $z = 4 - x^2 - 2y^2$ at the point $(1, 1, 1)$. Notice that $-2\vec{i} - 4\vec{j} - \vec{k}$ is not the only possible answer: any multiple of this vector will do.

(c)   We are looking for a 2-vector. Since grad $g = -2x\vec{i} + 0\vec{j}$, we have

$$\text{grad } g(1, 1) = -2\vec{i}.$$

Any multiple of this vector is perpendicular to the level curve also.

(d)   We are looking for a 3-vector. We rewrite $z = 4 - x^2$ as the level surface of the function G, where

$$G(x, y, z) = 4 - x^2 - z = 0.$$

Then

$$\text{grad } G = -2x\vec{i} - \vec{k}$$

So

$$\text{grad } G(1, 1, 3) = -2\vec{i} - \vec{k},$$

and any multiple of grad $G(1, 1, 3)$ is perpendicular to the surface $z = 4 - x^2$ at this point.

Example 4   (a)   A hiker on the surface $f(x, y) = 4 - x^2 - 2y^2$ at the point $(1, -1, 1)$ starts to climb along the path of steepest ascent. What is the relation between the vector grad $f(1, -1)$ and a vector tangent to the path at the point $(1, -1, 1)$ and pointing uphill?

(b)   Consider the surface $g(x, y) = 4 - x^2$. What is the relation between grad $g(-1, -1)$ and a vector tangent to the path of steepest ascent at $(-1, -1, 3)$?

(c)   At the point $(1, -1, 1)$ on the surface $f(x, y) = 4 - x^2 - 2y^2$, calculate a vector perpendicular to the surface and a vector, $\vec{T}$, tangent to the curve of steepest ascent.

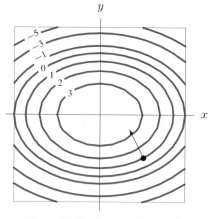

Figure 13.39: Contour diagram for $z = f(x, y) = 4 - x^2 - 2y^2$ showing direction of grad $f(1, -1)$

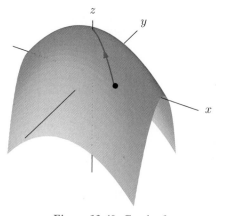

Figure 13.40: Graph of $f(x, y) = 4 - x^2 - 2y^2$ showing path of steepest ascent from the point $(1, -1, 1)$

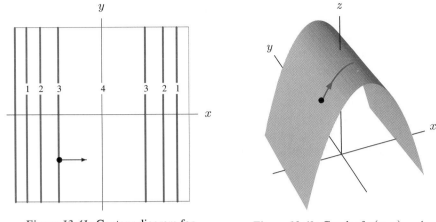

Figure 13.41: Contour diagram for $z = g(x, y) = 4 - x^2$ showing direction of grad $g(-1, -1)$

Figure 13.42: Graph of $g(x, y) = 4 - x^2$ showing path of steepest ascent from the point $(-1, -1, 3)$

Solution   (a)   The hiker at the point $(1, -1, 1)$ lies directly above the point $(1, -1)$ in the xy-plane. The vector grad $f(1, -1)$ lies in 2-space, pointing like a compass in the direction in which f increases most rapidly. Therefore, grad $f(1, -1)$ lies directly under a vector tangent to the hiker's path at $(1, -1, 1)$ and pointing uphill. (See Figures 13.39 and 13.40.)

(b)   The point $(-1, -1, 3)$ lies above the point $(-1, -1)$. The vector grad $g(-1, -1)$ points in the direction in which g increases most rapidly and lies directly under the path of steepest ascent. (See Figures 13.41 and 13.42.)

(c)   The surface is represented by $F(x, y, z) = 4 - x^2 - 2y^2 = 0$. Since grad $F = -2x\vec{i} - 4y\vec{j} - \vec{k}$, the normal, $\vec{N}$, to the surface is given by

$$\vec{N} = \text{grad } F(1, -1, 1) = -2(1)\vec{i} - 4(-1)\vec{j} - \vec{k} = -2\vec{i} + 4\vec{j} - \vec{k}.$$

We take the $\vec{i}$ and $\vec{j}$ components of $\vec{T}$ to be the vector grad $f(1, -1) = -2\vec{i} + 4\vec{j}$. Thus we have that, for some $a > 0$,

$$\vec{T} = -2\vec{i} + 4\vec{j} + a\vec{k}$$

We want $\vec{N} \cdot \vec{T} = 0$, so

$$\vec{N} \cdot \vec{T} = (-2\vec{i} + 4\vec{j} - \vec{k}) \cdot (-2\vec{i} + 4\vec{j} + a\vec{k}) = 4 + 16 - a = 0$$

So $a = 20$ and hence

$$\vec{T} = -2\vec{i} + 4\vec{j} + 20\vec{k}.$$

Example 5   Find the equation of the tangent plane to the sphere $x^2 + y^2 + z^2 = 14$ at the point $(1, 2, 3)$.

Solution   We write the sphere as a level surface as follows:

$$f(x, y, z) = x^2 + y^2 + z^2 = 14.$$

We have

$$\text{grad } f = 2x\vec{i} + 2y\vec{j} + 2z\vec{k},$$

so the vector

$$\text{grad } f(1, 2, 3) = 2\vec{i} + 4\vec{j} + 6\vec{k}$$

is perpendicular to the sphere at the point $(1, 2, 3)$. Since the vector grad $f(1, 2, 3)$ is normal to the tangent plane, the equation of the plane is

$$2x + 4y + 6z = 2(1) + 4(2) + 6(3) = 28 \quad \text{or} \quad x + 2y + 3z = 14.$$

Caution: Units and the Geometric Interpretation of the Gradient

When we interpreted the gradient of a function geometrically (page 123), we tacitly assumed that the scales along the x and y axes were the same. If they are not, the gradient vector may not look perpendicular to the contours. Consider the function $f(x, y) = x^2 + y$ with gradient vector given by grad $f = 2x\vec{i} + \vec{j}$. Figure 13.43 shows the gradient vector at $(1, 1)$ using the same scales in the x and y directions. As expected, the gradient vector is perpendicular to the contour line. Figure 13.44 shows contours of the same function with unequal scales on the two axes. Notice that the gradient vector no longer appears perpendicular to the contour lines. Thus we see that the geometric interpretation of the gradient vector requires that the same scale be used on the two axes.

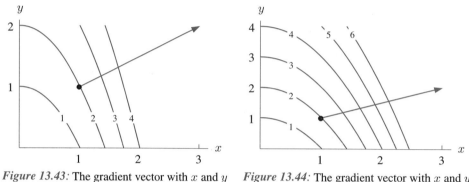

Figure 13.43: The gradient vector with x and y scales equal   *Figure 13.44:* The gradient vector with x and y scales unequal

Another problem arises when x and y are measured in different units. Consider, for example, the data relating corn production to rainfall, R, and temperature, T, on page 25. We plot the data twice, once using inches and Celsius and once using inches and Fahrenheit. The same scales are used along each axis as if units didn't matter. Figure 13.45 (inches and °C) shows the gradient at the point $(15, 25)$. Since $25°C = 77°F$, in Figure 13.46 (inches and °F), we show the gradient at the point $(15, 77)$. Notice that although the two points represent the same climate, the two contour diagrams and the two gradient vectors are different.

Gradients only make sense geometrically when x and y are measured in the same units. When x and y are both distances or both costs, for example, the size of the change $\sqrt{(\Delta x)^2 + (\Delta y)^2}$ is also a distance or a cost, so we can compare rates of change in all directions. However, if x and y are in different units, we cannot assign units to $\sqrt{(\Delta x)^2 + (\Delta y)^2}$. In this case, rates of change are only meaningful in directions parallel to the axes.

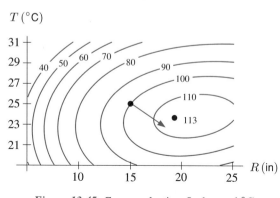

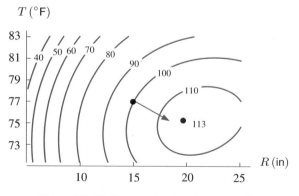

Figure 13.45: Corn production: Inches and °C   *Figure 13.46:* Corn production: Inches and °F

Problems for Section 13.5

1. If $f(x, y, z) = x^2 + 3xy + 2z$, find the directional derivative at the point $(2, 0, -1)$ in the direction of $2\vec{i} + \vec{j} - 2\vec{k}$.

2. Find the directional derivative of $f(x, y, z) = 3x^2y^2 + 2yz$ at the point $(-1, 0, 4)$ in the following directions (a) $\vec{i} - \vec{k}$ (b) $-\vec{i} + 3\vec{j} + 3\vec{k}$.

3. Find the equation of the tangent plane to $z = \sqrt{17 - x^2 - y^2}$ at the point $(3, 2, 2)$.

4. Find the equation of the tangent plane to $z = 8/(xy)$ at the point $(1, 2, 4)$.

5. Find an equation of the tangent plane and of a normal vector to the surface $x = y^3 z^7$ at the point $(1, -1, -1)$.

6. Let $f(x, y) = \cos x \sin y$ and let S be the surface $z = f(x, y)$.

   (a) Find a normal vector to the surface S at the point $(0, \pi/2, 1)$.
   (b) What is the equation of the tangent plane to the surface S at the point $(0, \pi/2, 1)$?

7. Consider the function $f(x, y) = (e^x - x) \cos y$. Suppose S is the surface $z = f(x, y)$.

   (a) Find a vector which is perpendicular to the level curve of f through the point $(2, 3)$ in the direction in which f decreases most rapidly.
   (b) Suppose $\vec{v} = 5\vec{i} + 4\vec{j} + a\vec{k}$ is a vector in 3-space which is tangent to the surface S at the point P lying on the surface above $(2, 3)$. What is a?

8. (a) Sketch the surface $z = f(x, y) = y^2$ in three dimensions.
   (b) Sketch the level curves of f in the xy plane.
   (c) If you are standing on the surface $z = y^2$ at the point $(2, 3, 9)$, in which direction should you move to climb the fastest? (Give your answer as a 2-vector.)

9. Suppose $z = y - \sin x$.

   (a) Sketch the contours for $z = -1, 0, 1, 2$.
   (b) A bug starts on the surface at the point $(\pi/2, 1, 0)$ and walks on the surface in the direction of the y-axis. Is the bug walking in a valley or on top of a ridge? Explain.
   (c) On the contour $z = 0$ in your sketch for part (a), draw the gradients of z at $x = 0$, $x = \pi/2$, and $x = \pi$.

10. Suppose that $F(x, y, z) = x^2 + y^4 + x^2 z^2$ gives the concentration of salt in a fluid at the point (x, y, z), and that you are at the point $(-1, 1, 1)$.

    (a) In which direction should you move if you want the concentration to increase the fastest?
    (b) Suppose you start to move in the direction you found in part (a) at a speed of 4 units/sec. How fast is the concentration changing? Explain your answer.

11. Consider S to be the surface represented by the equation $F = 0$, where

$$F(x, y, z) = x^2 - \left(\frac{y}{z^2}\right).$$

    (a) Find all points on S where a normal vector is parallel to the xy-plane.
    (b) Find the tangent plane to S at the points $(0, 0, 1)$ and $(1, 1, 1)$.
    (c) Find the unit vectors $\vec{u}_1$ and $\vec{u}_2$ pointing in the direction of maximum increase of F at the points $(0, 0, 1)$ and $(1, 1, 1)$ respectively.

12. At what point on the surface $z = 1 + x^2 + y^2$ is its tangent plane parallel to the following planes? (a) $z = 5$ (b) $z = 5 + 6x - 10y$.

13. A differentiable function $f(x, y)$ has the property that $f(4, 1) = 3$ and $f_x(4, 1) = 2$ and $f_y(4, 1) = -1$. Find the equation of the tangent plane at the point on the surface $z = f(x, y)$ where $x = 4$, $y = 1$.

14. A differentiable function $f(x, y)$ has the property that $f(1, 3) = 7$ and grad $f(1, 3) = 2\vec{i} - 5\vec{j}$.
    (a)   Find the equation of the tangent line to the level curve of f through the point $(1, 3)$.
    (b)   Find the equation of the tangent plane to the surface $z = f(x, y)$ at the point $(1, 3, 7)$.

15. Two surfaces are said to be *tangential* at a point P if they have the same tangent plane at P. Show that the surfaces $z = \sqrt{2x^2 + 2y^2 - 25}$ and $z = \frac{1}{5}(x^2 + y^2)$ are tangential at the point $(4, 3, 5)$.

16. Two surfaces are said to be *orthogonal* to each other at a point P if the normals to their tangent planes are perpendicular at P. Show that the surfaces $z = \frac{1}{2}(x^2 + y^2 - 1)$ and $z = \frac{1}{2}(1 - x^2 - y^2)$ are orthogonal at all points of intersection.

17. Let f and g be functions on 3-space. Suppose f is differentiable and that

$$\text{grad } f(x, y, z) = (x\vec{i} + y\vec{j} + z\vec{k})g(x, y, z).$$

Explain why f must be constant on any sphere centered at the origin.

18. Let $\vec{r}$ be the position vector of the point (x, y, z). If $\vec{\mu} = \mu_1\vec{i} + \mu_2\vec{j} + \mu_3\vec{k}$ is a constant vector, show that

$$\text{grad}(\vec{\mu} \cdot \vec{r}) = \vec{\mu}.$$

19. Let $\vec{r}$ be the position vector of the point (x, y, z). Show that, if a is a constant,

$$\text{grad}(\|\vec{r}\|^a) = a\|\vec{r}\|^{a-2}\vec{r}, \qquad \vec{r} \neq \vec{0}.$$

20. Suppose the earth has mass M and is located at the origin in 3-space, while the moon has mass m. Newton's Law of Gravitation states that if the moon is located at the point (x, y, z) then the attractive force exerted by the earth on the moon is given by the vector

$$\vec{F} = -GMm\frac{\vec{r}}{\|\vec{r}\|^3},$$

where $\vec{r} = x\vec{i} + y\vec{j} + z\vec{k}$. Show that $\vec{F} = \text{grad } \varphi$, where φ is the function given by

$$\varphi(x, y, z) = \frac{GMm}{\|\vec{r}\|}.$$

13.6 THE CHAIN RULE

Composition of Functions of Many Variables and Rates of Change

The chain rule enables us to differentiate *composite functions*. If we have a function of two variables $z = f(x, y)$ and we substitute $x = g(t), y = h(t)$ into $z = f(x, y)$, then we have a composite function in which z is a function of t:

$$z = f(g(t), h(t)).$$

If, on the other hand, we substitute $x = g(u, v), y = h(u, v)$, then we have a different composite function in which z is a function of u and v:

$$z = f(g(u, v), h(u, v)).$$

The next example shows how to calculate the rate of change of a composite function.

Example 1   Corn production, C, depends on annual rainfall, R, and average temperature, T, so $C = f(R, T)$. Global warming predicts that both rainfall and temperature depend on time. Suppose that according to a particular model of global warming, rainfall is decreasing at 0.2 cm per year and temperature is increasing at 0.1°C per year. Use the fact that at current levels of production, $f_R = 3.3$ and $f_T = -5$ to estimate the current rate of change, dC/dt.

Solution   By local linearity, we know that changes ΔR and ΔT generate a change, ΔC, in C given approximately by

$$\Delta C \approx f_R \Delta R + f_T \Delta T = 3.3 \Delta R - 5 \Delta T.$$

We want to know how ΔC depends on the time increment, Δt. The model of global warming tells us that

$$\frac{dR}{dt} = -0.2 \quad \text{and} \quad \frac{dT}{dt} = 0.1.$$

Thus, a time increment, Δt, generates changes of ΔR and ΔT given by

$$\Delta R \approx -0.2 \Delta t \quad \text{and} \quad \Delta T \approx 0.1 \Delta t.$$

Substituting for ΔR and ΔT in the expression for ΔC gives us

$$\Delta C \approx 3.3(-0.2 \Delta t) - 5(0.1 \Delta t) = -1.16 \Delta t.$$

Thus,

$$\frac{\Delta C}{\Delta t} \approx -1.16 \quad \text{and, therefore,} \quad \frac{dC}{dt} \approx -1.16$$

Thus, a change Δt causes changes ΔR and ΔT, which in turn cause a change ΔC. The relationship between ΔC and Δt, which gives the value of dC/dt, is an example of the *chain rule*. The argument in Example 1 can be used to generate more general statements of the chain rule.

The Chain Rule for $z = f(x, y),\ x = g(t),\ y = h(t)$

Since $z = f(g(t), h(t))$ is a function of t, we can consider the derivative dz/dt. The chain rule shows how dz/dt is related to the derivatives of $f, g,$ and h. Since dz/dt represents the rate of change of z with t, we look at how a small change, Δt, in t is propagated to z.

We substitute the local linearizations

$$\Delta x \approx \frac{dx}{dt} \Delta t \quad \text{and} \quad \Delta y \approx \frac{dy}{dt} \Delta t$$

into the local linearization

$$\Delta z \approx \frac{\partial z}{\partial x} \Delta x + \frac{\partial z}{\partial y} \Delta y,$$

yielding

$$\Delta z \approx \frac{\partial z}{\partial x}\frac{dx}{dt}\Delta t + \frac{\partial z}{\partial y}\frac{dy}{dt}\Delta t$$

$$= \left(\frac{\partial z}{\partial x}\frac{dx}{dt} + \frac{\partial z}{\partial y}\frac{dy}{dt}\right)\Delta t.$$

Thus,

$$\frac{\Delta z}{\Delta t} \approx \frac{\partial z}{\partial x}\frac{dx}{dt} + \frac{\partial z}{\partial y}\frac{dy}{dt}.$$

Taking the limit as $\Delta t \to 0$, we get the following result.

If f, g, and h are differentiable and if $z = f(x, y)$, and $x = g(t)$, and $y = h(t)$, then

$$\frac{dz}{dt} = \frac{\partial z}{\partial x}\frac{dx}{dt} + \frac{\partial z}{\partial y}\frac{dy}{dt}.$$

Visualizing the Chain Rule with a Tree Diagram

The "tree diagram" in Figure 13.47 provides a way of remembering the chain rule. It shows the chain of dependence: z depends on x and y, which in turn depend on t. Each line in the diagram is labeled with a derivative relating the variables at its ends.

The diagram keeps track of how a change in t propagates through the chain of composed functions. There are two paths from t to z, one through x and one through y. For each path, we multiply together the derivatives along the path. Then, to calculate dz/dt, we add up the contributions from the two paths.

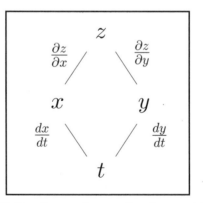

Figure 13.47: Tree diagram for $z = f(x, y)$, $x = g(t)$, $y = h(t)$

Example 2   Suppose that $z = f(x, y) = x \sin y$, where $x = t^2$ and $y = 2t + 1$. Let $z = g(t)$. Compute $g'(t)$ by two different methods.

Solution   Since $z = g(t) = f(t^2, 2t + 1) = t^2 \sin(2t + 1)$, it is possible to compute $g'(t)$ directly by one-variable methods:

$$g'(t) = t^2\frac{d}{dt}(\sin(2t+1)) + \left(\frac{d}{dt}(t^2)\right)\sin(2t+1) = 2t^2\cos(2t+1) + 2t\sin(2t+1).$$

The chain rule provides an alternative route to the same answer. We have

$$\frac{dz}{dt} = \frac{\partial z}{\partial x}\frac{dx}{dt} + \frac{\partial z}{\partial y}\frac{dy}{dt} = (\sin y)(2t) + (x\cos y)(2) = 2t\sin(2t+1) + 2t^2\cos(2t+1).$$

The Chain Rule in General

To find the rate of change of one variable with respect to another in a chain of composed differentiable functions:

- Draw a tree diagram expressing the relationship between the variables, and label each link in the diagram with the derivative relating the variables at its ends.
- For each path between the two variables, multiply together the derivatives from each step along the path.
- Add the contributions from each path.

A tree diagram keeps track of all the ways in which a change in one variable can cause a change in another; the diagram generates all the terms we would get from the appropriate substitutions into the local linearizations.

Example 3   Suppose that $z = f(x, y)$, with $x = g(u, v)$ and $y = h(u, v)$. Find formulas for $\partial z/\partial u$ and $\partial z/\partial v$.

Solution   Figure 13.48 shows the tree diagram for these variables. Adding the contributions for the two paths from z to u, we get

$$\frac{\partial z}{\partial u} = \frac{\partial z}{\partial x}\frac{\partial x}{\partial u} + \frac{\partial z}{\partial y}\frac{\partial y}{\partial u}.$$

Similarly, looking at the paths from z to v, we get

$$\frac{\partial z}{\partial v} = \frac{\partial z}{\partial x}\frac{\partial x}{\partial v} + \frac{\partial z}{\partial y}\frac{\partial y}{\partial v}.$$

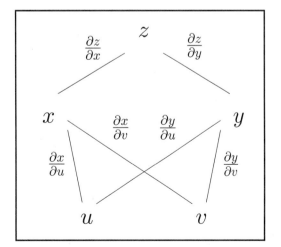

Figure 13.48: Tree diagram for $z = f(x, y)$, $x = g(u, v)$, $y = h(u, v)$

Example 4   Let $w = x^2 e^y$, $x = 4u$, and $y = 3u^2 - 2v$. Compute $\partial w/\partial u$ and $\partial w/\partial v$ using the chain rule.

Solution   Using the result of the previous example, we have

$$\frac{\partial w}{\partial u} = \frac{\partial w}{\partial x}\frac{\partial x}{\partial u} + \frac{\partial w}{\partial y}\frac{\partial y}{\partial u} = 2xe^y(4) + x^2 e^y(6u) = (8x + 6x^2 u)e^y$$

$$= (32u + 96u^3)e^{3u^2 - 2v}.$$

Similarly,

$$\frac{\partial w}{\partial v} = \frac{\partial w}{\partial x}\frac{\partial x}{\partial v} + \frac{\partial w}{\partial y}\frac{\partial y}{\partial v} = 2xe^y(0) + x^2 e^y(-2) = -2x^2 e^y$$

$$= -32u^2 e^{3u^2 - 2v}.$$

Example 5   A quantity z can be expressed either as a function of x and y, so that $z = f(x, y)$, or as a function of u and v, so that $z = g(u, v)$. The two coordinate systems are related by

$$x = u + v, \quad y = u - v.$$

(a)   Use the chain rule to express $\partial z/\partial u$ and $\partial z/\partial v$ in terms of $\partial z/\partial x$ and $\partial z/\partial y$.

(b)   Solve the equations in part (a) for $\partial z/\partial x$ and $\partial z/\partial y$.

(c)   Show that the expressions we get in part (b) are the same as we get by expressing u and v in terms of x and y and using the chain rule.

Solution   (a)   We have $\partial x/\partial u = 1$ and $\partial x/\partial v = 1$, and also $\partial y/\partial u = 1$ and $\partial y/\partial v = -1$. Thus,

$$\frac{\partial z}{\partial u} = \frac{\partial z}{\partial x}(1) + \frac{\partial z}{\partial y}(1) = \frac{\partial z}{\partial x} + \frac{\partial z}{\partial y}$$

and

$$\frac{\partial z}{\partial v} = \frac{\partial z}{\partial x}(1) + \frac{\partial z}{\partial y}(-1) = \frac{\partial z}{\partial x} - \frac{\partial z}{\partial y}.$$

(b)   Adding together the equations for $\partial z/\partial u$ and $\partial z/\partial v$ we get

$$\frac{\partial z}{\partial u} + \frac{\partial z}{\partial v} = 2\frac{\partial z}{\partial x}, \quad \text{so} \quad \frac{\partial z}{\partial x} = \frac{1}{2}\frac{\partial z}{\partial u} + \frac{1}{2}\frac{\partial z}{\partial v}.$$

Similarly, subtracting the equations for $\partial z/\partial u$ and $\partial z/\partial v$ yields

$$\frac{\partial z}{\partial y} = \frac{1}{2}\frac{\partial z}{\partial u} - \frac{1}{2}\frac{\partial z}{\partial v}.$$

(c)   Alternatively, we can solve the equations

$$x = u + v, \quad y = u - v$$

for u and v, which yields

$$u = \frac{1}{2}x + \frac{1}{2}y, \quad v = \frac{1}{2}x - \frac{1}{2}y.$$

Now we can think of z as a function of u and v, and u and v as functions of x and y, and apply the chain rule again. This gives us

$$\frac{\partial z}{\partial x} = \frac{\partial z}{\partial u}\frac{\partial u}{\partial x} + \frac{\partial z}{\partial v}\frac{\partial v}{\partial x} = \frac{1}{2}\frac{\partial z}{\partial u} + \frac{1}{2}\frac{\partial z}{\partial v}$$

and

$$\frac{\partial z}{\partial y} = \frac{\partial z}{\partial u}\frac{\partial u}{\partial y} + \frac{\partial z}{\partial v}\frac{\partial v}{\partial y} = \frac{1}{2}\frac{\partial z}{\partial u} - \frac{1}{2}\frac{\partial z}{\partial v}.$$

These are the same expressions we got in part (b).

An Example from Physical Chemistry

A chemist investigating the properties of a gas such as carbon dioxide may want to know how the internal energy U of a given quantity of the gas depends on its temperature, T, pressure, P, and volume, V. The three quantities T, P, and V are not independent, however. For instance, according to the ideal gas law, they satisfy the equation

$$PV = kT$$

where k is a constant which depends only upon the quantity of the gas. The internal energy can then be thought of as a function of any two of the three quantities T, P, and V:

$$U = U_1(T, P) = U_2(T, V) = U_3(P, V).$$

The chemist writes, for example, $\left(\frac{\partial U}{\partial T}\right)_P$ to indicate the partial derivative of U with respect to T *holding P constant*, signifying that for this computation U is viewed as a function of T and P. Thus, we interpret $\left(\frac{\partial U}{\partial T}\right)_P$ as

$$\left(\frac{\partial U}{\partial T}\right)_P = \frac{\partial U_1(T, P)}{\partial T}.$$

If U is to be viewed as a function of T and V, the chemist writes $\left(\frac{\partial U}{\partial T}\right)_V$ for the partial derivative of U with respect to T holding V constant: thus, $\left(\frac{\partial U}{\partial T}\right)_V = \frac{\partial U_2(T,V)}{\partial T}$.

Each of the functions U_1, U_2, U_3 gives rise to one of the following formulas for the differential dU:

$$dU = \left(\frac{\partial U}{\partial T}\right)_P dT + \left(\frac{\partial U}{\partial P}\right)_T dP \qquad \text{corresponds to } U_1$$

$$dU = \left(\frac{\partial U}{\partial T}\right)_V dT + \left(\frac{\partial U}{\partial V}\right)_T dV \qquad \text{corresponds to } U_2,$$

$$dU = \left(\frac{\partial U}{\partial P}\right)_V dP + \left(\frac{\partial U}{\partial V}\right)_P dV \qquad \text{corresponds to } U_3.$$

All the six partial derivatives appearing in formulas for dU have physical meaning, but they are not all equally easy to measure experimentally. A relationship among the partial derivatives, usually derived from the chain rule, may make it possible to evaluate one of the partials in terms of others that are more easily measured.

Example 6 Express $\left(\frac{\partial U}{\partial T}\right)_P$ in terms of $\left(\frac{\partial U}{\partial T}\right)_V$ and $\left(\frac{\partial U}{\partial V}\right)_T$ and $\left(\frac{\partial V}{\partial T}\right)_P$

Solution Since we are interested in the derivatives $\left(\frac{\partial U}{\partial T}\right)_V$ and $\left(\frac{\partial U}{\partial V}\right)_T$, we think of U as a function of T and V and use the formula

$$dU = \left(\frac{\partial U}{\partial T}\right)_V dT + \left(\frac{\partial U}{\partial V}\right)_T dV \qquad \text{corresponding to } U_2.$$

We want to find a formula for $\left(\frac{\partial U}{\partial T}\right)_P$, which means thinking of U as a function of T and P. Thus, we want to substitute for dV. Since V is a function of T and P, we have

$$dV = \left(\frac{\partial V}{\partial T}\right)_P dT + \left(\frac{\partial V}{\partial P}\right)_T dP.$$

Substituting for dV into the formula for dU corresponding to U_2 gives

$$dU = \left(\frac{\partial U}{\partial T}\right)_V dT + \left(\frac{\partial U}{\partial V}\right)_T \left(\left(\frac{\partial V}{\partial T}\right)_P dT + \left(\frac{\partial V}{\partial P}\right)_T dP\right).$$

Collecting the terms containing dT and the terms containing dP gives

$$dU = \left(\left(\frac{\partial U}{\partial T}\right)_V + \left(\frac{\partial U}{\partial V}\right)_T \left(\frac{\partial V}{\partial T}\right)_P\right) dT + \left(\frac{\partial U}{\partial V}\right)_T \left(\frac{\partial V}{\partial P}\right)_T dP.$$

But we also have the formula

$$dU = \left(\frac{\partial U}{\partial T}\right)_P dT + \left(\frac{\partial U}{\partial P}\right)_T dP \qquad \text{corresponding to } U_1.$$

We now have two formulas for dU in terms of dT and dP. The coefficients of dT must be identical, so we conclude

$$\left(\frac{\partial U}{\partial T}\right)_P = \left(\frac{\partial U}{\partial T}\right)_V + \left(\frac{\partial U}{\partial V}\right)_T \left(\frac{\partial V}{\partial T}\right)_P.$$

Example 6 expresses $\left(\frac{\partial U}{\partial T}\right)_P$ in terms of three other partial derivatives. Two of them, namely $\left(\frac{\partial U}{\partial T}\right)_V$, the constant-volume heat capacity, and $\left(\frac{\partial V}{\partial T}\right)_P$, the expansion coefficient, can be easily measured experimentally. The third, the internal pressure, $\left(\frac{\partial U}{\partial V}\right)_T$, cannot be measured directly but can be related to $\left(\frac{\partial P}{\partial T}\right)_V$, which is measurable. Thus, $\left(\frac{\partial U}{\partial T}\right)_P$ can be determined indirectly using this identity.

Problems for Section 13.6

For Problems 1–6, find dz/dt using the chain rule. Assume the variables are restricted to domains on which the functions are defined.

1. $z = xy^2$, $x = e^{-t}$, $y = \sin t$

2. $z = x \sin y + y \sin x$, $x = t^2$, $y = \ln t$

3. $z = \ln(x^2 + y^2)$, $x = 1/t$, $y = \sqrt{t}$

4. $z = \sin(x/y)$, $x = 2t$, $y = 1 - t^2$

5. $z = xe^y$, $x = 2t$, $y = 1 - t^2$

6. $z = (x + y)e^y$, $x = 2t$, $y = 1 - t^2$

For Problems 7–14, find $\partial z/\partial u$ and $\partial z/\partial v$. Assume the variables are restricted to domains on which the functions are defined.

7. $z = xe^{-y} + ye^{-x}$, $x = u \sin v$, $y = v \cos u$

8. $z = \cos(x^2 + y^2)$, $x = u \cos v$, $y = u \sin v$

9. $z = xe^y$, $x = \ln u$, $y = v$

10. $z = (x + y)e^y$, $x = \ln u$, $y = v$

11. $z = xe^y$, $x = u^2 + v^2$, $y = u^2 - v^2$

12. $z = (x + y)e^y$, $x = u^2 + v^2$, $y = u^2 - v^2$

13. $z = \sin(x/y)$, $x = \ln u$, $y = v$

14. $z = \tan^{-1}(x/y)$, $x = u^2 + v^2$, $y = u^2 - v^2$

15. Suppose $w = f(x, y, z)$ and that x, y, z are functions of u and v. Use a tree diagram to write down the chain rule formula for $\partial w / \partial u$ and $\partial w / \partial v$.

16. Suppose $w = f(x, y, z)$ and that x, y, z are all functions of t. Use a tree diagram to write down the chain rule for dw/dt.

17. Corn production, C, is a function of rainfall, R, and temperature, T. Figures 13.49 and 13.50 show how rainfall and temperature are predicted to vary with time because of global warming. Suppose we know that $\Delta C \approx 3.3\Delta R - 5\Delta T$. Use this to estimate the change in corn production between the year 2020 and the year 2021. Hence, estimate dC/dt when $t = 2020$.

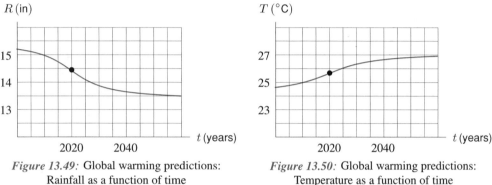

Figure 13.49: Global warming predictions: Rainfall as a function of time

Figure 13.50: Global warming predictions: Temperature as a function of time

18. The voltage, V, (in volts) across a circuit is given by Ohm's law: $V = IR$, where I is the current (in amps) flowing through the circuit and R is the resistance (in ohms). If we place two circuits, with resistance R_1 and R_2, in parallel, then their combined resistance, R, is given by

$$\frac{1}{R} = \frac{1}{R_1} + \frac{1}{R_2}.$$

Suppose the current is 2 amps and increasing at 10^{-2} amp/sec and R_1 is 3 ohms and increasing at 0.5 ohm/sec, while R_2 is 5 ohms and decreasing at 0.1 ohm/sec. Calculate the rate at which the voltage is changing.

19. A function $f(x, y)$ is *homogeneous of degree p* if $f(tx, ty) = t^p f(x, y)$ for all t. Show that any differentiable, homogeneous function of degree p satisfies Euler's Theorem:

$$x f_x(x, y) + y f_y(x, y) = p f(x, y).$$

[Hint: Define $g(t) = f(tx, ty)$ and compute $g'(1)$.]

Problems 20–22 are continuations of the physical chemistry example on page 139.

20. Write $\left(\frac{\partial U}{\partial P}\right)_V$ as a partial derivative of one of the functions $U_1, U_2,$ or U_3.

21. Express $\left(\frac{\partial U}{\partial P}\right)_T$ in terms of $\left(\frac{\partial U}{\partial V}\right)_T$ and $\left(\frac{\partial V}{\partial P}\right)_T$.

22. Use the result of Problem 21 to show that

$$\left(\frac{\partial U}{\partial V}\right)_P = \frac{\left(\frac{\partial U}{\partial T}\right)_V}{\left(\frac{\partial V}{\partial T}\right)_P} + \left(\frac{\partial U}{\partial V}\right)_T.$$

For Problems 23–24, suppose the quantity z can be expressed either as a function of Cartesian coordinates (x, y) or as a function of polar coordinates (r, θ), so that $z = f(x, y) = g(r, \theta)$. [Recall that $x = r \cos \theta, y = r \sin \theta$ and $r = \sqrt{x^2 + y^2}, \theta = \arctan(y/x)$]

23. (a) Use the chain rule to find $\partial z/\partial r$ and $\partial z/\partial \theta$ in terms of $\partial z/\partial x$ and $\partial z/\partial y$.
    (b) Solve the equations you have just written down for $\partial z/\partial x$ and $\partial z/\partial y$ in terms of $\partial z/\partial r$ and $\partial z/\partial \theta$.
    (c) Show that the expressions you get in part (b) are the same as you would get by using the chain rule to find $\partial z/\partial x$ and $\partial z/\partial y$ in terms of $\partial z/\partial r$ and $\partial z/\partial \theta$.

24. Show that

$$\left(\frac{\partial z}{\partial x}\right)^2 + \left(\frac{\partial z}{\partial y}\right)^2 = \left(\frac{\partial z}{\partial r}\right)^2 + \frac{1}{r^2}\left(\frac{\partial z}{\partial \theta}\right)^2.$$

25. Let $F(x, y, z)$ be a function and define a function $z = f(x, y)$ implicitly by letting $F(x, y, f(x, y)) = 0$. Use the chain rule to show that

$$\frac{\partial z}{\partial x} = -\frac{\partial F/\partial x}{\partial F/\partial z} \quad \text{and} \quad \frac{\partial z}{\partial y} = -\frac{\partial F/\partial y}{\partial F/\partial z}.$$

13.7   SECOND-ORDER PARTIAL DERIVATIVES

[handwritten note:]

$f(x,y) = xy^2 + 3x^2 e^y$

$f_x(x,y) = y^2 + 6xe^y$

$f_{xx}(x,y) = 6e^y$

(y part.) $f_{xy}(x,y) = 2y + 6xe^y$

take y partial of f_x part. →

- - - - - - - - - - -

$f_y(x,y) = 2xy + 3x^2 e^y$

$f_{yy}(x,y) = 2x + 3x^2 e^y$

take x partial of f_y part. → (x part.) $f_{yx}(x,y) = 2y + 6xe^y$

themselves functions, we can differentiate them, giving
$z = f(x, y)$ has two first-order partial derivatives, f_x
tives.

atives of $z = f(x, y)$

$\dfrac{\partial^2 z}{\partial x \partial y} = f_{yx} = (f_y)_x,$

$\dfrac{\partial^2 z}{\partial y^2} = f_{yy} = (f_y)_y.$

f_{xy} instead of $(f_x)_y$ and $\dfrac{\partial^2 z}{\partial y\, \partial x}$ instead of $\dfrac{\partial}{\partial y}\left(\dfrac{\partial z}{\partial x}\right)$.

Example 1   Compute the four second-order partial derivatives of $f(x, y) = xy^2 + 3x^2 e^y$.

Solution   From $f_x(x, y) = y^2 + 6xe^y$ we get (x partial)

$$f_{xx}(x, y) = \frac{\partial}{\partial x}(y^2 + 6xe^y) = 6e^y \quad \text{and} \quad f_{xy}(x, y) = \frac{\partial}{\partial y}(y^2 + 6xe^y) = 2y + 6xe^y.$$

From $f_y(x, y) = 2xy + 3x^2 e^y$ we get (y partial)

$$f_{yx}(x, y) = \frac{\partial}{\partial x}(2xy + 3x^2 e^y) = 2y + 6xe^y \quad \text{and} \quad f_{yy}(x, y) = \frac{\partial}{\partial y}(2xy + 3x^2 e^y) = 2x + 3x^2 e^y.$$

Observe that $f_{xy} = f_{yx}$ in this example.

Example 2   Use the values of the function $f(x, y)$ in Table 13.5 to estimate $f_{xy}(1, 2)$ and $f_{yx}(1, 2)$.

TABLE 13.5   *Values of $f(x, y)$*

$y \backslash x$	0.9	1.0	1.1
1.8	4.72	5.83	7.06
2.0	6.48	8.00	9.60
2.2	8.62	10.65	12.88

Solution   Since $f_{xy} = (f_x)_y$, we first estimate f_x

$$f_x(1, 2) \approx \frac{f(1.1, 2) - f(1, 2)}{0.1} = \frac{9.60 - 8.00}{0.1} = 16.0,$$

$$f_x(1, 2.2) \approx \frac{f(1.1, 2.2) - f(1, 2.2)}{0.1} = \frac{12.88 - 10.65}{0.1} = 22.3.$$

Thus,

$$f_{xy}(1, 2) \approx \frac{f_x(1, 2.2) - f_x(1, 2)}{0.2} = \frac{22.3 - 16.0}{0.2} = 31.5.$$

Similarly,

$$f_{yx}(1, 2) \approx \frac{f_y(1.1, 2) - f_y(1, 2)}{0.1} \approx \frac{1}{0.1} \left(\frac{f(1.1, 2.2) - f(1.1, 2)}{0.2} - \frac{f(1, 2.2) - f(1, 2)}{0.2} \right)$$

$$= \frac{1}{0.1} \left(\frac{12.88 - 9.60}{0.2} - \frac{10.65 - 8.00}{0.2} \right) = 31.5.$$

Observe that in this example also, $f_{xy} = f_{yx}$.

What Do the Second-Order Partial Derivatives Tell Us?

Example 3   Let us return to the guitar string of Example 13.2, page 107. The string is 1 meter long and at time t seconds, the point x meters from one end is displaced $f(x, t)$ meters from its rest position, where

$$f(x, t) = 0.003 \sin(\pi x) \sin(2765t).$$

Compute the four second-order partial derivatives of f at the point $(x, t) = (0.3, 1)$ and describe the meaning of their signs in practical terms.

Solution   First we compute $f_x(x, t) = 0.003\pi \cos(\pi x) \sin(2765t)$, from which we get

$$f_{xx}(x, t) = \frac{\partial}{\partial x}(f_x(x, t)) = -0.003\pi^2 \sin(\pi x) \sin(2765t), \quad \text{so} \quad f_{xx}(0.3, 1) \approx -0.01;$$

and

$$f_{xt}(x, t) = \frac{\partial}{\partial t}(f_x(x, t)) = (0.003)(2765)\pi \cos(\pi x) \cos(2765t), \quad \text{so} \quad f_{xt}(0.3, 1) \approx 14.$$

On page 107 we saw that $f_x(x, t)$ gives the slope of the string at any point and time. Therefore, $f_{xx}(x, t)$ measures the concavity of the string. The fact that $f_{xx}(0.3, 1) < 0$ means the string is concave down at the point $x = 0.3$ when $t = 1$. (See Figure 13.51.)

On the other hand, $f_{xt}(x, t)$ is the rate of change of the slope of the string with respect to time. Thus $f_{xt}(0.3, 1) > 0$ means that at time $t = 1$ the slope at the point $x = 0.3$ is increasing. (See Figure 13.52.)

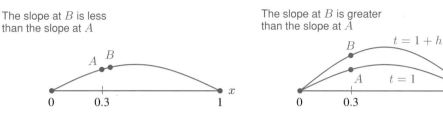

The slope at B is less
than the slope at A

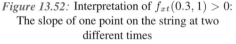

The slope at B is greater
than the slope at A

Figure 13.51: Interpretation of $f_{xx}(0.3, 1) < 0$:
The concavity of the string at $t = 1$

Figure 13.52: Interpretation of $f_{xt}(0.3, 1) > 0$:
The slope of one point on the string at two
different times

Now we compute $f_t(x, t) = (0.003)(2765)\sin(\pi x)\cos(2765t)$, from which we get

$$f_{tx}(x, t) = \frac{\partial}{\partial x}(f_t(x, t)) = (0.003)(2765)\pi \cos(\pi x)\cos(2765t), \quad \text{so} \quad f_{tx}(0.3, 1) \approx 14$$

and

$$f_{tt}(x, t) = \frac{\partial}{\partial t}(f_t(x, t)) = -(0.003)(2765)^2 \sin(\pi x)\sin(2765t), \quad \text{so} \quad f_{tt}(0.3, 1) \approx -7200.$$

On page 107 we saw that $f_t(x, t)$ gives the velocity of the string at any point and time. Therefore, $f_{tx}(x, t)$ and $f_{tt}(x, t)$ will both be rates of change of velocity. That $f_{tx}(0.3, 1) > 0$ means that at time $t = 1$ the velocities of points just to the right of $x = 0.3$ are greater than the velocity at $x = 0.3$. (See Figure 13.53.) That $f_{tt}(0.3, 1) < 0$ means that the velocity of the point $x = 0.3$ is decreasing at time $t = 1$. Thus, $f_{tt}(0.3, 1) = -7200$ m/sec^2 is the acceleration of this point. (See Figure 13.54.)

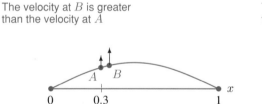

The velocity at B is greater
than the velocity at A

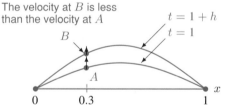

The velocity at B is less
than the velocity at A

Figure 13.53: Interpretation of $f_{tx}(0.3, 1) > 0$:
The velocity of different points on the string at
$t = 1$

Figure 13.54: Interpretation of $f_{tt}(0.3, 1) < 0$:
Negative acceleration. The velocity of one
point on the string at two different times

The Mixed Partial Derivatives Are Equal

It is not an accident that the estimates for $f_{xy}(1, 2)$ and $f_{yx}(1, 2)$ are equal in Example 2, because the same values of the function are used to calculate each one. The fact that $f_{xy} = f_{yx}$ in Examples 1 and 2 corroborates the following general result.

If f_{xy} and f_{yx} are continuous at (a, b), then

$$f_{xy}(a, b) = f_{yx}(a, b).$$

Most of the functions we will encounter not only have f_{xy} and f_{yx} continuous, but all their higher order partial derivatives (such as f_{xxy} or f_{xyyy}) will exist and be continuous. We call such functions *smooth*.

Problems for Section 13.7

In Problems 1–8, calculate all four second-order partial derivatives and show that $f_{xy} = f_{yx}$. Assume the variables are restricted to a domain on which the function is defined.

1. $f(x, y) = (x + y)^2$
2. $f(x, y) = (x + y)^3$
3. $f(x, y) = xe^y$
4. $f(x, y) = (x + y)e^y$
5. $f(x, y) = \sin(x^2 + y^2)$
6. $f(x, y) = \sqrt{x^2 + y^2}$
7. $f(x, y) = \sin(x/y)$
8. $f(x, y) = \tan^{-1}(x + y)$

9. If $z = f(x) + yg(x)$, what can you say about z_{yy}? Explain your answer.

10. If $z_{xy} = 4y$, what can you say about the value of (a) z_{yx}?   (b) z_{xyx}?   (c) z_{xyy}?

In Problems 11–19, use the level curves of the function $z = f(x, y)$ to decide the sign (positive, negative, or zero) of each of the following partial derivatives at the point P. Assume the x- and y-axes are in the usual positions.
(a) $f_x(P)$ (b) $f_y(P)$ (c) $f_{xx}(P)$ (d) $f_{yy}(P)$ (e) $f_{xy}(P)$

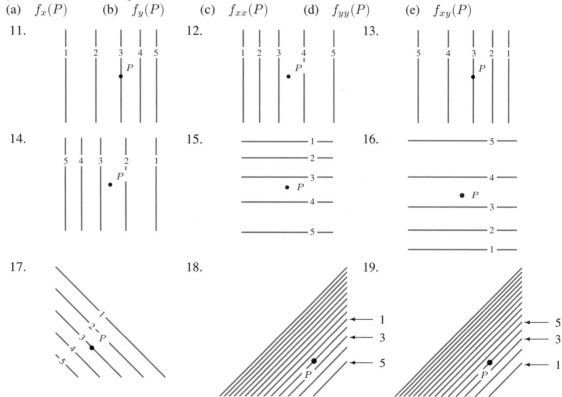

20. Give an explanation of why you might expect $f_{xy}(a, b) = f_{yx}(a, b)$ using the following steps.
    (a) Write the definition of $f_x(a, b)$.
    (b) Write a definition of $f_{xy}(a, b)$ as $(f_x)_y$.
    (c) Substitute the expression for f_x into the definition of f_{xy}.
    (d) Write an expression for f_{yx} similar to the one for f_{xy} you obtained in part (c).
    (e) Compare your answers to parts (c) and (d). What do you have to assume to conclude that f_{xy} and f_{yx} are equal?

13.8 PARTIAL DIFFERENTIAL EQUATIONS

Heat Flow

Imagine a room heated by a radiator along one wall. What happens after a window on the opposite wall is opened on a cold day? The temperature in the room begins to drop, quickly near the window, more slowly near the heater, and eventually stabilizes. The temperature $T = u(x, t)$ at any point in the room is a function of the distance x in meters from the heated wall and the time t in minutes since the window was opened. Figure 13.55 shows how the temperature, T, might look as a function of x at various values of t.

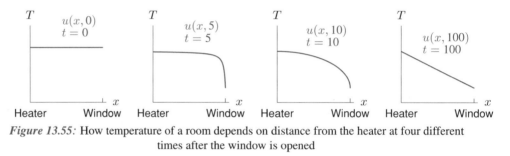

Figure 13.55: How temperature of a room depends on distance from the heater at four different times after the window is opened

Newton's Law of Cooling

The heat flow between two points is governed by Newton's Law of Cooling, which says that the rate of heat flow past a point x at time t is proportional to the partial derivative $u_x(x, t)$. This makes sense, since $u_x(x, t)$ measures the rate of change of u with respect to x at a fixed time t; in other words, it says that the greater the temperature gradient, the greater the rate of flow of heat. The temperature in a region of the room is increasing if the heat flowing into it is greater than the heat flowing out of it.

Example 1   Figure 13.56 shows the temperature $T = u(x, t)$ at a fixed time t. Use Newton's Law of Cooling to determine whether the temperature is increasing or decreasing:
(a)   At point p.   (b)   At point q.

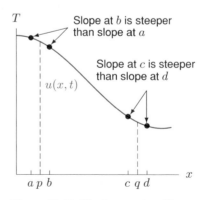

Figure 13.56: The temperature T as a function of x at a fixed time t

Solution (a) We must decide whether the small region $a < x < b$ that contains p is gaining or losing heat. (See Figure 13.56.) Since heat flows from hotter to colder regions, the heat flow is from left to right on the graph. In other words, heat flows downhill on a temperature graph. Newton's Law of Cooling asserts that the steeper the slope, $u_x(x, t)$, the greater the rate of flow of heat. Heat flows into the region $a < x < b$ at the point a and it flows out at b. The outflow is greater than the inflow because the tangent line at b is steeper than the tangent line at a. Thus, the region $a < x < b$ is losing heat and its temperature is decreasing. Thus, $u_t(p, t) < 0$.

(b) Heat flows into the small region $c < x < d$ at the point c and flows out at d, because the slopes $u_x(c, t)$ and $u_x(d, t)$ are negative. But the tangent line at c is steeper than the tangent line at d, so by Newton's Law of Cooling, the inflow is greater than the outflow. Thus, the region $c < x < d$ is gaining energy and so its temperature is increasing. Thus, $u_t(q, t) > 0$.

The Heat Equation

Example 1 shows that, for fixed t, the sign of u_t is determined by the concavity of the graph of $u(x, t)$. At $x = p$, we found that $u_t(p, t) < 0$ and the graph of u is concave down, so $u_{xx}(p, t) < 0$. At the point q, we have $u_t(q, t) > 0$ and $u_{xx}(q, t) > 0$. In fact, the two derivatives u_t and u_{xx} always have the same sign. In many situations the two derivatives u_t and u_{xx} are actually proportional, so the function $u(x, t)$ satisfies the following equation:

The One-Dimensional Heat (or Diffusion) Equation

$$u_t(x, t) = A\, u_{xx}(x, t), \qquad \text{where } A \text{ is a positive constant.}$$

The heat equation is an example of a *partial differential equation* (PDE), that is, an equation involving one or more partial derivatives of an unknown function.

Example 2 Which of the following two functions satisfies the heat equation $u_t = u_{xx}$?

(a) $u(x, t) = e^{-4t} \sin(2x)$ (b) $u(x, t) = \sin(x + t)$

Solution (a) Calculating partial derivatives of the function u gives

$$u_t = -4e^{-4t} \sin(2x), \quad u_x = 2e^{-4t} \cos(2x), \quad u_{xx} = -4e^{-4t} \sin(2x),$$

and so $u_t = u_{xx}$. Thus, $u(x, t) = e^{-4t} \sin(2x)$ is a solution.

(b) We have

$$u_t = \cos(x + t), \quad u_x = \cos(x + t), \quad u_{xx} = -\sin(x + t),$$

and so $u_t \neq u_{xx}$. Thus, $u(x, t) = \sin(x + t)$ is not a solution.

Example 3 A 10 cm metal rod is insulated so that heat can flow along the rod but cannot radiate into the air except at the ends. The temperature T ($^\circ$C) at x cm from one end and at time t seconds is a function $T = u(x, t)$ that satisfies the heat equation $u_t(x, t) = 0.1 u_{xx}(x, t)$. The initial temperature at several points is given in Table 13.6.

TABLE 13.6 *Temperature in metal rod at time $t = 0$*

x (cm)	0	2	4	6	8	10
$u(x, 0)$ ($^\circ$C)	50	52	56	62	70	80

(a) Is the temperature at the point $x = 6$ increasing or decreasing when $t = 0$?

(b) Make an estimate of the temperature $T = u(6, 1)$ at the point $x = 6$ at time $t = 1$.

Solution   (a)   The graph of $u(x,0)$ in Figure 13.57 suggests that $u(x,0)$ is concave up, so $u_{xx}(6,0) > 0$. Since $u_t(6,0) = 0.1u_{xx}(6,0)$ we have $u_t(6,0) > 0$ also. Since $u_t(6,0)$ gives the rate of change of temperature at $x = 6$ with respect to time, the fact that it is positive indicates that the temperature at $x = 6$ is increasing. We might have guessed this from the fact that the temperature $62°C$ at $x = 6$ is below the average $(56 + 70)/2 = 63°C$ of the temperatures of the neighboring points.

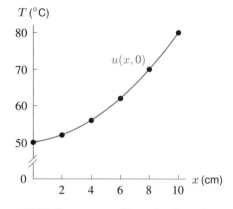

Figure 13.57: Temperature, $T = u(x,0)$, at time $t = 0$

(b)   To predict the temperature at $t = 1$ from the temperature at $t = 0$ we first estimate the rate of change of temperature with respect to time, $u_t(6,0)$. Since $u_t(6,0) = 0.1u_{xx}(6,0)$, we estimate $u_{xx}(6,0)$. Since $u_{xx}(6,0)$ is the rate of change of $u_x(x,0)$, we first estimate u_x at two points near $x = 6$:

$$u_x(5,0) \approx \frac{u(6,0) - u(4,0)}{6 - 4} = \frac{62 - 56}{2} = 3 \quad \text{and} \quad u_x(7,0) \approx \frac{u(8,0) - u(6,0)}{8 - 6} = 4.$$

Therefore,

$$u_{xx}(6,0) \approx \frac{u_x(7,0) - u_x(5,0)}{7 - 5} \approx \frac{4 - 3}{2} = 0.5.$$

Thus,

$$u_t(6,0) \approx 0.1u_{xx}(6,0) \approx (0.1)(0.5) = 0.05°C/\text{sec}.$$

Since the initial temperature is $62°C$ and we estimate that the temperature is increasing at $0.05°C/\text{sec}$, the temperature at $t = 1$ is approximately $62.05°C$.

Warning

Getting accurate numerical approximations of solutions to PDEs is generally quite difficult. Example 3 shows that quantitative information can be extracted from a PDE, but it is not a practical way to get accurate approximations to solutions.

Boundary Conditions

The heat equation $u_t = Au_{xx}$ has many solutions. Just as we needed initial conditions to get a unique solution to an ordinary differential equation, so more information is required in order to pick out a single solution to a PDE. In the case of the heated room, for example, we would need to know the temperature in the room when the window was opened, the outside temperature at all times, and the temperature near the heater (which tells us what the heater is doing). This information is referred to as the *boundary conditions*. Problems 10, 14, 15, 18 on page 152 show how boundary conditions are used.

A Traveling Wave

Think about a bottle that bobs up and down as a wave rolls through. The motion of the bottle depends on the shape and horizontal velocity of the wave; we will investigate a PDE which describes this motion.

Example 4   Suppose that the height y of the sea (and hence of the floating bottle) above normal at time t seconds and distance x meters from a reference point is given by the function $y = u(x, t)$, which is graphed at $t = 0$ in Figure 13.58. Suppose that the wave is moving in the direction of increasing x.
(a)   Determine whether $u_x(x, 0)$ and $u_t(x, 0)$ are positive or negative at the following points:
(i) $x = p$ (ii) $x = q$ (iii) $x = r$
(b)   Would $u_t(r, 0)$ be greater or smaller if the wave were traveling faster?

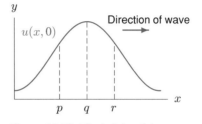

Figure 13.58: The height of the wave
at $t = 0$; wave moving to the right

Solution   (a)   (i)   The partial derivative $u_x(p, 0)$ gives the slope of the tangent line to the graph of $u(x, 0)$ at $x = p$. From Figure 13.58 it is clear that the slope is positive, so $u_x(p, 0) > 0$. On the other hand, $u_t(p, 0)$ equals the vertical velocity of the bottle at the point $x = p$ at the time $t = 0$. Since the crest of the wave has passed, the bottle is falling, so the velocity is negative. Thus, $u_t(p, 0) < 0$.

(ii)   We have $u_x(q, 0) = 0$, because the tangent to the wave at $x = q$ and $t = 0$ is horizontal and has slope zero. A bottle at $x = q$ and $t = 0$ is exactly at the crest of the wave, momentarily stopped with velocity zero. Thus, $u_t(q, 0) = 0$.

(iii)   At $t = 0$ the point r is on the back side of the wave, so $u_x(r, 0) < 0$. A bottle at $x = r$ would be rising at $t = 0$, so $u_t(r, 0) > 0$.
(b)   If the wave were moving faster, then a floating bottle would rise and fall faster, so the positive velocity $u_t(r, 0)$ would be greater than for the original, slower wave.

The Traveling Wave Equation

Notice that in Example 4 the derivatives $u_x(x, 0)$ and $u_t(x, 0)$ are of opposite sign or are both zero, for all x. This suggests that the two derivative functions u_x and u_t may be related.

To investigate further, let us suppose that a wave is moving to the right with velocity c and that its positions at two nearby times t and $t + \Delta t$ are shown in Figure 13.59. If Δt is small enough, $u_x(p, t)$, the slope of the graph at B, is well approximated by the slope of the secant line between

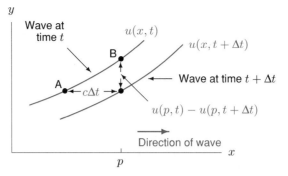

Figure 13.59: A traveling wave moving to the right at
speed c: the wave at two nearby times

the points A and B. Note that during time interval Δt the wave moves horizontally a distance of $c\,\Delta t$
(since distance = speed · time). Thus, we have

$$u_x(p,t) = \text{Slope of tangent at } B \approx \text{Slope of secant between } A \text{ and } B$$
$$= \frac{u(p,t) - u(p,t+\Delta t)}{c\,\Delta t}$$
$$= -\frac{1}{c}\frac{u(p,t+\Delta t) - u(p,t)}{\Delta t}$$
$$\approx -\frac{1}{c}u_t(p,t).$$

Therefore, $u(x,t)$ satisfies the following PDE:

The Traveling Wave Equation for a wave moving in the positive x-direction with speed c

$$u_t(x,t) = -cu_x(x,t), \qquad \text{where } c \text{ is a positive constant.}$$

A Formula For the Solution to the Traveling Wave Equation

Suppose that a traveling wave is moving in the positive x-direction with speed c and that the wave
has the shape $y = f(x)$ at time $t = 0$, which means that $u(x,0) = f(x)$. By time t the wave has
moved a distance ct to the right, so at time t the wave has shape $y = f(x - ct)$. In other words:

A wave of shape $f(x)$ traveling at speed c in the positive x-direction is represented by

$$u(x,t) = f(x - ct)$$

Example 5   (a)   Write a formula for the function $u(x,t)$ that describes a wave whose shape at time $t = 0$ is
$y = \sin x$ and that is moving in the positive x-direction with speed 0.5.
(b)   Show that the function $u(x,t)$ found in part (a) satisfies the traveling wave equation.
(c)   Sketch the graphs of $u(x,t)$ against x for $t = 0, 1, 2$.

Solution   (a)   Since the shape of the wave remains the same as it travels, we know that

$$u(x, t) = \sin(x - ct) = \sin(x - 0.5t).$$

(b)   Since $u_t(x, t) = -0.5\cos(x - 0.5t)$ and $u_x(x, t) = \cos(x - 0.5t)$, the function $u(x, t)$ satisfies the traveling wave equation with $c = 0.5$:

$$u_t(x, t) = -0.5u_x(x, t).$$

(c)   The graphs are in Figure 13.60. Notice that the forward motion of the wave is clearly visible.

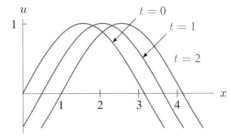

Figure 13.60: Graph of
$u(x, t) = \sin(x - 0.5t)$ at three times

A String Under Tension

How do we describe the motion of a vibrating string under tension, such as a plucked guitar string? Let $y = u(x, t)$ be the displacement from equilibrium at time t of the point on the string x units from one end. Then it can be shown that u satisfies the following equation:

The One-Dimensional Wave Equation:

$$u_{tt} = c^2 u_{xx}, \qquad \text{where } c \text{ is a positive constant.}$$

Let's see what this equation says. The function $u(x, t)$ describes the motion of a mass (the string) under the influence of a force (tension). Thus, Newton's Second Law of Motion applies, so we expect an equation of the type $F = ma$, where F is force, m is mass, and a is acceleration. The term $u_{tt}(x, t)$ in the wave equation is the acceleration of the point x at time t. The term $u_{xx}(x, t)$ is closely related to the force, for it measures the concavity of the string. The greater the concavity, the stronger the force back towards equilibrium, just as when shooting an arrow from a bow.

Problems for Section 13.8

1. Suppose $u(x, t)$ is as in Example 3.
   (a) Use Table 13.6 to make estimates for $u(4, 1)$ and $u(8, 1)$.
   (b) Use your answers to part (a) and the estimate for $u(6, 1)$ worked out in Example 3 to make an estimate for $u(6, 2)$.

2. Sketch the graph of $u(x, t) = 1 - (x - 2t)^2$ for $t = 0, 1, 2$. Explain how the graph represents a traveling wave. What is the speed and the direction of the wave?

3. We will model the spread of an epidemic through a region. Suppose $I(x, y, t)$ represents the density of sick people per unit area at the point (x, y) in the plane at time t. Suppose I satisfies the diffusion equation:

$$\frac{\partial I}{\partial t} = D\left(\frac{\partial^2 I}{\partial x^2} + \frac{\partial^2 I}{\partial y^2}\right),$$

where D is a constant. Suppose we know that for some particular epidemic

$$I = e^{ax+by+ct}.$$

What can you say about the relationship between a, b, and c?

4. For what values of the constants a and b will the function $u = (x + y)e^{ax+by}$ satisfy the equation

$$\frac{\partial^2 u}{\partial x \partial y} - \frac{\partial u}{\partial x} - \frac{\partial u}{\partial y} + u = 0?$$

Show that the functions in Problems 5–7 satisfy Laplace's equation, which is $F_{xx} + F_{yy} = 0$.

5. $F(x, y) = e^{-x} \sin y$ 6. $F(x, y) = \arctan(y/x)$
7. $F(x, y) = e^x \sin y + e^y \sin x$

8. The function $f(x, t) = 0.003 \sin(\pi x) \sin(2765t)$ describes a vibrating guitar string. Show that it is a solution of the wave equation, $f_{tt} = c^2 f_{xx}$, for some c. What is that value of c?

9. Suppose that f is any differentiable function of one variable. Define V, a function of two variables, by

$$V(x, t) = f(x + ct).$$

Show that V satisfies the equation

$$\frac{\partial V}{\partial t} = c\frac{\partial V}{\partial x}.$$

10. Consider the wave equation describing a vibrating string

$$\frac{\partial^2 y}{\partial t^2} = a^2 \frac{\partial^2 y}{\partial x^2},$$

where a is a constant and y is the displacement of the string at any point x and at any time t. Write the boundary conditions for a vibrating string of length L for which

(a) The ends at $x = 0$ and $x = L$ are fixed.
(b) The initial shape is given by $f(x)$.
(c) The initial velocity distribution is given by $g(x)$.

11. Figure 13.61 is the graph of the temperature versus position at one instant in a metal rod, where temperature $T = u(x, t)$ satisfies the heat equation $u_t = u_{xx}$. Determine for which x the temperature is increasing, for which it is decreasing, and use this information to sketch a graph of temperature at a slightly later time.

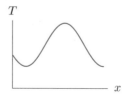

Figure 13.61

12. At any point (x, y, z) outside a spherically symmetric mass m located at the point (x_0, y_0, z_0), the gravitational potential, V, is defined by $V = -Gm/r$, where r is the distance from (x, y, z) to (x_0, y_0, z_0) and G is a constant. Show that, for all points outside the mass, V satisfies Laplace's equation:

$$\frac{\partial^2 V}{\partial x^2} + \frac{\partial^2 V}{\partial y^2} + \frac{\partial^2 V}{\partial z^2} = 0.$$

13. If $u(x, t) = e^{at} \sin(bx)$ satisfies the heat equation $u_t = u_{xx}$, find the relationship between a and b.

14. Assuming the solution to the wave equation is of the form

$$y = F(x + 2t) + G(x - 2t),$$

find a solution satisfying the boundary conditions

$$y(0, t) = y(5, t) = 0, \quad y(x, 0) = 0, \quad \left. \frac{\partial y}{\partial t} \right|_{t=0} = 5 \sin(\pi x), \quad 0 < x < 5, \quad t > 0.$$

15. (a) We study heat conduction in a 1 meter metal rod, $0 \le x \le 1$, whose sides are insulated and whose ends are maintained at $0°$ C at all times by being stuck into ice baths. The conditions at the ends of the rod represent a boundary condition on the possible functions $u(x, t)$ that could describe the temperature in the rod. State the boundary condition as a pair of equations.

(b) Determine all possible values of a and b such that $u(x, t) = e^{at} \sin(bx)$ satisfies both the PDE, $u_t = u_{xx}$, and the boundary conditions of part (a).

16. (a) Verify that the function

$$u(x, t) = \frac{1}{2\sqrt{\pi t}} e^{-x^2/(4t)}$$

satisfies the heat equation $u_t = u_{xx}$ for $t > 0$ and all x.

(b) Sketch the graphs of $u(x, t)$ against x for $t = 0.01, 0.1, 1, 10$. These graphs represent the temperature in an infinitely long insulated rod that at $t = 0$ is $0°$C everywhere except at the origin $x = 0$, and that is infinitely hot at $t = 0$ at the origin.

17. The temperature, T, of a metal plate can be described by a function $T = u(x, y, t)$ of three variables, the two space variables, x and y, and the time variable, t. The two-dimensional heat equation is the PDE

$$u_t = A(u_{xx} + u_{yy}), \qquad \text{where } A \text{ is a positive constant.}$$

Find conditions on a, b, and c such that $u(x, y, t) = e^{-at} \sin(bx) \sin(cy)$ satisfies this PDE.

18. (a) Verify that $u(x, t) = \sin(ax) \sin(at)$ satisfies the wave equation $u_{tt} = u_{xx}$. This solution represents a vibration with period $2\pi/a$, since $u(x, t + 2\pi/a) = u(x, t)$.

(b) Suppose that you wish to study the vibrations of a 1-meter string with fixed ends (such as a guitar string), so that $u(0, t) = 0$ and $u(1, t) = 0$ for all t, and such that $u_{tt} = u_{xx}$. The condition on the ends is a boundary condition. Find all $a > 0$ such that $u(x, t) = \sin(ax) \sin(at)$ satisfies both the PDE and the boundary condition. (It has been known to musicians since at least the time of Pythagoras that stretched strings can be made to vibrate at only special frequencies.)

19. You can generate a traveling wave on a string under tension by giving one end a snap. This suggests that there should be traveling wave solutions of the wave equation.

    (a) Show that if f is an arbitrary function, then $u(x, t) = f(x - ct)$ is a solution of the wave equation $u_{tt} = c^2 u_{xx}$.

    (b) Show that if g is an arbitrary function, then $u(x, t) = g(x + ct)$ is a solution of the wave equation $u_{tt} = c^2 u_{xx}$.

    (c) Show that if f and g are arbitrary functions, then $u(x, t) = f(x - ct) + g(x + ct)$ is a solution of the wave equation. (In fact, all solutions of the wave equation can be written in this form, as the sum of a forward traveling wave and a backward traveling wave.)

20. The vibration of a 2-dimensional object under tension, such as a drum head, is described by a function $u(x, y, t)$ of two space variables x and y and one time variable t. Such a function often satisfies the 2-dimensional wave equation $u_{tt} = c^2(u_{xx} + u_{yy})$. Find conditions on the constants a, b, and k such that $u(x, y, t) = \sin(ax)\sin(by)\sin(kt)$ satisfies this equation.

13.9   NOTES ON TAYLOR APPROXIMATIONS

Just as a function of one variable can usually be better approximated by a quadratic function than by a linear function, so can a function of several variables. In Section 13.3, we saw how to approximate $f(x, y)$ by a linear function (its local linearization). In this section, we see how to improve this approximation of $f(x, y)$ using a quadratic function.

Linear and Quadratic Approximations Near (0,0)

For a function of one variable, local linearity tells us that the best *linear* approximation is the degree 1 Taylor polynomial

$$f(x) \approx f(a) + f'(a)(x - a) \quad \text{for } x \text{ near } a.$$

A better approximation to $f(x)$ is given by the degree 2 Taylor polynomial:

$$f(x) \approx f(a) + f'(a)(x - a) + \frac{f''(a)}{2}(x - a)^2 \quad \text{for } x \text{ near } a.$$

For a function of two variables the local linearization for (x, y) near (a, b) is

$$f(x, y) \approx L(x, y) = f(a, b) + f_x(a, b)(x - a) + f_y(a, b)(y - b).$$

In the case $(a, b) = (0, 0)$, we have:

> **Taylor Polynomial of Degree 1 Approximating $f(x, y)$ for (x, y) near (0,0)**
> If f has continuous first-order partial derivatives, then
>
> $$f(x, y) \approx L(x, y) = f(0, 0) + f_x(0, 0)x + f_y(0, 0)y.$$

We get a better approximation to f by using a quadratic polynomial. We choose a quadratic polynomial $Q(x, y)$, with the same partial derivatives as the original function f. You can check that the following polynomial has this property.

> **Taylor Polynomial of Degree 2 Approximating $f(x, y)$ for (x, y) near (0,0)** If f has continuous second-order partial derivatives, then
>
> $$f(x, y) \approx Q(x, y)$$
>
> $$= f(0,0) + f_x(0,0)x + f_y(0,0)y + \frac{f_{xx}(0,0)}{2}x^2 + f_{xy}(0,0)xy + \frac{f_{yy}(0,0)}{2}y^2.$$

Example 1   Let $f(x, y) = \cos(2x + y) + 3\sin(x + y)$

    (a)   Compute the linear and quadratic Taylor polynomials, L and Q, approximating f near $(0, 0)$.

    (b)   Explain why the contour plots of L and Q for $-1 \le x \le 1$, $-1 \le y \le 1$ look the way they do.

Solution   (a)   We have $f(0, 0) = 1$. The derivatives we need are as follows:

$$f_x(x, y) = -2\sin(2x + y) + 3\cos(x + y) \quad \text{so} \quad f_x(0, 0) = 3,$$
$$f_y(x, y) = -\sin(2x + y) + 3\cos(x + y) \quad \text{so} \quad f_y(0, 0) = 3,$$
$$f_{xx}(x, y) = -4\cos(2x + y) - 3\sin(x + y) \quad \text{so} \quad f_{xx}(0, 0) = -4,$$
$$f_{xy}(x, y) = -2\cos(2x + y) - 3\sin(x + y) \quad \text{so} \quad f_{xy}(0, 0) = -2,$$
$$f_{yy}(x, y) = -\cos(2x + y) - 3\sin(x + y) \quad \text{so} \quad f_{yy}(0, 0) = -1.$$

Thus, the linear approximation, $L(x, y)$, to $f(x, y)$ at $(0, 0)$ is given by

$$f(x, y) \approx L(x, y) = f(0, 0) + f_x(0, 0)x + f_y(0, 0)y = 1 + 3x + 3y.$$

The quadratic approximation, $Q(x, y)$, to $f(x, y)$ near $(0, 0)$ is given by

$$f(x, y) \approx Q(x, y)$$

$$= f(0,0) + f_x(0,0)x + f_y(0,0)y + \frac{f_{xx}(0,0)}{2}x^2 + f_{xy}(0,0)xy + \frac{f_{yy}(0,0)}{2}y^2$$

$$= 1 + 3x + 3y - 2x^2 - 2xy - \frac{1}{2}y^2.$$

Notice that the linear terms in $Q(x, y)$ are the same as the linear terms in $L(x, y)$. The quadratic terms in $Q(x, y)$ can be thought of as "correction terms" to the linear approximation.

    (b)   The contour plots of $f(x, y)$, $L(x, y)$, and $Q(x, y)$ are in Figures 13.62–13.64.

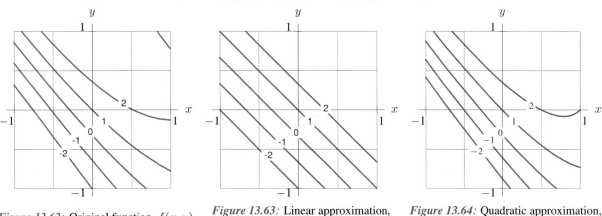

Figure 13.62: Original function, $f(x, y)$ *Figure 13.63:* Linear approximation, $L(x, y)$ *Figure 13.64:* Quadratic approximation, $Q(x, y)$

    Notice that the contour plot of Q is more similar to the contour plot of f than is the contour plot of L. Since L is linear, the contour plot of L consists of parallel, equally spaced lines.

An alternative, and much quicker, way to find the Taylor polynomial in the previous example is to use the single-variable approximations. For example, since

$$\cos u = 1 - \frac{u^2}{2!} + \frac{u^4}{4!}\dots \quad \text{and} \quad \sin v = v - \frac{v^3}{3!}\dots,$$

we can substitute $u = 2x + y$ and $v = x + y$ and expand. We discard terms beyond the second (since we want the quadratic polynomial) getting

$$\cos(2x + y) = 1 - \frac{(2x + y)^2}{2!} + \frac{(2x + y)^4}{4!}\dots \approx 1 - \frac{1}{2}(4x^2 + 4xy + y^2) = 1 - 2x^2 - 2xy - \frac{1}{2}y^2$$

and

$$\sin(x + y) = (x + y) - \frac{(x + y)^3}{3!}\dots \approx x + y.$$

Combining these results, we get

$$\cos(2x + y) + 3\sin(x + y) \approx 1 - 2x^2 - 2xy - \frac{1}{2}y^2 + 3(x + y) = 1 + 3x + 3y - 2x^2 - 2xy - \frac{1}{2}y^2.$$

Linear and Quadratic Approximations Near (a, b)

The local linearization for a function $f(x, y)$ at a point (a, b) is

---

Taylor Polynomial of Degree 1 Approximating $f(x, y)$ for (x, y) near (a, b)
If f has continuous first-order partial derivatives, then

$$f(x, y) \approx L(x, y) = f(a, b) + f_x(a, b)(x - a) + f_y(a, b)(y - b).$$

This suggests that a quadratic polynomial approximation $Q(x, y)$ for $f(x, y)$ near a point (a, b) should be written in terms of $(x - a)$ and $(y - b)$ instead of x and y. If we require that $Q(a, b) = f(a, b)$ and that the first- and second-order partial derivatives of Q and f at (a, b) be equal, then we get the following polynomial:

Taylor Polynomial of Degree 2 Approximating $f(x, y)$ for (x, y) near (a, b)
If f has continuous second-order partial derivatives, then

$$\begin{aligned}
f(x, y) \approx Q(x, y) \\
= f(a, b) + f_x(a, b)(x - a) + f_y(a, b)(y - b) \\
+ \frac{f_{xx}(a, b)}{2}(x - a)^2 + f_{xy}(a, b)(x - a)(y - b) + \frac{f_{yy}(a, b)}{2}(y - b)^2.
\end{aligned}$$

These coefficients are derived in exactly the same way as for $(a, b) = (0, 0)$.

Example 2   Find the Taylor polynomial of degree 2 at the point $(1, 2)$ for the function $f(x, y) = \dfrac{1}{xy}$.

Solution   Table 13.7 contains the partial derivatives and their values at the point $(1, 2)$.

TABLE 13.7   *Partial derivatives of $f(x, y) = 1/(xy)$*

Derivative	Formula	Value at $(1, 2)$	Derivative	Formula	Value at $(1, 2)$
$f(x, y)$	$1/(xy)$	$1/2$	$f_{xx}(x, y)$	$2/(x^3 y)$	1
$f_x(x, y)$	$-1/(x^2 y)$	$-1/2$	$f_{xy}(x, y)$	$1/(x^2 y^2)$	$1/4$
$f_y(x, y)$	$-1/(xy^2)$	$-1/4$	$f_{yy}(x, y)$	$2/(xy^3)$	$1/4$

So, the quadratic Taylor polynomial for f near $(1, 2)$ is

$$\frac{1}{xy} \approx Q(x, y)$$

$$= \frac{1}{2} - \frac{1}{2}(x - 1) - \frac{1}{4}(y - 2) + \frac{1}{2}(1)(x - 1)^2 + \frac{1}{4}(x - 1)(y - 2) + \left(\frac{1}{2}\right)\left(\frac{1}{4}\right)(y - 2)^2$$

$$= \frac{1}{2} - \frac{x - 1}{2} - \frac{y - 2}{4} + \frac{(x - 1)^2}{2} + \frac{(x - 1)(y - 2)}{4} + \frac{(y - 2)^2}{8}.$$

The Error in Linear and Quadratic Approximations

Let's return to the function $f(x, y) = \cos(2x + y) + \sin(x + y)$ and its linear and quadratic approximations, $L(x, y)$ and $Q(x, y)$. The contour plots in Example 1 are evidence that Q is a better approximation to f than L. Now we'll look at exactly how much better.

We begin by considering approximations about the point $(0, 0)$. We define the *error* in the linear approximation as the difference

$$E_L = f(x, y) - L(x, y).$$

The error in the quadratic approximation is defined similarly as

$$E_Q = f(x, y) - Q(x, y).$$

Table 13.8 shows how the magnitudes of these errors, $|E_L|$ and $|E_Q|$, depend on the distance, $d(x, y) = \sqrt{x^2 + y^2}$, of the point (x, y) from $(0, 0)$. The values in Table 13.8 suggest that, in this example,

$$E_L \text{ is proportional to } d^2 \quad \text{and} \quad E_Q \text{ is proportional to } d^3.$$

In general, the errors E_L and E_Q can be shown to be proportional to d^2 and d^3, respectively.

TABLE 13.8   *Magnitude of the error in the linear and quadratic approximations to $f(x, y) = \cos(2x + y) + \sin(x + y)$*

Point, (x, y)	Distance, d	Error, $|E_L|$	Error, $|E_Q|$
$x = y = 0$	0	0	0
$x = y = 10^{-1}$	$1.4 \cdot 10^{-1}$	$5 \cdot 10^{-2}$	$4 \cdot 10^{-3}$
$x = y = 10^{-2}$	$1.4 \cdot 10^{-2}$	$5 \cdot 10^{-4}$	$4 \cdot 10^{-6}$
$x = y = 10^{-3}$	$1.4 \cdot 10^{-3}$	$5 \cdot 10^{-6}$	$4 \cdot 10^{-9}$
$x = y = 10^{-4}$	$1.4 \cdot 10^{-4}$	$5 \cdot 10^{-8}$	$4 \cdot 10^{-12}$

To use these errors in practice, we need bounds on their magnitudes. If the distance between (x, y) and (a, b) is represented by $d(x, y) = \sqrt{(x - a)^2 + (y - b)^2}$, it can be shown that the following results hold:

Error Bound for Linear Approximation

Suppose $f(x, y)$ is a function with continuous second-order partial derivatives such that for $d(x, y) \leq d_0$,

$$|f_{xx}|, \; |f_{xy}|, \; |f_{yy}| \leq M_L.$$

Suppose

$$f(x, y) = L(x, y) + E_L(x, y)$$
$$= f(a, b) + f_x(a, b)(x - a) + f_y(a, b)(y - b) + E_L(x, y).$$

Then we have

$$|E_L(x, y)| \leq 2 M_L d(x, y)^2 \quad \text{for} \quad d(x, y) \leq d_0.$$

Note that the upper bound for the error term $E_L(x, y)$ has a form reminiscent of the second-order term in the Taylor formula for $f(x, y)$.

Error Bound for Quadratic Approximation

Suppose $f(x, y)$ is a function with continuous third-order partial derivatives such that for $d(x, y) \leq d_0$,

$$|f_{xxx}|, \; |f_{xxy}|, \; |f_{xyy}|, \; |f_{yyy}| \leq M_Q.$$

Suppose

$$f(x, y) = Q(x, y) + E_Q(x, y)$$
$$= f(a, b) + f_x(a, b)(x - a) + f_y(a, b)(y - b)$$
$$+ \frac{f_{xx}(a, b)}{2}(x - a)^2 + f_{xy}(a, b)(x - a)(y - b) + \frac{f_{yy}(a, b)}{2}(y - b)^2 + E_Q(x, y).$$

Then we have

$$|E_Q(x, y)| \leq \frac{4}{3} M_Q d(x, y)^3 \quad \text{for} \quad d(x, y) \leq d_0.$$

Problem 20 shows how these error estimates and the coefficients (2 and 4/3) are obtained. The important thing to notice is the fact that, for small d, the magnitude of E_L is much smaller than d and the magnitude of E_Q is much smaller than d^2. In other words we have the following result:

As $d(x, y) \to 0$:

$$\frac{E_L(x, y)}{d(x, y)} \to 0 \quad \text{and} \quad \frac{E_Q(x, y)}{(d(x, y))^2} \to 0.$$

This means that near the point (a, b), we can view the original function and the approximation as indistinguishable and behaving the same way.

Example 3 Suppose that the Taylor polynomial of degree 2 for f at $(0, 0)$ is $Q(x, y) = 5x^2 + 3y^2$. Suppose we are also told that

$$|f_{xxx}|, |f_{xxy}|, |f_{xyy}|, |f_{yyy}| \leq 9.$$

Notice that $Q(x, y) > 0$ for all (x, y) except $(0, 0)$. Show that, except at $(0, 0)$, we have

$$f(x, y) > 0 \quad \text{for all } (x, y) \text{ such that } \sqrt{x^2 + y^2} = d < 0.25.$$

Solution By the error bound for the Taylor polynomial of degree 2, we have

$$|E_Q(x, y)| = |f(x, y) - Q(x, y)| \leq \frac{4}{3}(9)d^3 = 12d^3$$

which can be written as

$$-12d^3 \leq f(x, y) - Q(x, y) \leq 12d^3.$$

Therefore we know that

$$Q(x, y) - 12d^3 \leq f(x, y).$$

Since $Q(x, y) = 5x^2 + 3y^2$, we have

$$3x^2 + 5y^2 - 12d^3 \leq f(x, y).$$

Since $3x^2 + 5y^2 \geq 3x^2 + 3y^2 = 3d^2$, we have

$$3d^2 - 12d^3 \leq f(x, y).$$

Now d^3 approaches 0 faster than d^2, so when d is small, we have

$$0 \leq 3d^2 - 12d^3 \leq f(x, y).$$

In fact, writing $3d^2 - 12d^3 = 3d^2(1 - 4d)$ shows that $d < 1/4$ ensures that $f(x, y) > 0$, except at $(0, 0)$ where $f = 0$. Thus, f has the same sign as Q for points near $(0, 0)$.

Problems for Section 13.9

Find the quadratic Taylor polynomials about $(0, 0)$ for the functions in Problems 1–3.

1. $e^{-2x^2 - y^2}$
2. $\sin 2x + \cos y$
3. $\ln(1 + x^2 - y)$

For each of the functions in Problems 4–11, find the linear and quadratic approximations valid near $(1, 1)$. Compare the values of the approximations at $(1.1, 1.1)$ with the exact value of the function.

4. $z = xe^y$
5. $z = (x + y)e^y$
6. $z = \sin(x^2 + y^2)$
7. $z = \sqrt{x^2 + y^2}$
8. $z = \arctan(x + y)$
9. $z = \dfrac{xe^y}{x + y}$
10. $z = \sin(x/y)$
11. $z = \arctan(x/y)$

12. Let $f(x, y) = \sqrt{x + 2y + 1}$.
    (a) Compute the local linearization of f at $(0, 0)$.
    (b) Compute the Taylor polynomial of degree 2 for f at $(0, 0)$.
    (c) Compare the values of the linear and quadratic approximations in part (a) and part (b) with the true values for $f(x, y)$ at the points $(0.1, 0.1), (-0.1, 0.1), (0.1, -0.1), (-0.1, -0.1)$. Which approximation gives the closest values?

13. Using a computer and your answer to Problem 12, draw the six contour diagrams of $f(x, y) = \sqrt{x + 2y + 1}$ and its linear and quadratic approximations, $L(x, y)$ and $Q(x, y)$, in the two windows $[-0.6, 0.6] \times [-0.6, 0.6]$ and $[-2, 2] \times [-2, 2]$. Explain the shape of the contours, their spacing, and the relationship between the contours of f, L, and Q.

14. Figure 13.65 shows the level curves of a function $f(x, y)$ around a maximum or minimum, M. One of the points P and Q has coordinates (x_1, y_1) and the other has coordinates (x_2, y_2). Suppose $b > 0$ and $c > 0$. Consider the two linear approximations to f given by

$$f(x, y) \approx a + b(x - x_1) + c(y - y_1)$$
$$f(x, y) \approx k + m(x - x_2) + n(y - y_2).$$

(a) What is the relationship between the values of a and k?
(b) What are the coordinates of P?
(c) Is M a maximum or a minimum?
(d) What can you say about the sign of the constants m and n?

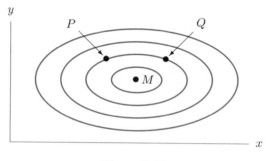

Figure 13.65

15. Consider the function $f(x, y) = (\sin x)(\sin y)$.

(a) Find the Taylor polynomials of degree 2 for f about the points $(0, 0)$ and $(\pi/2, \pi/2)$.
(b) Use the Taylor polynomials to sketch the contours of f close to each of the points $(0, 0)$ and $(\pi/2, \pi/2)$.

For Problems 16–19:

(a) Find the local linearization, $L(x, y)$, to the function $f(x, y)$ at the origin. Estimate the error $E_L(x, y) = f(x, y) - L(x, y)$ if $|x| \leq 0.1$ and $|y| \leq 0.1$.
(b) Find the degree 2 Taylor polynomial, $Q(x, y)$, for the function $f(x, y)$ at the origin. Estimate the error $E_Q(x, y) = f(x, y) - Q(x, y)$ if $|x| \leq 0.1$ and $|y| \leq 0.1$.
(c) Use a calculator to compute exactly $f(0.1, 0.1)$ and the errors $E_L(0.1, 0.1)$ and $E_Q(0.1, 0.1)$. How do these values compare with the errors predicted in parts (a) and (b)?

16. $f(x, y) = (\cos x)(\cos y)$

17. $f(x, y) = (e^x - x) \cos y$

18. $f(x, y) = e^{x+y}$

19. $f(x, y) = (x^2 + y^2)e^{x+y}$

20. It is known that if the derivatives of a one-variable function, $g(t)$, satisfy

$$|g^{(n+1)}(t)| \leq K \quad \text{for } |t| \leq d_0,$$

then the error, E_n, in the nth Taylor approximation, $P_n(x)$, is bounded as follows:

$$|E_n| = |g(t) - P_n(t)| \le \frac{K}{(n+1)!}|t|^{n+1} \quad \text{for } |t| \le d_0.$$

In this problem, we use this result for $g(t)$ to get the error bounds for the linear and quadratic Taylor approximations to $f(x, y)$. For a particular function $f(x, y)$, let $x = ht$ and $y = kt$ for fixed h and k, and define $g(t)$ as follows:

$$g(t) = f(ht, kt) \quad \text{for } 0 \le t \le 1.$$

(a)   Calculate $g'(t)$, $g''(t)$, and $g'''(t)$ using the chain rule.

(b)   Show that $L(ht, kt) = P_1(t)$ and that $Q(ht, kt) = P_2(t)$, where L is the linear approximation to f at $(0, 0)$ and Q is the Taylor polynomial of degree 2 for f at $(0, 0)$.

(c)   What is the relation between $E_L = f(x, y) - L(x, y)$ and E_1? What is the relation between $E_Q = f(x, y) - Q(x, y)$ and E_2?

(d)   Assuming that the second and third-order partial derivatives of f are bounded for $d(x, y) \le d_0$, show that $|E_L|$ and $|E_Q|$ are bounded as on page 158.

13.10 DIFFERENTIABILITY

Notes on Differentiability

In Section 13.3 we gave an informal introduction to the concept of differentiability. We called a function $f(x, y)$ *differentiable* at a point (a, b) if it is well-approximated by a linear function near (a, b). This section focuses on the precise meaning of the phrase "well-approximated." By looking at examples, we shall see that local linearity requires the existence of partial derivatives, but they do not tell the whole story. In particular, existence of partial derivatives at a point is not sufficient to guarantee local linearity at that point.

We begin by discussing the relation between continuity and differentiability. As an illustration, take a sheet of paper, crumple it into a ball and smooth it out again. Wherever there is a crease it would be difficult to approximate the surface by a plane — these are points of nondifferentiability of the function giving the height of the paper above the floor. Yet the sheet of paper models a graph which is continuous — there are no breaks. As in the case of one-variable calculus, continuity does not imply differentiability. But differentiability does *require* continuity: there cannot be linear approximations to a surface at points where there are abrupt changes in height.

Starting from the definition of differentiability for single-variable functions, we develop a definition of differentiability for two-variable functions.

Differentiability For Functions Of One Variable

We recall that a function $g(x)$ is *differentiable* at the point a if the limit

$$g'(a) = \lim_{h \to 0} \frac{g(a+h) - g(a)}{h}$$

exists. Geometrically, the definition means that the graph of $y = g(x)$ can be "well-approximated" by the line $y = L(x) = g(a) + g'(a)(x - a)$. How well does this line have to approximate the function $g(x)$ near the point a before we can say that g is differentiable at a? To answer this question, suppose g is differentiable at a and let $E(x)$ be the error between the function $g(x)$ and the line $L(x)$, so that

$$\begin{aligned} E(x) &= g(x) - L(x) \\ &= g(x) - g(a) - g'(a)(x - a). \end{aligned}$$

This means that at the point $x = a + h$ near a, the error $E(x)$ is given by

$$E(a + h) = g(a + h) - g(a) - g'(a)h.$$

Suppose we consider the *relative error* $E(a + h)/h$. We have

$$\frac{E(a + h)}{h} = \frac{g(a + h) - g(a)}{h} - g'(a).$$

Thus, in the limit as $h \to 0$, we have

$$\lim_{h \to 0} \frac{E(a + h)}{h} = \lim_{h \to 0} \frac{g(a + h) - g(a)}{h} - g'(a).$$

By the definition of the derivative, the right-hand side of the last equation is 0.

Therefore we see that if f is differentiable, the relative error tends to 0 as h tends to 0:

$$\lim_{h \to 0} \frac{E(a + h)}{h} = 0.$$

We will take "well approximated" to mean that this limit is zero. We use this idea to give a new definition of differentiability which can be generalized to functions of several variables

A function $g(x)$ is **differentiable at the point** a if there is a linear function $L(x) = g(a) + m(x - a)$ such that if the *error*, $E(x)$, is defined by

$$g(x) = L(x) + E(x)$$

and if $h = x - a$ then the *relative error* $E(a + h)/h$ satisfies

$$\lim_{h \to 0} \frac{E(a + h)}{h} = 0.$$

The function $L(x)$ is called the *local linearization* of $g(x)$ near a. The function g is *differentiable* if it is differentiable at each point of its domain.

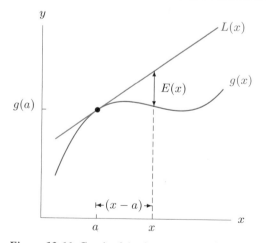

Figure 13.66: Graph of the function $y = g(x)$ and its local linearization $y = L(x)$ near the point a

This new definition tells us that the ratio $E(x)/(x-a)$ in Figure 13.66, the error divided by the distance from the point a, tends to 0 as $x \to a$. In addition, it can be shown that we must have $m = g'(a)$.

Differentiability For Functions Of Two Variables

Based on our new definition of differentiability, we define differentiability of a function of two variables at a point in terms of the error and the distance from the point. If the point is (a, b) and the nearby point is $(a+h, b+k)$, the distance is $\sqrt{h^2 + k^2}$. (See Figure 13.67.)

A function $f(x, y)$ is **differentiable at the point** (a, b) if there is a linear function $L(x, y) = f(a, b) + m(x - a) + n(y - b)$ such that if the *error* $E(x, y)$ is defined by

$$f(x, y) = L(x, y) + E(x, y),$$

and if $h = x - a, k = y - b$, then the *relative error* $E(a+h, b+k)/\sqrt{h^2 + k^2}$ satisfies

$$\lim_{\substack{h \to 0 \\ k \to 0}} \frac{E(a+h, b+k)}{\sqrt{h^2 + k^2}} = 0.$$

The function f is **differentiable** if it is differentiable at each point of its domain. The function $L(x, y)$ is called the *local linearization* of $f(x, y)$ near (a, b).

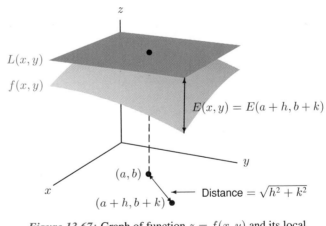

Figure 13.67: Graph of function $z = f(x, y)$ and its local linearization $z = L(x, y)$ near the point (a, b)

Partial Derivatives and Differentiability

In the next example, we show that this definition of differentiability is consistent with our previous notion — that is, that $m = f_x$ and $n = f_y$ and that the graph of $L(x, y)$ is the tangent plane.

Example 1   Show that if f is a differentiable function with local linearization $L(x, y) = f(a, b) + m(x - a) + n(y - b)$, then $m = f_x(a, b)$ and $n = f_y(a, b)$.

Solution   Since f is differentiable, we know that the relative error in $L(x, y)$ tends to 0 as we get close to (a, b). Suppose $h > 0$ and $k = 0$. Then we know that

$$0 = \lim_{h \to 0} \frac{E(a + h, b)}{\sqrt{h^2 + k^2}} = \lim_{h \to 0} \frac{E(a + h, b)}{h} = \lim_{h \to 0} \frac{f(a + h, b) - L(a + h, b)}{h}$$

$$= \lim_{h \to 0} \frac{f(a + h, b) - f(a, b) - mh}{h}$$

$$= \lim_{h \to 0} \left(\frac{f(a + h, b) - f(a, b)}{h} \right) - m = f_x(a, b) - m.$$

A similar result holds if $h < 0$, so we have $m = f_x(a, b)$. The result $n = f_y(a, b)$ is found in a similar manner.

The previous example shows that if a function is differentiable at a point, it has partial derivatives there. Therefore, if any of the partial derivatives fail to exist, then the function cannot be differentiable. This is what happens in the following example of a cone.

Example 2   Consider the function $f(x, y) = \sqrt{x^2 + y^2}$. Is f differentiable at the origin?

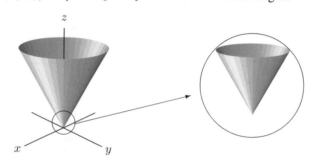

Figure 13.68: The function $f(x, y) = \sqrt{x^2 + y^2}$ is not locally linear at $(0, 0)$: Zooming in around $(0, 0)$ does not make the graph look like a plane

Solution   If we zoom in on the graph of the function $f(x, y) = \sqrt{x^2 + y^2}$ at the origin, as shown in Figure 13.68, the sharp point remains; the graph never flattens out to look like a plane. Near its vertex, the graph does not look like it is well approximated (in any reasonable sense) by any plane.

Judging from the graph of f, we would not expect f to be differentiable at $(0, 0)$. Let us check this by trying to compute the partial derivatives of f at $(0, 0)$:

$$f_x(0, 0) = \lim_{h \to 0} \frac{f(h, 0) - f(0, 0)}{h} = \lim_{h \to 0} \frac{\sqrt{h^2 + 0} - 0}{h} = \lim_{h \to 0} \frac{|h|}{h}.$$

Since $|h|/h = \pm 1$, depending on whether h approaches 0 from the left or right, this limit does not exist and so neither does the partial derivative $f_x(0, 0)$. Thus, f cannot be differentiable at the origin. If it were, both of the partial derivatives, $f_x(0, 0)$ and $f_y(0, 0)$, would exist.

Alternatively, we could show directly that there is no linear approximation near $(0, 0)$ that satisfies the small relative error criterion for differentiability. Any plane passing through the point $(0, 0, 0)$ has the form $L(x, y) = mx + ny$ for some constants m and n. If $E(x, y) = f(x, y) - L(x, y)$, then

$$E(x, y) = \sqrt{x^2 + y^2} - mx - ny.$$

Then for f to be differentiable at the origin, we would need to show that

$$\lim_{\substack{h \to 0 \\ k \to 0}} \frac{\sqrt{h^2 + k^2} - mh - nk}{\sqrt{h^2 + k^2}} = 0.$$

Taking $k = 0$ gives

$$\lim_{h \to 0} \frac{|h| - mh}{|h|} = 1 - m \lim_{h \to 0} \frac{h}{|h|}.$$

This limit exists only if $m = 0$ for the same reason as before. But then the value of the limit is 1 and not 0 as required. Thus, we again conclude f is not differentiable.

In Example 2 the partial derivatives f_x and f_y did not exist at the origin and this was sufficient to establish nondifferentiability there. We might expect that if both partial derivatives do exist, then f *is* differentiable. But the next example shows that this not necessarily true: the existence of both partial derivatives at a point is *not* sufficient to guarantee differentiability.

Example 3   Consider the function $f(x, y) = x^{1/3}y^{1/3}$. Show that the partial derivatives $f_x(0, 0)$ and $f_y(0, 0)$ exist, but that f is not differentiable at $(0, 0)$.

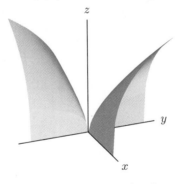

Figure 13.69: Graph of $z = x^{1/3}y^{1/3}$ for $z \geq 0$

Solution   See Figure 13.69 for the part of the graph of $z = x^{1/3}y^{1/3}$ when $z \geq 0$. We have $f(0, 0) = 0$ and we compute the partial derivatives using the definition:

$$f_x(0, 0) = \lim_{h \to 0} \frac{f(h, 0) - f(0, 0)}{h} = \lim_{h \to 0} \frac{0 - 0}{h} = 0,$$

and similarly

$$f_y(0, 0) = 0.$$

So, if there did exist a linear approximation near the origin, it would have to be $L(x, y) = 0$. But we can show that this choice of $L(x, y)$ doesn't result in the small relative error that is required for differentiability. In fact, since $E(x, y) = f(x, y) - L(x, y) = f(x, y)$, we need to look at the limit

$$\lim_{\substack{h \to 0 \\ k \to 0}} \frac{h^{1/3}k^{1/3}}{\sqrt{h^2 + k^2}}.$$

If this limit exists, we get the same value no matter how h and k approach 0. Suppose we take $k = h > 0$. Then the limit becomes

$$\lim_{h \to 0} \frac{h^{1/3}h^{1/3}}{\sqrt{h^2 + h^2}} = \lim_{h \to 0} \frac{h^{2/3}}{h\sqrt{2}} = \lim_{h \to 0} \frac{1}{h^{1/3}\sqrt{2}}.$$

But this limit does not exist, since small values for h will make the fraction arbitrarily large. So the only possible candidate for a linear approximation at the origin does not have a sufficiently small relative error. Thus, this function is *not* differentiable at the origin, even though the partial derivatives $f_x(0,0)$ and $f_y(0,0)$ exist. Figure 13.69 confirms that near the origin the graph of $z = f(x, y)$ is not well approximated by any plane.

In summary,

- If a function is differentiable at a point, then both partial derivatives exist there.
- Having both partial derivatives at a point does not guarantee that a function is differentiable there.

Continuity and Differentiability

We know that differentiable functions of one variable are continuous. Similarly, it can be shown that if a function of two variables is differentiable at a point, then the function is continuous there.

In Example 3 the function f was continuous at the point where it was not differentiable. Example 4 shows that even if the partial derivatives of a function exist at a point, the function is not necessarily continuous at that point if it is not differentiable there.

Example 4   Suppose that f is the function of two variables defined by

$$f(x, y) = \begin{cases} \dfrac{xy}{x^2 + y^2}, & (x, y) \neq (0, 0), \\ 0, & (x, y) = (0, 0). \end{cases}$$

Problem 4 on page 55 showed that $f(x, y)$ is not continuous at the origin. Show that the partial derivatives $f_x(0,0)$ and $f_y(0,0)$ exist. Could f be differentiable at $(0, 0)$?

Solution   From the definition of the partial derivative we see that

$$f_x(0,0) = \lim_{h \to 0} \frac{f(h, 0) - f(0, 0)}{h} = \lim_{h \to 0} \left(\frac{1}{h} \cdot \frac{0}{h^2 + 0^2} \right) = \lim_{h \to 0} \frac{0}{h} = 0,$$

and similarly

$$f_y(0,0) = 0.$$

So, the partial derivatives $f_x(0,0)$ and $f_y(0,0)$ exist. However, f cannot be differentiable at the origin since it is not continuous there.

In summary,

- If a function is differentiable at a point, then it is continuous there.
- Having both partial derivatives at a point does not guarantee that a function is continuous there.

How Do We Know If a Function Is Differentiable?

Can we use partial derivatives to tell us if a function is differentiable? As we see from Examples 3 and 4, it is not enough that the partial derivatives exist. However, the following condition *does* guarantee differentiability:

> If the partial derivatives, f_x and f_y, of a function f exist and are continuous on a small disk centered at the point (a, b), then f is differentiable at (a, b).

We will not prove this fact, although it provides a criterion for differentiability which is often simpler to use than the definition. It turns out that the requirement of continuous partial derivatives is more stringent than that of differentiability, so there exist differentiable functions which do not have continuous partial derivatives. However, most functions we encounter will have continuous partial derivatives. The class of functions with continuous partial derivatives is given the name C^1.

Example 5 Show that the function $f(x, y) = \ln(x^2 + y^2)$ is differentiable everywhere in its domain.

Solution The domain of f is all of 2-space except for the origin. We shall show that f has continuous partial derivatives everywhere in its domain (that is, the function f is in C^1). The partial derivatives are

$$f_x = \frac{2x}{x^2 + y^2} \quad \text{and} \quad f_y = \frac{2y}{x^2 + y^2}.$$

Since each of f_x and f_y is the quotient of continuous functions, the partial derivatives are continuous everywhere except the origin (where the denominators are zero). Thus, f is differentiable everywhere in its domain.

Most functions built up from elementary functions have continuous partial derivatives, except perhaps at a few obvious points. Thus, in practice, we can often identify functions as being C^1 without explicitly computing the partial derivatives.

Problems for Section 13.10

For the functions f in Problems 1–4 answer the following questions. Justify your answers.
(a) Use a computer to draw a contour diagram for f.
(b) Is f differentiable at all points $(x, y) \neq (0, 0)$?
(c) Do the partial derivatives f_x and f_y exist and are they continuous at all points $(x, y) \neq (0, 0)$?
(d) Is f differentiable at $(0, 0)$?
(e) Do the partial derivatives f_x and f_y exist and are they continuous at $(0, 0)$?

1. $f(x, y) = \begin{cases} \dfrac{x}{y} + \dfrac{y}{x}, & x \neq 0 \text{ and } y \neq 0, \\ 0, & x = 0 \text{ or } y = 0. \end{cases}$

2. $f(x, y) = \begin{cases} \dfrac{2xy}{(x^2 + y^2)^2}, & (x, y) \neq (0, 0), \\ 0, & (x, y) = (0, 0). \end{cases}$

3. $f(x, y) = \begin{cases} \dfrac{xy}{\sqrt{x^2 + y^2}}, & (x, y) \neq (0, 0), \\ 0, & (x, y) = (0, 0). \end{cases}$

4. $f(x, y) = \begin{cases} \dfrac{x^2 y}{x^4 + y^2}, & (x, y) \neq (0, 0), \\ 0, & (x, y) = (0, 0). \end{cases}$

5. Consider the function

$$f(x, y) = \begin{cases} \dfrac{xy^2}{x^2 + y^2}, & (x, y) \neq (0, 0), \\ 0, & (x, y) = (0, 0). \end{cases}$$

(a) Use a computer to draw the contour diagram for f.

(b) Is f differentiable for $(x, y) \neq (0, 0)$?

(c) Show that $f_x(0, 0)$ and $f_y(0, 0)$ exist.

(d) Is f differentiable at $(0, 0)$?

(e) Suppose $x(t) = at$ and $y(t) = bt$, where a and b are constants, not both zero. If $g(t) = f(x(t), y(t))$, show that

$$g'(0) = \frac{ab^2}{a^2 + b^2}.$$

(f) Show that

$$f_x(0, 0)x'(0) + f_y(0, 0)y'(0) = 0.$$

Does the chain rule hold for the composite function $g(t)$ at $t = 0$? Explain.

(g) Show that the directional derivative $f_{\vec{u}}(0, 0)$ exists for each unit vector $\vec{u}$. Does this imply that f is differentiable at $(0, 0)$?

6. Consider the function

$$f(x, y) = \begin{cases} \dfrac{xy^2}{x^2 + y^4}, & (x, y) \neq (0, 0), \\ 0, & (x, y) = (0, 0). \end{cases}$$

(a) Use a computer to draw the contour diagram for f.

(b) Show that the directional derivative $f_{\vec{u}}(0, 0)$ exists for each unit vector $\vec{u}$.

(c) Is f continuous at $(0, 0)$? Is f differentiable at $(0, 0)$? Explain.

7. Consider the function $f(x, y) = \sqrt{|xy|}$.

(a) Use a computer to draw the contour diagram for f. Does the contour diagram look like that of a plane when we zoom in on the origin?

(b) Use a computer to draw the graph of f. Does the graph look like a plane when we zoom in on the origin?

(c) Is f differentiable for $(x, y) \neq (0, 0)$?

(d) Show that $f_x(0, 0)$ and $f_y(0, 0)$ exist.

(e) Is f differentiable at $(0, 0)$? [Hint: Consider the directional derivative $f_{\vec{u}}(0, 0)$ for $\vec{u} = (\vec{i} + \vec{j})/\sqrt{2}$.]

8. Suppose a function f is differentiable at the point (a, b). Show that f is continuous at (a, b).

9. Suppose $f(x, y)$ is a function such that $f_x(0,0) = 0$ and $f_y(0,0) = 0$, and $f_{\vec{u}}(0,0) = 3$ for $\vec{u} = (\vec{i} + \vec{j})/\sqrt{2}$.

    (a) Is f differentiable at $(0,0)$? Explain.

    (b) Give an example of a function f defined on 2-space which satisfies these conditions. [Hint: The function f does not have to be defined by a single formula valid over all of 2-space.]

10. Consider the following function:

$$f(x, y) = \begin{cases} \dfrac{xy(x^2 - y^2)}{x^2 + y^2}, & (x, y) \neq (0,0), \\ 0, & (x, y) = (0,0). \end{cases}$$

The graph of f is shown in Figure 13.70, and the contour diagram of f is shown in Figure 13.71.

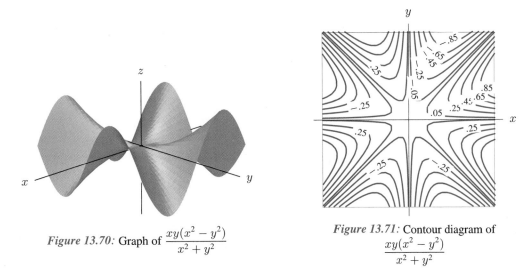

Figure 13.70: Graph of $\dfrac{xy(x^2 - y^2)}{x^2 + y^2}$

Figure 13.71: Contour diagram of $\dfrac{xy(x^2 - y^2)}{x^2 + y^2}$

    (a) Find $f_x(x, y)$ and $f_y(x, y)$ for $(x, y) \neq (0,0)$.
    (b) Show that $f_x(0,0) = 0$ and $f_y(0,0) = 0$.
    (c) Are the functions f_x and f_y continuous at $(0,0)$?
    (d) Is f differentiable at $(0,0)$?

REVIEW PROBLEMS FOR CHAPTER THIRTEEN

For Problems 1–4, find the indicated partial derivatives. Assume the variables are restricted to a domain on which the function is defined.

1. $\dfrac{\partial z}{\partial x}$ and $\dfrac{\partial z}{\partial y}$ if $z = (x^2 + x - y)^7$

2. $\dfrac{\partial F}{\partial L}$ if $F(L, K) = 3\sqrt{LK}$

3. $\dfrac{\partial f}{\partial p}$ and $\dfrac{\partial f}{\partial q}$ if $f(p, q) = e^{p/q}$

4. $\dfrac{\partial f}{\partial x}$ if $f(x, y) = e^{xy}(\ln y)$

Find both partial derivatives for the functions in Problems 5–8. Assume the variables are restricted to a domain on which the function is defined.

5. $z = x^4 - x^7 y^3 + 5xy^2$

6. $z = \tan(\theta)/r$

7. $w = s \ln(s + t)$

8. $w = \arctan(ue^{-v})$

9. If $f(x, y) = x^2 y$ and $\vec{v} = 4\vec{i} - 3\vec{j}$, find the directional derivative at the point $(2, 6)$ in the direction of $\vec{v}$.

Assume that $f(x, y)$ is a differentiable function. Are the statements in Problems 10–16 true or false? Explain your answer.

10. $f_{\vec{u}}(x_0, y_0)$ is a scalar.

11. $f_{\vec{u}}(a, b) = \|\nabla f(a, b)\|$

12. If $\vec{u}$ is tangent to the level curve of f at some point, then grad $f \cdot \vec{u} = 0$ there.

13. Suppose that f is differentiable at (a, b). Then there is always a direction in which the rate of change of f at (a, b) is 0.

14. There is a function with a point in its domain where $\|\operatorname{grad} f\| = 0$ and where there is a nonzero directional derivative.

15. There is a function with $\|\operatorname{grad} f\| = 4$ and $f_{\vec{i}} = 5$ at some point.

16. There is a function with $\|\operatorname{grad} f\| = 4$ and $f_{\vec{j}} = -3$ at some point.

17. Let $f(w, z) = w^2 z + 3z^2$.

    (a) Use difference quotients with $h = 0.01$ to approximate $f_w(2, 2)$ and $f_z(2, 2)$.
    (b) Now evaluate $f_w(2, 2)$ and $f_z(2, 2)$ exactly.

18. Figure 13.72 shows a contour diagram for the temperature T (in °C) along a wall in a heated room as a function of distance x along the wall and time t in minutes. Estimate $\partial T / \partial x$ and $\partial T / \partial t$ at the given points. Give units for your answers and say what the answers mean.
    (a) $x = 15, t = 20$ (b) $x = 5, t = 12$

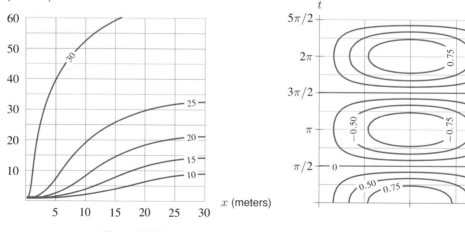

Figure 13.72   Figure 13.73

19. Figure 13.73 shows a contour diagram for a vibrating string function, $f(x, t)$.

    (a) Is $f_t(\pi/2, \pi/2)$ positive or negative? How about $f_t(\pi/2, \pi)$? What does the sign of $f_t(\pi/2, b)$ tell you about the motion of the point on the string at $x = \pi/2$ when $t = b$?
    (b) Find all t for which f_t is positive, for $0 \le t \le 5\pi/2$.
    (c) Find all x and t such that f_x is positive.

20. The quantity Q (in pounds) of beef that a certain community buys during a week is a function $Q = f(b, c)$ of the prices of beef, b, and chicken, c, during the week. Do you expect $\partial Q/\partial b$ to be positive or negative? What about $\partial Q/\partial c$?

21. Suppose the cost of producing one unit of a certain product is given by

$$c = a + bx + ky,$$

    where x is the amount of labor used (in man hours) and y is the amount of raw material used (by weight) and a and b and k are constants. What does $\partial c/\partial x = b$ mean? What is the practical interpretation of b?

22. People commuting to a city can choose to go either by bus or by train. The number of people who choose either method depends in part upon the price of each. Let $f(P_1, P_2)$ be the number of people who take the bus when P_1 is the price of a bus ride and P_2 is the price of a train ride. What can you say about the signs of $\partial f/\partial P_1$ and $\partial f/\partial P_2$? Explain your answers.

23. The acceleration g due to gravity, at a distance r from the center of a planet of mass m, is given by

$$g = \frac{Gm}{r^2},$$

    where G is the universal gravitational constant.

    (a) Find $\partial g/\partial m$ and $\partial g/\partial r$.
    (b) Interpret each of the partial derivatives you found in part (a) as the slope of a graph in the plane and sketch the graph.

24. Suppose that the function $P = f(K, L)$ expresses the production of a firm as a function of the capital invested, K, and its labor costs, L.

    (a) Suppose $f(K, L) = 60K^{1/3}L^{2/3}$. Find the relationship between K and L if the marginal productivity of capital (that is, the rate of change of production with capital) equals the marginal productivity of labor cost (that is, the rate of change of production with labor cost). Simplify your answer.
    (b) Now suppose $f(K, L) = cK^aL^b$, with a, b, c positive constants. Answer the same question as in part (a).

25. In analyzing a factory and deciding whether or not to hire more workers, it is useful to know under what circumstances productivity increases. Suppose $P = f(x_1, x_2, x_3)$ is the total quantity produced as a function of x_1, the number of workers, and any other variables x_2, x_3. We define the average productivity of a worker as P/x_1. Show that the average productivity increases as x_1 increases when marginal production, $\partial P/\partial x_1$, is greater than the average productivity, P/x_1.

26. For the Cobb-Douglas production function $P = 40L^{0.25}K^{0.75}$, find the differential dP when $L = 2$ and $K = 16$.

27. The area of a triangle can be calculated from the formula $S = \frac{1}{2}ab\sin C$. Show that if an error of $10'$ (or $\pi/1080$ radians) is made in measuring C then the error in S is approximately $\pi S/(1080\tan C)$. [Note: $10'$ means 10 minutes, where 1 minute$= 1/60$ degree.]

28. The gas equation for one mole of oxygen relates its pressure, P (in atmospheres), its temperature, T (in K), and its volume, V (in cubic decimeters, dm^3):

$$T = 16.574\frac{1}{V} - 0.52754\frac{1}{V^2} + 0.3879P + 12.187VP.$$

    (a) Find the temperature T and differential dT if the volume is 25 dm^3 and the pressure is 1 atmosphere.
    (b) Use your answer to part (a) to estimate how much the volume would have to change if the pressure increased by 0.1 atmosphere and the temperature remained constant.

29. Find the rate of change of $f(x, y) = xe^y$ at the point $(1, 1)$ in the direction of $\vec{i} + 2\vec{j}$.

30. Find the directional derivative of $z = x^2 - y^2$ at the point $(3, -1)$ in the direction making an angle $\theta = \pi/4$ with the x-axis. In which direction is the directional derivative the largest?

31. Figure 13.74 shows the level curves of a function $f(x, y)$. Give the approximate value of $f_{\vec{u}}(3, 1)$ with $\vec{u} = (-2\vec{i} + \vec{j})/\sqrt{5}$. Explain your answer.

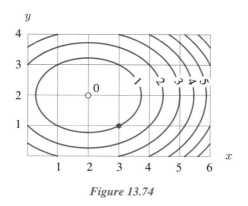

Figure 13.74

32. Figure 13.75 shows the monthly payment, m, on a 5-year car loan if you borrow P dollars at r percent interest. Find a formula for a linear function which approximates m. What is the practical significance of the constants in your formula?

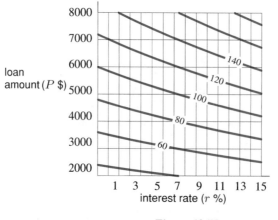

Figure 13.75

33. Find the point(s) on $x^2 + y^2 + z^2 = 8$ where the tangent plane is parallel to the plane $x - y + 3z = 0$.

34. Suppose the temperature at a point (x, y) is given by the function $T(x, y) = 100 - x^2 - y^2$. In which direction should a heat-seeking bug move from the point (x, y) to increase its temperature fastest?

35. Suppose that the values of the function $f(x, y)$ near the point $x = 2$, $y = 3$ are given in Table 13.9. Estimate the following.

(a) $\dfrac{\partial f}{\partial x}\bigg|_{(2,3)}$ and $\dfrac{\partial f}{\partial y}\bigg|_{(2,3)}$.

(b) The rate of change of f at $(2, 3)$ in the direction of the vector $\vec{i} + 3\vec{j}$.

(c) The maximum possible rate of change of f as you move away from the point $(2, 3)$. In which direction should you move to obtain this rate of change?

(d) Write an equation for the level curve through the point $(2, 3)$.

(e) Find a vector tangent to the level curve of f through the point $(2, 3)$.

(f) Find the differential of f at the point $(2, 3)$. If $dx = 0.03$ and $dy = 0.04$, find df. What does df represent in this case?

TABLE 13.9

		x	
		2.00	2.01
y	3.00	7.56	7.42
	3.02	7.61	7.47

36. The function $g(x, y)$ is differentiable and has the property that $g(1, 3) = 4$ and $g_x(1, 3) = -1$ and $g_y(1, 3) = 2$.

(a) Find the equation of the level curve of g through the point $(1, 3)$.

(b) Find the coordinates of the point on the surface $z = g(x, y)$ above the point $(1, 3)$.

(c) Find the equation of the tangent plane to the surface $z = g(x, y)$ at the point you found in part (b).

37. Suppose $w = f(x, y, z) = 3xy + yz$ and that x, y, z are functions of u and v such that

$$x = \ln u + \cos v, \quad y = 1 + u \sin v, \quad z = uv.$$

(a) Find $\partial w / \partial u$ and $\partial w / \partial v$ at $(u, v) = (1, \pi)$.

(b) Suppose now that u and v are also functions of t such that

$$u = 1 + \sin(\pi t), \quad v = \pi t^2.$$

Use your answer to part (a) to find dw/dt at $t = 1$.

38. A circular city has radius r km and an average population density of ρ people/km^2. In 1997 the population was 3 million, the radius was 25 km and growing at 0.1 km/year. If the density was increasing at 200 people/km^2/year, find the rate at which the total population of the city was growing.

39. Show that if F is any differentiable function of one variable, then $V(x, y) = xF(2x + y)$ satisfies the equation

$$x\frac{\partial V}{\partial x} - 2x\frac{\partial V}{\partial y} = V.$$

40. Find a particular solution to the differential equation in Problem 39 satisfying

$$V(1, y) = y^2.$$

41. Find the quadratic Taylor polynomial about $(0, 0)$ for $f(x, y) = \cos{(x + 2y)} \sin{(x - y)}$.

42. Suppose $f(x, y) = e^{(x-1)^2 + (y-3)^2}$.

    (a) Find the first-order Taylor polynomial about $(0, 0)$.
    (b) Find the second-order (quadratic) Taylor polynomial about the point $(1, 3)$.
    (c) Find a 2-vector perpendicular to the level curve through $(0, 0)$.
    (d) Find a 3-vector perpendicular to the surface $z = f(x, y)$ at the point $(0, 0)$.

43. The function $T(x, y, z, t)$ is a solution to the *heat equation*

$$T_t = K(T_{xx} + T_{yy} + T_{zz}),$$

and gives the temperature at the point (x, y, z) in 3-space and time t. The constant K is the *thermal conductivity* of the medium through which the heat is flowing.

    (a) Show that the function

$$T(x, y, z, t) = \frac{1}{(4\pi K t)^{3/2}} e^{-(x^2 + y^2 + z^2)/4Kt}$$

    is a solution to the heat equation for all (x, y, z) in 3-space and $t > 0$.
    (b) For each fixed time t, what are the level surfaces of the function $T(x, y, z, t)$ in 3-space?
    (c) Regard t as fixed and compute $\operatorname{grad} T(x, y, z, t)$. What does $\operatorname{grad} T(x, y, z, t)$ tell us about the direction and magnitude of the heat flow?

44. Each diagram (I) – (IV) in Figure 13.76 represents the level curves of a function $f(x, y)$. For each function f, consider the point above P on the surface $z = f(x, y)$ and choose from the lists which follow:

    (a) A vector which could be the normal to the surface at that point;
    (b) An equation which could be the equation of the tangent plane to the surface at that point.

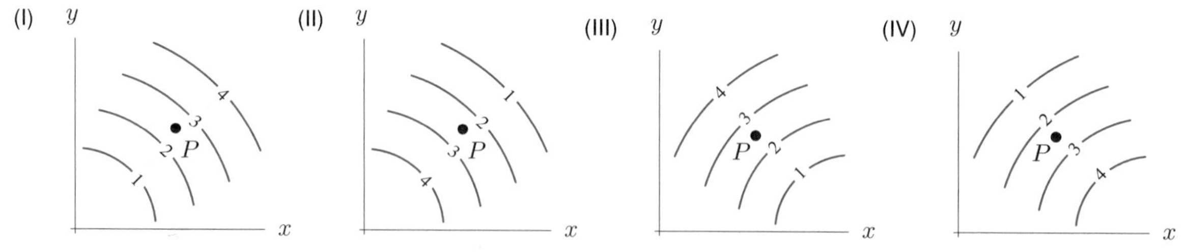

Figure 13.76

Vectors
(E) $2\vec{i} + 2\vec{j} - 2\vec{k}$
(F) $2\vec{i} + 2\vec{j} + 2\vec{k}$
(G) $2\vec{i} - 2\vec{j} + 2\vec{k}$
(H) $-2\vec{i} + 2\vec{j} + 2\vec{k}$

Equations
(J) $x + y + z = 4$
(K) $2x - 2y - 2z = 2$
(L) $-3x - 3y + 3z = 6$
(M) $-\dfrac{x}{2} + \dfrac{y}{2} - \dfrac{z}{2} = -7$

CHAPTER FOURTEEN

OPTIMIZATION: LOCAL AND GLOBAL EXTREMA

In one-variable calculus we saw how to find the maximum and minimum values of a function of one variable. In practice, there are often several variables in an optimization problem. For example, you may have $10,000 to invest in new equipment and advertising for your business. What combination of equipment and advertising will yield the greatest profit? Or, what combination of drugs will lower a patient's temperature the most? In this chapter we consider optimization problems, where the variables are completely free to vary (unconstrained optimization) and where there is a constraint on the variables (for example, a budget constraint).

14.1 LOCAL EXTREMA

Functions of several variables, like functions of one variable, can have *local* and *global* extrema. (That is, local and global maxima and minima.) A function has a local extremum at a point where it takes on the largest or smallest values in a small region around the point. Global extrema are the largest or smallest values anywhere on the domain under consideration. (See Figures 14.1 and 14.2.)

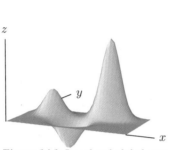

Figure 14.1: Local and global extrema for a function of two variables on $0 \leq x \leq a$, $0 \leq y \leq b$

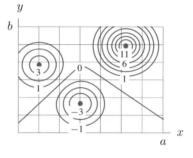

Figure 14.2: Contour map of the function in Figure 14.1

More precisely, considering only points at which f is defined, we say:

- f has a **local maximum** at the point P_0 if $f(P_0) \geq f(P)$ for all points P near P_0.
- f has a **local minimum** at the point P_0 if $f(P_0) \leq f(P)$ for all points P near P_0.

How Do We Detect a Local Maximum or Minimum?

Recall that if the gradient vector of a function is defined and nonzero, then it points in a direction in which the function increases. Suppose that a function f has a local maximum at a point P_0 which is not on the boundary of the domain. If the vector grad $f(P_0)$ were defined and nonzero, then we could increase f by moving in the direction of grad $f(P_0)$. Since f has a local maximum at P_0, there is no direction in which f is increasing. Thus, if grad $f(P_0)$ is defined, we must have

$$\text{grad } f(P_0) = \vec{0}.$$

Similarly, suppose f has a local minimum at the point P_0. If grad $f(P_0)$ were defined and nonzero, then we could decrease f by moving in the direction opposite to grad $f(P_0)$, and so we must again have grad $f(P_0) = \vec{0}$ Therefore, we make the following definition:

Points where the gradient is either $\vec{0}$ or undefined are called **critical points** of the function. If a function has a local maximum or minimum at a point P_0, not on the boundary of its domain, then P_0 is a critical point.

For a function of two variables, we can also see that the gradient vector must be zero or undefined at a local maximum by looking at its contour diagram and a plot of its gradient vectors. (See Figures 14.3 and 14.4.) Around the maximum the vectors are all pointing inward, perpendicularly to the contours. At the maximum the gradient vector must be zero or undefined. A similar argument shows that the gradient must be zero at a local minimum.

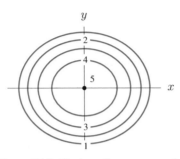

Figure 14.3: Contour diagram around a local maximum of a function

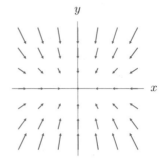

Figure 14.4: Gradients pointing toward the local maximum of the function in Figure 14.3

Finding Critical Points

To find critical points we set grad $f = f_x\vec{i} + f_y\vec{j} + f_z\vec{k} = \vec{0}$, which means setting all the partial derivatives of f equal to zero. We must also look for the points where one or more of the partial derivatives is undefined.

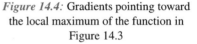

Example 1   Find and analyze the critical points of $f(x, y) = x^2 - 2x + y^2 - 4y + 5$.

Solution   To find the critical points, we set both partial derivatives equal to zero:

$$f_x = 2x - 2 = 0$$
$$f_y = 2y - 4 = 0.$$

Solving these equations gives $x = 1$, $y = 2$. Hence, f has only one critical point, namely $(1, 2)$. To see the behavior of f near $(1, 2)$, look at the values of the function in Table 14.1.

TABLE 14.1   *Values of $f(x, y)$ near the point $(1, 2)$*

			x		
		0.8	0.9	1.0	1.1	1.2
	1.8	0.08	0.05	0.04	0.05	0.08
	1.9	0.05	0.02	0.01	0.02	0.05
y	2.0	0.04	0.01	0.00	0.01	0.04
	2.1	0.05	0.02	0.01	0.02	0.05
	2.2	0.08	0.05	0.04	0.05	0.08

The table suggests that the function has a local minimum value of 0 at $(1, 2)$. We can verify this by completing the square:

$$f(x, y) = x^2 - 2x + y^2 - 4y + 5 = (x - 1)^2 + (y - 2)^2.$$

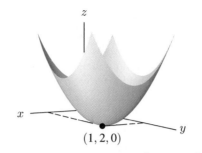

$(1, 2, 0)$

Figure 14.5: The graph of $f(x, y) = x^2 - 2x + y^2 - 4y + 5$
with a local minimum at the point $(1, 2)$

Figure 14.5 shows that the graph of f is a parabolic bowl with vertex at the point $(1, 2, 0)$. It is the same shape as the graph of $z = x^2 + y^2$ (shown in Figure 11.18 on page 15), except that the vertex has been shifted to $(1, 2)$. So the point $(1, 2)$ is a local minimum of f (as well as a global minimum).

Example 2   Find and analyze any critical points of $f(x, y) = -\sqrt{x^2 + y^2}$.

Solution   We look for points where grad $f = \vec{0}$ or is undefined. The partial derivatives are given by

$$\frac{\partial f}{\partial x} = -\frac{x}{\sqrt{x^2 + y^2}},$$

$$\frac{\partial f}{\partial y} = -\frac{y}{\sqrt{x^2 + y^2}}.$$

These are never both zero; but they are both undefined at $x = 0$, $y = 0$. Thus, $(0, 0)$ is a critical point and a possible extreme point. The graph of f (see Figure 14.6) is a cone, with vertex at $(0, 0)$. So f has a local and global maximum at $(0, 0)$.

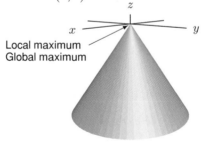

Local maximum
Global maximum

Figure 14.6: Graph of $f(x, y) = -\sqrt{x^2 + y^2}$

Example 3   Find the local extrema of the function $f(x, y) = 8y^3 + 12x^2 - 24xy$.

Solution   We begin by looking for critical points:

$$f_x = 24x - 24y,$$
$$f_y = 24y^2 - 24x.$$

Setting these expressions equal to zero gives the system of equations

$$x = y, \qquad x = y^2,$$

which has two solutions, $(0, 0)$ and $(1, 1)$. Are these maxima, minima or neither? Let's look at the contours near the points: Figure 14.7 shows the contour diagram of this function. Notice that

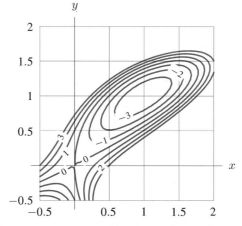

Figure 14.7: Contour diagram of $f(x, y) = 8y^3 + 12x^2 - 24xy$ showing
critical points at $(0, 0)$ and $(1, 1)$

$f(1, 1) = -4$ and that there is no other -4 contour. The contours near $P = (1, 1)$ appear oval in shape and show that f is increasing in value no matter in which direction you move away from P. This suggests that f has a local minimum at the point $(1, 1)$.

The level curves near $Q = (0, 0)$ show a very different behavior. While $f(0, 0) = 0$, we see that f takes on both positive and negative values at nearby points. Thus, the point $(0, 0)$ is a critical point which is neither a local maximum nor a local minimum.

Saddle Points

The previous example shows that critical points can occur at local maxima or minima, or at points which are neither — the value of the function is larger in some directions and smaller in others. We make the following definition:

> A function, f, has a **saddle point** at P_0 if P_0 is a critical point of f and within any distance of P_0, no matter how small, there are points, P_1 and P_2, with
>
> $$f(P_1) > f(P_0) \quad \text{and} \quad f(P_2) < f(P_0).$$

Thus, we see from Figure 14.7 that the function $f(x, y) = 8y^3 + 12x^2 - 24xy$ in Example 3 has a saddle point at the origin.

For another example, look at the graph of $g(x, y) = x^2 - y^2$ in Figure 14.8. The origin is a critical point and $g(0, 0) = 0$. Since there are positive values on the x-axis and negative values on the y-axis, the origin is a saddle point. Notice that the graph of g looks like a saddle there.

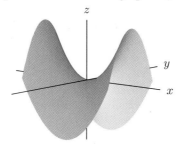

Figure 14.8: Graph of $g(x, y) = x^2 - y^2$, showing a saddle point at the origin

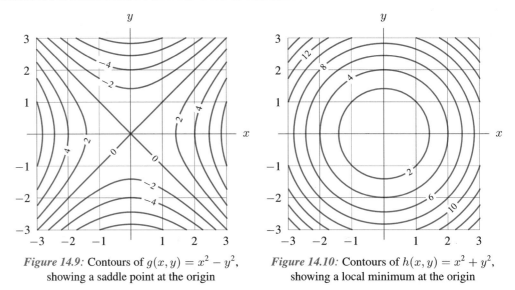

Figure 14.9: Contours of $g(x, y) = x^2 - y^2$, showing a saddle point at the origin

Figure 14.10: Contours of $h(x, y) = x^2 + y^2$, showing a local minimum at the origin

Figure 14.9 shows the level curves of g near the saddle point $(0, 0)$. They are hyperbolas showing both positive and negative values of g near $(0, 0)$. Contrast this with the appearance of level curves near a local maximum or minimum. For example, Figure 14.10 shows $h(x, y) = x^2 + y^2$ near $(0, 0)$.

Is a Critical Point a Local Maximum, Local Minimum, or Saddle Point?

We can see whether a critical point of a function, f, is a maximum, minimum, or saddle point by looking at the contour diagram. There is also a simple analytic method for making the distinction if the critical point is one at which the partial derivatives of f are zero. Near most critical points, a function has the same behavior as its quadratic Taylor approximation about that point, so we must first understand quadratic functions.

Quadratic Functions of the form $f(x, y) = ax^2 + bxy + cy^2$

We begin by looking at what can happen at critical points of quadratic functions of the form $f(x, y) = ax^2 + bxy + cy^2$, where a, b and c are constants.

Example 4   Find and analyze the local extrema of the function $f(x, y) = x^2 + xy + y^2$.

Solution   To find critical points, we set

$$f_x = 2x + y = 0,$$
$$f_y = x + 2y = 0.$$

The only critical point is $(0, 0)$, and the value of the function there is $f(0, 0) = 0$. If f is always positive or zero near $(0, 0)$, then $(0, 0)$ is a local minimum; if f is always negative or zero near $(0, 0)$, it is a local maximum; if f takes both positive and negative values it is a saddle point. The graph in Figure 14.11 suggests that $(0, 0)$ is a local minimum.

How can we be sure that $(0, 0)$ is a local minimum? The algebraic way to determine if a quadratic function is always negative, always positive, or neither, is to complete the square. Writing

$$f(x, y) = x^2 + xy + y^2 = \left(x + \frac{1}{2}y\right)^2 + \frac{3}{4}y^2,$$

shows that $f(x, y)$ is a sum of two squares, so it must always be greater than or equal to zero. Thus, the critical point is both a local and a global minimum.

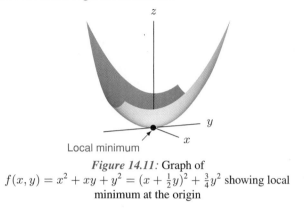

Local minimum

Figure 14.11: Graph of
$f(x, y) = x^2 + xy + y^2 = (x + \frac{1}{2}y)^2 + \frac{3}{4}y^2$ showing local
minimum at the origin

The Shape of the Graph of $f(x, y) = ax^2 + bxy + cy^2$

In general, a function of the form $f(x, y) = ax^2 + bxy + cy^2$ has one critical point at $(0, 0)$. To analyze its graph, we complete the square. Assuming $a \neq 0$, we write

$$ax^2 + bxy + cy^2 = a\left[x^2 + \frac{b}{a}xy + \frac{c}{a}y^2\right]$$

$$= a\left[\left(x + \frac{b}{2a}y\right)^2 + \left(\frac{c}{a} - \frac{b^2}{4a^2}\right)y^2\right]$$

$$= a\left[\left(x + \frac{b}{2a}y\right)^2 + \left(\frac{4ac - b^2}{4a^2}\right)y^2\right].$$

The shape of the graph of f depends on whether the coefficient of y^2 is positive, negative, or zero. The sign of $D = 4ac - b^2$, called the *discriminant*, determines the sign of the coefficient of y^2.

- If $D > 0$, then the expression inside the brackets is positive or zero, so the function has a local maximum or a local minimum.
    - If $a > 0$, the function has a local minimum, since the graph is a right-side-up paraboloid, like $z = x^2 + y^2$. (See Figure 14.12.)
    - If $a < 0$, the function has a local maximum, since the graph is an upside-down paraboloid, like $z = -x^2 - y^2$. (See Figure 14.13.)
- If $D < 0$, then the function goes up in some directions and goes down in others, like $z = x^2 - y^2$. Hence it has a saddle point. (See Figure 14.14.)
- If $D = 0$, then the quadratic function is $a(x + by/2a)^2$, whose graph is a parabolic cylinder. (See Figure 14.15.)

Figure 14.12: Concave up:
$D > 0$ and $a > 0$

Figure 14.13: Concave
down: $D > 0$ and $a < 0$

Figure 14.14: Saddle-shaped:
$D < 0$

Figure 14.15: Parabolic
cylinder: $D = 0$

More generally, the graph of $g(x, y) = a(x - x_0)^2 + b(x - x_0)(y - y_0) + c(y - y_0)^2$ has exactly the same shape as the graph of $f(x, y) = ax^2 + bxy + cy^2$, except that the critical point is at (x_0, y_0) rather than $(0, 0)$. Therefore, the discriminant test[1] gives the same results for the behavior of g near (x_0, y_0).

Classifying the Critical Points of a Function

Now, suppose that f is any function with $f(0, 0) = 0$ and grad $f(0, 0) = \vec{0}$. Recall from page 156 that f can be approximated by its quadratic Taylor polynomial near $(0, 0)$:

$$f(x, y) \approx f(0, 0) + f_x(0, 0)x + f_y(0, 0)y$$
$$+ \frac{1}{2} f_{xx}(0, 0)x^2 + f_{xy}(0, 0)xy + \frac{1}{2} f_{yy}(0, 0)y^2.$$

Since $f(0, 0) = 0$ and $f_x(0, 0) = f_y(0, 0) = 0$, the quadratic polynomial simplifies to

$$f(x, y) \approx \frac{1}{2} f_{xx}(0, 0)x^2 + f_{xy}(0, 0)xy + \frac{1}{2} f_{yy}(0, 0)y^2.$$

The discriminant is

$$D = 4ac - b^2 = 4 \left(\frac{1}{2} f_{xx}(0, 0) \right) \left(\frac{1}{2} f_{yy}(0, 0) \right) - \left(f_{xy}(0, 0) \right)^2,$$

which simplifies to

$$D = f_{xx}(0, 0) f_{yy}(0, 0) - \left(f_{xy}(0, 0) \right)^2.$$

There is a similar formula for D if $f(0, 0) \neq 0$ or if the critical point is at (x_0, y_0). Thus, we get the following test:

Second Derivative Test for Functions of Two Variables

Suppose (x_0, y_0) is a point where grad $f(x_0, y_0) = \vec{0}$. Let

$$D = f_{xx}(x_0, y_0) f_{yy}(x_0, y_0) - \left(f_{xy}(x_0, y_0) \right)^2.$$

- If $D > 0$ and $f_{xx}(x_0, y_0) > 0$, then f has a local minimum at (x_0, y_0).
- If $D > 0$ and $f_{xx}(x_0, y_0) < 0$, then f has a local maximum at (x_0, y_0).
- If $D < 0$, then f has a saddle point at (x_0, y_0).
- If $D = 0$, anything can happen: f can have a local maximum, or a local minimum or a saddle point at (x_0, y_0).

Example 5   Find the local maxima, minima, and saddle points of the function

$$f(x, y) = \frac{x^2}{2} + 3y^3 + 9y^2 - 3xy + 9y - 9x.$$

Solution   The partial derivatives of f are $f_x = x - 3y - 9$ and $f_y = 9y^2 + 18y - 3x + 9$. The equations $f_x = 0$ and $f_y = 0$ give

$$9y^2 + 18y + 9 - 3x = 0,$$
$$x - 3y - 9 = 0.$$

Eliminating x gives

$$9y^2 + 9y - 18 = 0,$$

which has solutions $y = -2$ and $y = 1$. We find the corresponding values of x, so the critical points of f are $(3, -2)$ and $(12, 1)$. The discriminant is

$$D(x, y) = f_{xx}f_{yy} - f_{xy}^2 = (1)(18y + 18) - (-3)^2 = 18y + 9.$$

---

[1]We assumed that $a \neq 0$. If $a = 0$ and $c \neq 0$, the same argument works. If both $a = 0$ and $c = 0$, then $f(x, y) = bxy$, which is a saddle.

Since $D(3, -2) = -36 + 9 < 0$, we know that $(3, -2)$ is a saddle point of f. Since $D(12, 1) = 18 + 9 > 0$ and $f_{xx}(12, 1) = 1 > 0$, we know that $(12, 1)$ is a local minimum of f.

The second derivative test does not give any information in the case $D = 0$. However, as the following example illustrates, we can still classify the critical points by looking at the graph of the function.

Example 6 Classify the critical point $(0, 0)$ of the functions $f(x, y) = x^4 + y^4$, and $g(x, y) = -x^4 - y^4$, and $h(x, y) = x^4 - y^4$.

Solution Each of these functions has a critical point at $(0, 0)$. However, all the second partial derivatives are 0 there, so each function has $D = 0$. Near the origin, the graphs of f, g and h look like the surfaces in Figures 14.12–14.14, respectively, and so we see that f has a minimum at $(0, 0)$, and g has a maximum at $(0, 0)$, and h has a saddle point at $(0, 0)$.

We can get the same results algebraically. Since $f(0, 0) = 0$ and $f(x, y) > 0$ elsewhere, f must have a minimum at the origin. Since $g(0, 0) = 0$ and $g(x, y) < 0$ elsewhere, g has a maximum at the origin. Lastly, h has a saddle point at the origin since $h(0, 0) = 0$ and $h(x, y) > 0$ on the x-axis and $h(x, y) < 0$ on the y-axis.

Problems for Section 14.1

1. Consider the points marked A, B, C in the contour plot in Figure 14.16. Which of these appear to be critical points? Classify those that are critical points.

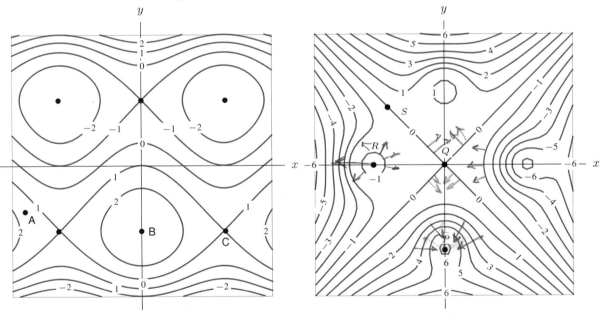

Figure 14.16 *Figure 14.17*

For Problems 2–4, use Figure 14.17, which shows level curves of some function $f(x, y)$.

2. Decide whether you think each point is a local maximum, local minimum, saddle point, or none of these.

(a) P (b) Q (c) R (d) S

3. Sketch the direction of ∇f at several points around each of P, Q, and R.

4. Put arrows showing the direction of ∇f at the points where $\|\nabla f\|$ is largest.

For Problems 5–11, find the local maxima, local minima, and saddle points of the given function.

5. $f(x, y) = x^3 - 3x + y^3 - 3y$

6. $f(x, y) = x^3 + e^{-y^2}$

7. $f(x, y) = (x + y)(xy + 1)$

8. $f(x, y) = 8xy - \frac{1}{4}(x + y)^4$

9. $E(x, y) = 1 - \cos x + y^2/2$

10. $f(x, y) = \sin x \sin y$

11. $P(x, y) = 400 - 3x^2 - 4x + 2xy - 5y^2 + 48y$

12. Suppose $f(x, y) = A - (x^2 + Bx + y^2 + Cy)$. What values of A, B, and C give $f(x, y)$ a local maximum value of 15 at the point $(-2, 1)$?

Each function in Problems 13–15 has a critical point at $(0, 0)$. What sort of critical point is it?

13. $f(x, y) = x^6 + y^6$

14. $g(x, y) = x^4 + y^3$

15. $h(x, y) = \cos x \cos y$

16. (a) Find all critical points of
$$f(x, y) = e^x(1 - \cos y).$$
    (b) Are these critical points local maxima, local minima, or saddle points?

17. Suppose $f_x = f_y = 0$ at $(1, 3)$ and $f_{xx} > 0$, $f_{yy} > 0$, $f_{xy} = 0$.
    (a) What can you conclude about the behavior of the function near the point $(1, 3)$?
    (b) Sketch a possible contour diagram.

18. Suppose that for some function $f(x, y)$ at the point (a, b), we have $f_x = f_y = 0$, $f_{xx} > 0$, $f_{yy} = 0$, $f_{xy} > 0$.
    (a) What can you conclude about the shape of the graph of f near the point (a, b)?
    (b) Sketch a possible contour diagram.

19. The behavior of a function can be complicated near a critical point where $D = 0$. Suppose that
$$f(x, y) = x^3 - 3xy^2.$$
    Show that there is one critical point at $(0, 0)$ and that $D = 0$ there. Then show that the contour for $f(x, y) = 0$ consists of three lines intersecting at the origin and that these lines divide the plane into six regions around the origin where f alternates from positive to negative. Sketch a contour map for f near $(0, 0)$. The graph of this function is called a *monkey saddle*.

20. On a computer, draw contour diagrams for the family of functions
$$f(x, y) = k(x^2 + y^2) - 2xy$$
    for $k = -2, -1, 0, 1, 2$. Use these figures to classify the critical point at $(0, 0)$ for each value of k. Explain your observations using the discriminant, D.

14.2 GLOBAL EXTREMA: UNCONSTRAINED OPTIMIZATION

Suppose we want to find the highest and the lowest points in some region of the country. First of all, it makes a difference what the region is, the whole United States, an individual state, or a county.

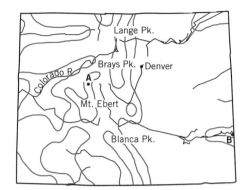

Figure 14.18: The highest and lowest
points in the state of Colorado

Let's suppose the region is Colorado (a contour map is shown in Figure 14.18). The highest point is the top of a mountain peak (point A on the map, Mt. Elbert, 14,431 feet high). What about the lowest point? Colorado does not have large pits without drainage, like Death Valley in California. A drop of rain falling at any point in Colorado will flow eventually out of the state to either the Pacific or the Atlantic Ocean. If there is no local minimum inside the state, where is the lowest point? It must be on the state boundary at a point where a river is flowing out of the state (point B where the Arkansas River leaves the state, 3,400 feet high). The highest point in Colorado is a global maximum for the elevation function in Colorado and the lowest point is the global minimum.

In general, if we are given a function f defined on a region R, we say:

> - f has a **global maximum on** R at the point P_0 if $f(P_0) \geq f(P)$ for all points P in R.
> - f has a **global minimum on** R at the point P_0 if $f(P_0) \leq f(P)$ for all points P in R.

The process of finding a global maximum or minimum for a function f on a region R is called *optimization*. If the region R is the entire xy-plane, we speak of *unconstrained optimization*; if the region R is not the entire xy-plane, that is, if x or y is restricted in some way, then we speak of *constrained optimization*. If the region R is not stated explicitly, it is understood to be the whole xy-plane.

How Do We Find Global Maxima and Minima?

As the Colorado example illustrates, a global extremum can occur either at a critical point inside the region or at a point on the boundary of the region. This is analogous to single-variable calculus, where a function achieves its global extrema on an interval either at a critical point inside the interval or at an endpoint of the interval. Optimization for functions of more than one variable, however, is more difficult because regions in 2-space can have very complicated boundaries.

> ### For an Unconstrained Optimization Problem
> - Find the critical points.
> - Investigate whether the critical points give global maxima or minima.

Not all functions have a global maximum or minimum: it depends on the function and the region. For the remainder of this section we consider applications in which global extrema are expected from practical considerations. In general, the fact that a function has a single local maximum or minimum does not guarantee that the point is the global maximum or minimum. (See Problem 23.) An exception is if the function is quadratic, in which case the local maximum or minimum is the global maximum or minimum. See Example 1 on page 177.

Economic Example: Maximizing Profit

In planning production, a company is concerned with how much of a particular item to manufacture and the price at which to sell the item. In general, the higher the price, the less that can be sold. To determine how much to produce, the company often chooses the combination of price and quantity that maximizes the profit. To calculate the maximum we use the fact that

$$\text{Profit} = \text{Revenue} - \text{Cost},$$

and, provided the price is constant,

$$\text{Revenue} = \text{Price} \times \text{Quantity} = pq.$$

In addition, we need to know how the cost and price depend on quantity.

Example 1   A company manufactures two items which are sold in two separate markets. The quantities, q_1 and q_2, demanded by consumers, and the prices, p_1 and p_2 (in dollars), of each item are related by

$$p_1 = 600 - 0.3q_1 \quad \text{and} \quad p_2 = 500 - 0.2q_2.$$

Thus, if the price for either item increases, the demand for it decreases. The company's total production cost is given by

$$C = 16 + 1.2q_1 + 1.5q_2 + 0.2q_1q_2.$$

If the company wants to maximize its total profits, how much of each product should it produce? What will be the maximum profit? [2]

Solution   The total revenue, R, is the sum of the revenues, p_1q_1 and p_2q_2, from each market. Substituting for p_1 and p_2, we get

$$R = p_1q_1 + p_2q_2$$
$$= (600 - 0.3q_1)q_1 + (500 - 0.2q_2)q_2$$
$$= 600q_1 - 0.3q_1^2 + 500q_2 - 0.2q_2^2.$$

Thus, the total profit P is given by

$$P = R - C$$
$$= \overset{R}{(600q_1 - 0.3q_1^2 + 500q_2 - 0.2q_2^2)} - \overset{C}{(16 + 1.2q_1 + 1.5q_2 + 0.2q_1q_2)}$$
$$= -16 + 598.8q_1 - 0.3q_1^2 + 498.5q_2 - 0.2q_2^2 - 0.2q_1q_2.$$

To maximize P, we compute partial derivatives and set them equal to 0:

$$\frac{\partial P}{\partial q_1} = 598.8 - 0.6q_1 - 0.2q_2 = 0,$$

$$\frac{\partial P}{\partial q_2} = 498.5 - 0.4q_2 - 0.2q_1 = 0.$$

Since grad P is defined everywhere, the only critical points of P are those where grad $P = \vec{0}$. Thus, solving for q_1, q_2, we find that

$$q_1 = 699.1 \quad \text{and} \quad q_2 = 896.7.$$

The corresponding prices are

$$p_1 = 390.27 \quad \text{and} \quad p_2 = 320.66.$$

---

[2] Adapted from M. Rosser, *Basic Mathematics for Economists*, p. 316 (New York: Routledge, 1993).

To see whether or not we have found a maximum, we compute second partial derivatives:

$$\frac{\partial^2 P}{\partial q_1^2} = -0.6, \qquad \frac{\partial^2 P}{\partial q_2^2} = -0.4, \qquad \frac{\partial^2 P}{\partial q_1 \partial q_2} = -0.2,$$

so,

$$D = \frac{\partial^2 P}{\partial q_1^2}\frac{\partial^2 P}{\partial q_2^2} - \left(\frac{\partial^2 P}{\partial q_1 \partial q_2}\right)^2 = (-0.6)(-0.4) - (-0.2)^2 = 0.2.$$

Therefore we have found a local maximum. The graph of P is an upside-down paraboloid and so $(699.1, 896.7)$ is in fact a global maximum. The company should produce 699.1 units of the first item priced at \$390.27 per unit, and 896.7 units of the second item priced at \$320.66 per unit. The maximum profit $P(699.1, 896.7) \approx \$433{,}000$.

Fitting a Line to Data

An important application of optimization is to the problem of fitting the "best" line to some data. Suppose the data is plotted in the plane. We measure the distance from a line to the data points by adding the squares of the vertical distances from each point to the line. The smaller this sum of squares is, the better the line fits the data. The line with the minimum sum of square distances is called the *least squares line*, or the *regression line*. If the data is nearly linear, the least squares line will be a good fit; otherwise it may not be. (See Figure 14.19.)

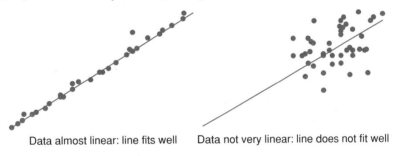

Data almost linear: line fits well   Data not very linear: line does not fit well

Figure 14.19: Fitting lines to data points

Example 2   Find a least squares line for the following data points: $(1, 1)$, $(2, 1)$, and $(3, 3)$.

Solution   Suppose the line has equation $y = b + mx$. If we find b and m then we have found the line. So, for this problem, b and m are the two variables. We want to minimize the function $f(b, m)$ that gives the sum of the three squared vertical distances from the points to the line in Figure 14.20.

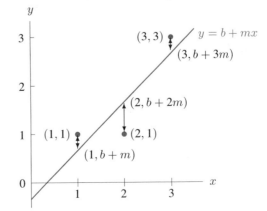

Figure 14.20: The least squares line minimizes the sum of the squares of these vertical distances

The vertical distance from the point $(1, 1)$ to the line is the difference in the y-coordinates $1 - (b + m)$; similarly for the other points. Thus, the sum of squares is

$$f(b, m) = (1 - (b + m))^2 + (1 - (b + 2m))^2 + (3 - (b + 3m))^2.$$

To minimize f we look for critical points. First we differentiate f with respect to b:

$$\begin{aligned}
\frac{\partial f}{\partial b} &= -2(1 - (b + m)) - 2(1 - (b + 2m)) - 2(3 - (b + 3m)) \\
&= -2 + 2b + 2m - 2 + 2b + 4m - 6 + 2b + 6m \\
&= -10 + 6b + 12m.
\end{aligned}$$

Now we differentiate with respect to m:

$$\begin{aligned}
\frac{\partial f}{\partial m} &= 2(1 - (b + m))(-1) + 2(1 - (b + 2m))(-2) + 2(3 - (b + 3m))(-3) \\
&= -2 + 2b + 2m - 4 + 4b + 8m - 18 + 6b + 18m \\
&= -24 + 12b + 28m.
\end{aligned}$$

The equations $\dfrac{\partial f}{\partial b} = 0$ and $\dfrac{\partial f}{\partial m} = 0$ give a system of two linear equations in two unknowns:

$$-10 + 6b + 12m = 0,$$
$$-24 + 12b + 28m = 0.$$

The solution to this pair of equations is the critical point $b = -1/3$ and $m = 1$. Since

$$D = f_{bb}f_{mm} - (f_{mb})^2 = (6)(28) - 12^2 = 24 \quad \text{and} \quad f_{bb} = 6 > 0,$$

we have found a local minimum. The graph of $f(b, m)$ is a parabolic bowl, so the local minimum is the global minimum of f. Thus, the least squares line is

$$y = x - \frac{1}{3}.$$

As a check, notice that the line $y = x$ passes through the points $(1, 1)$ and $(3, 3)$. It is reasonable that introducing the point $(2, 1)$ moves the y-intercept down from 0 to $-1/3$.

The general formulas for the slope and y-intercept of a least squares line are in Problem 18 at the end of this section. Many calculators have these formulas built in, so that when you enter the data, out come the values of b and m. At the same time, you get the *correlation coefficient*, which measures how close the data points actually come to fitting the least squares line.

Gradient Search for Finding Local Extrema

So far we have searched for values that maximize or minimize a function $f(x, y)$ by first finding the critical points of f. Finding the critical points amounts to solving the equation grad $f = \vec{0}$, which is really a pair of simultaneous equations for x and y:

$$\frac{\partial f}{\partial x}(x_0, y_0) = 0 \quad \text{and} \quad \frac{\partial f}{\partial y}(x_0, y_0) = 0.$$

However, solving such equations can be very difficult. In practice, most optimization problems are solved by numerical methods such as the *gradient search*. The gradient search method can be explained by analogy with a mountain climber who wishes to maximize his elevation by getting to the top of the highest mountain. All he has to do is keep going up and eventually he will get to the top of some mountain. If he starts near the highest mountain, that is probably the mountain he will conquer. If not, he may go up a lower mountain, ending up at a local rather than a global maximum.

The gradient search method is illustrated in the next example. It is a minimization problem, so imagine a hiker seeking the bottom of the lowest valley by always going down.

Example 3 Twenty cubic meters of gravel are to be delivered to a landfill by a trucker. She plans to purchase an open-top box in which to transport the gravel in numerous trips. The cost to her is the cost of the box plus $2 per trip. The box must have height 0.5 m, but she can choose the length and width. The cost of the box will be $20/m^2 for the ends and $10/m^2 for the bottom and sides. Notice the tradeoff she faces: A smaller box is cheaper to buy but requires more trips. What size box should she buy to minimize her over-all costs? [3]

Solution We first get an algebraic expression for the trucker's cost. Let the length of the box be x meters and the width be y meters and let the height be 0.5 m (See Figure 14.21.)

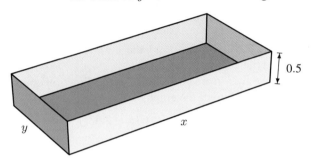

0.5

y

x

TABLE 14.2 *Trucker's itemized cost*

$20/(0.5xy)$ at $2/trip	$80/(xy)$
2 ends at $20/m^2 × $0.5y$ m^2	$20y$
2 sides at $10/m^2 × $0.5x$ m^2	$10x$
1 bottom at $10/m^2 × xy m^2	$10xy$
Total cost	$f(x, y)$

Figure 14.21: The box for transporting gravel

The volume of the box is $0.5xy$ m^3, so delivery of 20 m^3 of gravel will require $20/(0.5xy)$ trips. The trucker's cost is itemized in Table 14.2. The problem is to choose x and y to minimize

$$\text{Total cost} = f(x, y) = \frac{80}{xy} + 20y + 10x + 10xy.$$

We pick a starting point (x_0, y_0) that may not minimize f but which we hope is not too far from the minimum point. In this example we start with $(x_0, y_0) = (5, 5)$, which is definitely not a critical point of f because

$$\text{grad } f(5, 5) = 59.4\vec{i} + 69.4\vec{j} \neq \vec{0}.$$

We plan to move from (x_0, y_0) to a new point (x_1, y_1) in such a way that f decreases, that is $f(x_1, y_1) < f(x_0, y_0)$. Moving in the direction of grad $f(x_0, y_0)$ increases f as rapidly as possible, so we move in the opposite direction, namely $-$ grad $f(x_0, y_0)$. We continue to move in this direction until the f values begin to increase again. If we move parallel to $-$ grad $f(x_0, y_0)$, our displacement from the original point is of the form $-t$ grad $f(x_0, y_0)$, where t is a scalar to be determined. Since grad $f(x_0, y_0) = 59.4\vec{i} + 69.4\vec{j}$, the coordinates of our final point are

$$(x_0 - 59.4t, y_0 - 69.4t).$$

We want to find the minimum value of the function f as t increases. Figure 14.22 gives the graph of

$$f\big((x_0, y_0) - t \text{ grad } f(x_0, y_0)\big) = f(5 - 59.4t, 5 - 69.4t)$$

for positive t. Zooming in shows that the local minimum is at $t \approx 0.0554$. So we move to the point given by

$$(x_1, y_1) = (5 - (59.4)(0.0554), 5 - (69.4)(0.0554)) \approx (1.71, 1.16).$$

---

[3] Adapted from Claude McMillan, Jr., *Mathematical Programming*, 2nd ed., p. 156-157 (New York: Wiley, 1978).

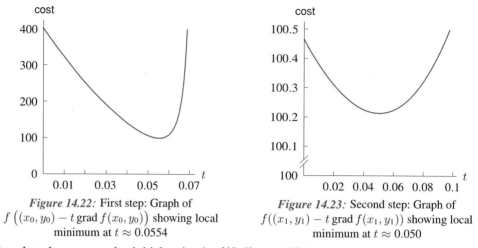

Figure 14.22: First step: Graph of
$f\left((x_0, y_0) - t \operatorname{grad} f(x_0, y_0)\right)$ showing local
minimum at $t \approx 0.0554$

Figure 14.23: Second step: Graph of
$f((x_1, y_1) - t \operatorname{grad} f(x_1, y_1))$ showing local
minimum at $t \approx 0.050$

Notice that the cost at the initial point is $f(5, 5) = 403.2$ and the cost at the new point is $f(1.71, 1.16) = 100.47$, so we have decreased the cost considerably.

We can further decrease the cost by moving away from $(1.71, 1.16)$ in the direction opposite to $\operatorname{grad} f(1.71, 1.16) = -1.99\vec{i} + 2.33\vec{j}$. Figure 14.23 gives the graph of

$$f\left((x_1, y_1) - t \operatorname{grad} f(x_1, y_1)\right) = f(1.71 + 1.99t, 1.16 - 2.33t)$$

which achieves a local minimum at $t \approx 0.050$. So we take

$$(x_2, y_2) = (1.71 + (1.99)(0.050), 1.16 - (2.33)(0.050)) \approx (1.81, 1.04).$$

Notice that $f(1.81, 1.04) = 100.22$ whereas $f(1.71, 1.16) = 100.47$. The move from (x_1, y_1) to (x_2, y_2) decreased the cost f by a rather small amount, only \$0.25. Thus, although we have not actually achieved a minimum, we may feel that for practical purposes we are probably close enough. The trucker will round off, buying a box of dimensions about 1.8m × 1m × 0.5m.

How Do We Know Whether a Function Has a Global Maximum or Minimum?

Under what circumstances does a function of two variables have a global maximum or minimum? The next example shows that a function may have both a global maximum and a global minimum on a region, or just one, or neither.

Example 4   Investigate the global maxima and minima of the following functions:

(a) $h(x, y) = 1 + x^2 + y^2$ on the disk $x^2 + y^2 \leq 1$.
(b) $f(x, y) = x^2 - 2x + y^2 - 4y + 5$ on the xy-plane.
(c) $g(x, y) = x^2 - y^2$ on the xy-plane.

Solution   (a) The graph of $h(x, y) = 1 + x^2 + y^2$ is a bowl shaped paraboloid with a global minimum of 1 at $(0, 0)$, and a global maximum of 2 on the edge of the region, $x^2 + y^2 = 1$.

(b) The graph of f in Figure 14.5 on page 178 shows that f has a global minimum at the point $(1, 2)$ and no global maximum (because the value of f increases without bound as $x \to \infty$, $y \to \infty$).

(c) The graph of g in Figure 14.8 on page 179 shows that g has no global maximum because $g(x, y) \to \infty$ as $x \to \infty$ if y is constant. Similarly, g has no global minimum because $g(x, y) \to -\infty$ as $y \to \infty$ if x is constant.

There are, however, conditions that guarantee that a function has a global maximum and minimum. For $h(x)$, a function of one variable, the function must be continuous on a closed interval $a \leq x \leq b$. If h is continuous on a non-closed interval, such as $a \leq x < b$ or $a < x < b$, or on an interval which is not bounded, such as $a < x < \infty$, then h need not have a maximum or minimum value. What is the situation for functions of two variables? As it turns out, a similar result is true for continuous functions defined on regions which are closed and bounded, analogous to the closed and bounded interval $a \leq x \leq b$. In everyday language we say

- A **closed** region is one which contains its boundary;
- A **bounded** region is one which does not stretch to infinity in any direction.

More precise definitions are as follows. Suppose R is a region in 2-space. A point (x_0, y_0) is a *boundary point* of R if, for every $r > 0$, the disk $(x - x_0)^2 + (y - y_0)^2 < r^2$ with center (x_0, y_0) and radius r contains both points which are in R and points which are not in R. See Figure 14.24. A point (x_0, y_0) can be a boundary point of the region R without actually belonging to R. A point (x_0, y_0) in R is an *interior point* if it is not a boundary point; thus, for small enough $r > 0$, the disk of radius r centered at (x_0, y_0) lies entirely in the region R. See Figure 14.25. The collection of all the boundary points is the *boundary of* R and the collection of all the interior points is the *interior* of R. The region R is *closed* if it contains its boundary, while it is *open* if every point in R is an interior point.

A region R in 2-space is *bounded* if the distance between every point (x, y) in R and the origin is less than or equal to some constant number K. Closed and bounded regions in 3-space are defined in the same way.

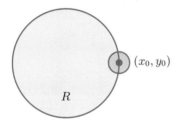

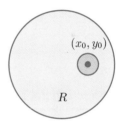

Figure 14.24: Boundary point (x_0, y_0) of R *Figure 14.25:* Interior point (x_0, y_0) of R

Example 5 (a)   The square $-1 \leq x \leq 1$, $-1 \leq y \leq 1$ is closed and bounded.
            (b)   The first quadrant $x \geq 0$, $y \geq 0$ is closed but is not bounded.
            (c)   The disk $x^2 + y^2 < 1$ is open and bounded, but is not closed.
            (d)   The half-plane $y > 0$ is open, but is neither closed nor bounded.

The reason that closed and bounded regions are useful is the following result[4]:

If f is a continuous function on a closed and bounded region R, then f has a global maximum at some point (x_0, y_0) in R and a global minimum at some point (x_1, y_1) in R.

The result is also true for functions of three or more variables.

[4]For a proof, see W. Rudin, *Principles of Mathematical Analysis*, 2nd ed., p. 89, (New York: McGraw-Hill, 1976)

If f is not continuous or the region R is not closed and bounded, there is no guarantee that f will achieve a global maximum or global minimum on R. In Example 4, the function g is continuous but does not achieve a global maximum or minimum in 2-space, a region which is closed but not bounded. The following example illustrates what can go wrong when the region is bounded but not closed.

Example 6 Does the following function have a global maximum or minimum on the region R given by $0 < x^2 + y^2 \leq 1$?

$$f(x, y) = \frac{1}{x^2 + y^2}$$

Solution The region R is bounded, but it is not closed since it does not contain the boundary point $(0, 0)$. We see from the graph of $z = f(x, y)$ in Figure 14.26 that f has a global minimum on the circle $x^2 + y^2 = 1$. However, $f(x, y) \to \infty$ as $(x, y) \to (0, 0)$, so f has no global maximum.

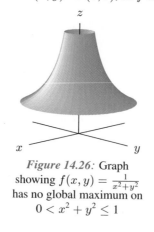

Figure 14.26: Graph showing $f(x, y) = \frac{1}{x^2+y^2}$ has no global maximum on $0 < x^2 + y^2 \leq 1$

Problems for Section 14.2

1. By looking at the weather map in Figure 11.1 on page 2, find the maximum and minimum daily high temperatures in the states of Mississippi, Alabama, Pennsylvania, New York, California, Arizona, and Massachusetts.

2. Does the function $f(x, y) = -2x^2 - 7y^2$ have global maxima and minima? Explain.

3. Does the function $f(x, y) = x^2/2 + 3y^3 + 9y^2 - 3x + 9y - 9$ have global maxima and minima? Explain.

4. Which of the following functions have both a global maximum and global minimum? Which have neither?

$$f(x, y) = 5 + x^2 - 2y^2, \quad g(x, y) = x^2y^2, \quad h(x, y) = x^3 + y^3.$$

In Problems 5–7, find the global maximum and minimum of the given function over the square $-1 \leq x \leq 1, -1 \leq y \leq 1$, and say whether it occurs on the boundary of the square. (Hint: Consider the graph of the function.)

5. $z = x^2 + y^2$

6. $z = x^2 - y^2$

7. $z = -x^2 - y^2$

8. The quantity of a product demanded by consumers is a function of its price. The quantity of one product demanded may also depend on the price of other products. For example, the demand for tea is affected by the price of coffee; the demand for cars is affected by the price of gas. Suppose the quantities demanded, q_1 and q_2, of two products depend on the prices, p_1 and p_2, as follows

$$q_1 = 150 - 2p_1 - p_2$$

$$q_2 = 200 - p_1 - 3p_2.$$

   (a) What does the fact that the coefficients of p_1 and p_2 are negative tell you? Give an example of two products that might be related this way.

   (b) Suppose one manufacturer sells both of these products. How should the manufacturer set prices to earn the maximum possible revenue? What is that maximum possible revenue?

9. A company operates two plants which manufacture the same item and whose total cost functions are

$$C_1 = 8.5 + 0.03q_1^2 \quad \text{and} \quad C_2 = 5.2 + 0.04q_2^2,$$

   where q_1 and q_2 are the quantities produced by each plant. The total quantity demanded, $q = q_1 + q_2$, is related to the price, p, by

$$p = 60 - 0.04q.$$

   How much should each plant produce in order to maximize the company's profit? [5]

10. Assume that two products are manufactured in quantities q_1 and q_2 and sold at prices of p_1 and p_2 respectively, and that the cost of producing them is given by

$$C = 2q_1^2 + 2q_2^2 + 10.$$

   (a) Find the maximum profit that can be made, assuming the prices are fixed.

   (b) Find the rate of change of that maximum profit as p_1 increases.

11. A missile has a remote guidance device which is sensitive to both temperature and humidity. If t is the temperature in °C and h is percent humidity, the range over which the missile can be controlled is given by:

$$\text{Range in km} = 27{,}800 - 5t^2 - 6ht - 3h^2 + 400t + 300h,$$

   What are the optimal atmospheric conditions for controlling the missile?

12. Some items are sold at different prices to different groups of people. For example, there are sometimes discounts for senior citizens or for children. The reason is that these groups may be more sensitive to price, so a discount will have greater impact on their purchasing decisions. The seller faces an optimization problem: How large a discount to offer in order to maximize profits?

   A theater can sell q_c child tickets and q_a adult tickets at prices p_c and p_a, according to the following demand functions:

$$q_c = rp_c^{-4} \quad \text{and} \quad q_a = sp_a^{-2},$$

   and has operating costs proportional to the total number of tickets sold. What should be the relative price of children's and adults' tickets?

---

[5] Adapted from M. Rosser, *Basic Mathematics for Economists*, p. 318 (New York: Routledge, 1993).

13. Show analytically that the function $f(x, y)$ in Example 3 has a local minimum at $(2, 1)$.

14. Design a rectangular milk carton box of width w, length l, and height h which holds 512 cm³ of milk. The sides of the box cost 1 cent/cm² and the top and bottom cost 2 cent/cm². Find the dimensions of the box that minimize the total cost of materials used.

15. An international airline has a regulation that each passenger can carry a suitcase having the sum of its width, length and height less than or equal to 135 cm. Find the dimensions of the suitcase of maximum volume that a passenger may carry under this regulation.

16. A company manufactures a product which requires capital and labor to produce. The quantity, Q, of the product manufactured is given by the Cobb-Douglas production function

$$Q = AK^a L^b,$$

where K is the quantity of capital and L is the quantity of labor used and A, a, and b are positive constants with $0 < a < 1$ and $0 < b < 1$. Suppose one unit of capital costs $\$k$ and one unit of labor costs $\$\ell$. The price of the product is fixed at $\$p$ per unit.

   (a) If $a + b < 1$, how much capital and labor should the company use to maximize its profit?
   (b) Is there a maximum profit in the case $a + b = 1$? What about $a + b \geq 1$? Explain.

   [Note: See page 29 for a discussion of the Cobb-Douglas production function. The three cases considered above, namely $a + b < 1$, $a + b = 1$, and $a + b > 1$ are, respectively, the cases of *decreasing returns to scale*, *constant returns to scale*, and *increasing returns to scale*.]

17. Compute the regression line for the points $(-1, 2)$, $(0, -1)$, $(1, 1)$ using least squares.

18. In this problem you will derive the general formulas for the slope and y-intercept of a least squares line. Assume that you have n data points $(x_1, y_1), (x_2, y_2), \cdots, (x_n, y_n)$. Let the equation of the least squares line be $y = b + mx$.

   (a) For each data point (x_i, y_i), show that the corresponding point directly above or below it on the least squares line has y-coordinate $b + mx_i$.
   (b) For each data point (x_i, y_i), show that the square of the vertical distance from it to the point found in part (a) is $(y_i - (b + mx_i))^2$.
   (c) Form the function $f(b, m)$ which is the sum of all of the n squared distances found in part (b). That is,

$$f(b, m) = \sum_{i=1}^{n} (y_i - (b + mx_i))^2.$$

   Show that the partial derivatives $\dfrac{\partial f}{\partial b}$ and $\dfrac{\partial f}{\partial m}$ are given by

$$\frac{\partial f}{\partial b} = -2 \sum_{i=1}^{n} (y_i - (b + mx_i))$$

   and

$$\frac{\partial f}{\partial m} = -2 \sum_{i=1}^{n} (y_i - (b + mx_i)) \cdot x_i.$$

   (d) Show that the critical point equations $\dfrac{\partial f}{\partial b} = 0$ and $\dfrac{\partial f}{\partial m} = 0$ lead to a pair of simultaneous linear equations in b and m:

$$nb + \left(\sum x_i \right) m = \sum y_i$$
$$\left(\sum x_i \right) b + \left(\sum x_i^2 \right) m = \sum x_i y_i$$

(e)   Solve the equations in part (d) for b and m, getting

$$b = \left(\sum_{i=1}^{n} x_i^2 \sum_{i=1}^{n} y_i - \sum_{i=1}^{n} x_i \sum_{i=1}^{n} x_i y_i \right) \bigg/ \left(n \sum_{i=1}^{n} x_i^2 - \left(\sum_{i=1}^{n} x_i \right)^2 \right)$$

$$m = \left(n \sum_{i=1}^{n} x_i y_i - \sum_{i=1}^{n} x_i \sum_{i=1}^{n} y_i \right) \bigg/ \left(n \sum_{i=1}^{n} x_i^2 - \left(\sum_{i=1}^{n} x_i \right)^2 \right)$$

(f)   Apply these formulas to the data points $(1,1), (2,1), (3,3)$ to verify that you get the same result as in Example 2.

When data is not linear, it can sometimes be transformed in such a way that it looks more linear. For example, suppose we expect that data points (x, y) lie approximately on an exponential curve, say

$$y = Ce^{ax},$$

where a and C are constants. Taking the natural log of both sides, we find that $\ln y$ is a linear function of x.

$$\ln y = ax + \ln C.$$

To find a and C, we use least squares for $\ln y$ against x. Use this method in Problems 19–20.

19.   The population of the United States was about 180 million in 1960, grew to 206 million in 1970, and 226 million in 1980.

(a)   Based on this data and assuming that the population was growing at an exponential rate, use the method of least squares to estimate the population in 1990.

(b)   According to the national census, the 1990 population was 249 million. What does this say about the assumption of exponential growth?

(c)   Predict the population in the year 2010.

20.   The data in Table 14.3 shows the cost of a first class stamp in the US over the last 70 years.

TABLE 14.3   *Cost of a first class stamp*

Year	1920	1932	1958	1963	1968	1971	1974
Postage	0.02	0.03	0.04	0.05	0.06	0.08	0.10
Year	1975	1978	1981	1985	1988	1991	1995
Postage	0.13	0.15	0.20	0.22	0.25	0.29	0.32

(a)   Find the line of best fit through the data. Using this line, predict the cost of a postage stamp in the year 2010.

(b)   Plot the data. Does it look linear?

(c)   Plot the year against the natural logarithm of the price: Does this look linear? If it is linear, what does that tell you about the price of a stamp as a function of time? Find the line of best fit through this data, and use your answer to again predict the cost of a postage stamp in the year 2010.

21.   We wish to find the minimum value of

$$f(x, y) = (x + 1)^4 + (y - 1)^4 + \frac{1}{x^2 y^2 + 1}.$$

(a)   Use a computer to plot the contour diagram for f.
(b)   Minimize f using the gradient search method.

22.   The government wants to build a pipe that will pump water up from a dam to a reservoir, as in Figure 14.27. The cost, C, (in millions of dollars) will depend on the diameter, d, of the pipe (in meters) and the number, n, of pumping stations, according to the following formula[6]:

$$C = 0.15n + 3 \left(\frac{4d}{5} \right)^{-4.87} + \left(\frac{4d}{5} \right)^{1.8} + 3 \left(\frac{4d}{5} \right)^{1.8} n^{-1}.$$

Using the gradient search method, find the optimal number of pumping stations and pipe diameter.

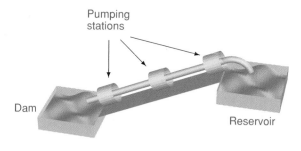

Figure 14.27

23.   Consider the function given by $f(x, y) = x^2(y + 1)^3 + y^2$. Show that f has only one critical point, namely $(0, 0)$, and that point is a local minimum but not a global minimum. Contrast this with the case of a function with a single local minimum in one-variable calculus.

14.3   CONSTRAINED OPTIMIZATION: LAGRANGE MULTIPLIERS

Many, perhaps most, real optimization problems are constrained by external circumstances. For example, a city wanting to build a public transportation system has only a limited number of tax dollars it can spend on the project. In this section, we will see how to find an optimum value under such constraints.

Graphical Approach: Maximizing Production Subject to a Budget Constraint

Suppose we want to maximize the production of a firm under a budget constraint. Suppose production, f, is a function of two variables, x and y, which are quantities of two raw materials, and that

$$f(x, y) = x^{2/3} y^{1/3}.$$

If x and y are purchased at prices of p_1 and p_2 thousands of dollars per unit, what is the maximum production f that can be obtained with a budget of c thousand dollars?

[6]From Douglass J. Wilde, *Globally Optimal Design*, (New York: John Wiley & Sons, 1978).

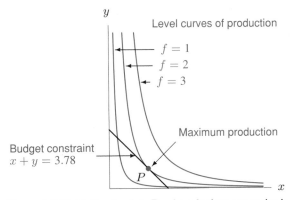

Figure 14.28: Optimal point, P, where budget constraint is
tangent to a level of production function

To maximize f without regard to the budget, we simply increase x and y. However, the budget constraint prevents us from increasing x and y beyond a certain point. Exactly how does the budget constrain us? With prices of p_1 and p_2, the amount spent on x is $p_1 x$ and the amount spent on y is $p_2 y$, so we must have

$$g(x, y) = p_1 x + p_2 y \leq c,$$

where $g(x, y)$ is the total cost of the raw materials x and y and c is the budget in thousands of dollars. Let's look at the case when $p_1 = p_2 = 1$ and $c = 3.78$. Then

$$x + y \leq 3.78.$$

Figure 14.28 shows some contours of f and the budget constraint represented by the line $x + y = 3.78$. Any point on or below the line represents a pair of values of x and y that we can afford. A point on the line completely exhausts the budget, a point below the line represents values of x and y which can be bought without using up the budget. Any point above the line represents a pair of values that we cannot afford. To maximize f, we find the point which lies on the level curve with the largest possible value of f *and* which lies within the budget. The point must lie on the budget constraint because we should spend all the available money. Unless we are at the point where the budget constraint is tangent to the contour $f = 2$, we can increase f by moving along the line representing the budget constraint in Figure 14.28. For example, if we are on the line to the left of the point of tangency, moving right will increase f; if we are on the line to the right of the point of tangency, moving left will increase f. Thus, the maximum value of f on the budget constraint occurs at the point where the budget constraint is tangent to the contour $f = 2$.

Analytical Solution: Lagrange Multipliers

We know that maximum production is achieved at the point where the budget constraint is tangent to a level curve of the production function. The method of Lagrange multipliers uses this fact in algebraic form. Figure 14.29 shows that at the optimum point, P, the gradient of f and the normal to the budget line $g(x, y) = 3.78$ are parallel. Thus, at P, grad f and grad g are parallel, so for some scalar λ, called the *Lagrange multiplier*

$$\text{grad } f = \lambda \text{ grad } g$$

Since grad $f = \left(\dfrac{2}{3}x^{-1/3}y^{1/3}\right) \vec{i} + \left(\dfrac{1}{3}x^{2/3}y^{-2/3}\right) \vec{j}$ and grad $g = \vec{i} + \vec{j}$, we have, by equating components,

$$\frac{2}{3}x^{-1/3}y^{1/3} = \lambda \quad \text{and} \quad \frac{1}{3}x^{2/3}y^{-2/3} = \lambda.$$

Eliminating λ gives

$$\frac{2}{3}x^{-1/3}y^{1/3} = \frac{1}{3}x^{2/3}y^{-2/3}, \quad \text{which leads to} \quad 2y = x.$$

Since we must also satisfy the constraint $x + y = 3.78$, we have $x = 2.52$ and $y = 1.26$. For these values,

$$f(2.52, 1.26) = (2.52)^{2/3}(1.26)^{1/3} \approx 2.$$

Thus, as before, we see that the maximum value of f is approximately 2; we also learn that this maximum occurs at $x = 2.52$ and $y = 1.26$.

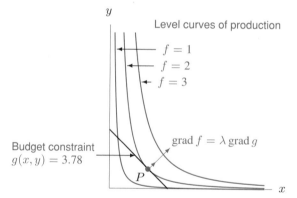

Figure 14.29: At the point, P, of maximum production, the vectors grad f and grad g are parallel

Lagrange Multipliers in General

Suppose we want to optimize an *objective function* $f(x, y)$ subject to a *constraint* $g(x, y) = c$. We consider only those points which satisfy the constraint and look for extrema among them. We make the following definition.

Suppose P_0 is a point satisfying the constraint $g(x, y) = c$.

- f has a **local maximum** at P_0 **subject to the constraint** if $f(P_0) \geq f(P)$ for all points P near P_0 satisfying the constraint.

- f has a **global maximum** at P_0 **subject to the constraint** if $f(P_0) \geq f(P)$ for all points P satisfying the constraint.

Local and global minima are defined similarly.

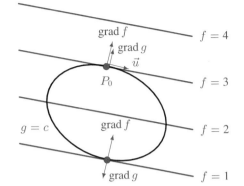

Figure 14.30: Maximum and minimum values
of $f(x, y)$ on $g(x, y) = c$ are at points where
grad f is parallel to grad g

As we saw in the production example, constrained extrema occur at points where the contours of f are tangent to the contours of g; they can also occur at endpoints of the constraint. We sometimes find these extrema by substituting from the constraint into the objective function. However, the method of Lagrange multipliers works when substitution is not possible.

At any point, grad f points in the direction in which f increases most rapidly. Suppose $\vec{u}$ is a unit vector tangent to the constraint. If grad $f \cdot \vec{u} > 0$, then the directional derivative, $f_{\vec{u}}$, is positive and moving in the direction of $\vec{u}$ increases f. If grad $f \cdot \vec{u} < 0$, then $f_{\vec{u}}$ is negative and moving in the direction of $-\vec{u}$ increases f. Thus, at a point P_0 where f has a constrained local maximum, we must have grad $f \cdot \vec{u} = 0$. Therefore, at P_0, both grad f and grad g are perpendicular to $\vec{u}$ so grad f and grad g are parallel. (See Figure 14.30.) Thus, provided grad $g \neq \vec{0}$ at P_0, we can use the following method:

To optimize f subject to the constraint $g = c$, solve the equations

$$\text{grad } f = \lambda \text{ grad } g \quad \text{and} \quad g = c$$

where λ is called the **Lagrange multiplier**.

If f and g are functions of two variables, the Lagrange method gives us three equations for three unknowns, x, y, λ:

$$f_x = \lambda g_x, \quad f_y = \lambda g_y, \quad g(x, y) = c.$$

If f and g are functions of three variables, the Lagrange method gives us four equations for four unknowns, x, y, z, λ:

$$f_x = \lambda g_x, \quad f_y = \lambda g_y, \quad f_z = \lambda g_z, \quad g(x, y, z) = c$$

Example 1   Find the maximum and minimum values of $x + y$ on the circle $x^2 + y^2 = 4$.

Solution   The objective function is

$$f(x, y) = x + y,$$

and the constraint is

$$g(x, y) = x^2 + y^2 = 4.$$

Since grad $f = f_x \vec{i} + f_y \vec{j} = \vec{i} + \vec{j}$ and grad $g = g_x \vec{i} + g_y \vec{j} = 2x\vec{i} + 2y\vec{j}$, then grad $f = \lambda$ grad g

$$(\vec{i} + \vec{j}) = \lambda(2x + 2y)$$

gives

$$1 = 2\lambda x; \qquad\qquad 0 = x - y$$
$$1 = 2\lambda y; \qquad\qquad x = y$$

so

$$x = y.$$

We also know that

$$x^2 + y^2 = 4,$$

giving $x = y = \sqrt{2}$ or $x = y = -\sqrt{2}$.

Since $f(x, y) = x + y$, the maximum value of f is $f(\sqrt{2}, \sqrt{2}) = 2\sqrt{2}$, and occurs when $x = y = \sqrt{2}$; the minimum value is $f(-\sqrt{2}, -\sqrt{2}) = -2\sqrt{2}$, and occurs when $x = y = -\sqrt{2}$. (See Figure 14.31.)

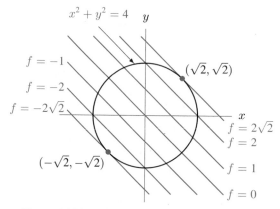

Figure 14.31: Maximum and minimum values of $f(x, y) = x + y$ on the circle $x^2 + y^2 = 4$ are at points where contours of f are tangent to the circle

How to Distinguish Maxima from Minima

There is a second derivative test[7] for classifying the critical points of constrained optimization problems, but it is more complicated than the test in Section 14.1. However, as you see from the examples, a graph of the constraint and some level curves can usually make it clear which points are maxima, which points are minima, and which are neither.

The Meaning of λ in the Production Example

In our previous examples, we never found (or needed) the value of λ. However, λ does have a practical interpretation.

Let's look back at the production problem where we wanted to maximize

$$f(x, y) = x^{2/3} y^{1/3}$$

subject to the constraint

$$g(x, y) = x + y = 3.78.$$

We solved the equations

$$\frac{2}{3} x^{-1/3} y^{1/3} = \lambda,$$

$$\frac{1}{3} x^{2/3} y^{-2/3} = \lambda,$$

$$x + y = 3.78,$$

---

[7]See J. E. Marsden and A. J. Tromba, *Vector Calculus*, 2nd ed., pp. 224–230 (San Francisco: W.H. Freeman, 1981).

to get $x = 2.52, y = 1.26$ and $f(2.52, 1.26) \approx 2$. Continuing to find λ gives us

$$\lambda \approx 0.53.$$

Suppose now we did another, apparently unrelated calculation. Suppose our budget is increased slightly, from 3.78 to 4.78, giving a new budget constraint of $x + y = 4.78$. Then the corresponding solution is at $x = 3.19$ and $y = 1.59$ and the new maximum value (instead of $f = 2$) is

$$f = (3.19)^{2/3}(1.59)^{1/3} \approx 2.53.$$

Notice that the amount by which f has increased is 0.53, the value of λ. Thus, in this example, the value of λ represents the extra production achieved by increasing the budget by one — in other words, the extra "bang" you get for an extra "buck" of budget. In fact, this is true in general:

- The value of λ is approximately the increase in the optimum value of f when the budget is increased by 1 unit.

More precisely:

- The value of λ represents the rate of change of the optimum value of f as the budget increases.

The Meaning of λ in General

To interpret λ, we look at how the optimum value of the objective function f changes as the value c of the constraint function g is varied. The optimum point (x_0, y_0) will, in general, depend on the constraint value c. So, provided x_0 and y_0 are differentiable functions of c, we can use the chain rule to differentiate the optimum value $f(x_0(c), y_0(c))$ with respect to c:

$$\frac{df}{dc} = \frac{\partial f}{\partial x}\frac{dx_0}{dc} + \frac{\partial f}{\partial y}\frac{dy_0}{dc}.$$

At the optimum point (x_0, y_0), we have $f_x = \lambda g_x$ and $f_y = \lambda g_y$, and therefore

$$\frac{df}{dc} = \lambda \left(\frac{\partial g}{\partial x}\frac{dx_0}{dc} + \frac{\partial g}{\partial y}\frac{dy_0}{dc} \right) = \lambda \frac{dg}{dc}.$$

But, as $g(x_0(c), y_0(c)) = c$, we see that $dg/dc = 1$, and so $df/dc = \lambda$. Thus, we have the following interpretation of the Lagrange multiplier λ:

The value of λ is the rate of change of the optimum value of f as c increases (where $g(x, y) = c$). If the optimum value of f is written as $f(x_0(c), y_0(c))$, then we have

$$\frac{d}{dc} f(x_0(c), y_0(c)) = \lambda.$$

Example 2   Suppose the quantity of goods produced according to the function $f(x, y) = x^{2/3}y^{1/3}$ is maximized subject to the budget constraint $x + y \leq 3.78$. Suppose the budget is increased to allow for a small increase in production. What price must the product sell for if it is to be worth the increased budget for its production?

Solution   We know that $\lambda = 0.53$, which tells us that $df/dc = 0.53$. Therefore increasing the budget by $1 increases production by about 0.53 unit. In order to make the increase in budget profitable, the extra goods produced must sell for more than $1. Thus, if p is the price of each unit of the good, then $0.53p$ is the revenue from the extra 0.53 units sold. Thus, we need $0.53p \geq 1$ so $p \geq 1/0.53 = \$1.89$.

Optimization with Inequality Constraints

The production problem that we looked at first was to maximize production $f(x, y)$ subject to a budget constraint

$$g(x, y) = p_1 x + p_2 y \leq c.$$

This budget constraint is an inequality constraint, which restricts (x, y) to a region of the plane rather than to a curve in the plane. In principle, we should first check to see whether or not $f(x, y)$ has any critical points in the interior defined by

$$p_1 x + p_2 y < c.$$

However, in the case of a budget constraint, we can see that the maximum of f must occur when the budget is exhausted, and so we look for the maximum value of f on the boundary line:

$$p_1 x + p_2 y = c.$$

Strategy for Optimizing $f(x, y)$ Subject to the Constraint $g(x, y) \leq c$

- Find all points in the interior $g(x, y) < c$ where grad f is zero or undefined.
- Use Lagrange multipliers to find the local extrema of f on the boundary $g(x, y) = c$.
- Evaluate f at the points found in the previous two steps and compare the values.

From Section 14.2 we know that if f is continuous on a closed and bounded region, R, then f is guaranteed to attain its global maximum and minimum values on R.

Example 3   Find the maximum and minimum values of $f(x, y) = (x - 1)^2 + (y - 2)^2$ subject to the constraint $x^2 + y^2 \leq 45$.

Solution   First, we look for all critical points for f in the interior of the region. Setting

$$f_x = 2(x - 1) = 0$$
$$f_y = 2(y - 2) = 0$$

we find f has exactly one critical point at $x = 1$, $y = 2$. Since $1^2 + 2^2 < 45$, that critical point is in the interior of the region.

Next, we find the local extrema of f on the boundary curve $x^2 + y^2 = 45$. To do this, we use Lagrange multipliers with constraint $g(x, y) = x^2 + y^2 = 45$. Setting grad $f = \lambda$ grad g, we get

$$2(x - 1) = \lambda \cdot 2x,$$
$$2(y - 2) = \lambda \cdot 2y.$$

If $\lambda = 0$, then $x = 1$, $y = 2$, the interior critical point. Thus, on the boundary we have

$$\frac{x}{x - 1} = \frac{y}{y - 2}$$

and so

$$y = 2x.$$

Combining this with the constraint $x^2 + y^2 = 45$, we get

$$5x^2 = 45$$

so

$$x = \pm 3.$$

Since $y = 2x$, we have possible local extrema at $x = 3$, $y = 6$ and $x = -3$, $y = -6$.

We conclude that the only candidates for the maximum and minimum values of f in the region occur at $(1, 2)$, $(3, 6)$, and $(-3, -6)$. Evaluating f at these three points we find

$$f(1, 2) = 0, \qquad f(3, 6) = 20, \qquad f(-3, -6) = 80.$$

Therefore, the minimum value of f is 0 at $(1, 2)$ and the maximum value is 80 at $(-3, -6)$.

Optimization Problems With Two Constraints

In applications we encounter optimization problems where the objective function f is a function of three or more variables and where there are two or more constraint functions. In this case the contours of f, g_1, and g_2 are surfaces. In order to optimize f subject to the constraints $g_1 = c_1$ and $g_2 = c_2$, we solve the system of equations,[8]

$$\operatorname{grad} f(x, y, z) = \lambda_1 \operatorname{grad} g_1(x, y, z) + \lambda_2 \operatorname{grad} g_2(x, y, z),$$
$$g_1(x, y, z) = c_1,$$
$$g_2(x, y, z) = c_2,$$

for the five unknowns $x, y, z, \lambda_1, \lambda_2$.

Example 4   The plane $x + y + z = 1$ cuts the cylinder $x^2 + y^2 = 2$ in a curve C. Find the points on C of minimum and maximum height above the xy-plane.

Solution   Since z is the distance of a point above the xy-plane, we want to maximize the objective function $f(x, y, z) = z$ subject to the constraints

$$g_1(x, y, z) = x^2 + y^2 = 2 \quad \text{and} \quad g_2(x, y, z) = x + y + z = 1.$$

We solve the equations $\operatorname{grad} f = \lambda_1 \operatorname{grad} g_1 + \lambda_2 \operatorname{grad} g_2$, and $g_1(x, y, z) = 2$ and $g_2(x, y, z) = 1$:

$$0 = 2\lambda_1 x + \lambda_2,$$
$$0 = 2\lambda_1 y + \lambda_2,$$
$$1 = \lambda_2,$$
$$x^2 + y^2 = 2,$$
$$x + y + z = 1.$$

From the first two equations, we get $x = y$. Using the third as well gives $x = y = -1/(2\lambda_1)$. Substituting these values for x and y in the fourth, we get

$$\lambda_1 = \pm 1/2.$$

This gives $x = y = \pm 1$ and the last equation gives $z = -1$ and $z = 3$. Therefore, $P_1 = (-1, -1, 3)$ is the point on C of maximum height above the xy-plane and $P_2 = (1, 1, -1)$ is the point of minimum height.

---

[8]A justification of the method of Lagrange multipliers requires the *implicit function theorem*. See, for example, J. E. Marsden and M. H. Hoffman, *Elementary Classical Analysis*, 2nd ed. (New York: W. H. Freeman, 1993).

The Lagrangian Function

Constrained optimization problems are frequently solved using a *Lagrangian function*, $\mathcal{L}$. For example, to optimize the function $f(x, y)$ subject to the constraint $g(x, y) = c$, we use the Lagrangian function

$$\mathcal{L}(x, y, \lambda) = f(x, y) - \lambda(g(x, y) - c).$$

To see why the function $\mathcal{L}$ is useful, compute the partial derivatives of $\mathcal{L}$:

$$\frac{\partial \mathcal{L}}{\partial x} = \frac{\partial f}{\partial x} - \lambda \frac{\partial g}{\partial x},$$

$$\frac{\partial \mathcal{L}}{\partial y} = \frac{\partial f}{\partial y} - \lambda \frac{\partial g}{\partial y},$$

$$\frac{\partial \mathcal{L}}{\partial \lambda} = -(g(x, y) - c).$$

Notice that if (x_0, y_0) is a critical point of $f(x, y)$ subject to the constraint $g(x, y) = c$ and λ_0 is the corresponding Lagrange multiplier, then at the point (x_0, y_0, λ_0) we have

$$\frac{\partial \mathcal{L}}{\partial x} = 0 \quad \text{and} \quad \frac{\partial \mathcal{L}}{\partial y} = 0 \quad \text{and} \quad \frac{\partial \mathcal{L}}{\partial \lambda} = 0.$$

In other words, (x_0, y_0, λ_0) is a critical point for the unconstrained problem of optimization of the Lagrangian, $\mathcal{L}(x, y, \lambda)$. Thus, the Lagrangian enables us to convert a constrained optimization problem to an unconstrained problem.

The Lagrangian used to optimize the function $f(x, y, z)$ subject to two constraints $g_1(x, y, z) = c_1$ and $g_2(x, y, z) = c_2$ is

$$\mathcal{L}(x, y, z, \lambda_1, \lambda_2) = f(x, y, z) - \lambda_1(g_1(x, y, z) - c_1) - \lambda_2(g_2(x, y, z) - c_2).$$

Example 5   A company has a production function with three inputs x, y, and z given by

$$f(x, y, z) = 50x^{2/5}y^{1/5}z^{1/5}.$$

The total budget is \$24,000 and the company can buy x, y, and z at \$80, \$12, and \$10 per unit, respectively. What combination of inputs will maximize production? [9]

Solution   We need to maximize the objective function

$$f(x, y, z) = 50x^{2/5}y^{1/5}z^{1/5},$$

subject to the constraint

$$g(x, y, z) = 80x + 12y + 10z = 24{,}000.$$

Therefore, the Lagrangian function is

$$\mathcal{L}(x, y, z) = 50x^{2/5}y^{1/5}z^{1/5} - \lambda(80x + 12y + 10z - 24{,}000),$$

and so we look for solutions to the system of equations we get from grad $\mathcal{L} = 0$:

$$\frac{\partial \mathcal{L}}{\partial x} = 20x^{-3/5}y^{1/5}z^{1/5} - 80\lambda = 0,$$

$$\frac{\partial \mathcal{L}}{\partial y} = 10x^{2/5}y^{-4/5}z^{1/5} - 12\lambda = 0,$$

$$\frac{\partial \mathcal{L}}{\partial z} = 10x^{2/5}y^{1/5}z^{-4/5} - 10\lambda = 0,$$

$$\frac{\partial \mathcal{L}}{\partial \lambda} = -(80x + 12y + 10z - 24{,}000) = 0.$$

[9]Adapted from M. Rosser, *Basic Mathematics for Economists*, p. 363 (New York: Routledge, 1993).

We simplify this system to give

$$\lambda = \frac{1}{4}x^{-3/5}y^{1/5}z^{1/5},$$

$$\lambda = \frac{5}{6}x^{2/5}y^{-4/5}z^{1/5},$$

$$\lambda = x^{2/5}y^{1/5}z^{-4/5},$$

$$80x + 12y + 10z = 24{,}000.$$

Eliminating z from the first two equations gives $x = 0.3y$. Eliminating x from the second and third equations gives $z = 1.2y$. Substituting for x and z into $80x + 12y + 10z = 24{,}000$ gives

$$80(0.3y) + 12y + 10(1.2y) = 24{,}000,$$

so $y = 500$. Hence we get $x = 150$ and $z = 600$, and the corresponding value of f is $f(150, 500, 600) = 4{,}622$ units.

The graph of the constraint is a plane in 3-space. Since the inputs x, y, z must be nonnegative, the constraint is a triangle in the first quadrant, with edges on the coordinate planes. On the boundary of the triangle, one (or more) of the variables x, y, z is zero and so the function f is zero. Thus, $x = 150$, $y = 500$, $z = 600$ is a maximum.

Problems for Section 14.3

In Problems 1–17, use Lagrange multipliers to find the maximum and minimum values of $f(x, y)$ subject to the given constraints.

1. $f(x, y) = x + y, \quad x^2 + y^2 = 1$

2. $f(x, y) = 3x - 2y, \quad x^2 + 2y^2 = 44$

3. $f(x, y) = x^2 + y, \quad x^2 - y^2 = 1$

4. $f(x, y) = xy, \quad 4x^2 + y^2 = 8$

5. $f(x, y) = x^2 + y^2, \quad x^4 + y^4 = 2$

6. $f(x, y) = x^2 - xy + y^2, \quad x^2 - y^2 = 1$

7. $f(x, y, z) = x + 3y + 5z, \quad x^2 + y^2 + z^2 = 1$

8. $f(x, y, z) = 2x + y + 4z, \quad x^2 + y + z^2 = 16$

9. $f(x, y, z) = x^2 - y^2 - 2z, \quad x^2 + y^2 = z$

10. $f(x, y, z) = x^2 - 2y + 2z^2, \quad x^2 + y^2 + z^2 = 1$

11. $f(x, y, z) = x + y + z$, subject to $x^2 + y^2 + z^2 = 1$ and $x - y = 1$

12. $f(x, y) = x^2 + 2y^2, \quad x^2 + y^2 \le 4$

13. $f(x, y) = xy, \quad x^2 + 2y^2 \le 1$

14. $f(x, y) = x^2 - y^2, \quad x^2 \ge y$

15. $f(x, y) = x + 3y, \quad x^2 + y^2 \le 2$

16. $f(x, y) = x^3 + y, \quad x + y \ge 1$

17. $f(x, y) = x^3 - y^2, \quad x^2 + y^2 \le 1$

18. A company manufactures a product using inputs x, y, and z according to the production function

$$Q(x, y, z) = 20x^{1/2}y^{1/4}z^{2/5}.$$

The prices per unit are \$20 for x, and \$10 for y, and \$5 for z. What quantity of each input should the company use in order to manufacture 1,200 products at minimum cost?[10]

---

[10]Adapted from M. Rosser, *Basic Mathematics for Economists*, p.363 (New York: Routledge, 1993).

19. Consider a firm which manufactures a commodity at two different factories. The total cost of manufacturing depends on the quantities, q_1 and q_2, supplied by each factory, and is expressed by the *joint cost function*, $C = f(q_1, q_2)$. Suppose the joint cost function is approximated by

$$f(q_1, q_2) = 2q_1^2 + q_1 q_2 + q_2^2 + 500$$

and that the company's objective is to produce 200 units, at the same time minimizing production costs. How many units should be supplied by each factory?

20. An industry manufactures a product from two raw materials. The quantity produced, Q, can be given by the Cobb-Douglas function:

$$Q = cx^a y^b,$$

where x and y are quantities of each of the two raw materials used and a, b, and c are positive constants. Suppose the first raw material costs $\$P_1$ per unit and the second costs $\$P_2$ per unit. Find the maximum production possible if no more than $\$K$ can be spent on raw materials.

21. Each person tries to balance his or her time between leisure and work. The tradeoff is that as you work less your income falls. Therefore each person has *indifference curves* which connect the number of hours of leisure, l, and income, s. If, for example, you are indifferent between 0 hours of leisure and an income of $\$1125$ a week on the one hand, and 10 hours of leisure and an income of $\$750$ a week on the other hand, then the points $l = 0$, $s = 1125$, and $l = 10$, $s = 750$ both lie in the same indifference curve. Table 14.4 gives information on three indifference curves, I, II, and III.

TABLE 14.4

Weekly Income			Weekly Leisure Hours		
I	II	III	I	II	III
1125	1250	1375	0	20	40
750	875	1000	10	30	50
500	625	750	20	40	60
375	500	625	30	50	70
250	375	500	50	70	90

(a) Sketch the three indifference curves on graph paper.

(b) Suppose you have 100 hours a week available for work and leisure combined, and that you earn $\$10$/hour. Write an equation in terms of l and s which represents this constraint.

(c) On the same graph paper, sketch a graph of this constraint.

(d) Estimate from the graph what combination of leisure hours and income you would choose under these circumstances. Give the corresponding number of hours per week you would work. Explain how you made this estimate.

22. Figure 14.32 shows ∇f for a function $f(x, y)$ and two curves $g(x, y) = 1$ and $g(x, y) = 2$. Notice that $g = 1$ is the inside curve and $g = 2$ is the outside curve. Mark the following points on a copy of the figure.

(a) The point(s) A where f has a local maximum.

(b) The point(s) B where f has a saddle point.

(c) The point C where f has a maximum on $g = 1$.

(d)   The point D where f has a minimum on $g = 1$.

(e)   If you used Lagrange multipliers to find C, what would the sign of λ be? Why?

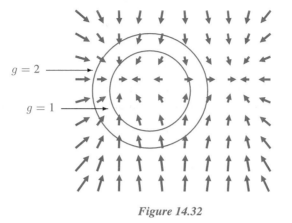

Figure 14.32

23.   Design a closed cylindrical container which holds 100 cm^3 and has the minimal possible surface area. What should its dimensions be?

24.   A company manufactures x units of one item and y units of another. The total cost in dollars, C, of producing these two items is approximated by the function

$$C = 5x^2 + 2xy + 3y^2 + 800.$$

(a)   If the production quota for the total number of items (both types combined) is 39, find the minimum production cost.

(b)   Estimate the additional production cost or savings if the production quota is raised to 40 or lowered to 38.

25.   A mountain climber at the summit of a mountain wants to descend to a lower altitude as fast as possible. Suppose the altitude of the mountain is given approximately by

$$h(x, y) = 3000 - \frac{1}{10000}(5x^2 + 4xy + 2y^2) \quad \text{meters},$$

where x, y are horizontal coordinates on the earth (in meters), with the mountain summit located above the origin. In thirty minutes, the climber can reach any point (x, y) on a circle of radius 1000 m. In which direction should she travel in order to descend as far as possible?

26.   Suppose the quantity, q, of a product manufactured depends on the number of workers, W, and the amount of capital invested, K, and is represented by the Cobb-Douglas function

$$q = 6W^{3/4}K^{1/4}.$$

In addition, labor costs are \$10 per worker and capital costs are \$20 per unit, and the budget is \$3000.

(a)   What are the optimum number of workers and the optimum number of units of capital?

(b)   Show that at the optimum values of W and K, the ratio of the marginal productivity of labor $(\partial q/\partial W)$ to the marginal productivity of capital $(\partial q/\partial K)$ is the same as the ratio of the cost of a unit of labor to the cost of a unit of capital.

(c)   Recompute the optimum values of W and K when the budget is increased by one dollar. Check that increasing the budget by \$1 allows the production of λ extra units of the good, where λ is the Lagrange multiplier.

27. The director of a neighborhood health clinic has an annual budget of $600,000. He wants to allocate his budget so as to maximize the number of patient visits, V, which is given as a function of the number of doctors, D, and the number of nurses, N, by

$$V = 1000D^{0.6}N^{0.3}.$$

Doctors receive a salary of $40,000, while nurses get $10,000.

(a) Set up the director's constrained optimization problem.
(b) Describe, in words, the conditions which must be satisfied by $\partial V/\partial D$ and $\partial V/\partial N$ for V to have an optimum value.
(c) Solve the problem formulated in part (a).
(d) Find the value of the Lagrange multiplier and interpret its meaning in this problem.
(e) At the optimum point, what is the marginal cost of a patient visit (that is, the cost of an additional visit)? Will that marginal cost rise or fall with the number of visits? Why?

28. Minimize

$$f(x, y, z) = \sqrt{(x-a)^2 + (y-b)^2 + (z-c)^2},$$

subject to the constraint $Ax + By + Cz + D = 0$. What is the geometric meaning of your solution?

29. Let $f(x, y)$ be a linear function, so that $f(x, y) = ax + by + c$ where a, b and c are constants, and let R be a region in the xy-plane.

(a) If R is any disk, show that the maximum and minimum values of f on R occur on the boundary of the disk.
(b) If R is any rectangle, show that the maximum and minimum values of f on R occur at the corners of the rectangle. They may occur at other points of the rectangle as well.
(c) Explain, with the aid of a graph of the plane $z = f(x, y)$, why you expect the answers you obtained in parts (a) and (b).

30. (a) In Problem 26, does the value of λ change if the budget changes from $3000 to $4000?
(b) In Problem 27, does the value of λ change if the budget changes from $600,000 to $700,000?
(c) What condition must a Cobb-Douglas production function

$$Q = cK^aL^b$$

satisfy to ensure that the marginal increase of production (that is, the rate of increase of production with budget) is not affected by the size of the budget?

REVIEW PROBLEMS FOR CHAPTER FOURTEEN

1. Find the local maxima, minima, and saddle points of the function
$$f(x, y) = \sin x + \sin y + \sin(x + y), \quad 0 < x < \pi, \quad 0 < y < \pi.$$

For Problems 2–4, find the local maxima, local minima, and saddle points of the functions given. Decide if the local maxima or minima are global maxima or minima. Explain.

2. $f(x, y) = x^2 + y^3 - 3xy$

3. $f(x, y) = xy + \ln x + y^2 - 10$ $(x > 0)$

4. $f(x, y) = x + y + \dfrac{1}{x} + \dfrac{4}{y}$

5. Suppose $f_x = f_y = 0$ at $(1, 3)$ and $f_{xx} < 0$, $f_{yy} < 0$, $f_{xy} = 0$. Draw a possible contour diagram.

6. Find the least squares line for the data points $(0, 4)$, $(1, 3)$, $(2, 1)$.

7. Find the minimum and maximum of the function $z = 4x^2 - xy + 4y^2$ over the closed disk $x^2 + y^2 \leq 2$.

8. What are the maximum and minimum values of $f(x, y) = -3x^2 - 2y^2 + 20xy$ on the line $x + y = 100$?

9. A company sells two products which are partial substitutes for each other, such as coffee and tea. If the price of one product rises, then the demand for the other product rises. The quantities demanded, q_1 and q_2, are given as a function of the prices, p_1 and p_2, by

$$q_1 = 517 - 3.5p_1 + 0.8p_2 \quad \text{and} \quad q_2 = 770 - 4.4p_2 + 1.4p_1.$$

What prices should the company charge in order to maximize the total sales revenue? [11]

10. A biological rule of thumb states that as the area A of an island increases tenfold, the number of animal species, N, living on it doubles. Table 14.5 shows the area (in square km) of several islands in the West Indies and the number of species living on each one. Assume that N is a power function of A. Using the biological rule of thumb, find

(a) N as a function of A.

(b) $\ln N$ as a function of $\ln A$.

(c)   Using the data given, tabulate $\ln N$ against $\ln A$ and find the line of best fit. Does your answer agree with the biological rule of thumb?

TABLE 14.5   *Number of species on various islands*

Island	Area (sq km)	Number
Redonda	3	5
Saba	20	9
Montserrat	192	15
Puerto Rico	8858	75
Jamaica	10854	70
Hispaniola (Haiti & Dominican Rep.)	75571	130
Cuba	113715	125

11. Suppose that the quantity, Q, manufactured of a certain product depends on the quantity of labor, L, and of capital, K, used according to the function

$$Q = 900L^{1/2}K^{2/3}.$$

Suppose that labor costs $100 per unit and that capital costs $200 per unit. What combination

---

[11] Adapted from M. Rosser, *Basic Mathematics for Economists*, p. 318 (New York: Routledge, 1993).

of labor and capital should be used to produce 36,000 units of the goods at minimum cost? What is that minimum cost?

12. An international organization must decide how to spend the $2000 they have been allotted for famine relief in a remote area. They expect to divide the money between buying rice at $5/sack and beans at $10/sack. The number, P, of people who would be fed if they buy x sacks of rice and y sacks of beans is given by

$$P = x + 2y + \frac{x^2 y^2}{2 \cdot 10^8}.$$

What is the maximum number of people that can be fed, and how should the organization allocate its money?

13. The quantity, Q, of a product manufactured by a company is given by

$$Q = aK^{0.6}L^{0.4},$$

where a is a positive constant, K is the quantity of capital and L is the quantity of labor used. Capital costs are $20 per unit, labor costs are $10 per unit, and the company wants costs for capital and labor combined to be no higher than $150. Suppose you are asked to consult for the company, and learn that 5 units each of capital and labor are being used.

   (a) What do you advise? Should the plant use more or less labor? More or less capital? If so, by how much?

   (b) Write a one sentence summary that could be used to sell your advice to the board of directors.

14. A doctor wants to schedule visits for two patients who have been operated on for tumors so as to minimize the expected delay in detecting a new tumor. Visits for patients 1 and 2 are scheduled at intervals of x_1 and x_2 weeks, respectively. A total of m visits per week is available for both patients combined.

   The recurrence rates for tumors for patients 1 and 2 are judged to be v_1 and v_2 tumors per week, respectively. Thus, $v_1/(v_1 + v_2)$ and $v_2/(v_1 + v_2)$ are the probabilities that patient 1 and patient 2, respectively, will have the next tumor. It is known that the expected delay in detecting a tumor for a patient checked every x weeks is $x/2$. Hence, the expected detection delay for both patients combined is given by[12]

$$f(x_1, x_2) = \frac{v_1}{v_1 + v_2} \cdot \frac{x_1}{2} + \frac{v_2}{v_1 + v_2} \cdot \frac{x_2}{2}.$$

Find the values of x_1 and x_2 in terms of v_1 and v_2 that minimize $f(x_1, x_2)$ subject to the fact that m, the number of visits per week, is fixed.

15. What is the value of the Lagrange multiplier in Problem 14? What are the units of λ? What is its practical significance to the doctor?

16. The Cobb-Douglas equation models the total quantity, q, of a commodity produced as a function of the number of workers, W, and the amount of capital invested, K, by the production function

$$q = cW^{1-a}K^a$$

where a and c are positive constants. Assume labor costs are $\$p_1$ per worker, capital costs are $\$p_2$ per unit, and there is a fixed budget of $\$b$. Show that when W and K are at their optimal levels, the ratio of marginal productivity of labor to marginal productivity of capital equals the ratio of the cost of one unit of labor to one unit of capital.

---

[12]Adapted from Daniel Kent, Ross Shachter, *et al.*, Efficient Scheduling of Cystoscopies in Monitoring for Recurrent Bladder Cancer in *Medical Decision Making* (Philadelphia: Hanley and Belfus, 1989).

17. An irrigation canal has a trapezoidal cross-section of area 50 m^2, as in Figure 14.33. The average flow rate in the canal is inversely proportional to the wetted perimeter, p, of the canal, that is, to the perimeter of the trapezoid in Figure 14.33, excluding the top. Thus, to maximize the flow rate we must minimize p. Find the depth d, base width w, and angle θ that give the maximum flow rate.[13]

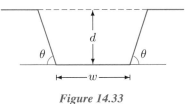

Figure 14.33

18. The energy required to compress a gas from pressure p_1 to pressure p_{N+1} in N stages is proportional to

$$E = \left(\frac{p_2}{p_1}\right)^2 + \left(\frac{p_3}{p_2}\right)^2 + \cdots + \left(\frac{p_{N+1}}{p_N}\right)^2 - N.$$

Show how to choose the intermediate pressures $p_2, \ldots, p_N$ so as to minimize the energy requirement. [14]

19. A family wants to move to a house at a point that is better situated with respect to the children's school and both the parents' places of work. See Figure 14.34.

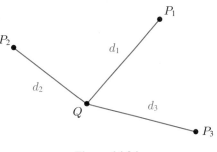

Figure 14.34

Currently they live at Q, the school is at P_1, the mother's work is at P_2, and the father's work is at P_3. They want to minimize $d = d_1 + d_2 + d_3$, where

$$d_1 = \text{distance to school,}$$
$$d_2 = \text{distance to mother's work,}$$
$$d_3 = \text{distance to father's work.}$$

(a) Show that grad d_i is a unit vector pointing directly away from P_i, for $i = 1, 2, 3$. [Hint: Draw contours of d_i, paying careful attention to the spacing of the contours.]

(b) Use your answer to part (a) to draw grad d at the point Q in Figure 14.34. In what direction should the family move to decrease d?

(c) Find as best you can the point on the diagram where grad $d = 0$. What geometric condition characterizes this location?

---

[13] Adapted from Robert M. Stark and Robert L. Nichols, *Mathematical Foundations of Design: Civil Engineering Systems*, (New York: McGraw-Hill, 1972).

[14] Adapted from Rutherford, Aris, *Discrete Dynamic Programming*, p. 35 (New York: Blaisdell, 1964).

20. Consider the function $f(x, y) = 2x^2 - 3xy + 8y^2 + x - y$.

    (a) Compute the critical points of f and classify them.
    (b) By completing the square, plot the contour diagram of f and show that the local extremum found in part (a) is a global one.
    (c) Starting at the point $(1, 1)$, minimize f using the gradient search method and compare your answer after two iterations with your answer from part (a).

21. A light ray crossing the boundary between two different media (for example, vacuum and glass, or air and water) undergoes a change in direction or is *refracted* by an amount which depends on the properties of the media. Suppose that a light ray travels from point A to point B, as shown in Figure 14.35, with velocity v_1 in medium 1 and velocity v_2 in medium 2.

    (a) Find the time $T(\theta_1, \theta_2)$ taken for the ray to travel from A to B in terms of the angles θ_1, θ_2 and the constants a, b, v_1, and v_2.
    (b) Show that the angles θ_1, θ_2 satisfy the constraint

$$a \tan \theta_1 + b \tan \theta_2 = d.$$

    (c) What is the effect on the time, T, of letting $\theta_1 \to -\pi/2$ (that is, moving R far to the left of A') or of letting $\theta_1 \to \pi/2$ (that is, moving R far to the right of B')?
    (d) *Fermat's Principle* states that the light ray follows a path such that the time taken T is minimized. Use the method of Lagrange multipliers to show that $T(\theta_1, \theta_2)$ is minimized when *Snell's Law of Refraction* holds:

$$\frac{\sin \theta_1}{\sin \theta_2} = \frac{v_1}{v_2}.$$

(The constant v_1/v_2 is called the *index of refraction* of medium 2 with respect to medium 1. For example, the indices of air, water, and glass with respect to a vacuum are approximately 1.0003, 1.33, and 1.52, respectively. The lenses of modern reading "glasses" are made of plastics with a high index of refraction in order to reduce weight and thickness when the prescription is strong.)

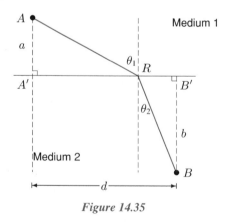

Figure 14.35

22. Twenty-six teams compete for the Stanley cup in hockey. At the beginning of the season an experienced fan estimates that the probability that team i will win is some number p_i, where $0 \le p_i \le 1$ and

$$\sum_{i=1}^{26} p_i = 1.$$

Exactly one team will actually win, so the probabilities have to add up to 1. If one of the teams, say team i, is certain to win then p_i is equal to 1 and all the other p_j must be equal to zero. Another extreme case occurs if all the teams are equally likely to win, so all the p_i are equal to $1/26$, and the outcome of the hockey season is completely unpredictable. Thus, the *uncertainty* in the outcome of the hockey season depends on the probabilities $p_1, \ldots, p_{26}$. In this problem we measure this uncertainty quantitatively using the following function:

$$S(p_1, \ldots, p_{26}) = - \sum_{i=1}^{26} p_i \frac{\ln p_i}{\ln 2}.$$

Note that as $p_i \le 1$, we have $-\ln p_i \ge 0$ and hence $S \ge 0$.

(a)   Show that $\lim_{p \to 0} p \ln p = 0$. (This means that S is a continuous function of the p_i, where $0 \le p_i \le 1$ and $1 \le i \le 26$, if we set $p \ln p|_{p=0}$ equal to zero. Since S is then a continuous function on a closed and bounded region, it attains a maximum and a minimum value on this region.)

(b)   Find the maximum value of $S(p_1, \ldots, p_{26})$ subject to the constraint $p_1 + \cdots + p_{26} = 1$. What are the values of p_i in this case? What does your answer mean in terms of the uncertainty in the outcome of the hockey season?

(c)   Find the minimum value of $S(p_1, \ldots, p_{26})$, subject to the constraint $p_1 + \cdots + p_{26} = 1$. What are the values of p_i in this case? What does your answer mean in terms of the uncertainty in the outcome of the hockey season?

[Note: The function S is an example of an *entropy* function; the concept of entropy is used in information theory, statistical mechanics, and thermodynamics when measuring the uncertainty in an experiment (the hockey season in this problem) or physical system.]

23.   Find the minimum distance from the point $(1, 2, 10)$ to the paraboloid given by the equation $z = x^2 + y^2$. Give a geometric justification for your answer.

24.   The cone $z^2 = x^2 + y^2$ is cut by the plane $z = 1 + x + y$. Find the points lying on the intersection of the cone and the plane which are nearest to and farthest from the origin. Give a geometric justification for your answer.

CHAPTER FIFTEEN

INTEGRATING FUNCTIONS OF MANY VARIABLES

A definite integral is the limit of a sum. We use a definite integral to calculate the total population of a region given the population density as a function of position. If the population density is a function of one variable only (for example, the distance from the center of a city), we have an ordinary one-variable definite integral. If the density depends on more than one variable, we need a multivariable integral to calculate the total population. In this chapter we develop double and triple integrals in Cartesian and polar coordinates.

15.1 THE DEFINITE INTEGRAL OF A FUNCTION OF TWO VARIABLES

In this section we see how to estimate total population from population density in the plane. This leads to the definition of the definite integral of a function of two variables.

Population Density of Foxes in England

The fox population in parts of England is important to public health officials concerned about the disease rabies, which is spread by animals. The contour diagram in Figure 15.1 shows the population density $D = f(x, y)$ of foxes in the southwestern corner of England, where x and y are in kilometers from the southwest corner of the map and D is in foxes per square kilometer.[1] The bold contour is the coastline (approximately), and may be thought of as the $D = 0$ contour; clearly the density is zero outside it.

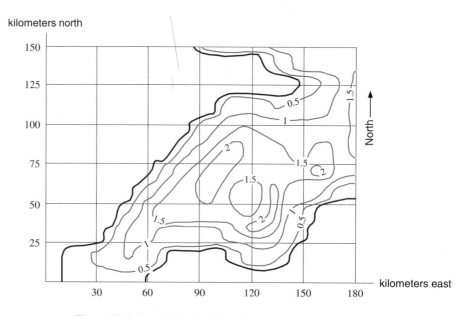

Figure 15.1: Population density of foxes in southwestern England

---

Example 1   Estimate the total fox population in the region represented by the map in Figure 15.1.

Solution   We want to find upper and lower bounds for the population. We subdivide the map into the 36 rectangles shown in Figure 15.1 and estimate the population in each rectangle. We find an upper bound for the population density in each rectangle, multiply it by the area of the rectangle to get an upper bound for the population in that rectangle, then add these upper bounds to get an upper bound for the total population. The bottom left rectangle includes a small region between the 0.5 and 1 contours, so we estimate the density of foxes in this rectangle to be at most 0.6 foxes per square kilometer. The next rectangle northwards is all sea and so there are no foxes. However, the rectangle

[1] Adapted from J. D. Murray, *Mathematical Biology*, Springer-Verlag, 1989

TABLE 15.1 *Upper estimates of the fox population density*

0	0	0	0.8	1.5	1.5
0	0	0.5	1.5	1.5	1.5
0	0	1.5	2.5	1.9	2.3
0	1.2	2.2	2.5	2.5	2.5
0	1.3	2	2.2	2.5	0.5
0.6	1.3	1	1	1.3	0

TABLE 15.2 *Lower estimates of the fox population density*

0	0	0	0	0	0.1
0	0	0	0	0	0.1
0	0	0	0.5	1.2	1.2
0	0	0	1	1	0
0	0	0.5	0.5	0	0
0	0	0	0	0	0

to the east includes a region between the contours labeled 1 and 1.5, so we estimate the maximum density to be 1.3 foxes per square kilometer. Continuing in this way we obtain the upper bounds in Table 15.1. Similarly, we obtain the lower bounds shown in Table 15.2.

Each rectangle has an area $30 \times 25 = 750 \text{ km}^2$, so we obtain

$$\text{Lower estimate} = (0.1 + 0.1 + 0.5 + 1.2 + 1.2 + 1 + 1 + 0.5 + 0.5) \cdot 750 = 4575 \quad \text{foxes.}$$

Similarly, we obtain an upper bound of

$$\text{Upper estimate} = 41.6 \cdot 750 = 31{,}200 \quad \text{foxes}$$

The average of our two estimates is about 18,000, so we take that as our estimate. Notice that there is a wide discrepancy between the upper and lower estimates; by taking finer subdivisions we could make the upper and lower estimates closer.

Definition of the Definite Integral

The sums used to approximate the fox population are similar to the Riemann sums used to define the definite integral of a function of one variable. We now define the definite integral for a function f of two variables on a rectangular region.[2] Given a continuous function $f(x, y)$ defined on a region $a \leq x \leq b$ and $c \leq y \leq d$, we construct a Riemann sum by subdividing the region into smaller rectangles. We do this by subdividing each of the intervals $a \leq x \leq b$ and $c \leq y \leq d$ into n and m equal subintervals respectively, giving nm subrectangles. (See Figure 15.2.)

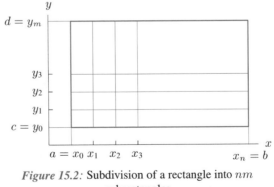

Figure 15.2: Subdivision of a rectangle into nm subrectangles

The area of each subrectangle is ΔA, where $\Delta A = \Delta x \, \Delta y$ and $\Delta x = (b - a)/n$ is the width of each subdivision along the x-axis, and $\Delta y = (d - c)/m$ is the width of each subdivision along the y-axis. To compute the Riemann sum, we multiply the area of each subrectangle by the value of the function at a point in the rectangle and add all the resulting numbers. Choosing the point which gives the maximum value, M_{ij}, of the function on each rectangle, we get the *upper sum*, $\sum_{i,j} M_{ij} \Delta x \Delta y$.

---

[2] A review of the one-variable definite integral is given in Appendix D.

The *lower sum,* $\sum_{i,j} L_{ij}\Delta x \Delta y$, is obtained by taking the minimum value on each rectangle. Thus, any other Riemann sum satisfies

$$\sum_{i,j} L_{ij}\Delta x \Delta y \leq \sum_{i,j} f(x_i, y_j)\, \Delta x\, \Delta y \leq \sum_{i,j} M_{ij}\Delta x \Delta y$$

where (x_i, y_i) is any point in the ij-th subrectangle. We define the definite integral by taking the limit as the numbers of subdivisions, n and m, tend to infinity. By comparing upper and lower sums, as we did for the fox population, it can be shown that the limit exists when the function, f, is continuous. We get the same limit by letting Δx and Δy tend to 0. Thus, we have the following definition:

Suppose the function f is continuous on R, the rectangle $a \leq x \leq b$, $c \leq y \leq d$. We define the **definite integral** of f over R

$$\int_R f\, dA = \lim_{\Delta x, \Delta y \to 0} \sum_{i,j} f(x_i, y_j)\Delta x \Delta y.$$

Such an integral is called a **double integral**.

Sometimes we think of dA as being the area of an infinitesimal rectangle of length dx and height dy, so that $dA = dx\, dy$. Then we use the notation[3]

$$\int_R f\, dA = \int_R f(x, y)\, dx\, dy.$$

The Riemann sum used in the definition, with equal-sized rectangular subdivisions, is just one type of Riemann sum. For a general Riemann sum, the subdivisions do not all have to be the same size.

Example 2   Let R be the rectangle $0 \leq x \leq 1$ and $0 \leq y \leq 1$. Use Riemann sums to estimate $\displaystyle\int_R e^{-(x^2+y^2)}\, dA$.

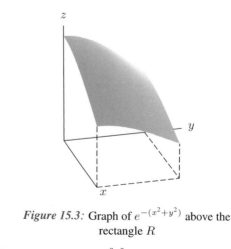

Figure 15.3: Graph of $e^{-(x^2+y^2)}$ above the rectangle R

---

[3]Another common notation for the double integral is $\int \int_R f\, dA$.

Solution   We divide R into 16 subrectangles by dividing each edge into 4 parts. Figure 15.3 shows that $f(x, y) = e^{-(x^2+y^2)}$ decreases as we move away from the origin. Thus, to get an upper sum we evaluate f on each subrectangle at the corner nearest the origin. For example, in the rectangle $0 \le x \le 0.25, 0 \le y \le 0.25$, we evaluate f at $(0, 0)$.

Using Table 15.3 we find that

$$
\begin{aligned}
\text{Upper sum} = [(\ 1\ +0.9394 + 0.7788 + 0.5698) \\
+(0.9394 + 0.8825 + 0.7316 + 0.5353) \\
+(0.7788 + 0.7316 + 0.6065 + 0.4437) \\
+(0.5698 + 0.5353 + 0.4437 + 0.3247)](0.0625) = 0.68.
\end{aligned}
$$

To get a lower sum, we must evaluate f at the opposite corner of each rectangle because the surface slopes down in both the x and y directions. This yields a lower sum of 0.44. Thus,

$$ 0.44 \le \int_R e^{-(x^2+y^2)}\, dA \le 0.68. $$

To get a better approximation, we compute lower and upper estimates with more subdivisions. The results for several cases with equal numbers of subdivisions in the x and y-directions are shown in Table 15.4.

TABLE 15.3   *Values of $f(x, y) = e^{-(x^2+y^2)}$ on the rectangle R*

		0.0	0.25	0.50	0.75	1.00
				y		
	0.0	1	0.9394	0.7788	0.5698	0.3679
	0.25	0.9394	0.8825	0.7316	0.5353	0.3456
x	0.50	0.7788	0.7316	0.6065	0.4437	0.2865
	0.75	0.5698	0.5353	0.4437	0.3247	0.2096
	1.00	0.3679	0.3456	0.2865	0.2096	0.1353

TABLE 15.4   *Riemann sum approximations to $\int_R e^{-(x^2+y^2)}\, dA$*

Number of subdivisions in x and y directions

	8	16	32	64
Upper	0.6168	0.5873	0.5725	0.5651
Lower	0.4989	0.5283	0.5430	0.5504

The true value of the double integral, $0.5577\ldots$, is trapped between the lower and upper sums. Notice that the lower sum increases and the upper sum decreases as the number of subdivisions grows. (Why is this?) However, even with 64 subdivisions, leading to $64^2 = 4096$ terms in the Riemann sum, the lower and upper sums agree with the true value of the integral only in the first decimal place. Hence, we should look for better ways to approximate the integral.

The Region R

In our definition of the definite integral $\int_R f(x, y)\, dA$, the region R is a rectangle. However, the definite integral can be defined for regions of other shapes, including triangles, circles, and regions bounded by the graphs of piecewise continuous functions.

To approximate the definite integral over a region, R, which is not rectangular, we use a grid of rectangles approximating the region. We obtain this grid by enclosing R in a large rectangle and subdividing that rectangle; we consider just the subrectangles which are inside R.

As before, we pick a point (x_i, y_j) in each subrectangle and form a Riemann sum

$$\sum_{i,j} f(x_i, y_j) \Delta x \Delta y.$$

This time, however, the sum is over only those subrectangles within R. For example, in the case of the fox population we can use the rectangles which are entirely on land. As the subdivisions become finer, the grid resembles the region R more closely. For a function, f, which is continuous on R, we define the definite integral as follows:

$$\int_R f \, dA = \lim_{\Delta x, \Delta y \to 0} \sum_{i,j} f(x_i, y_j) \Delta x \Delta y$$

where the Riemann sum is taken over the subrectangles inside R.

You may wonder why we can leave out the rectangles which cover the edge of R — if we included them might we get a different value for the integral? The answer is that for any region that we are likely to meet, the area of the subrectangles covering the edge tends to 0 as the grid becomes finer. Therefore, omitting these rectangles does not affect the limit.

Interpretations of the Double Integral

Interpretation as an Area

Suppose $f(x, y) = 1$ for all points (x, y) in the region R. Then each term in the Riemann sum is of the form $1 \cdot \Delta A = \Delta A$ and the double integral gives the area of the region R.

$$\text{Area}(R) = \int_R 1 \, dA = \int_R dA$$

Interpretation as a Volume

Just as the definite integral of a positive one-variable function can be interpreted as an area under the graph of the function, so the definite integral of a positive two-variable function can be interpreted as a volume under its graph. In the one-variable case we visualize the Riemann sums as the total area of rectangles above the subdivisions. In the two-variable case we get solid bars instead of rectangles. As the number of subdivisions grows, the tops of the bars approximate the surface better, and the volume of the bars gets closer to the volume under the surface and above the region R. (See Figure 15.4.)

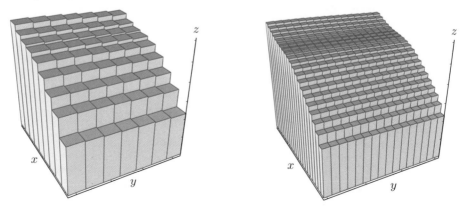

Figure 15.4: Approximating volume under a graph with finer and finer Riemann sums

Example 3   Find the volume under the graph of $f(x, y) = 2 - x^2 - y^2$ lying above the rectangle $-1 \leq x \leq 1$ and $-1 \leq y \leq 1$. (See Figure 15.5.)

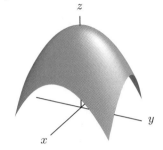

Figure 15.5: Graph of $f(x, y) = 2 - x^2 - y^2$ above
$-1 \leq x \leq 1$ and $-1 \leq y \leq 1$

Solution   If R is the rectangle $-1 \leq x \leq 1$, $-1 \leq y \leq 1$, the volume we want is given by

$$\text{Volume} = \int_R (2 - x^2 - y^2) \, dA$$

Table 15.5 contains values of Riemann sums, S_n, calculated by subdividing the rectangle into n^2 subrectangles and evaluating f at the point with minimum x and y-values. The table suggests the value of the integral is about 5.3.

TABLE 15.5   *Riemann sums for*
$\int_R (2 - x^2 - y^2) \, dA$

n	5	10	20	40
S_n	5.12	5.28	5.32	5.33

Interpretation of the Definite Integral When f is a Density Function

A function of two variables can represent a density per unit area, for example, the fox population density (in foxes per unit area), or the mass density of a thin metal plate. Then the integral $\int_R f \, dA$ represents the total population or total mass in the region R.

Interpretation of the Definite Integral as an Average Value

As in the one-variable case, the definite integral can be used to compute the average value of a function:

$$\frac{\text{Average value of } f}{\text{on the region } R} = \frac{1}{\text{Area of } R} \int_R f \, dA$$

We can rewrite this as

$$\text{Average value} \times \text{Area of } R = \int_R f \, dA.$$

Thus, if we interpret the integral as the volume under the graph of f, then we can think of the average value of f as the height of the box with the same volume that is on the same base. (See Figure 15.6.)

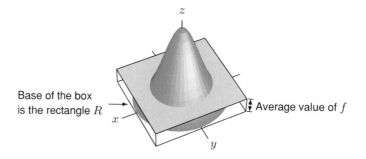

Figure 15.6: Volume and average value

One way to think of this is to imagine that the volume under the graph is made out of wax; if the wax melted and leveled out within walls erected on the perimeter of R, then it would end up box-shaped with height equal to the average value of f.

Problems for Section 15.1

1. A function $f(x, y)$ has values in Table 15.6. Let R be the rectangle $1 \leq x \leq 1.2, 2 \leq y \leq 2.4$. Find the Riemann sums which are reasonable over- and under-estimates for $\int_R f(x, y)\, dA$ with $\Delta x = 0.1$ and $\Delta y = 0.2$.

TABLE 15.6

		x		
		1.0	1.1	1.2
	2.0	5	7	10
y	2.2	4	6	8
	2.4	3	5	4

2. A solid is formed above the rectangle R with $0 \leq x \leq 2, \quad 0 \leq y \leq 4$ by the graph of $f(x, y) = 2 + xy$. Using Riemann sums with four subdivisions, find upper and lower bounds for the volume of this solid.

3. Let R be the rectangle with vertices $(0, 0)$, $(4, 0)$, $(4, 4)$, and $(0, 4)$ and let $f(x, y) = \sqrt{xy}$.

   (a) Find reasonable upper and lower bounds for $\int_R f\, dA$ without subdividing R.

   (b) Estimate $\int_R f\, dA$ by partitioning R into four subrectangles and evaluating f at its maximum and minimum values on each subrectangle.

4. Figure 15.7 shows the distribution of temperature, in °C, in a 5 meter by 5 meter heated room. Using Riemann sums, estimate the average temperature in the room.

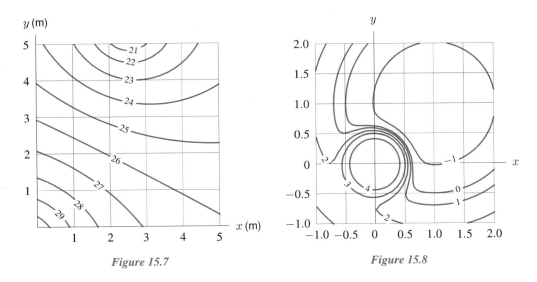

Figure 15.7   **Figure 15.8**

5. Figure 15.8 shows the contour diagram of a function $z = f(x, y)$. Let R be the square $-0.5 \leq x \leq 1$, $-0.5 \leq y \leq 1$. Is the integral $\int_R f \, dA$ positive or negative? Explain your reasoning.

6. A biologist studying insect populations measures the population density of flies and mosquitos at various points in a rectangular study region. The graphs of the two population densities for the region are shown in Figures 15.9 and 15.10. Assuming that the units along the corresponding axes are the same in the two graphs, are there more flies or more mosquitos in the region?

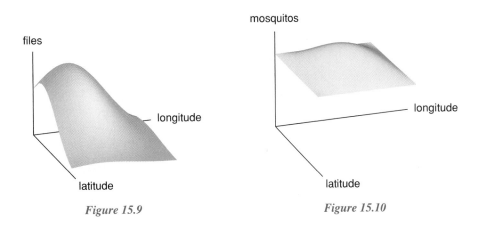

Figure 15.9   **Figure 15.10**

In Problems 7–8, use a computer or calculator program that finds two-dimensional Riemann sums to estimate the given integral.

7. If R is the rectangle $1 \leq x \leq 2$, $1 \leq y \leq 3$, estimate $\int_R (x^2 + y^2) \, dA$.

8. If R is the rectangle $-\pi \leq x \leq 0$, $0 \leq y \leq \pi/2$, estimate $\int_R \sin(xy) \, dA$.

9. The hull of a certain boat has width $w(x, y)$ feet at a point x feet from the front and y feet below the water line. A table of values of w follows. Set up a definite integral that gives the volume of the hull below the waterline, and then estimate the value of the integral.

TABLE 15.7

Front of boat $\longrightarrow$ Back of boat

		0	10	20	30	40	50	60
	0	2	8	13	16	17	16	10
Depth	2	1	4	8	10	11	10	8
below	4	0	3	4	6	7	6	4
waterline	6	0	1	2	3	4	3	2
(in feet)	8	0	0	1	1	1	1	1

10. Let $f(x, y)$ be a function of x and y which is independent of y, that is, $f(x, y) = g(x)$ for some one-variable function g.

   (a) What does the graph of f look like?

   (b) Let R be the rectangle $a \leq x \leq b, c \leq y \leq d$. By interpreting the integral as a volume, and using your answer to part (a), express $\int_R f \, dA$ in terms of a one-variable integral.

11. Figure 15.11 shows contours for the annual frequency of tornados per 10,000 square miles in the U.S.[4] Each grid square is 100 miles on a side. Use the map to estimate the total number of tornados per year in (a) Texas   (b) Florida   (c) Arizona.

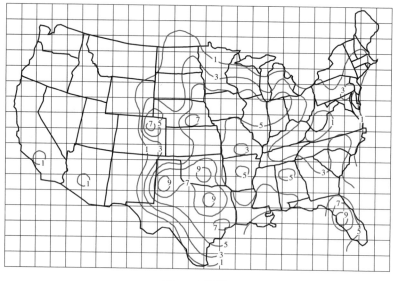

Figure 15.11

---

[4]From *Modern Physical Geography*, Alan H. Strahler and Arthur H. Strahler, Fourth Edition, John Wiley & Sons, New York, 1992, p. 128

12. Figure 15.12 shows contours of annual rainfall (in centimeters) in Oregon.[5] Use it to estimate how much rain falls in Oregon in one year. Each grid square is 100 kilometers on a side.

Figure 15.12

13. Let D be the region inside the unit circle centered at the origin, let R be the right half of D and let B be the bottom half of D. Decide (without calculating the value of any of the integrals) whether each integral is positive, negative, or zero.

   (a) $\int_D dA$

   (b) $\int_B dA$

   (c) $\int_R 5x \, dA$

   (d) $\int_B 5x \, dA$

   (e) $\int_D 5x \, dA$

   (f) $\int_D (y^3 + y^5) \, dA$

   (g) $\int_B (y^3 + y^5) \, dA$

   (h) $\int_R (y^3 + y^5) \, dA$

   (i) $\int_B (y - y^3) \, dA$

   (j) $\int_D (y - y^3) \, dA$

   (k) $\int_D \sin y \, dA$

   (l) $\int_D \cos y \, dA$

   (m) $\int_D e^x \, dA$

   (n) $\int_D xe^x \, dA$

   (o) $\int_D xy^2 \, dA$

   (p) $\int_B x \cos y \, dA$

14. For any numbers a and b, assume that $|a + b| \le |a| + |b|$. Use this to explain why

$$\left| \int_R f \, dA \right| \le \int_R |f| \, dA.$$

15.2 ITERATED INTEGRALS

In Section 15.1 we approximated double integrals using Riemann sums. In this section we see how to compute double integrals exactly using ordinary one-variable integrals.

---

[5]From *Physical Geography of the Global Environment*, H. J. de Blij and Peter O. Muller, John Wiley & Sons, New York, 1993, p. 133

The Fox Population Again: Expressing a Double Integral as an Iterated Integral

To estimate the fox population, we computed a sum of the form

$$\text{Total population} \approx \sum_{i,j} f(x_i, y_j)\Delta x\, \Delta y,$$

where $1 \leq i \leq n$ and $1 \leq j \leq m$ and the values $f(x_i, y_j)$ can be arranged as in Table 15.8.

TABLE 15.8   *Upper bounds for fox population densities for $n = m = 6$*

0	0	0	0.8	1.5	1.5
0	0	0.5	1.5	1.5	1.5
0	0	1.5	2.5	1.9	2.3
0	1.2	2.2	2.5	2.5	2.5
0	1.3	2	2.2	2.5	0.5
0.6	1.3	1	1	1.3	0

For any values of n and m, there are two ways to compute this sum: one is to add across the rows first, the other is to add down the columns first. If we add rows first, we can write the sum in the form

$$\text{Total population} \approx \sum_{j=1}^{m}\left(\sum_{i=1}^{n} f(x_i, y_j)\Delta x\right)\Delta y.$$

approximates the integral $\int_0^{180} f(x, y_j)\, dx$. Thus, we have

$$\approx \sum_{j=1}^{m}\left(\int_0^{180} f(x, y_j)\, dx\right)\Delta y.$$

sum approximating another integral, this time with integration of y. Thus, we can write the total population in terms of

$$= \int_0^{150}\left(\int_0^{180} f(x, y)\, dx\right)dy.$$

ented by $\int_R f\, dA$, we have discovered a way of expressing

$$\iint f(x,y)\, dx\, dy$$

take first (inner) with respect to x; take outer with respect to y.

If R is the rectangle $a \leq x \leq b$, $c \leq y \leq d$ and if $f(x,y)$ is a continuous function of two variables, then

$$\int_R f\, dA = \int_c^d\left(\int_a^b f(x,y)\, dx\right)dy.$$

The expression $\int_c^d\left(\int_a^b f(x,y)\, dx\right)dy$, or simply $\int_c^d \int_a^b f(x,y)\, dx\, dy$, is called an **iterated integral.**

The inside integral is performed with respect to x, holding y constant, and then the result is integrated with respect to y.

The Parallel Between Repeated Summation and Repeated Integration

It is helpful to notice how the summation and integral notation parallel each other. Viewing a Riemann sum as a sum of sums,

$$\sum_{i,j} f(x_i, y_j)\, \Delta x\, \Delta y = \sum_{j}\left(\sum_{i} f(x_i, y_j)\, \Delta x\right)\Delta y$$

leads us to see a double integral as an integral of integrals:

$$\int_R f(x, y)\, dA = \int_c^d \left(\int_a^b f(x, y)\, dx\right) dy,$$

where R is the rectangle $a \le x \le b$ and $c \le y \le d$.

Example 1   A building is 8 feet wide and 16 feet long. It has a flat roof that is 12 feet high at one corner, and 10 feet high at each of the adjacent corners. What is the volume of the building?

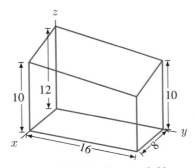

Figure 15.13: A slant-roofed hut

Solution   If we put the high corner on the z-axis, the long side along the y-axis, and the short side along the x-axis, as in Figure 15.13, then the roof is a plane with z-intercept 12, and x slope $(-2)/8 = -1/4$, and y slope $(-2)/16 = -1/8$. Hence, the equation of the roof is

$$z = 12 - \tfrac{1}{4}x - \tfrac{1}{8}y.$$

To calculate the volume, we will integrate over the rectangle $0 \le x \le 8, 0 \le y \le 16$. Setting up an iterated integral, we get

$$\text{Volume} = \int_0^{16}\int_0^8 (12 - \tfrac{1}{4}x - \tfrac{1}{8}y)\, dx\, dy.$$

The inside integral is

$$\int_0^8 (12 - \tfrac{1}{4}x - \tfrac{1}{8}y)\, dx = \left(12x - \tfrac{1}{8}x^2 - \tfrac{1}{8}xy\right)\Big|_0^8 = 88 - y.$$

Then the outside integral is $(88y)$ (16) $-(0)$

$$\int_0^{16} (88 - y)\, dy = (88y - \tfrac{1}{2}y^2)\Big|_0^{16} = 1280.$$

So the volume of the building is 1280 cubic feet.

The Order of Integration

In computing the fox population, we could have chosen to add columns (fixed x) first, instead of the rows. This leads to an iterated integral where x is constant in the inner integral instead of y. Thus,

$$\int_R f(x, y)\, dA = \int_a^b \left(\int_c^d f(x, y)\, dy \right) dx$$

where R is the rectangle $a \leq x \leq b$ and $c \leq y \leq d$.

For any function we are likely to meet, it doesn't matter in which order we integrate over a rectangular region R; we get the same value for the double integral either way.

$$\int_R f\, dA = \int_c^d \left(\int_a^b f(x, y)\, dx \right) dy = \int_a^b \left(\int_c^d f(x, y)\, dy \right) dx$$

Example 2   Compute the volume of Example 1 as an iterated integral with y fixed in the inner integral.

Solution   Rewriting the integral, we have

$$\text{Volume} = \int_0^8 \left(\int_0^{16} (12 - \tfrac{1}{4}x - \tfrac{1}{8}y)\, dy \right) dx = \int_0^8 \left((12y - \tfrac{1}{4}xy - \tfrac{1}{16}y^2)\Big|_0^{16} \right) dx$$

$$= \int_0^8 (176 - 4x)\, dx = (176x - 2x^2)\Big|_0^8 = 1280.$$

Iterated Integrals Over Non-Rectangular Regions

Example 3   The density at the point (x, y) of a right triangular metal plate, as shown in Figure 15.14, is $\delta(x, y)$. Express its mass as an iterated integral.

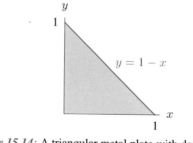

Figure 15.14: A triangular metal plate with density $\delta(x, y)$ at the point (x, y)

Solution   Using a grid, divide the region into small rectangles of side Δx and Δy. Thus, the mass of one piece is given by

$$\text{Mass of rectangle} \approx \text{Density} \cdot \text{Area} \approx \delta(x, y)\Delta x \Delta y.$$

Summing over all rectangles gives a Riemann sum which approximates a double integral:

$$\text{Mass} = \int_R \delta(x, y)\, dA,$$

where R is the triangle. The sloping edge of the triangle is the line $y = 1 - x$. We want to compute this integral using an iterated integral. Think about how an iterated integral over a rectangle, such as

$$\int_a^b \int_c^d f(x, y)\, dy\, dx,$$

works. This integral is over the rectangle $a \le x \le b$, $c \le y \le d$. The inside integral with respect to y is along vertical strips from $y = c$ to $y = d$. There is one such strip for each x value between $x = a$ and $x = b$. Thus, the value of the inside integral depends on the value of x. After computing the inside integral with respect to y, we compute the outside integral with respect to x, which means adding the contributions from the individual vertical strips that make up the rectangle. (See Figure 15.15).

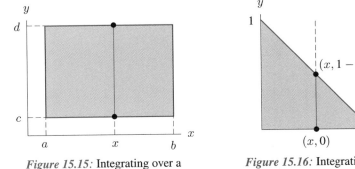

Figure 15.15: Integrating over a rectangle using vertical strips

Figure 15.16: Integrating over a triangle using vertical strips

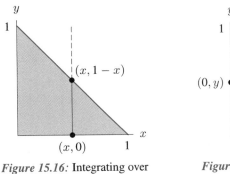

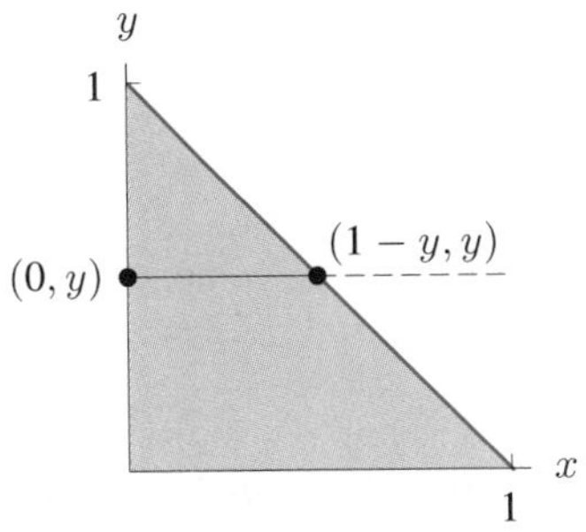

Figure 15.17: Integrating over a triangle using horizontal strips

For the triangular region in Figure 15.14, the idea is the same. The only difference is that the individual vertical strips no longer all go from $y = c$ to $y = d$. The vertical strip that enters the triangle at the point $(x, 0)$ leaves it at the point $(x, 1 - x)$, because the top edge of the triangle is the line $y = 1 - x$. See Figure 15.16. Thus, on this vertical strip, y goes from 0 to $1 - x$. Hence, the inside integral is

$$\int_0^{1-x} \delta(x, y)\, dy.$$

Finally, since there is one of these integrals for each x value between 0 and 1, the outside integral goes from 0 to 1. Thus, the iterated integral we want is

$$\text{Mass} = \int_0^1 \int_0^{1-x} \delta(x, y)\, dy\, dx.$$

We could have chosen to integrate in the opposite order, keeping y fixed in the inner integral instead of x. The limits are formed by looking at horizontal strips instead of vertical ones, and expressing the x-values at the end points in terms of y. A typical horizontal strip goes from $x = 0$ to $x = 1 - y$, and since y-values overall range between 0 and 1, the iterated integral is

$$\text{Mass} = \int_0^1 \int_0^{1-y} \delta(x, y) \, dx \, dy.$$

Limits on Iterated Integrals

- The limits on the outer integral must be constants.
- If the inner integral is with respect to x, its limits should be constants or expressions in terms of y, and vice versa.

Example 4   Find the mass M of a metal plate R bounded by $y = x$ and $y = x^2$, with density given by $\delta(x, y) = 1 + xy$ kg/meter^2. (See Figure 15.18.)

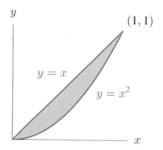

Figure 15.18: A metal plate
with density $\delta(x, y)$

Solution   The mass is given by

$$M = \int_R \delta(x, y) \, dA.$$

We integrate along vertical strips first; this means we do the y integral first, which goes from the bottom boundary $y = x^2$ to the top boundary $y = x$. The left edge of the region is at $x = 0$ and the right edge is at the intersection point of $y = x$ and $y = x^2$, which is $(1, 1)$. Thus, the x-coordinate of the vertical strips can vary from $x = 0$ to $x = 1$, and so the mass is given by

$$M = \int_0^1 \int_{x^2}^x \delta(x, y) \, dy \, dx = \int_0^1 \int_{x^2}^x (1 + xy) \, dy \, dx.$$

Calculating the inner integral first gives

$$M = \int_0^1 \left(\int_{x^2}^x (1 + xy) \, dy \right) dx = \int_0^1 \left((y + x\frac{y^2}{2}) \Big|_{y=x^2}^{y=x} \right) dx$$

$$= \int_0^1 (x - x^2 + \frac{x^3}{2} - \frac{x^5}{2}) \, dx = \left(\frac{x^2}{2} - \frac{x^3}{3} + \frac{x^4}{8} - \frac{x^6}{12} \right) \Big|_0^1$$

$$= \frac{5}{24} \text{ kg.}$$

Example 5   A city is in the form of a semicircular region of radius 3 km bordering on the ocean. Find the average distance from any point in the city to the ocean.

Solution   Think of the ocean as everything below the x-axis in the xy-plane and think of the city as the upper half of the circular disk of radius 3 bounded by $x^2 + y^2 = 9$. (See Figure 15.19).

The distance from any point (x, y) in the city to the ocean is the vertical distance to the x-axis, namely y. Thus, we want to compute

$$\text{Average distance} = \frac{1}{\text{Area}(R)} \int_R y \, dA,$$

where R is the region between the upper half of the circle $x^2 + y^2 = 9$ and the x-axis. The area of R is $\pi 3^2 / 2 = 9\pi/2$. To compute the integral, let's take the inner integral with respect to y. Then a typical vertical strip goes from the x-axis, namely $y = 0$, to the semicircle. The upper limit must be expressed in terms of x so we solve $x^2 + y^2 = 9$ to get $y = \sqrt{9 - x^2}$. Since x varies from -3 to 3 throughout the region, the integral is:

$$\int_R y \, dA = \int_{-3}^{3} \left(\int_0^{\sqrt{9-x^2}} y \, dy \right) dx = \int_{-3}^{3} \left(\left. \frac{y^2}{2} \right|_0^{\sqrt{9-x^2}} \right) dx$$

$$= \int_{-3}^{3} \frac{1}{2}(9 - x^2) \, dx = \frac{1}{2}\left(9x - \frac{x^3}{3}\right) \Big|_{-3}^{3} = \frac{1}{2}(18 - (-18)) = 18.$$

Therefore, the average distance is $18/(9\pi/2) = 4/\pi$ km.

What if we choose the inner integral to be with respect to x? Then we get the limits by looking at horizontal strips, not vertical, and we solve $x^2 + y^2 = 9$ for x in terms of y. We get $x = -\sqrt{9 - y^2}$ at the left end of the strip and $x = \sqrt{9 - y^2}$ at the right. Now y varies from 0 to 3 so the integral becomes:

$$\int_R y \, dA = \int_0^3 \left(\int_{-\sqrt{9-y^2}}^{\sqrt{9-y^2}} y \, dx \right) dy = \int_0^3 \left(\left. yx \right|_{x=-\sqrt{9-y^2}}^{x=\sqrt{9-y^2}} \right) dy = \int_0^3 2y\sqrt{9 - y^2} \, dy$$

$$= -\frac{2}{3}(9 - y^2)^{3/2} \Big|_0^3 = -\frac{2}{3}(0 - 27) = 18.$$

We get the same result as before. Thus, the average distance to the ocean is $(2/(9\pi))18 = 4/\pi$ km.

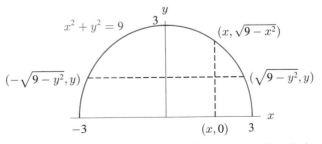

Figure 15.19: The city by the ocean showing a typical vertical strip and a typical horizontal strip

In the examples so far, the region was given and the problem was to determine the limits for an iterated integral. Sometimes only the limits are known and we want to determine the region.

Example 6   Sketch the region of integration for the iterated integral $\int_0^6 \int_{x/3}^2 x\sqrt{y^3 + 1}\, dy\, dx$.

Solution   The inner integral is with respect to y, so we imagine vertical strips crossing the region of integration. The bottom of each strip is $y = x/3$, a line through the origin, and the top is $y = 2$, a horizontal line. Since the limits of the outer integral are 0 and 6, the whole region is contained between the vertical lines $x = 0$ and $x = 6$. Notice that the lines $y = 2$ and $y = x/3$ meet where $x = 6$. The region is shown in Figure 15.20 .

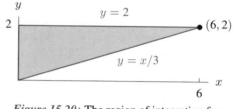

Figure 15.20: The region of integration for Example 6

Reversing the Order of Integration

It can sometimes be helpful to reverse the order of integration in an iterated integral. Surprisingly enough, an integral which is difficult or impossible with the limits in one order can be quite straightforward in the other. The next example is such a case.

Example 7   Evaluate $\int_0^6 \int_{x/3}^2 x\sqrt{y^3 + 1}\, dy\, dx$ using the region sketched in Figure 15.20.

Solution   Since $\sqrt{y^3 + 1}$ has no elementary antiderivative, we cannot calculate the inner integral symbolically. We try reversing the order of integration. From Figure 15.20, we see that horizontal strips go from $x = 0$ to $x = 3y$. For the whole region, y varies from 0 to 2. Thus, when we change the order of integration we get

$$\int_0^6 \int_{x/3}^2 x\sqrt{y^3 + 1}\, dy\, dx = \int_0^2 \int_0^{3y} x\sqrt{y^3 + 1}\, dx\, dy.$$

Now we can at least do the inner integral because we know the antiderivative of x. What about the outer integral?

$$\int_0^2 \int_0^{3y} x\sqrt{y^3 + 1}\, dx\, dy = \int_0^2 \left(\frac{x^2}{2} \sqrt{y^3 + 1} \right) \Big|_{x=0}^{x=3y} dy = \int_0^2 \frac{9y^2}{2}(y^3 + 1)^{1/2}\, dy$$

$$= (y^3 + 1)^{3/2} \Big|_0^2 = 27 - 1 = 26.$$

Thus, reversing the order of integration made the integral in the previous problem much easier. Notice that to reverse the order it is essential first to sketch the region over which the integration is being performed.

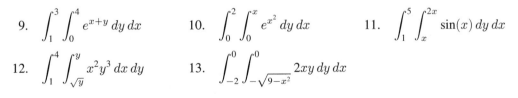

Problems for Section 15.2

For Problems 1–4, evaluate the given integral.

1. $\int_R \sqrt{x+y}\, dA$, where R is the rectangle $0 \le x \le 1, 0 \le y \le 2$.

2. Calculate the integral in Problem 1 using the other order of integration.

3. $\int_R (5x^2 + 1) \sin 3y\, dA$, where R is the rectangle $-1 \le x \le 1, 0 \le y \le \pi/3$.

4. $\int_R (2x + 3y)^2\, dA$, where R is the triangle with vertices at $(-1, 0)$, $(0, 1)$, and $(1, 0)$.

For each of the regions R in Problems 5–8, write $\int_R f\, dA$ as an iterated integral.

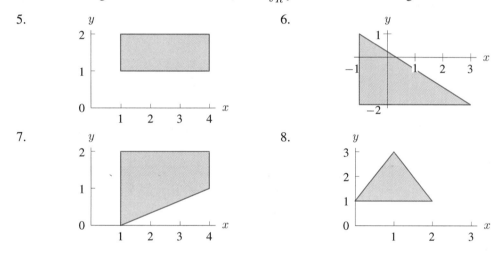

For Problems 9–13, sketch the region of integration and evaluate the integral.

9. $\displaystyle\int_1^3 \int_0^4 e^{x+y}\, dy\, dx$

10. $\displaystyle\int_0^2 \int_0^x e^{x^2}\, dy\, dx$

11. $\displaystyle\int_1^5 \int_x^{2x} \sin(x)\, dy\, dx$

12. $\displaystyle\int_1^4 \int_{\sqrt{y}}^y x^2 y^3\, dx\, dy$

13. $\displaystyle\int_{-2}^0 \int_{-\sqrt{9-x^2}}^0 2xy\, dy\, dx$

14. Consider the integral $\displaystyle\int_0^4 \int_0^{-(y-4)/2} g(x, y)\, dx\, dy$.

    (a) Sketch the region over which the integration is being performed.

    (b) Write the integral with the order of the integration reversed.

Evaluate the integral in Problems 15–17 by reversing the order of integration.

15. $\displaystyle\int_0^1 \int_y^1 e^{x^2}\, dx\, dy$

16. $\displaystyle\int_0^3 \int_{y^2}^9 y \sin(x^2)\, dx\, dy$

17. $\displaystyle\int_0^1 \int_{\sqrt{y}}^1 \sqrt{2 + x^3}\, dx\, dy$

18.   Reverse the order of integration: $\displaystyle\int_{-4}^{0}\int_{0}^{2x+8} f(x,y)\,dy\,dx + \int_{0}^{4}\int_{0}^{-2x+8} f(x,y)\,dy\,dx.$

For Problems 19–21 set up, but do not evaluate, an iterated integral for the volume of the solid.

19.   Under the graph of $f(x,y) = 25 - x^2 - y^2$ and above the xy-plane.

20.   Below the graph of $f(x,y) = 25 - x^2 - y^2$ and above the plane $z = 16$.

21.   The three-sided pyramid whose base is on the xy-plane and whose three sides are the vertical planes $y = 0$ and $y - x = 4$, and the slanted plane $2x + y + z = 4$.

For Problems 22–24, find the volume of the given region.

22.   Under the graph of $f(x,y) = xy$ and above the square $0 \le x \le 2, 0 \le y \le 2$ in the xy-plane.

23.   The solid between the planes $z = 3x + 2y + 1$ and $z = x + y$, and above the triangle with vertices $(1,0,0)$, $(2,2,0)$, and $(0,1,0)$ in the xy-plane. See Figure 15.21.

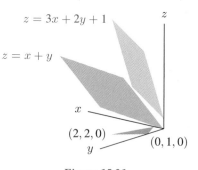

Figure 15.21

24.   The region R bounded by the graph of $ax + by + cz = 1$ and the coordinate planes. Assume a, b, and c are positive.

25.   Find the average distance to the x-axis for points in the region bounded by the x-axis and the graph of $y = x - x^2$.

26.   Prove that for a right triangle the average distance from any point in the triangle to one of the legs is one-third the length of the other leg. (The legs of a right triangle are the two sides that are not the hypotenuse.)

27.   Evaluate $\displaystyle\int_{0}^{1}\int_{y}^{1} \sin(x^2)\,dx\,dy$ 28.   Evaluate $\displaystyle\int_{0}^{1}\int_{e^y}^{e} \frac{x}{\ln x}\,dx\,dy$

29.   At airports, departure gates are often lined up in a terminal like points along a line. If you arrive at one gate and proceed to another gate for a connecting flight, what proportion of the length of the terminal will you have to walk, on average? This can be modeled by randomly choosing two numbers, $0 \le x \le 1$ and $0 \le y \le 1$, and calculating the average value of $|x - y|$. Use a double integral to show that, on average, you have to walk $1/3$ the length of the terminal.

30. In Problem 29, the terminal gates are not actually located continuously from 0 to 1, as we assumed. There are only a finite number of gates and they are likely to be equally spaced. Suppose there are $n + 1$ gates located $1/n$ units apart from one end of the terminal ($x_0 = 0$) to the other ($x_n = 1$). Assuming that all pairs (i, j) of arrival and departure gates are equally likely, show that

$$\text{Average distance between gates} = \frac{1}{(n + 1)^2} \cdot \sum_{i=0}^{n} \sum_{j=0}^{n} \left| \frac{i}{n} - \frac{j}{n} \right|.$$

Identify this sum as approximately (but not exactly) a Riemann sum with n subdivisions for the integrand used in Problem 29. Compute this sum for $n = 5$ and $n = 10$ and compare to the answer of $1/3$ obtained in Problem 29.

15.3 TRIPLE INTEGRALS

A continuous function of three variables can be integrated over a solid region W in 3-space in the same way as a function of two variables is integrated over a flat region in 2-space. Again, we start with a Riemann sum. First we subdivide W into smaller regions, then we multiply the volume of each region by a value of the function in that region, and then we add the results. For example, if W is the box $a \leq x \leq b$, $c \leq y \leq d$, $p \leq z \leq q$, then we subdivide each side into l, m, and n pieces, thereby chopping W into lmn smaller boxes, as shown in Figure 15.22.

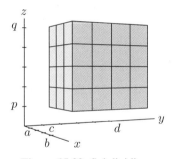

Figure 15.22: Subdividing a three-dimensional box

The volume of each smaller box is

$$\Delta V = \Delta x \Delta y \Delta z,$$

where $\Delta x = (b - a)/l$, and $\Delta y = (d - c)/m$, and $\Delta z = (q - p)/n$. Using this subdivision, we pick a point (x_i, y_j, z_k) in the ijk-th small box and construct a Riemann sum

$$\sum_{i,j,k} f(x_i, y_j, z_k) \, \Delta V.$$

If f is continuous, as Δx, Δy, and Δz approach 0, this Riemann sum approaches the definite integral, $\int_W f \, dV$, called a *triple integral*, which is defined as

$$\int_W f \, dV = \lim_{l,m,n \to \infty} \sum_{i,j,k} f(x_i, y_j, z_k) \Delta V.$$

As in the case of a double integral, we can evaluate this integral as an iterated integral:

Triple integral as an iterated integral

$$\int_W f \, dV = \int_p^q \left(\int_c^d \left(\int_a^b f(x, y, z) \, dx \right) dy \right) dz,$$

where y and z are treated as constants in the innermost (dx) integral, and z is treated as a constant in the middle (dy) integral. The integration can be performed in any order.

Example 1   A cube C has sides of length 4 cm and is made of a material of variable density. If one corner is at the origin and the adjacent corners are on the positive x, y, and z axes, then the density (in gm/cm^3) at the point (x, y, z) is $\delta(x, y, z) = 1 + xyz$ gm/cm^3. Find the mass of the cube.

Solution   Consider a small piece ΔV of the cube, small enough so that the density remains close to constant over the piece. Then

$$\text{Mass of small piece} = \text{Density} \cdot \text{Volume} \approx \delta(x, y, z) \, \Delta V.$$

To get the total mass, we add the masses of the small pieces and take the limit as $\Delta V \to 0$. Thus, the mass is the triple integral

$$M = \int_C \delta \, dV = \int_0^4 \int_0^4 \int_0^4 (1 + xyz) \, dx \, dy \, dz = \int_0^4 \int_0^4 \left[x + \frac{1}{2} x^2 yz \right]_{x=0}^{x=4} dy \, dz$$

$$= \int_0^4 \int_0^4 (4 + 8yz) \, dy \, dz = \int_0^4 \left[4y + 4y^2 z \right]_{y=0}^{y=4} dz = \int_0^4 (16 + 64z) \, dz = 576 \, \text{gm}.$$

Example 2   Express the volume of the building described in Example 1 on page 227 as a triple integral.

Solution   The building is given by $0 \le x \le 8$, $0 \le y \le 16$, and $0 \le z \le 12 - x/4 - y/8$. (See Figure 15.23.) To find its volume, we divide it into small cubes of volume $\Delta V = \Delta x \, \Delta y \, \Delta z$ and add. First we form a vertical stack of cubes above the point $(x, y, 0)$. This stack goes from $z = 0$ to $z = 12 - x/4 - y/8$, and

$$\text{Volume of vertical stack} \approx \sum_z \Delta V = \sum_z \Delta x \, \Delta y \, \Delta z = \left(\sum_z \Delta z \right) \Delta x \, \Delta y.$$

Next we line up these stacks parallel to the y-axis to form a slice from $y = 0$ to $y = 16$. So

$$\text{Volume of slice} \approx \left(\sum_y \sum_z \Delta z \, \Delta y \right) \Delta x.$$

Finally we line up the slices along the x-axis from $x = 0$ to $x = 8$ and add up their volumes, to get

$$\text{Volume of building} \approx \sum_x \sum_y \sum_z \Delta z \, \Delta y \, \Delta x.$$

Thus, in the limit,

$$\text{Volume of building} = \int_0^8 \int_0^{16} \int_0^{12 - x/4 - y/8} 1 \, dz \, dy \, dx.$$

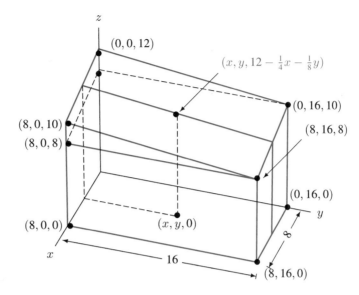

Figure 15.23: Volume of building as triple integral

Example 3   Set up an iterated integral to compute the mass of the solid cone bounded by $z = \sqrt{x^2 + y^2}$ and $z = 3$, if the density is given by $\delta(x, y, z) = z$.

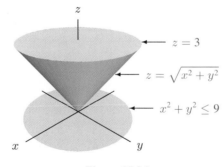

Figure 15.24

Solution   The cone is shown in Figure 15.24. We break the cone into small cubes of volume $\Delta V = \Delta x \, \Delta y \, \Delta z$, on which the density is approximately constant, and approximate the mass of each cube $\delta(x, y, z) \, \Delta x \, \Delta y \, \Delta z$. Stacking the cubes vertically above the point $(x, y, 0)$ starting on the cone at height $z = \sqrt{x^2 + y^2}$ and going up to $z = 3$ tells us that the inner integral is

$$\int_{\sqrt{x^2+y^2}}^{3} \delta(x, y, z) \, dz = \int_{\sqrt{x^2+y^2}}^{3} z \, dz.$$

There is a stack for every point in the xy-plane in the shadow cast by the cone. Since the cone $z = \sqrt{x^2 + y^2}$ intersects the horizontal plane $z = 3$ in the circle $x^2 + y^2 = 9$, this means that there is a stack for all (x, y) in the region $x^2 + y^2 \leq 9$. Lining up the stacks parallel to the y-axis gives a slice from $y = -\sqrt{9 - x^2}$ to $y = \sqrt{9 - x^2}$, for each fixed value of x. Thus, the limits on the middle integral are

$$\int_{-\sqrt{9-x^2}}^{\sqrt{9-x^2}} \int_{\sqrt{x^2+y^2}}^{3} z \, dz \, dy.$$

Finally, there is a slice for each x between -3 and 3, so the integral we want is

$$\text{Mass} = \int_{-3}^{3} \int_{-\sqrt{9-x^2}}^{\sqrt{9-x^2}} \int_{\sqrt{x^2+y^2}}^{3} z \, dz \, dy \, dx.$$

Notice that setting up the limits on the two outer integrals was just like setting up the limits for a double integral over the region $x^2 + y^2 \leq 9$.

As the previous example illustrates, for a region W contained between two surfaces, the inner most limits correspond to these surfaces. The middle and outer limits ensure that we integrate over the "shadow" of W in the xy-plane.

Limits on Triple Integrals

- The limits for the outer integral are constants.
- The limits for the middle integral can involve only one variable (that in the outer integral).
- The limits for the inner integral can involve two variables (those on the two outer integrals).

Problems for Section 15.3

In Problems 1–4, find the triple integrals of the given functions over the given regions.

1. $f(x, y, z) = x^2 + 5y^2 - z$, W is the rectangular box $0 \leq x \leq 2, -1 \leq y \leq 1, 2 \leq z \leq 3$.
2. $f(x, y, z) = \sin x \cos(y + z)$, W is the cube $0 \leq x \leq \pi, 0 \leq y \leq \pi, 0 \leq z \leq \pi$.
3. $h(x, y, z) = ax + by + cz$, W is the rectangular box $0 \leq x \leq 1, 0 \leq y \leq 1, 0 \leq z \leq 2$.
4. $f(x, y, z) = e^{-x-y-z}$, W is the rectangular box with corners at $(0, 0, 0)$, $(a, 0, 0)$, $(0, b, 0)$, and $(0, 0, c)$.

For Problems 5–11 describe or sketch the region of integration for the triple integrals. If the limits do not make sense, say why.

5. $\displaystyle\int_{0}^{6} \int_{0}^{3-x/2} \int_{0}^{6-x-2y} f(x, y, z) \, dz \, dy \, dx$
6. $\displaystyle\int_{0}^{1} \int_{0}^{x} \int_{0}^{x} f(x, y, z) \, dz \, dy \, dx$

7. $\displaystyle\int_{0}^{1} \int_{0}^{z} \int_{0}^{x} f(x, y, z) \, dz \, dy \, dx$
8. $\displaystyle\int_{0}^{3} \int_{-\sqrt{9-y^2}}^{0} \int_{\sqrt{x^2+y^2}}^{3} f(x, y, z) \, dz \, dx \, dy$

9. $\displaystyle\int_{1}^{3} \int_{1}^{x+y} \int_{0}^{y} f(x, y, z) \, dz \, dx \, dy$
10. $\displaystyle\int_{0}^{1} \int_{0}^{2-x} \int_{0}^{3} f(x, y, z) \, dz \, dy \, dx$

11. $\displaystyle\int_{-1}^{1}\int_{0}^{\sqrt{1-x^2}}\int_{0}^{\sqrt{2-x^2-y^2}} f(x,y,z)\,dz\,dy\,dx$

12. Find the volume of the pyramid with base in the plane $z = -6$ and sides formed by the three planes $y = 0$ and $y - x = 4$ and $2x + y + z = 4$.

13. Find the mass of the solid bounded by the xy-plane, yz-plane, xz-plane, and the plane $(x/3) + (y/2) + (z/6) = 1$, if the density of the solid is given by $\delta(x, y, z) = x + y$.

14. Find the average value of the sum of the squares of three numbers x, y, z, where each number is between 0 and 2.

15. Set up, but do not evaluate, an iterated integral for the volume of the solid formed by the intersections of the cylinders $x^2 + z^2 = 1$ and $y^2 + z^2 = 1$.

The motion of a solid object can be analyzed by thinking of the mass as concentrated at a single point, the *center of mass*. If the object has density $\rho(x, y, z)$ at the point (x, y, z) and occupies a region W, then the coordinates $(\bar{x}, \bar{y}, \bar{z})$ of the center of mass are given by

$$\bar{x} = \frac{1}{m}\int_W x\rho\,dV \qquad \bar{y} = \frac{1}{m}\int_W y\rho\,dV \qquad \bar{z} = \frac{1}{m}\int_W z\rho\,dV$$

where $m = \int_W \rho\,dV$ is the total mass of the body. Use these definitions for Problems 16–17.

16. A solid is bounded below by the square $z = 0, 0 \le x \le 1, 0 \le y \le 1$ and above by the surface $z = x + y + 1$. Find the total mass and the coordinates of the center of mass if the density is 1 gm/cm^3 and x, y, z are measured in centimeters.

17. Find the center of mass of the tetrahedron that is bounded by the x, y, and z planes and the plane $x + y/2 + z/3 = 1$. Assume the density is 1 gm/cm^3.

The *moment of inertia* of a solid body about an axis in 3-space gives the angular acceleration about this axis for a given torque (force twisting the body). The moments of inertia about the coordinate axes of a body of constant density and mass m occupying a region W of volume V are defined to be

$$I_x = \frac{m}{V}\int_W (y^2 + z^2)\,dV \qquad I_y = \frac{m}{V}\int_W (x^2 + z^2)\,dV \qquad I_z = \frac{m}{V}\int_W (x^2 + y^2)\,dV$$

Use these definitions for Problems 18–20.

18. Find the moment of inertia about the z-axis of the rectangular solid of mass m given by $0 \le x \le 1, 0 \le y \le 2, 0 \le z \le 3$.

19. Find the moment of inertia about the x-axis of the rectangular solid $-a \le x \le a, -b \le y \le b$ and $-c \le z \le c$ of mass m.

20. Let a, b, and c denote the moments of inertia of a homogeneous solid object about the x, y and z-axes respectively. Explain why $a + b > c$.

15.4 NUMERICAL INTEGRATION: THE MONTE CARLO METHOD

There are many one-variable definite integrals in which the integrand has no elementary antiderivative. A familiar example is $\int_0^1 e^{-x^2}\,dx$. There are also intractable double and triple integrals. They can be approximated by Riemann sums or by a variant of Simpson's rule (see Problem 10). In this section, we give an alternative method called the Monte Carlo method (after the gambling resort).

A One-Variable Example

Let us consider the integral $\int_0^1 x^2\, dx$, which we know has value $1/3$. We now approach it probabilistically. We graph the function $y = x^2$ in the square $0 \leq x \leq 1$, $0 \leq y \leq 1$ and throw darts at the square. We expect that some darts will land above the curve, some below. The fraction of the darts which land below the curve gives an estimate of the ratio of the area under the curve to the area of the square. This is the basis of the Monte Carlo method.

Example 1   Approximate the integral $\displaystyle\int_0^1 x^2\, dx$ using the Monte Carlo method.

Solution   If we choose points from the unit square in Figure 15.25 at random, we expect that the ratio of the number of points in region R, say N_R, to the total number of points, N, will approximate the integral:

$$\frac{N_R}{N} \approx \frac{\int_0^1 x^2\, dx}{\text{Area of unit square}} = \int_0^1 x^2\, dx.$$

Since we are selecting the points at random we cannot expect to get the same ratio every time, but as the number of points increases, the approximation should improve. Table 15.9 shows the values of N_R/N for six different trials each with $N = 50$ points. These, and all subsequent trials, were obtained by using a computer program to generate random points in the region and counting how many fall inside R.

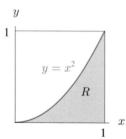

Figure 15.25: Region whose area is $\int_0^1 x^2\, dx$ as fraction of the unit square

TABLE 15.9   *Six trials each with $N = 50$ points*

$N = 50$	1	2	3	4	5	6
N_R/N	0.2	0.24	0.52	0.36	0.38	0.28

These approximations are not particularly good. Their average is 0.33, which has two digit accuracy. Repeating this process again for $N = 50$ gives the results in Table 15.10.

TABLE 15.10   *Six more trials for $N = 50$ points*

$N = 50$	1	2	3	4	5	6
N_R/N	0.44	0.42	0.28	0.34	0.28	0.32

Notice that the average of these is 0.347. This is not as close to the true value of $1/3$ as before, but remember that this is a random process. Each time it is repeated we expect a different result. To continue with this example we now approximate $\int_0^1 x^2\, dx$ by taking increasingly large values of N, say $N = 10, 100, 1000$, and $10,000$. The results of one computer experiment are given in

Table 15.11, but if you carry out a similar experiment your results will probably be slightly different. However, it appears that as N increases, the ratio gets closer to the exact value of $1/3$.

TABLE 15.11   *Value of N_R/N as N increases*

N	10	100	1000	10000
N_R/N	0.2000	0.3400	0.3250	0.3343

The basis of the Monte Carlo method is the generation of random numbers. Fortunately, nearly all computer programming languages have a random number generator built in. Any two random numbers, x and y, between 0 and 1 give a point (x, y) in the unit square. Then we check to see if $y \leq x^2$. If this is true, the point is in the region under the parabola. We assume that every point is equally likely to be chosen, enabling us to compute the area of the region under the parabola by the following method.

Monte Carlo Method for Estimating an Integral

Suppose the integral, $\int_a^b f(x)dx$, is given by the area of a region, R. Enclose the region in a rectangle of area A. If N random points are chosen in A and N_R of them fall into the region R, then we expect

$$\frac{N_R}{N} \approx \frac{\text{Area}(R)}{\text{Area}(A)} = \frac{\int_a^b f(x)dx}{\text{Area}(A)}.$$

A Two-Variable Example

We can extend the idea of the Monte Carlo method to evaluating integrals of more than one variable.

Example 2   Use a Monte Carlo method to approximate the double integral

$$\int_0^1 \int_0^1 e^{-(x^2+y^2)} \, dx \, dy.$$

Solution   This integral gives the volume of the region W above the unit square and below the graph of $z = e^{-(x^2+y^2)}$. Since the volume we are considering is contained in the cube C given by $0 \leq x \leq 1$, $0 \leq y \leq 1$, and $0 \leq z \leq 1$, we count points of the form (x, y, z) which lie in the cube and which satisfy the condition

$$0 \leq z \leq e^{-(x^2+y^2)}.$$

If N_R of the N randomly chosen points satisfy this condition, then, since $\text{Vol}(C) = 1$, we have

$$\frac{N_R}{N} \approx \frac{\text{Vol}(W)}{\text{Vol}(C)} = \text{Vol}(W) = \int_0^1 \int_0^1 e^{-(x^2+y^2)} \, dx \, dy.$$

TABLE 15.12   *Ten trials each with $N = 100$*

$N = 100$	1	2	3	4	5	6	7	8	9	10
N_R/N	0.54	0.60	0.57	0.60	0.51	0.53	0.59	0.56	0.56	0.57

Table 15.12 shows the value of N_R/N for ten trials of $N = 100$ points each. The average of the ten values N_R/N is 0.563. We take this as an approximate value of the integral. Taking $N = 10,000$ gives

$$\int_0^1 \int_0^1 e^{-(x^2+y^2)}\, dx\, dy \approx \frac{N_R}{N} \approx 0.5654.$$

This is consistent with the estimate made in Example 2 on page 218.

When using the Monte Carlo method, it is important to choose a small box C that entirely contains the region W. Intuitively, the better the fit between the two volumes, the fewer random points are needed to obtain a reasonable approximation. In fact, the biggest problem with the Monte Carlo method is finding a sufficiently small rectangular box which encloses the volume.

Problems for Section 15.4

Problems 1–9 require a computer or calculator that generates random numbers.

1. Use a Monte Carlo method to approximate the integral $\int_0^1 \sqrt{1-x^2}\, dx$. Explain geometrically why your answer gives an approximation for $\pi/4$.

2. Use a Monte Carlo method to approximate the integral $\int_0^1 e^{-x^2} dx$.

3. Approximate the integral $\int_0^1 \int_0^1 e^{-xy}\, dx\, dy$ to four decimal places of accuracy.

4. Approximate the integral $\int_0^1 \int_0^1 xy^{xy}\, dx\, dy$ to two decimal places of accuracy.

5. Approximate the integral $\int_0^\pi \int_0^2 x \sin y\, dx\, dy$ and compare your result to the exact answer.

6. Explain why the Monte Carlo method fails to approximate the integral $\int_0^1 \int_0^1 x^{-y}\, dx\, dy$.

In the Monte Carlo method described in the text, we generated triplets of random numbers and evaluated the function at the first two numbers to estimate a double integral. Here is another Monte Carlo method that requires only pairs of random numbers. Recall that

$$\begin{array}{c} \text{Average value of} \\ f(x,y) \text{ on } R \end{array} = \frac{1}{\text{Area}(R)} \int_R f(x,y)\, dx\, dy.$$

We can also estimate the average value of f by choosing N points, (x_i, y_i), at random in R, summing the values of f at those points and dividing by N, giving

$$A = \frac{1}{N} \sum_{i=1}^N f(x_i, y_i).$$

Thus, we have the approximation

$$\int_R f\, dA \approx \text{Area}(R) \cdot A$$

In Problems 7–9, use this method to approximate the integrals to two decimal places of accuracy,

7. The integral in Prob-
lem 3.

8. The integral in Prob-
lem 4.

9. The integral in Problem 5.

10. Here is a way to use Simpson's rule twice to approximate a definite integral. Suppose the integral is $\int_1^5 \int_2^6 \sqrt{x^2 + y^2} \, dy \, dx$. Use Simpson's rule with $\Delta y = 1$ to approximate the inner integral when x is fixed at 1. Repeat for $x = 1.5, 2, 2.5, 3, 3.5, 4, 4.5, 5$. You now have approximations for the inner integral at nine different values of x. Now use Simpson's rule again with $\Delta x = 1$, using the nine different values for the inner integral, to approximate the outer integral (and therefore the whole double integral).

15.5 DOUBLE INTEGRALS IN POLAR COORDINATES

Integration in Polar Coordinates

We started this chapter by putting a rectangular grid on the fox population density map, to estimate the total population using a Riemann sum. However, sometimes a polar grid is more appropriate. A review of polar coordinates is in Appendix G.

Example 1   A biologist studying insect populations around a circular lake divides the area into polar sectors as in Figure 15.26. The population density in each sector is shown in millions per square km. Estimate the total insect population around the lake.

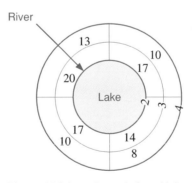

Figure 15.26: An insect infested lake
showing the insect population density
by sector

Solution   To get our estimate we will multiply the population density in each sector by the area of that sector. Unlike the rectangles in a rectangular grid, the sectors in this grid do not all have the same area. The inner sectors have area

$$\frac{1}{4}(\pi 3^2 - \pi 2^2) = \frac{5\pi}{4} \approx 3.93 \text{ km}^2,$$

and the outer sectors have area

$$\frac{1}{4}(\pi 4^2 - \pi 3^2) = \frac{7\pi}{4} \approx 5.50 \text{ km}^2,$$

so we estimate

$$\text{Population} \approx (20)(3.93) + (17)(3.93) + (14)(3.93) + (17)(3.93) +$$
$$(13)(5.50) + (10)(5.50) + (8)(5.50) + (10)(5.50)$$
$$\approx 493 \text{ million insects.}$$

What is dA in Polar Coordinates?

The previous example used a polar grid rather than a rectangular grid. A rectangular grid is constructed from vertical and horizontal lines corresponding to $x = k$ (a constant) and $y = l$ (another constant), respectively. In polar coordinates, setting $r = k$ gives a circle of radius k centered at the origin and setting $\theta = l$ gives a ray emanating from the origin (at angle l with the x-axis). A polar grid is built out of these circles and rays. Figure 15.27 shows a subdivision of the polar region $a \le r \le b$, $\alpha \le \theta \le \beta$, using n subdivisions each way. This bent rectangle is the sort of region that is naturally represented in polar coordinates.

In general, dividing R as shown in Figure 15.27, gives a Riemann sum:

$$\sum_{i,j} f(r_i, \theta_j) \, \Delta A.$$

However, calculating the area ΔA is more complicated in polar coordinates than in Cartesian coordinates. Figure 15.28 shows ΔA. If Δr and $\Delta \theta$ are small, the shaded region is approximately a rectangle with sides $r \, \Delta \theta$ and Δr, so

$$\Delta A \approx r \Delta \theta \Delta r.$$

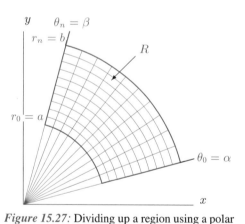

Figure 15.27: Dividing up a region using a polar grid

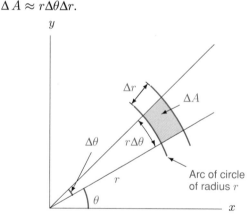

Figure 15.28: Calculating area ΔA in polar coordinates

Thus, the Riemann sum is approximately

$$\sum_{i,j} f(r_i, \theta_j) \, r_i \, \Delta \theta \, \Delta r.$$

If we take the limit as Δr and $\Delta \theta$ approach 0, we obtain

$$\int_R f \, dA = \int_\alpha^\beta \int_a^b f(r, \theta) \, r \, dr \, d\theta.$$

When computing integrals in polar coordinates, put $dA = r \, dr \, d\theta$ or $dA = r \, d\theta \, dr$.

Example 2   Compute the integral of $f(x, y) = 1/(x^2 + y^2)^{3/2}$ over the region R shown in Figure 15.29.

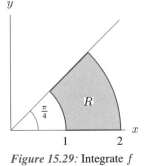

Figure 15.29: Integrate f
over the polar region

Solution   The region R is described by the inequalities $1 \leq r \leq 2$, $0 \leq \theta \leq \pi/4$. In polar coordinates, $r = \sqrt{x^2 + y^2}$, so we can write f as

$$f(x, y) = \frac{1}{(r^2)^{3/2}} = \frac{1}{r^3}.$$

Then

$$\int_R f \, dA = \int_0^{\pi/4} \int_1^2 \frac{1}{r^3} r \, dr \, d\theta = \int_0^{\pi/4} \left(\int_1^2 r^{-2} \, dr \right) d\theta$$

$$= \int_0^{\pi/4} \left[-\frac{1}{r} \right]_{r=1}^{r=2} d\theta = \int_0^{\pi/4} \frac{1}{2} \, d\theta = \frac{\pi}{8}.$$

Example 3   For each region in Figure 15.30, decide whether to integrate using polar or Cartesian coordinates. On the basis of its shape, write an iterated integral of an arbitrary function $f(x, y)$ over the region.

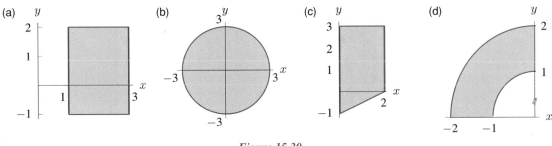

Figure 15.30

Solution   (a)   Since this is a rectangular region, Cartesian coordinates are likely to be a better choice. The rectangle is described by the inequalities $1 \leq x \leq 3$ and $-1 \leq y \leq 2$, so the integral is

$$\int_{-1}^2 \int_1^3 f(x, y) \, dx \, dy.$$

(b)   A circle is best described in polar coordinates. The radius is 3, so r goes from 0 to 3, and to describe the whole circle, θ goes from 0 to 2π. The integral is

$$\int_0^{2\pi} \int_0^3 f(r \cos \theta, r \sin \theta) r \, dr \, d\theta.$$

(c)   The bottom boundary of this trapezoid is the line $y = (x/2) - 1$ and the top is the line $y = 3$, so we use Cartesian coordinates. If we integrate with respect to y first, the lower limit of the integral is $(x/2) - 1$ and the upper limit is 3. The x limits are $x = 0$ to $x = 2$. So the integral is

$$\int_0^2 \int_{(x/2)-1}^3 f(x,y)\, dy\, dx.$$

(d)   This is another polar region: it is a piece of a ring in which r goes from 1 to 2. Since it is in the second quadrant, θ goes from $\pi/2$ to π. The integral is

$$\int_{\pi/2}^\pi \int_1^2 f(r\cos\theta, r\sin\theta)\, r\, dr\, d\theta.$$

Problems for Section 15.5

For each of the regions R in Problems 1–4, write $\int_R f\, dA$ as an iterated integral in polar coordinates.

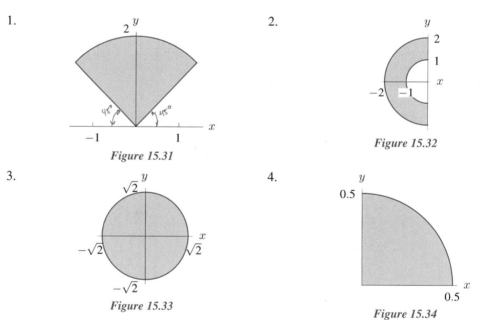

1.

Figure 15.31

2.

Figure 15.32

3.

Figure 15.33

4.

Figure 15.34

Sketch the region over which the integrals in Problems 5–11 are computed.

5. $\displaystyle\int_0^{2\pi}\int_1^2 f(r,\theta)\, r\, dr\, d\theta.$

6. $\displaystyle\int_{\pi/2}^\pi\int_0^1 f(r,\theta)\, r\, dr\, d\theta.$

7. $\displaystyle\int_{\pi/6}^{\pi/3}\int_0^1 f(r,\theta)\, r\, dr\, d\theta.$

8. $\displaystyle\int_3^4\int_{3\pi/4}^{3\pi/2} f(r,\theta)\, r\, d\theta\, dr.$

9. $\displaystyle\int_0^{\pi/4}\int_0^{1/\cos\theta} f(r,\theta)\, r\, dr\, d\theta.$

10. $\displaystyle\int_{\pi/4}^{\pi/2}\int_0^{2/\sin\theta} f(r,\theta)\, r\, dr\, d\theta.$

11. $\displaystyle\int_0^4 \int_{-\pi/2}^{\pi/2} f(r,\theta)\, r\, d\theta\, dr.$

12. Evaluate $\displaystyle\int_R \sin(x^2 + y^2)\, dA$, where R is the disk of radius 2 centered at the origin.

13. Evaluate $\displaystyle\int_R (x^2 - y^2)\, dA$, where R is the first quadrant region between the circles of radius 1 and radius 2.

14. Consider the integral $\displaystyle\int_0^3 \int_{x/3}^1 f(x,y)\, dy\, dx.$

     (a) Sketch the region R over which the integration is being performed.
     (b) Rewrite the integral with the order of integration reversed.
     (c) Rewrite the integral in polar coordinates.

Convert the integrals in Problems 15–17 to polar coordinates and evaluate.

15. $\displaystyle\int_{-1}^0 \int_{-\sqrt{1-x^2}}^{\sqrt{1-x^2}} x\, dy\, dx$ 16. $\displaystyle\int_0^{\sqrt{2}} \int_y^{\sqrt{4-y^2}} xy\, dx\, dy$ 17. $\displaystyle\int_0^{\sqrt{6}} \int_{-x}^x dy\, dx$

18. Find the volume of the region between the graph of $f(x,y) = 25 - x^2 - y^2$ and the xy plane.

19. An ice cream cone can be modeled by the region bounded by the hemisphere $z = \sqrt{8 - x^2 - y^2}$ and the cone $z = \sqrt{x^2 + y^2}$. Find its volume.

20. A disk of radius 5 cm has density 10 gm/cm^2 at its center, density 0 at its edge, and its density is a linear function of the distance from the center. Find the mass of the disk.

21. A city by the ocean surrounds a bay as shown in Figure 15.35. The population density of the city (in thousands of people per square km) is given by the function $\delta(r,\theta)$, where r and θ are polar coordinates with respect to the x and y-axes shown, and the distances indicated on the y-axis are in km.

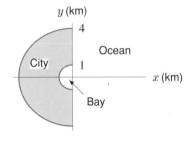

Figure 15.35

     (a) Set up an iterated integral in polar coordinates giving the total population of the city.
     (b) The population density decreases the farther you live from the shoreline of the bay; it also decreases the farther you live from the ocean. Which of the following functions best describes this situation?

          (i) $\delta(r,\theta) = (4 - r)(2 + \cos\theta)$ (ii) $\delta(r,\theta) = (4 - r)(2 + \sin\theta)$
          (iii) $\delta(r,\theta) = (r + 4)(2 + \cos\theta)$

     (c) Estimate the population using your answers to parts (a) and (b).

22. A watch spring lies flat on the table. It is made of a coiled steel strip standing a height of 0.2 inches above the table. The inner edge is the spiral $r = 0.25 + 0.04\theta$, where $0 \leq \theta \leq 4\pi$ (so the spiral makes two complete turns). The outer edge is given by $r = 0.26 + 0.04\theta$. Find the volume of the spring.

15.6 INTEGRALS IN CYLINDRICAL AND SPHERICAL COORDINATES

Some double integrals are easier to evaluate in polar, rather than Cartesian, coordinates. Similarly, some triple integrals are easier in non-Cartesian coordinates.

Cylindrical Coordinates

The cylindrical coordinates of a point (x, y, z) in 3-space are obtained by representing the x and y coordinates in polar coordinates and letting the z-coordinate be the z-coordinate of the Cartesian coordinate system. (See Figure 15.36.)

Relation between Cartesian and Cylindrical Coordinates

Each point in 3-space is represented using $0 \leq r < \infty, 0 \leq \theta \leq 2\pi, -\infty < z < \infty$.

$$x = r \cos \theta,$$
$$y = r \sin \theta,$$
$$z = z.$$

As with polar coordinates in the plane, note that $x^2 + y^2 = r^2$.

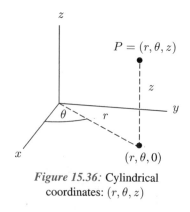

Figure 15.36: Cylindrical coordinates: (r, θ, z)

A useful way to visualize cylindrical coordinates is to sketch the surfaces obtained by setting one of the coordinates equal to a constant. See Figures 15.37–15.39.

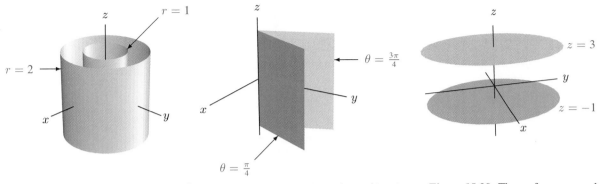

Figure 15.37: The surfaces $r = 1$ and $r = 2$ *Figure 15.38:* The surfaces $\theta = \pi/4$ and $\theta = 3\pi/4$ *Figure 15.39:* The surfaces $z = -1$ and $z = 3$

Setting $r = c$ (where c is constant) gives a cylinder around the z-axis whose radius is c. Setting $\theta = c$ gives a half-plane perpendicular to the xy plane, with one edge along the z-axis, making an angle c with the x-axis. Setting $z = c$ gives a horizontal plane $|c|$ units from the xy-plane. We call these *fundamental surfaces*.

The regions that can most easily be described in cylindrical coordinates are those regions whose boundaries are such fundamental surfaces. (For example, vertical cylinders, or wedge-shaped parts of vertical cylinders.)

Example 1   Describe in cylindrical coordinates a wedge of cheese cut from a cylinder 4 cm high and 6 cm in radius; this wedge subtends an angle of $\pi/6$ at the center. (See Figure 15.40.)

Solution   The wedge is described by the inequalities $0 \leq r \leq 6$, and $0 \leq z \leq 4$, and $0 \leq \theta \leq \pi/6$.

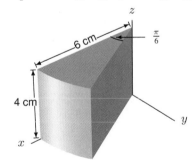

Figure 15.40: A wedge of cheese

Integration in Cylindrical Coordinates

To integrate in polar coordinates, we had to express the area element dA in terms of polar coordinates: $dA = r\, dr\, d\theta$. To evaluate a triple integral $\int_W f\, dV$ in cylindrical coordinates, we need to express the volume element dV in cylindrical coordinates.

Consider the volume element ΔV, shown in Figure 15.41. It is a wedge bounded by fundamental surfaces. The area of the base is $\Delta A \approx r\Delta r\Delta\theta$. Since the height is Δz, the volume element is given approximately by $\Delta V \approx r\, \Delta r\, \Delta\theta\, \Delta z$.

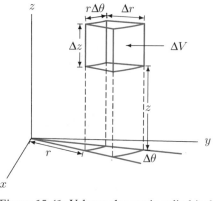

Figure 15.41: Volume element in cylindrical coordinates

When computing integrals in cylindrical coordinates, put $dV = r\, dr\, d\theta\, dz$. Other orders of integration are also possible.

Example 2   Find the mass of the wedge of cheese in Example 1, if its density is 1.2 grams/cm^3.

Solution   If the wedge is W, its mass is

$$\int_W 1.2\, dV.$$

In cylindrical coordinates this integral is

$$\int_0^4 \int_0^{\pi/6} \int_0^6 1.2\, r\, dr\, d\theta\, dz = \int_0^4 \int_0^{\pi/6} 0.6r^2 \Big|_0^6 d\theta\, dz = 21.6 \int_0^4 \int_0^{\pi/6} d\theta\, dz$$

$$= 21.6(\frac{\pi}{6})4 \approx 45.24 \text{ grams.}$$

Example 3   A water tank in the shape of a hemisphere has radius a; its base is its plane face. Find the volume, V, of water in the tank as a function of h, the depth of the water.

Solution   In Cartesian coordinates a sphere of radius a has the equation $x^2 + y^2 + z^2 = a^2$. (See Figure 15.42.) In cylindrical coordinates, $r^2 = x^2 + y^2$, so this becomes

$$r^2 + z^2 = a^2.$$

Thus, if we want to describe the amount of water in the tank in cylindrical coordinates, we let r go from 0 to $\sqrt{a^2 - z^2}$, we let θ go from 0 to 2π, and we let z go from 0 to h, giving

$$\begin{aligned}
\text{Volume} &= \int_W dV = \int_0^{2\pi} \int_0^h \int_0^{\sqrt{a^2-z^2}} r\, dr\, dz\, d\theta = \int_0^{2\pi} \int_0^h \frac{r^2}{2}\Big|_{r=0}^{r=\sqrt{a^2-z^2}} dz\, d\theta \\
\text{of water} \\
&= \int_0^{2\pi} \int_0^h \frac{1}{2}(a^2 - z^2)\, dz\, d\theta = \int_0^{2\pi} \frac{1}{2}\left(a^2 z - \frac{z^3}{3}\right)\Big|_{z=0}^{z=h} d\theta \\
&= \int_0^{2\pi} \frac{1}{2}\left(a^2 h - \frac{h^3}{3}\right) d\theta = \pi\left(a^2 h - \frac{h^3}{3}\right).
\end{aligned}$$

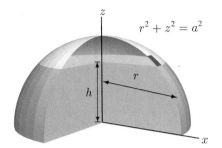

$$r^2 + z^2 = a^2$$

Figure 15.42: Hemispherical water tank with radius a and water of depth h

Spherical Coordinates

In Figure 15.43, the point P has coordinates (x, y, z) in the Cartesian coordinate system. We define spherical coordinates ρ, ϕ, and θ for P as follows: $\rho = \sqrt{x^2 + y^2 + z^2}$ is the distance of P from the origin; ϕ is the angle between the positive z-axis and the line through the origin and the point P; and θ is the same as in cylindrical coordinates.

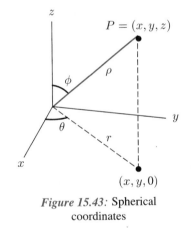

Figure 15.43: Spherical coordinates

In cylindrical coordinates,

$$x = r \cos \theta, \quad \text{and} \quad y = r \sin \theta, \quad \text{and} \quad z = z.$$

From Figure 15.43 we have $z = \rho \cos \phi$ and $r = \rho \sin \phi$, giving the following relationship:

Relation between Cartesian and Spherical Coordinates

Each point in 3-space is represented using $0 \le \rho < \infty, 0 \le \phi \le \pi$, and $0 \le \theta \le 2\pi$.

$$x = \rho \sin \phi \cos \theta$$
$$y = \rho \sin \phi \sin \theta$$
$$z = \rho \cos \phi.$$

Also, $\rho^2 = x^2 + y^2 + z^2$.

This system of coordinates is useful when there is spherical symmetry with respect to the origin, either in the region of integration or in the integrand. The fundamental surfaces in spherical coordinates are $\rho = k$ (a constant), which is a sphere of radius k centered at the origin, $\theta = k$ (a constant), which is the half-plane with its edge along the z-axis, and $\phi = k$ (a constant), which is a cone if $k \neq \pi/2$ and the xy-plane if $k = \pi/2$. (See Figures 15.44-15.46.)

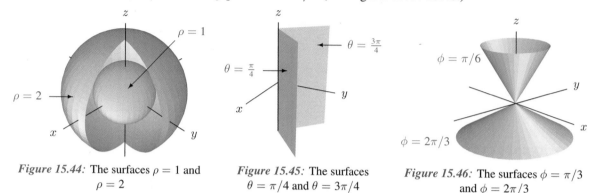

Figure 15.44: The surfaces $\rho = 1$ and $\rho = 2$

Figure 15.45: The surfaces $\theta = \pi/4$ and $\theta = 3\pi/4$

Figure 15.46: The surfaces $\phi = \pi/3$ and $\phi = 2\pi/3$

Integration in Spherical Coordinates

To use spherical coordinates in triple integrals we need to express the volume element, dV, in spherical coordinates. From Figure 15.47, we see that the volume element can be approximated by a box with curved edges. One edge has length $\Delta\rho$. The edge parallel to the xy-plane is an arc of a circle made from rotating the cylindrical radius r ($= \rho \sin \phi$) through an angle $\Delta\theta$, and so has length $\rho \sin \phi \, \Delta\theta$. The remaining side comes from rotating the radius ρ through an angle $\Delta\phi$, and so has length $\rho \, \Delta\phi$. Therefore, $\Delta V \approx \Delta\rho(\rho \, \Delta\phi)(\rho \sin \phi \, \Delta\theta) = \rho^2 \sin \phi \, \Delta\rho \, \Delta\phi \, \Delta\theta$.

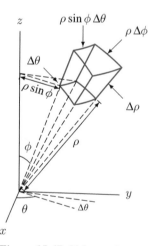

Figure 15.47: Volume element in spherical coordinates

Thus,

When computing integrals in spherical coordinates, put $dV = \rho^2 \sin \phi \, d\rho \, d\phi \, d\theta$. Other orders of integration are also possible.

Example 4   Use spherical coordinates to derive the formula for the volume of a solid sphere of radius a.

Solution   In spherical coordinates, a sphere of radius a is described by the inequalities $0 \leq \rho \leq a, 0 \leq \theta \leq 2\pi$, and $0 \leq \phi \leq \pi$. Note that θ goes all the way around the circle, whereas ϕ only goes from 0 to π. We find the volume by integrating the constant density function 1 over the sphere:

$$\text{Volume} = \int_R 1 \, dV = \int_0^{2\pi} \int_0^{\pi} \int_0^a \rho^2 \sin \phi \, d\rho \, d\phi \, d\theta = \int_0^{2\pi} \int_0^{\pi} \frac{1}{3} a^3 \sin \phi \, d\phi \, d\theta$$

$$= \frac{1}{3} a^3 \int_0^{2\pi} -\cos \phi \Big|_0^{\pi} d\theta = \frac{2}{3} a^3 \int_0^{2\pi} d\theta = \frac{4\pi a^3}{3}.$$

Example 5   Find the magnitude of the gravitational force exerted by a solid hemisphere of radius a and constant density δ on a unit mass located at the center of the base of the hemisphere.

Solution   Assume the base of the hemisphere rests on the xy-plane with center at the origin. (See Figure 15.48.) Newton's law of gravitation says that the force between two masses m_1 and m_2 at a distance r apart is $F = Gm_1m_2/r^2$. In this example, symmetry shows that the net component of the force on the particle at the origin due to the hemisphere is in the z direction only. Any force in the x or y direction from some part of the hemisphere will be canceled by the force from another part of the hemisphere directly opposite the first. To compute the net z component of the gravitational force, we imagine a small piece of the hemisphere with volume ΔV, located at spherical coordinates (ρ, θ, ϕ). This piece has mass $\delta \Delta V$, and exerts a force of magnitude F on the unit mass at the origin. The z-component of this force is given by its projection onto the z-axis, which can be seen from the figure to be $F \cos \phi$. The distance from the mass $\delta \Delta V$ to the unit mass at the origin is the spherical coordinate ρ. Therefore the z-component of the force due to the small piece ΔV is

$$\begin{array}{c} z\text{-component} \\ \text{of force} \end{array} = \frac{G(\delta \Delta V)(1)}{\rho^2} \cos \phi.$$

Adding the contributions of the small pieces, we get a vertical force with magnitude

$$F = \int_0^{2\pi} \int_0^{\pi/2} \int_0^a \left(\frac{G\delta}{\rho^2} \right) (\cos \phi) \rho^2 \sin \phi \, d\rho \, d\phi \, d\theta = \int_0^{2\pi} \int_0^{\pi/2} G\delta (\cos \phi \sin \phi) \rho \Big|_{\rho=0}^{\rho=a} d\phi \, d\theta$$

$$= \int_0^{2\pi} \int_0^{\pi/2} G\delta a \cos \phi \sin \phi \, d\phi \, d\theta = \int_0^{2\pi} G\delta a \left(-\frac{(\cos \phi)^2}{2} \right) \Big|_{\phi=0}^{\phi=\pi/2} d\theta$$

$$= \int_0^{2\pi} G\delta a (\frac{1}{2}) \, d\theta = G\delta a\pi.$$

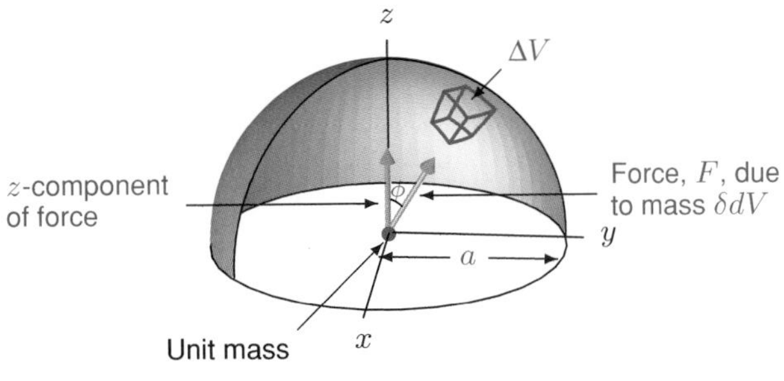

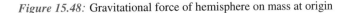

Figure 15.48: Gravitational force of hemisphere on mass at origin

Problems for Section 15.6

In Problems 1–2, evaluate the triple integrals in cylindrical coordinates.

1. $f(x, y, z) = x^2 + y^2 + z^2$, W is the region $0 \le r \le 4$, $\pi/4 \le \theta \le 3\pi/4$, $-1 \le z \le 1$.
2. $f(x, y, z) = \sin(x^2 + y^2)$, W is the solid cylinder with height 4 and with base of radius 1 centered on the z axis at $z = -1$.

In Problems 3–4, evaluate the triple integrals in spherical coordinates.

3. $f(x, y, z) = 1/(x^2 + y^2 + z^2)^{1/2}$ over the bottom half of the sphere of radius 5 centered at the origin.
4. $f(\rho, \theta, \phi) = \sin \phi$, over the region $0 \le \theta \le 2\pi$, $0 \le \phi \le \pi/4$, $1 \le \rho \le 2$.

For Problems 5–9, choose a set of coordinate axes, and then set up the three-variable integral in an appropriate coordinate system for integrating a density function δ over the given region.

5.

6.

7.   A piece of a sphere; angle at the center is $\pi/3$.

8.

9.

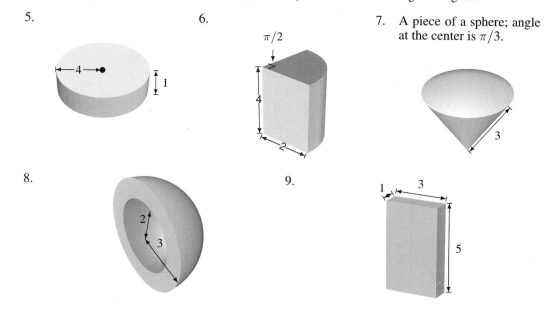

10.   Sketch the region R over which the integration is being performed:

$$\int_0^{\pi/2} \int_{\pi/2}^{\pi} \int_0^1 f(\rho, \phi, \theta) \rho^2 \sin \phi \, d\rho \, d\phi \, d\theta.$$

Evaluate the integrals in Problems 11–12.

11. $\displaystyle \int_0^1 \int_{-\sqrt{1-x^2}}^{\sqrt{1-x^2}} \int_{-\sqrt{1-x^2-z^2}}^{\sqrt{1-x^2-z^2}} \frac{1}{(x^2 + y^2 + z^2)^{1/2}} \, dy \, dz \, dx$

12. $\displaystyle \int_0^1 \int_{-1}^1 \int_{-\sqrt{1-x^2}}^{\sqrt{1-x^2}} \frac{1}{(x^2 + y^2)^{1/2}} \, dy \, dx \, dz$

Without performing the integration, decide whether each of the integrals in Problems 13–14 is positive, negative, or zero. Give reasons for your decision.

13. W_1 is the unit ball, $x^2 + y^2 + z^2 \leq 1$.

    (a) $\int_{W_1} \sin \phi \, dV$ (b) $\int_{W_1} \cos \phi \, dV$

14. W_2 is the top half of the unit ball, $0 \leq z \leq \sqrt{1 - x^2 - y^2}$.

    (a) $\int_{W_2} (z^2 - z) \, dV$ (b) $\int_{W_2} (-xz) \, dV$

15. Write a triple integral representing the volume of a slice of the cylindrical cake of height 2 and radius 5 between the planes $\theta = \pi/6$ and $\theta = \pi/3$. Evaluate this integral.

16. Find the mass M of the solid region W given in spherical coordinates by

    $$W = \{(\rho, \phi, \theta) : 0 \leq \rho \leq 3, 0 \leq \theta < 2\pi, 0 \leq \phi \leq \pi/4\}$$

    if the density, $\delta(P)$, at any point P is given by the distance of P from the origin.

17. A particular spherical cloud of gas of radius 3 km is more dense at the center than towards the edge. The density, D, of the gas at a distance ρ km from the center is given by $D(\rho) = 3 - \rho$. Write an integral representing the total mass of the cloud of gas, and evaluate it.

18. Find the volume that remains after a cylindrical hole of radius R is bored through a sphere of radius a, where $0 < R < a$, passing through the center of the sphere along the pole.

19. Use appropriate coordinates to find the average distance to the origin for points in the ice cream cone region bounded by the hemisphere $z = \sqrt{8 - x^2 - y^2}$ and the cone $z = \sqrt{x^2 + y^2}$. [Hint: The volume of this region is computed in Problem 19 on page 247.]

20. Compute the force of gravity exerted by a solid cylinder of radius R, height H, and constant density δ on a unit mass at the center of the base of the cylinder.

For Problems 21–24, use the definition of center of mass given on page 239.

21. Let C be a solid cone with both height and radius 1 and contained between the surfaces $z = \sqrt{x^2 + y^2}$ and $z = 1$. If C has constant mass density of 1 gm/cm³, find the z-coordinate of C's center of mass.

22. Suppose that the density of the cone C in Problem 21 is given by $\rho(z) = z^2$ gm/cm³. Find

    (a) The mass of C. (b) The z-coordinate of C's center of mass.

23. For $a > 0$, consider the family of solids bounded below by the paraboloid $z = a(x^2 + y^2)$ and above by the plane $z = 1$. If the solids all have constant mass density 1 gm/cm³, show that the z-coordinate of the center of mass is $2/3$, and hence is independent of the parameter a.

24. Find the location of the center of mass of a hemisphere of radius a and density b gm/cm³.

For Problems 25–26, use the definition of moment of inertia given on page 239.

25. The moment of inertia of a solid homogeneous ball B of mass 1 and radius a centered at the origin is the same about any of the coordinate axes (due to the symmetry of the ball). It is easier to evaluate the sum of the three integrals involved in computing the moment of inertia about each of the axes than to compute them individually. Find the sum of the moments of inertia about the x, y and z-axes and thus find the individual moments of inertia.

26. Find the moment of inertia about the z axis of the solid "fat ice cream cone" given in spherical coordinates by $0 \leq \rho \leq a, 0 \leq \phi \leq \frac{\pi}{3}$ and $0 \leq \theta \leq 2\pi$. Assume that the solid is homogeneous with mass m.

15.7 APPLICATIONS OF INTEGRATION TO PROBABILITY

To represent how a quantity such as height or weight is distributed throughout a population, we use a density function. A review of single variable density functions is in Appendix F. To study two or more quantities at the same time and see how they are related, we use a multivariable density function.

Density Functions

Distribution of Weight and Height in Expectant Mothers

Table 15.13 shows the distribution of weight and height in a survey of expectant mothers. The histogram in Figure 15.49 is constructed in such a way that the volume of each bar represents the percentage in the corresponding weight and height range. For example, the bar representing the mothers who weighed 60–70 kg and were 160–165 cm tall has base of area 10 kg · 5 cm = 50 kg cm. The volume of this bar is 12%, so its height is 12%/50 kg cm = 0.24%/ kg cm. Notice that the units on the vertical axis are percent/ kg cm. Thus, volumes under the histogram are in units of %. The total volume is 100% = 1.

TABLE 15.13   *Distribution of weight and height in a survey of expectant mothers*

	45–50 kg	50–60 kg	60–70 kg	70–80 kg	80–105 kg	Totals by height
150–155 cm	2	4	4	2	1	13
155–160 cm	0	12	8	2	1	23
160–165 cm	1	7	12	4	3	27
165–170 cm	0	8	12	6	2	28
170–180 cm	0	1	3	4	1	9
Totals by weight	3	32	39	18	8	100

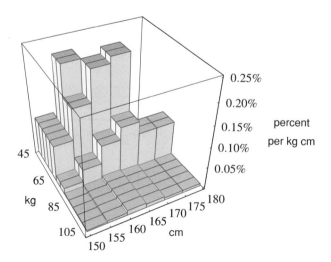

Figure 15.49: Histogram representing the data in Table 15.13.

Example 1   Find the percentage of mothers in the survey with height between 170 and 180 cm.

Solution   We add the percentages across the row corresponding to the 170–180 cm height range; this is equivalent to adding the volumes of the corresponding rectangular solids in the histogram.

$$\text{Percentage of mothers} = 0 + 1 + 3 + 4 + 1 = 9\%.$$

Smoothing the Histogram

If we have smaller weight and height groups (and a larger sample), we can draw a smoother histogram and get finer estimates. In the limit, we replace the histogram with a smooth surface, in such a way that the volume under the surface above a rectangle is the percentage of mothers in that rectangle. We define a *density function*, $p(w, h)$, to be the function whose graph is the smooth surface. It has the property that

$$\begin{array}{ccc} \text{Fraction of sample with} & & \text{Volume under graph of } p \\ \text{weight between } a \text{ and } b \text{ and} & = & \text{over the rectangle} \\ \text{height between } c \text{ and } d & & a \le w \le b, c \le h \le d \end{array} = \int_a^b \int_c^d p(w, h) \, dh \, dw.$$

Joint Density Functions

We generalize this idea to represent any two characteristics, x and y, distributed throughout a population.

A function $p(x, y)$ is called a **joint density function** for x and y if

$$\begin{array}{ccc} \text{Fraction of population with} & & \text{Volume under graph of } p \\ x \text{ between } a \text{ and } b \text{ and} & = & \text{above the rectangle} \\ y \text{ between } c \text{ and } d & & a \le x \le b, c \le y \le d \end{array} = \int_a^b \int_c^d p(x, y) \, dy \, dx$$

where

$$\int_{-\infty}^{\infty} \int_{-\infty}^{\infty} p(x, y) \, dy \, dx = 1 \quad \text{and} \quad p(x, y) \ge 0 \text{ for all } x \text{ and } y.$$

A joint density function need not be continuous, as in Example 2 which follows. In addition, as in Example 4, the integrals involved may be improper and must be computed by methods similar to those used for improper one-variable integrals.

Example 2   Let $p(x, y)$ be defined on the square $0 \le x \le 1, 0 \le y \le 1$ by $p(x, y) = x + y$; let $p(x, y) = 0$ if (x, y) is outside this square. Verify that p is a joint density function. In terms of the distribution of x and y in the population, what does it mean that $p(x, y) = 0$ outside the square?

Solution   First, we have $p(x, y) \geq 0$ for all x and y. To verify that p is a joint density function, we check that the total volume under the graph is 1:

$$\int_{-\infty}^{\infty} \int_{-\infty}^{\infty} p(x, y) \, dy \, dx = \int_{0}^{1} \int_{0}^{1} (x + y) \, dy \, dx$$

$$= \int_{0}^{1} \left(xy + \frac{y^2}{2} \right) \Big|_{0}^{1} dx = \int_{0}^{1} \left(x + \frac{1}{2} \right) dx = \left(\frac{x^2}{2} + \frac{x}{2} \right) \Big|_{0}^{1} = 1$$

The fact that $p(x, y) = 0$ outside the square means that the variables x and y never take values outside the interval $[0, 1]$, that is, the value of x and y for any individual in the population is always between 0 and 1.

Example 3   Suppose two variables x and y are distributed in a population according to the density function of Example 2. Find the fraction of the population with $x \leq 1/2$, the fraction with $y \leq 1/2$, and the fraction with both $x \leq 1/2$ and $y \leq 1/2$.

Solution   The fraction with $x \leq 1/2$ is the volume under the graph to the left of the line $x = 1/2$:

$$\int_{0}^{1/2} \int_{0}^{1} (x + y) \, dy \, dx = \int_{0}^{1/2} \left(xy + \frac{y^2}{2} \right) \Big|_{0}^{1} dx = \int_{0}^{1/2} \left(x + \frac{1}{2} \right) dx$$

$$= \left(\frac{x^2}{2} + \frac{x}{2} \right) \Big|_{0}^{1/2} = \frac{1}{8} + \frac{1}{4} = \frac{3}{8}.$$

Since the function is symmetric in x and y, the fraction with $y \leq 1/2$ is also $3/8$. Finally, the fraction with both $x \leq 1/2$ and $y \leq 1/2$ is

$$\int_{0}^{1/2} \int_{0}^{1/2} (x + y) \, dy \, dx = \int_{0}^{1/2} \left(xy + \frac{y^2}{2} \right) \Big|_{0}^{1/2} dx = \int_{0}^{1/2} \left(\frac{1}{2}x + \frac{1}{8} \right) dx$$

$$= \left(\frac{1}{4}x^2 + \frac{1}{8}x \right) \Big|_{0}^{1/2} = \frac{1}{16} + \frac{1}{16} = \frac{1}{8}$$

Recall that a one-variable density function $p(x)$ is a function such that $p(x) \geq 0$ for all x, and $\int_{-\infty}^{\infty} p(x) \, dx = 1$. (See Appendix F.)

Example 4   Let p_1 and p_2 be one-variable density functions for x and y, respectively. Verify that $p(x, y) = p_1(x)p_2(y)$ is a joint density function.

Solution   Since both p_1 and p_2 are density functions, they are nonnegative everywhere. Thus, their product $p_1(x)p_2(x) = p(x, y)$ is nonnegative everywhere. Now we must check that the volume under the graph of p is 1. Since $\int_{-\infty}^{\infty} p_2(y) \, dy = 1$ and $\int_{-\infty}^{\infty} p_1(x) \, dx = 1$, we have

$$\int_{-\infty}^{\infty} \int_{-\infty}^{\infty} p(x, y) \, dy \, dx = \int_{-\infty}^{\infty} \int_{-\infty}^{\infty} p_1(x)p_2(y) \, dy \, dx = \int_{-\infty}^{\infty} p_1(x) \left(\int_{-\infty}^{\infty} p_2(y) \, dy \right) dx$$

$$= \int_{-\infty}^{\infty} p_1(x)(1) \, dx = \int_{-\infty}^{\infty} p_1(x) \, dx = 1.$$

Density Functions and Probability

What is the probability that an expectant mother weighs 60–70 kg and is 155–160 cm tall? Table 15.13 shows that 8% of mothers fall in this group, so the probability that a randomly chosen mother falls in this group is 0.08.

$$
\begin{array}{ccc}
\begin{array}{c}
\text{Probability that a mother} \\
\text{has weight between } a \text{ and } b \\
\text{and height between } c \text{ and } d
\end{array}
& = &
\begin{array}{c}
\text{Volume under graph of } p \\
\text{above the rectangle} \\
a \le w \le b, c \le h \le d
\end{array}
& = \int_a^b \int_c^d p(w, h)\, dh\, dw.
\end{array}
$$

For a joint density function $p(x, y)$, the probability that x falls in an interval of width Δx around x_0 and y falls in an interval of width Δy around y_0 is approximately $p(x_0, y_0)\Delta x \Delta y$. Thus, p is often called a *probability density function*.

Example 5   A machine in a factory is set to produce components 10 cm long and 5 cm in diameter. In fact, there is a slight variation from one component to the next. A component is usable if its length and diameter deviate from the correct values by less than 0.1 cm. If the length is x cm and the diameter is y cm, the probability density function for the variation in x and y is

$$
p(x, y) = \frac{50\sqrt{2}}{\pi} e^{-100(x-10)^2} e^{-50(y-5)^2}.
$$

What is the probability that a component will be usable? (See Figure 15.50.)

Figure 15.50: The density function
$p(x, y) = \frac{50\sqrt{2}}{\pi} e^{-100(x-10)^2} e^{-50(y-5)^2}.$

Solution   We know that

$$
\begin{array}{c}
\text{Probability that } x \text{ and } y \text{ satisfy} \\
x_0 - \Delta x \le x \le x_0 + \Delta x \\
y_0 - \Delta y \le y \le y_0 + \Delta y
\end{array}
= \frac{50\sqrt{2}}{\pi} \int_{y_0 - \Delta y}^{y_0 + \Delta y} \int_{x_0 - \Delta x}^{x_0 + \Delta x} e^{-100(x-10)^2} e^{-50(y-5)^2}\, dx\, dy.
$$

Thus,

$$
\begin{array}{c}
\text{Probability that} \\
\text{component is usable}
\end{array}
= \frac{50\sqrt{2}}{\pi} \int_{4.9}^{5.1} \int_{9.9}^{10.1} e^{-100(x-10)^2} e^{-50(y-5)^2}\, dx\, dy.
$$

The double integral must be evaluated numerically. This yields

$$\text{Probability that component is usable} = \frac{50\sqrt{2}}{\pi}(0.02556) \approx 0.57530.$$

Thus, there is a 57.5% chance that the component will be usable.

Dependence and Independence of Variables

If we study the distribution of height and weight in a population, we expect to see a relation between them. All other things being equal, tall people are more likely to be heavy than short ones. On the other hand, if we study the distribution of height and annual income, we do not expect to see much of a relationship; tall people and short people probably earn the same, on average.

How Can We Determine Dependence from the Joint Density Function?

Let's see how the dependence between weight and height shows up in the data in Table 15.13. Look at the column corresponding to expectant mothers weighing 70–80 kg. This group is 18% of the whole sample. The subset of this group with height between 170 and 180 cm is 4% of the whole sample. So the probability that a woman in this weight group is 170–180 cm tall is

$$\frac{\text{Probability that height is 170–180 cm and weight is 70–80 kg}}{\text{Probability that weight is 70–80 kg}} = \frac{4}{18} = 0.22.$$

Now look at a lighter group, the women who weigh 60–70 kg. This group forms 39% of the total and the subset with height 170–180 cm is 3% of the total. So the probability that a woman in this group is 170–180 cm tall is

$$\frac{\text{Probability that height is 170–180 cm and weight is 60–70 kg}}{\text{Probability that weight is 60–70 kg}} = \frac{3}{39} = 0.08.$$

This is a lower probability than for the 70–80 kg group. This is because a light woman is less likely to be tall than a heavy woman. In this situation, we say that the two variables w and h appear *dependent*, because to a certain extent they depend on each other.

Conditional Probability

We can generalize these ideas to any joint density function. We want to calculate the probability that y falls in a certain range, given that x falls in a certain range.

If $p(x, y)$ is a probability density function, we define the **conditional probability** by

$$\text{Conditional probability that } a \leq x \leq b \text{ given } c \leq y \leq d = \frac{\text{Probability that } a \leq x \leq b \text{ and } c \leq y \leq d}{\text{Probability that } c \leq y \leq d} = \frac{\int_a^b \int_c^d p(x, y)\, dy\, dx}{\int_{-\infty}^{\infty} \int_c^d p(x, y)\, dy\, dx}.$$

Example 6   For the probability density function in Example 5, calculate the probability that the length is between 9.9 and 10.1 cm, given that the diameter is between

(a) 4.9 and 5.1 cm.   (b) 5.3 and 5.5 cm.

Solution   (a) We have that

$$
\frac{\text{Probability that } 9.9 \leq x \leq 10.1}{\text{given } 4.9 \leq y \leq 5.1} = \frac{\text{Probability that } 9.9 \leq x \leq 10.1 \text{ and } 4.9 \leq y \leq 5.1}{\text{Probability that } 4.9 \leq y \leq 5.1}
$$

$$
= \frac{\dfrac{50\sqrt{2}}{\pi} \displaystyle\int_{9.9}^{10.1} \int_{4.9}^{5.1} e^{-100(x-10)^2} e^{-50(y-5)^2} \, dy \, dx}{\dfrac{50\sqrt{2}}{\pi} \displaystyle\int_{-\infty}^{\infty} \int_{4.9}^{5.1} e^{-100(x-10)^2} e^{-50(y-5)^2} \, dy \, dx}
$$

$$
\approx \frac{0.57}{0.68} \approx 0.84.
$$

(b) Similarly, we find that

$$
\frac{\text{Probability that } 9.9 \leq x \leq 10.1}{\text{given } 5.3 \leq y \leq 5.5} = \frac{\text{Probability that } 9.9 \leq x \leq 10.1 \text{ and } 5.3 \leq y \leq 5.5}{\text{Probability that } 5.3 \leq y \leq 5.5}
$$

$$
= \frac{\dfrac{50\sqrt{2}}{\pi} \displaystyle\int_{9.9}^{10.1} \int_{5.3}^{5.5} e^{-100(x-10)^2} e^{-50(y-5)^2} \, dy \, dx}{\dfrac{50\sqrt{2}}{\pi} \displaystyle\int_{-\infty}^{\infty} \int_{5.3}^{5.5} e^{-100(x-10)^2} e^{-50(y-5)^2} \, dy \, dx}
$$

$$
\approx \frac{0.00114}{0.00135} \approx 0.84.
$$

Look at the denominators in the ratios used to calculate the conditional probabilities. Notice that it is much less likely that $5.3 \leq y \leq 5.5$ than that $4.9 \leq y \leq 5.1$ (a probability of 0.00135 as opposed to 0.68). However, the probability for the length x to fall in the range $9.9 \leq x \leq 10.1$ is the same in both cases, about 0.84. Thus, the variation in the length appears to be independent of the variation in the diameter. We say that the variables x and y are *independent*.

Example 7   For the density function in Example 2, find the probability that $x \leq 1/2$ given that $y \leq 1/2$.

Solution   We have that

$$
\frac{\text{Probability that}}{x \leq \tfrac{1}{2} \text{ given } y \leq \tfrac{1}{2}} = \frac{\text{Probability that } x \leq \tfrac{1}{2} \text{ and } y \leq \tfrac{1}{2}}{\text{Probability that } y \leq \tfrac{1}{2}} = \frac{1/8}{3/8} = \frac{1}{3}.
$$

Since $1/3 < 3/8$, we see that having $y \leq 1/2$ makes $x \leq 1/2$ less likely. In this case the variables do not appear to be independent.

How Can We Tell If Two Variables are Independent?

Two events are said to be independent if the probability that they both happen is the product of the probabilities that they would happen individually. For example, if we throw two dice, the probability of a double four is $(1/6) \cdot (1/6) = 1/36$. This is because the face showing on the first die is independent of the face showing on the second and the probability of a four on either die is $1/6$. We use this idea to find the joint probability distribution function for two qualities x and y that vary independently in a population.

Suppose x has density function $p_1(x)$ and y has density function $p_2(y)$. If x and y are independent, then we expect

$$
\begin{matrix}
\text{Probability that} \\
x_0 \leq x \leq x_0 + \Delta x \\
\text{and } y_0 \leq y \leq y_0 + \Delta y
\end{matrix}
=
\begin{matrix}
\text{Probability that} \\
x_0 \leq x \leq x_0 + \Delta x
\end{matrix}
\cdot
\begin{matrix}
\text{Probability that} \\
y_0 \leq y \leq y_0 + \Delta y
\end{matrix}
$$

$$
\approx (p_1(x_0)\,\Delta x) \cdot (p_2(y_0)\,\Delta y)
$$

$$
= p_1(x_0)p_2(y_0)\,\Delta x\,\Delta y.
$$

On the other hand, if $p(x, y)$ is the joint density function for x and y,

$$
\begin{matrix}
\text{Probability that} \\
x_0 \leq x \leq x_0 + \Delta x \\
\text{and } y_0 \leq y \leq y_0 + \Delta y
\end{matrix}
=
\begin{matrix}
\text{Volume under the graph of } p \\
\text{above } x_0 \leq x \leq x_0 + \Delta x, \, y_0 \leq y \leq y_0 + \Delta y
\end{matrix}
$$

$$
\approx p(x_0, y_0)\,\Delta x\,\Delta y.
$$

Thus,

$$
p_1(x_0)p_2(y_0)\,\Delta x\,\Delta y \approx p(x_0, y_0)\,\Delta x\,\Delta y.
$$

Dividing by $\Delta x\,\Delta y$, we arrive at the conclusion that:

> If x has probability density $p_1(x)$ and y has probability density $p_2(y)$, and if x and y are independent, then the joint density for x and y is $p(x, y) = p_1(x)p_2(y)$.

Conversely, if the joint density function $p(x, y)$ can be written as a product of one-variable density functions $p_1(x)$ and $p_2(y)$, then

$$
\begin{matrix}
\text{Probability that} \\
a \leq x \leq b \text{ and } c \leq y \leq d
\end{matrix}
= \int_a^b \int_c^d p_1(x)p_2(y)\,dy\,dx
$$

$$
= \int_a^b p_1(x) \left(\int_c^d p_2(y)\,dy \right) dx
$$

$$
= \int_a^b p_1(x)\,dx \cdot (\text{Probability that } c \leq y \leq d)
$$

$$
= \text{Probability that } a \leq x \leq b \cdot \text{Probability that } c \leq y \leq d.
$$

Hence the variables are independent. Thus, we conclude that:

> If the joint density function p of x and y can be expressed as a product $p(x,y) = p_1(x)p_2(y)$, where p_1 and p_2 are density functions, then x and y are independent.

The one-variable *normal* probability density function with mean μ and standard deviation σ is defined by

$$p(x) = \frac{1}{\sigma\sqrt{2\pi}}e^{-(x-\mu)^2/(2\sigma^2)}.$$

The normal density function arises frequently in applications and is one of the most widely used probability density functions.

Example 8   Show that the length and diameter of the components in Example 5 are independent.

Solution   The joint density function may be written as

$$\frac{50\sqrt{2}}{\pi}e^{-100(x-10)^2}e^{-50(y-5)^2} = \left(\frac{10\sqrt{2}}{\sqrt{2\pi}}e^{-(x-10)^2/(2(\frac{1}{10\sqrt{2}})^2)}\right)\left(\frac{10}{\sqrt{2\pi}}e^{-(y-5)^2/(2(\frac{1}{10})^2)}\right),$$

which is a product of a normal distribution in x with mean 10 and standard deviation $1/(10\sqrt{2})$ and a normal distribution in y with mean 5 and standard deviation $1/10$.

Problems for Section 15.7

1. Let x and y have joint density function

$$p(x,y) = \begin{cases} \frac{2}{3}(x+2y) & \text{for } 0 \le x \le 1, 0 \le y \le 1, \\ 0 & \text{otherwise.} \end{cases}$$

Find the probability that   (a) $x > 1/3$.   (b) $x < (1/3) + y$.

2. Table 15.14 gives some values of the joint density function for two variables x and y. We assume x can take the values 1, 2, 3 and 4 and y can take the values 1, 2 and 3.
   (a) Explain why this table defines a joint density function.
   (b) What is the probability that $x = 2$?
   (c) Find the probability that $y \le 2$.
   (d) Find the probability that $x \le 3$ and $y \le 2$.

TABLE 15.14

		y		
		1	2	3
x	1	0.3	0.2	0.1
	2	0.2	0.1	0
	3	0.1	0	0
	4	0	0	0

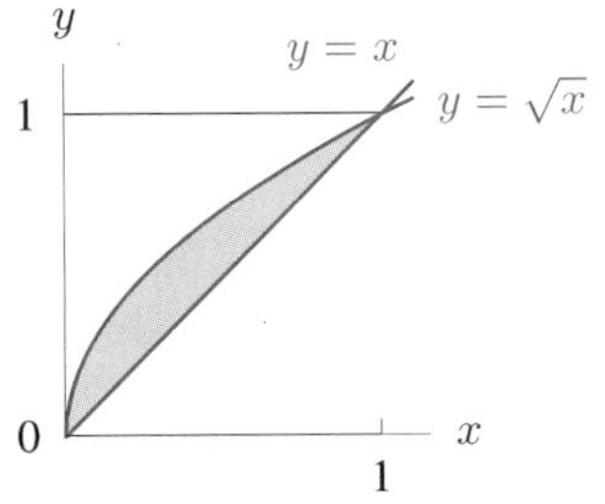

Figure 15.51

3. Assume that the joint density function for x, y is given by

$$f(x, y) = \begin{cases} kxy & \text{for } 0 \leq x \leq y \leq 1, \\ 0 & \text{otherwise.} \end{cases}$$

   (a) Determine the value of k.
   (b) Find the probability that (x, y) lies in the shaded region in Figure 15.51.

4. A joint density function is given by

$$f(x, y) = \begin{cases} kx^2 & \text{for } 0 \leq x \leq 2 \text{ and } 0 \leq y \leq 1, \\ 0 & \text{otherwise.} \end{cases}$$

   (a) Find the value of the constant k.
   (b) Find the probability that a point (x, y) satisfies $x + y \leq 2$.
   (c) Find the probability that a point (x, y) satisfies $x \leq 1$ and $y \leq 1/2$.

5. A health insurance company wants to know what proportion of its policies are going to cost them a lot of money because the insured people are over 65 and sick. In order to compute this proportion, the company defines a *disability index*, x, with $0 \leq x \leq 1$, where $x = 0$ represents perfect health and $x = 1$ represents total disability. In addition, the company uses a density function, $f(x, y)$, defined in such a way that the quantity

$$f(x, y)\, \Delta x\, \Delta y$$

   approximates the fraction of the population with disability index between x and $x + \Delta x$, and aged between y and $y + \Delta y$. The company knows from experience that a policy no longer covers its costs if the insured person is over 65 and has a disability index exceeding 0.8. Write an expression for the fraction of the company's policies held by people meeting these criteria.

6. Assume that a point is chosen at random from the region S in the xy-plane containing all points (x, y) such that $-1 \leq x \leq 1, -2 \leq y \leq 2$ and $x - y \geq 0$ (at random means that the density function is constant on S).
   (a) Determine the joint density function for x and y.
   (b) If T is a subset of S with area α, then find the probability that a point (x, y) is in T.

7. Give the joint density of x and y where x and y are independent, x has a normal distribution with mean 5 and standard deviation $1/10$, and y has a normal distribution with mean 15 and standard deviation $1/6$.

8. The probability that a radioactive substance will decay at time t is modeled by the density function

$$p(t) = \lambda e^{-\lambda t}$$

   for $t \geq 0$, and $p(t) = 0$ for $t < 0$. The positive constant λ depends on the material, and is called the decay rate.
   (a) Verify that p is a density function.
   (b) Consider two materials with decay rates λ and μ, which decay independently of each other. Write the joint density function for the probability that the first material decays at time t and the second at time s.
   (c) Find the probability that the first substance decays before the second.

9. Suppose Figure 15.52 represents a baseball field, with the bases at $(1,0)$, $(1,1)$, $(0,1)$, and home plate at $(0,0)$. The outer bound of the outfield is a piece of a circle about the origin with radius 4. Whenever a ball is hit by a batter we can record the spot on the field (i.e., in the plane) where the ball is caught.

     Let $p(r, \theta)$ be a function in the plane that gives the density of the distribution of such spots. Write an expression that represents the probability that a hit will be caught in

(a) The right field (region R).   (b) The center field (region C).

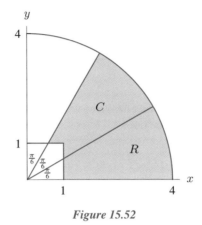

Figure 15.52

15.8 NOTES ON CHANGE OF VARIABLES IN A MULTIPLE INTEGRAL

In the previous sections, we used polar, cylindrical, and spherical coordinates to simplify iterated integrals. In this section, we discuss more general changes of variable. In the process, we will see where the extra factor of r comes from when we change from Cartesian to polar coordinates and the factor $\rho^2 \sin \phi$ when we change from Cartesian to spherical coordinates.

Polar Change of Variables Revisited

Consider the integral $\int_R (x + y)\, dA$ where R is the region in the first quadrant bounded by the circle $x^2 + y^2 = 16$ and the x and y-axes. Writing the integral in Cartesian and polar coordinates we have

$$\int_R (x+y)\, dA = \int_0^4 \int_0^{\sqrt{16-x^2}} (x+y)\, dy\, dx = \int_0^{\pi/2} \int_0^4 (r\cos\theta + r\sin\theta)r\, dr d\theta.$$

This is an integral over the rectangle in the $r\theta$-space given by $0 \leq r \leq 4$, $0 \leq \theta \leq \pi/2$. The conversion from polar to Cartesian coordinates changes this rectangle into a quarter-disk. Figure 15.53 shows how a typical rectangle (shaded) in the $r\theta$-plane with sides of length Δr and $\Delta\theta$ corresponds to a curved rectangle in the xy-plane with sides of length Δr and $r\Delta\theta$. The extra r is needed because the correspondence between r, θ and x, y not only curves the lines $r = 1, 2, 3 \ldots$ into circles, it also stretches those lines around larger and larger circles.

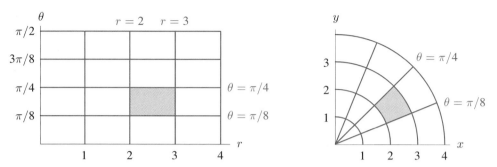

Figure 15.53: A grid in the $r\theta$-plane and the corresponding curved grid in the xy-plane

General Change of Variables

We now consider a general change of variable, where x, y coordinates are related to s, t coordinates by the differentiable functions

$$x = x(s, t) \quad y = y(s, t).$$

Just as a rectangular region in the $r\theta$-plane corresponds to a circular region in the xy-plane, a rectangular region, T, in the st-plane corresponds to a curved region, R, in the xy-plane. We assume that the change of coordinates is one-to-one, that is, that each point R corresponds to one point in T.

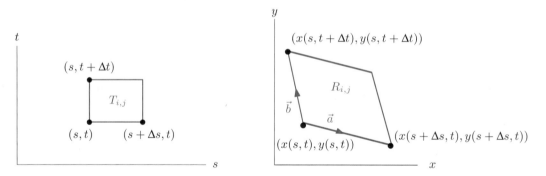

Figure 15.54: A small rectangle $T_{i,j}$ in the st-plane and the corresponding region $R_{i,j}$ of the xy-plane

We divide T into small rectangles $T_{i,j}$ with sides of length Δs and Δt. (See Figure 15.54.) The corresponding piece $R_{i,j}$ of the xy-plane is a quadrilateral with curved sides. If we choose Δs and Δt very small, then by local linearity, $R_{i,j}$ is approximately a parallelogram.

Recall from Chapter 12 that the area of the parallelogram with sides $\vec{a}$ and $\vec{b}$ is $\|\vec{a} \times \vec{b}\|$. Thus, we need to find the sides of $R_{i,j}$ as vectors. The side of $R_{i,j}$ corresponding to the bottom side of $T_{i,j}$ has endpoints $(x(s, t), y(s, t))$ and $(x(s + \Delta s, t), y(s + \Delta s, t))$, so in vector form that side is

$$\vec{a} = (x(s + \Delta s, t) - x(s, t))\vec{i} + (y(s + \Delta s, t) - y(s, t))\vec{j} + 0\vec{k} \approx \left(\frac{\partial x}{\partial s}\Delta s\right)\vec{i} + \left(\frac{\partial y}{\partial s}\Delta s\right)\vec{j} + 0\vec{k}.$$

Similarly, the side of $R_{i,j}$ corresponding to the left edge of $T_{i,j}$ is given by

$$\vec{b} \approx \left(\frac{\partial x}{\partial t}\Delta t\right)\vec{i} + \left(\frac{\partial y}{\partial t}\Delta t\right)\vec{j} + 0\vec{k}.$$

Computing the cross product, we get

$$\text{Area } R_{i,j} \approx \|\vec{a} \times \vec{b}\| \approx \left|\left(\frac{\partial x}{\partial s}\Delta s\right)\left(\frac{\partial y}{\partial t}\Delta t\right) - \left(\frac{\partial x}{\partial t}\Delta t\right)\left(\frac{\partial y}{\partial s}\Delta s\right)\right|$$

$$= \left|\frac{\partial x}{\partial s}\cdot\frac{\partial y}{\partial t} - \frac{\partial x}{\partial t}\cdot\frac{\partial y}{\partial s}\right|\Delta s\Delta t.$$

Using determinant notation, we define the *Jacobian*, $\dfrac{\partial(x,y)}{\partial(s,t)}$, as follows

$$\frac{\partial(x,y)}{\partial(s,t)} = \frac{\partial x}{\partial s}\cdot\frac{\partial y}{\partial t} - \frac{\partial x}{\partial t}\cdot\frac{\partial y}{\partial s} = \begin{vmatrix} \frac{\partial x}{\partial s} & \frac{\partial y}{\partial s} \\ \frac{\partial x}{\partial t} & \frac{\partial y}{\partial t} \end{vmatrix}.$$

Thus, we can write

$$\text{Area } R_{i,j} \approx \left|\frac{\partial(x,y)}{\partial(s,t)}\right|\Delta s\,\Delta t.$$

To compute $\int_R f(x,y)\,dA$, where f is a continuous function, we look at the Riemann sum obtained by dividing the region R into the small curved regions $R_{i,j}$, giving

$$\int_R f(x,y)\,dA \approx \sum_{i,j} f(x_i,y_j)\cdot \text{Area of } R_{i,j} \approx \sum_{i,j} f(x_i,y_j)\left|\frac{\partial(x,y)}{\partial(s,t)}\right|\Delta s\,\Delta t.$$

Each point (x_i,y_j) corresponds to a point (s_i,t_j), so the sum can be written in terms of s and t:

$$\sum_{i,j} f(x(s_i,t_j),y(s_i,t_j))\left|\frac{\partial(x,y)}{\partial(s,t)}\right|\Delta s\,\Delta t.$$

This is a Riemann sum in terms of s and t, so as Δs and Δt approach 0, we get

$$\int_R f(x,y)\,dA = \int_T f(x(s,t),y(s,t))\left|\frac{\partial(x,y)}{\partial(s,t)}\right|ds\,dt.$$

> To convert an integral from x,y to s,t coordinates we make three changes:
> 1. Substitute for x and y in the integrand in terms of s and t.
> 2. Change the xy region R into an st region T.
> 3. Introduce the absolute value of the Jacobian, $\left|\dfrac{\partial(x,y)}{\partial(s,t)}\right|$, representing the change in the area element.

Example 1   Verify that the Jacobian $\dfrac{\partial(x,y)}{\partial(r,\theta)} = r$ for polar coordinates $x = r\cos\theta$, $y = r\sin\theta$.

Solution $\dfrac{\partial(x,y)}{\partial(r,\theta)} = \begin{vmatrix} \frac{\partial x}{\partial r} & \frac{\partial y}{\partial r} \\ \frac{\partial x}{\partial\theta} & \frac{\partial y}{\partial\theta} \end{vmatrix} = \begin{vmatrix} \cos\theta & \sin\theta \\ -r\sin\theta & r\cos\theta \end{vmatrix} = r\cos^2\theta + r\sin^2\theta = r.$

Example 2   Find the area of the ellipse $\dfrac{x^2}{a^2} + \dfrac{y^2}{b^2} = 1$.

Solution   Let $x = as$, $y = bt$. Then the ellipse $x^2/a^2 + y^2/b^2 = 1$ in the xy-plane corresponds to the circle $s^2 + t^2 = 1$ in the st-plane. The Jacobian is $\begin{vmatrix} a & 0 \\ 0 & b \end{vmatrix} = ab$. Thus, if we let R be the ellipse in the xy-plane and T the unit circle in the st-plane, we get

$$\text{Area of } xy\text{-ellipse} = \int_R 1 \, dA = \int_T 1ab \, ds \, dt = ab \int_T ds \, dt = ab \cdot \text{Area of } st\text{-circle} = \pi ab.$$

Change of Variables in Triple Integrals

For triple integrals, there is a similar formula. Suppose the differentiable functions

$$x = x(s, t, u), \quad y = y(s, t, u), \quad z = z(s, t, u)$$

define a change of variables from a region S in stu-space to a region W in xyz-space. Then, the Jacobian of this change of variables is given by the determinant

$$\frac{\partial(x, y, z)}{\partial(s, t, u)} = \begin{vmatrix} \frac{\partial x}{\partial s} & \frac{\partial y}{\partial s} & \frac{\partial z}{\partial s} \\ \frac{\partial x}{\partial t} & \frac{\partial y}{\partial t} & \frac{\partial z}{\partial t} \\ \frac{\partial x}{\partial u} & \frac{\partial y}{\partial u} & \frac{\partial z}{\partial u} \end{vmatrix}.$$

Just as the Jacobian in two dimensions gives us the change in the area element, the Jacobian in three dimensions represents the change in the volume element. Thus, we have

$$\int_W f(x, y, z) \, dx \, dy \, dz = \int_S f(x(s, t, u), y(s, t, u), z(s, t, u)) \left| \frac{\partial(x, y, z)}{\partial(s, t, u)} \right| ds \, dt \, du.$$

Problem 3 at the end of this section asks you to verify that the Jacobian for the change of variables for spherical coordinates is $\rho^2 \sin \phi$. The next example generalizes Example 2 to ellipsoids.

Example 3   Find the volume of the ellipsoid $\dfrac{x^2}{a^2} + \dfrac{y^2}{b^2} + \dfrac{z^2}{c^2} = 1$.

Solution   Let $x = as$, $y = bt$, $z = cu$. The Jacobian is computed to be abc. The xyz-ellipsoid corresponds to the stu-sphere $s^2 + t^2 + u^2 = 1$. Thus, as in Example 2,

$$\text{Volume of } xyz\text{-ellipsoid} = abc \cdot \text{Volume of } stu\text{-sphere} = abc \frac{4}{3}\pi = \frac{4}{3}\pi abc.$$

Problems for Section 15.8

1. Find the region R in the xy-plane corresponding to the region $T = \{(s,t) \mid 0 \le s \le 3, 0 \le t \le 2\}$ under the change of variables $x = 2s - 3t$, $y = s - 2t$. Check that

$$\int_R dx\,dy = \int_T \left| \frac{\partial(x,y)}{\partial(s,t)} \right| ds\,dt.$$

2. Find the region R in the xy-plane corresponding to the region $T = \{(s,t) \mid 0 \le s \le 2, s \le t \le 2\}$ under the change of variables $x = s^2$, $y = t$. Check that

$$\int_R dx\,dy = \int_T \left| \frac{\partial(x,y)}{\partial(s,t)} \right| ds\,dt.$$

3. Compute the Jacobian for the change of variables into spherical coordinates:

$$x = \rho \sin \phi \cos \theta, \quad y = \rho \sin \phi \sin \theta, \quad z = \rho \cos \phi.$$

4. For the change of variables $x = 3s - 4t$, $y = 5s + 2t$, show that

$$\frac{\partial(x,y)}{\partial(s,t)} \cdot \frac{\partial(s,t)}{\partial(x,y)} = 1$$

5. Use the change of variables $x = 2s + t$, $y = s - t$ to compute the integral $\int_R (x + y)\,dA$, where R is the parallelogram formed by $(0,0)$, $(3,-3)$, $(5,-2)$, and $(2,1)$.

6. Use the change of variables $x = \frac{1}{2}s$, $y = \frac{1}{3}t$ to compute the integral $\int_R (x^2 + y^2)\,dA$, where R is the region bounded by the curve $4x^2 + 9y^2 = 36$.

7. Use the change of variables $s = xy$, $t = xy^2$ to compute $\int_R xy^2\,dA$, where R is the region bounded by $xy = 1$, $xy = 4$, $xy^2 = 1$, $xy^2 = 4$.

8. Evaluate the integral $\int_R \cos\left(\frac{x-y}{x+y}\right) dx\,dy$ where R is the triangle bounded by $x + y = 1$, $x = 0$, and $y = 0$.

REVIEW PROBLEMS FOR CHAPTER FIFTEEN

1. Figure 15.55 shows contours of average annual rainfall (in inches) in South America.[6] Each grid square is 500 miles on a side. Estimate the total volume of rain that falls on the considered area in a year.

2. Figure 15.56 gives isotherms for low winter temperature in Washington, DC.[7] The grid squares are one mile on a side. Find the average low temperature over the whole city (the city is the shaded region).

---

[6]From *Modern Physical Geography*, Alan H. Strahler and Arthur H. Strahler, Fourth Edition, John Wiley & Sons, New York, 1992, p. 144

[7]From *Physical Geography of the Global Environment*, H. J. de Blij and Peter O. Muller, John Wiley & Sons, New York, 1993, p. 220

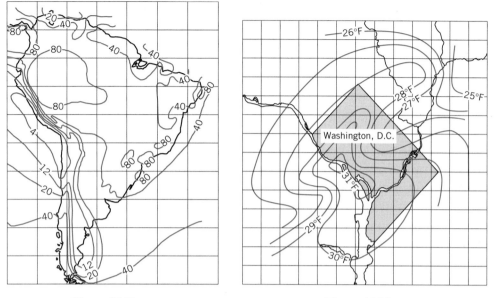

Figure 15.55   Figure 15.56

Sketch the regions over which the integrals in Problems 3–6 are being performed.

3. $\int_1^4 \int_{-\sqrt{y}}^{\sqrt{y}} f(x, y)\, dx\, dy$

4. $\int_0^1 \int_0^{\sin^{-1} y} f(x, y)\, dx\, dy$

5. $\int_{-1}^1 \int_{-\sqrt{1-x^2}}^{\sqrt{1-x^2}} f(x, y)\, dy\, dx$

6. $\int_0^2 \int_{-\sqrt{4-y^2}}^0 f(x, y)\, dx\, dy$

Calculate exactly the integrals in Problems 7–12. (Your answer may contain e, π, $\sqrt{2}$, and so on.)

7. $\int_0^1 \int_0^z \int_0^2 (y + z)^7\, dx\, dy\, dz$

8. $\int_0^1 \int_3^4 (\sin(2 - y)) \cos(3x - 7)\, dx\, dy$

9. $\int_0^{10} \int_0^{0.1} x e^{xy}\, dy\, dx$

10. $\int_0^1 \int_0^y (\sin^3 x)(\cos x)(\cos y)\, dx\, dy$

11. $\int_3^4 \int_0^1 x^2 y \cos(xy)\, dy\, dx$

12. $\int_0^1 \int_{-\sqrt{1-x^2}}^{\sqrt{1-x^2}} e^{-(x^2 + y^2)}\, dy\, dx$

13. Write $\int_R f(x, y)\, dA$ as an iterated integral if R is the region in Figure 15.57.

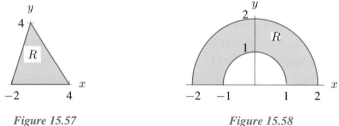

Figure 15.57   Figure 15.58

14. Evaluate $\int_R \sqrt{x^2 + y^2}\, dA$ where R is the region in Figure 15.58.

15. Set up $\int_R f\, dV$ as an iterated integral in all six possible orders of integration, where R is the hemisphere bounded by the upper half of $x^2 + y^2 + z^2 = 1$ and the xy-plane.

Evaluate the integrals in Problems 16–18 by changing them to cylindrical or spherical coordinates as appropriate.

16. $\displaystyle\int_{-\sqrt{3}}^{\sqrt{3}} \int_{-\sqrt{3-x^2}}^{\sqrt{3-x^2}} \int_{1}^{4-x^2-y^2} \frac{1}{z^2}\, dz\, dy\, dx$ 17. $\displaystyle\int_{0}^{3} \int_{-\sqrt{9-z^2}}^{\sqrt{9-z^2}} \int_{-\sqrt{9-y^2-z^2}}^{\sqrt{9-y^2-z^2}} x^2\, dx\, dy\, dz$

18. $\displaystyle\int_{0}^{1} \int_{0}^{\sqrt{1-x^2}} \int_{0}^{\sqrt{x^2+y^2}} (z + \sqrt{x^2+y^2})\, dz\, dy\, dx$

19. If $W = \{(x,y,z) : 1 \leq x^2 + y^2 \leq 4, 0 \leq z \leq 4\}$ evaluate the integral $\displaystyle\int_{W} \frac{z}{(x^2+y^2)^{3/2}}\, dV$.

20. Write an integral representing the mass of a sphere of radius 3 if the density of the sphere at any point is twice the distance of that point from the center of the sphere.

21. A forest next to a road has the shape in Figure 15.59. The population density of rabbits is proportional to the distance from the road. It is 0 at the road, and 10 rabbits per square mile at the opposite edge of the forest. Find the total rabbit population in the forest.

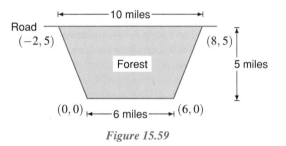

Figure 15.59

For Problems 22–23, use the definition of moment of inertia on Page 239.

22. Consider a rectangular brick with length 5, width 3, and height 1, and of uniform density 1. Compute the moment of inertia about each of the three axes passing through the center of the brick, perpendicular to one of the sides.

23. Compute the moment of inertia of a ball of radius R about an axis passing through its center. Assume that the ball has a constant density of 1.

24. In this problem you will derive one of the remarkable formulas of mathematics, namely that

$$\int_{-\infty}^{\infty} e^{-x^2}\, dx = \sqrt{\pi}.$$

(a) Change the following double integral into polar coordinates and evaluate it:

$$\int_{-\infty}^{\infty} \int_{-\infty}^{\infty} e^{-(x^2+y^2)}\, dx\, dy.$$

(b) Explain why

$$\int_{-\infty}^{\infty} \int_{-\infty}^{\infty} e^{-(x^2+y^2)}\, dx\, dy = \left(\int_{-\infty}^{\infty} e^{-x^2}\, dx\right)^2.$$

(c) Explain why the answers to parts (a) and (b) give the formula we want.

25. A particle of mass m is placed at the center of one base of a circular cylindrical shell of inner radius r_1, outer radius r_2, height h, and constant density δ. Find the force of gravitational attraction exerted by the cylinder on the particle.

26. Find the area of the half-moon shape with circular arcs as edges and the dimensions shown in Figure 15.60.

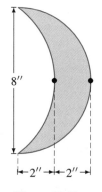

Figure 15.60

27. Find the area of the metal frames with one or four cutouts shown in Figure 15.61. Start with Cartesian coordinates x, y aligned along one side. Consider slanted coordinates $u = x - y$, $v = y$ in which the frame is "straightened". [Hint: First describe the shape of the cut-out in the uv-plane; second, calculate its area in the uv-plane; third, using Jacobians, calculate its area in the xy-plane.]

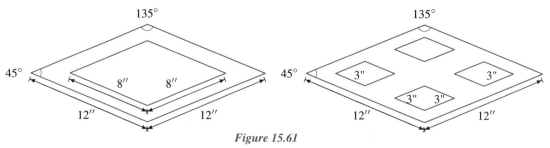

Figure 15.61

28. A river follows the path $y = f(x)$ where x, y are in kilometers. Near the sea, it widens into a lagoon, then narrows again at its mouth. See Figure 15.62. At the point (x, y), the depth, $d(x, y)$, of the lagoon is given by

$$d(x, y) = 40 - 160(y - f(x))^2 - 40x^2 \text{ meters.}$$

The lagoon itself is described by $d(x, y) \geq 0$. What is the volume of the lagoon in cubic meters? [Hint: Use new coordinates $u = x/2$, $v = y - f(x)$ and Jacobians.]

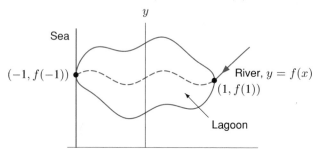

Figure 15.62

CHAPTER SIXTEEN

PARAMETERIZED CURVES AND SURFACES

In single-variable calculus, we study the motion of a particle along a line. For example, we represent the motion of an object thrown straight up into the air by a single function $h(t)$, the height of the object above the ground at time t.

To study the motion of a particle in space, we must express all the coordinates of the particle in terms of t, giving $x(t)$, $y(t)$, and $z(t)$ if the motion is in 3-space. This is called the *parametric representation* of the path of motion, which is a curve. The parametric representation enables us to find the velocity and acceleration of the particle. In addition, we use parameterization to study surfaces in 3-dimensional space.

16.1 PARAMETERIZED CURVES

How Do We Represent Motion?

To represent the motion of a particle in the xy-plane we use two equations, one for the x-coordinate of the particle, $x = f(t)$, and another for the y-coordinate, $y = g(t)$. Thus at time t the particle is at the point $(f(t), g(t))$. The equation for x describes the right-left motion; the equation for y describes the up-down motion. The two equations for x and y are called *parametric equations* with *parameter* t.

Example 1   Describe the motion of the particle whose coordinates at time t are $x = \cos t, y = \sin t$.

Solution   Since $(\cos t)^2 + (\sin t)^2 = 1$, we have $x^2 + y^2 = 1$. That is, at any time t the particle is at a point (x, y) somewhere on the unit circle $x^2 + y^2 = 1$. We plot points at different times to see how the particle moves on the circle. (See Figure 16.1 and Table 16.1.) The particle moves at a uniform speed, completing one full trip counterclockwise around the circle every 2π units of time. Notice how the x-coordinate goes repeatedly back and forth from -1 to 1 while the y-coordinate goes repeatedly up and down from -1 to 1. The two motions combine to trace out a circle.

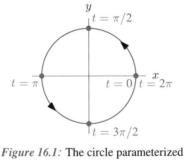

Figure 16.1: The circle parameterized by $x = \cos t, y = \sin t$

TABLE 16.1   *Points on the circle with* $x = \cos t, y = \sin t$

t	x	y
0	1	0
$\pi/2$	0	1
π	-1	0
$3\pi/2$	0	-1
2π	1	0

Example 2   Figure 16.2 shows the graphs of two functions, $f(t)$ and $g(t)$. Describe the motion of the particle whose coordinates at time t are $x = f(t), \quad y = g(t)$.

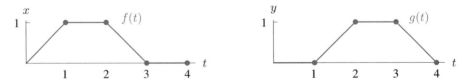

Figure 16.2: Graphs of $x = f(t)$ and $y = g(t)$ used to trace out the path $(f(t), g(t))$ in Figure 16.3

Solution   Between times $t = 0$ and $t = 1$, the x-coordinate goes from 0 to 1, while the y-coordinate stays fixed at 0. So the particle moves along the x-axis from $(0, 0)$ to $(1, 0)$. Then, between times $t = 1$ and $t = 2$, the x-coordinate stays fixed at $x = 1$, while the y-coordinate goes from 0 to 1. Thus, the particle moves along the vertical line from $(1, 0)$ to $(1, 1)$. Similarly, between times $t = 2$ and $t = 3$, it moves horizontally back to $(0, 1)$, and between times $t = 3$ and $t = 4$ it moves down the y-axis to $(0, 0)$. Thus, it traces out the square in Figure 16.3.

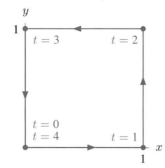

Figure 16.3: The square parameterized by $(f(t), g(t))$

Different Motions Along the Same Path

Example 3   Describe the motion of the particle whose x and y coordinates at time t are given by the equations

$$x = \cos 3t, \quad y = \sin 3t.$$

Solution   Since $(\cos 3t)^2 + (\sin 3t)^2 = 1$, we have $x^2 + y^2 = 1$, giving motion around the unit circle. But if we plot points at different times, we see that in this case the particle is moving three times as fast as in Example 1 on page 274. (See Figure 16.4 and Table 16.2.)

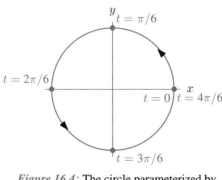

Figure 16.4: The circle parameterized by
$x = \cos 3t$, $y = \sin 3t$

TABLE 16.2   *Points on circle with $x = \cos 3t$, $y = \sin 3t$*

t	x	y
0	1	0
$\pi/6$	0	1
$2\pi/6$	-1	0
$3\pi/6$	0	-1
$4\pi/6$	1	0

Example 3 is obtained from Example 1 by replacing t by $3t$; this is called a *change in parameter*. If we make a change in parameter, the particle traces out the same curve (or a part of it) but at a different speed or in a different direction. Section 16.2 shows how to compute the speed of a moving particle.

Example 4   Describe the motion of the particle whose x and y coordinates at time t are

$$x = \cos(e^{-t^2}), \quad y = \sin(e^{-t^2}).$$

Solution   As in Examples 1 and 3, we have $x^2 + y^2 = 1$ so the motion lies on the unit circle. As time t goes from $-\infty$ (way back in the past) to 0 (the present) to ∞ (way off in the future), e^{-t^2} goes from near 0 to 1 back to near 0. So $(x, y) = (\cos(e^{-t^2}), \sin(e^{-t^2}))$ goes from near $(1, 0)$ to $(\cos 1, \sin 1)$ and back to near $(1, 0)$. The particle does not actually reach the point $(1, 0)$. (See Figure 16.5 and Table 16.3.)

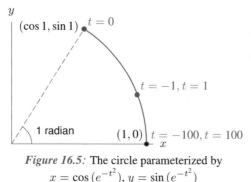

Figure 16.5: The circle parameterized by
$x = \cos\left(e^{-t^2}\right)$, $y = \sin\left(e^{-t^2}\right)$

TABLE 16.3   *Points on circle with*
$x = \cos(e^{-t^2})$, $y = \sin(e^{-t^2})$

t	x	y
-100	~ 1	~ 0
-1	0.93	0.36
0	0.54	0.84
1	0.93	0.36
100	~ 1	~ 0

Parametric Representations of Curves in the Plane

Sometimes we are more interested in the curve traced out by the particle than we are in the motion itself. In that case we will call the parametric equations a *parameterization* of the curve. As we can see by comparing Examples 1 and 3, two different parameterizations can describe the same curve in 2-space. Though the parameter, which we usually denote by t, may not have physical meaning it is still helpful to think of it as time.

Example 5   Give a parameterization of the semicircle of radius 1 shown in Figure 16.6.

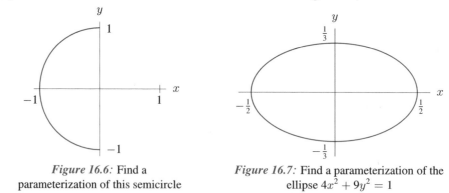

Figure 16.6: Find a
parameterization of this semicircle

Figure 16.7: Find a parameterization of the
ellipse $4x^2 + 9y^2 = 1$

Solution   We can use the equations $x = \cos t$ and $y = \sin t$ for counterclockwise motion in a circle, from Example 1 on page 274. The particle passes $(0, 1)$ at $t = \pi/2$, moves counterclockwise around the circle, and reaches $(0, -1)$ at $t = 3\pi/2$. So a parameterization is

$$x = \cos t, \ y = \sin t, \quad \frac{\pi}{2} \le t \le \frac{3\pi}{2}.$$

Example 6   Give a parameterization of the ellipse $4x^2 + 9y^2 = 1$ shown in Figure 16.7.

Solution   Since $(2x)^2 + (3y)^2 = 1$, we adapt the parameterization of the circle in Example 1. Replacing x by $2x$ and y by $3y$ gives the equations $2x = \cos t, 3y = \sin t$. A parameterization of the ellipse is thus

$$x = \tfrac{1}{2}\cos t, \qquad y = \tfrac{1}{3}\sin t, \qquad 0 \le t \le 2\pi.$$

We usually require that the parameterization of a curve go from one end of the curve to the other without retracing any portion of the curve. This is different from parameterizing the motion of a particle, where, for example, a particle may move around the same circle many times.

Parameterizing the Graph of a Function

The graph of any function $y = f(x)$ can be parameterized by letting the parameter t be x:

$$x = t, \quad y = f(t).$$

Example 7   Give parametric equations for the curve $y = x^3 - x$. In which direction does this parameterization trace out the curve?

Solution   Let $x = t, y = t^3 - t$. Thus, $y = t^3 - t = x^3 - x$. Since $x = t$, as time increases the x-coordinate moves from left to right, so the particle traces out the curve $y = x^3 - x$ from left to right.

Curves Given Parametrically

Some complicated curves can be graphed more easily using parametric equations; the next example shows such a curve.

Example 8   Assume t is time in seconds. Sketch the curve traced out by the particle whose motion is given by

$$x = \cos 3t, \quad y = \sin 5t.$$

Solution   The x-coordinate oscillates back and forth between 1 and -1, completing 3 oscillations every 2π seconds. The y-coordinate oscillates up and down between 1 and -1, completing 5 oscillations every 2π seconds. Since both the x- and y-coordinates return to their original values every 2π seconds, the curve is retraced every 2π seconds. The result is a pattern called a Lissajous figure. (See Figure 16.8.) Problems 35–38 concern Lissajous figures $x = \cos at, y = \sin bt$ for other values of a and b.

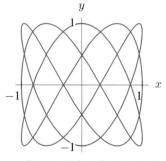

Figure 16.8: A Lissajous figure:
$x = \cos 3t, y = \sin 5t$

Parametric Equations in Three Dimensions

To describe a motion in 3-dimensional space parametrically, we need a third equation giving z in terms of t.

Example 9   Describe in words the motion given parametrically by

$$x = \cos t, \quad y = \sin t, \quad z = t.$$

Solution   The particle's x- and y-coordinates are the same as in Example 1, which gives circular motion in the xy-plane, while the z-coordinate increases steadily. Thus, the particle traces out a rising spiral, like a coiled spring. (See Figure 16.9.) This curve is called a *helix*.

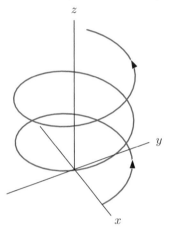

Figure 16.9: The helix $x = \cos t, y = \sin t, z = t$

Example 10   Find parametric equations for the line through the point $(1, 5, 7)$ and parallel to the vector $2\vec{i} + 3\vec{j} + 4\vec{k}$.

Solution   Let's imagine a particle at the point $(1, 5, 7)$ at time $t = 0$ and moving through a displacement of $2\vec{i} + 3\vec{j} + 4\vec{k}$ for each unit of time, t. When $t = 0$, $x = 1$ and x increases by 2 units for every unit of time. Thus, at time t, the x-coordinate of the particle is given by

$$x = 1 + 2t.$$

Similarly, the y-coordinate starts at $y = 5$ and increases at a rate of 3 units for every unit of time. The z-coordinate starts at $y = 7$ and increases by 4 units for every unit of time. Thus, the parametric equations of the line are

$$x = 1 + 2t, \quad y = 5 + 3t, \quad z = 7 + 4t.$$

We can generalize the previous example as follows:

> **Parametric Equations of a Line** through the point (x_0, y_0, z_0) and parallel to the vector $a\vec{i} + b\vec{j} + c\vec{k}$ are
> $$x = x_0 + at, \quad y = y_0 + bt, \quad z = z_0 + ct.$$

Notice that the parameterization of a line given above expresses the coordinates x, y, and z as linear functions of the parameter t.

Example 11 (a) Describe in words the curve given by these parametric equations:

$$x = 3 + t, \quad y = 2t, \quad z = 1 - t.$$

(b) Find parametric equations for the line through the points $(1, 2, -1)$ and $(3, 3, 4)$.

Solution (a) The curve is a line through the point $(3, 0, 1)$ and parallel to the vector $\vec{i} + 2\vec{j} - \vec{k}$.
(b) The line is parallel to the displacement vector between the points $P = (1, 2, -1)$ and $Q = (3, 3, 4)$.

$$\overrightarrow{PQ} = (3 - 1)\vec{i} + (3 - 2)\vec{j} + (4 - (-1))\vec{k} = 2\vec{i} + \vec{j} + 5\vec{k}.$$

Thus, the parametric equations are

$$x = 1 + 2t, \quad y = 2 + t, \quad z = -1 + 5t.$$

Note that the equations $x = 3 + 2t, y = 3 + t, z = 4 + 5t$ represent the same line.

Where Does a Curve Pierce a Surface?

Parametric equations for a curve enable us to find where the curve intersects a given surface.

Example 12 Find the points at which the line $x = t, y = 2t, z = 1 + t$ pierces the sphere of radius 10 centered at the origin.

Solution The equation for the sphere of radius 10 and center at the origin is

$$x^2 + y^2 + z^2 = 100.$$

To find the intersection points of the line and the sphere, substitute the parametric equations of the line into the equation of the sphere, giving

$$t^2 + 4t^2 + (1 + t)^2 = 100,$$

so

$$6t^2 + 2t - 99 = 0,$$

which has the two solutions at approximately $t = -4.23$ and $t = 3.90$. Using the parametric equation for the line, $(x, y, z) = (t, 2t, 1 + t)$, we see that the line cuts the sphere at the two points

$$(x, y, z) = (-4.23, 2(-4.23), 1 + (-4.23)) = (-4.23, -8.46, -3.23),$$

and

$$(x, y, z) = (3.90, 2(3.90), 1 + 3.90) = (3.90, 7.80, 4.90).$$

Problems for Section 16.1

For Problems 1–4, describe the motion of a particle whose position at time t is $x = f(t)$, $y = g(t)$, where the graphs of f and g are as shown.

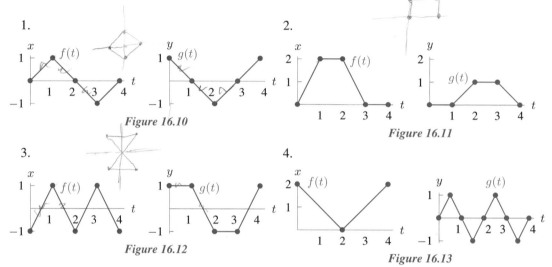

Figure 16.10

Figure 16.11

Figure 16.12

Figure 16.13

Problems 5–10 give parameterizations of the unit circle or a part of it. In each case, describe in words how the circle is traced out, including when and where the particle is moving clockwise and when and where the particle is moving counterclockwise.

5. $x = \cos t$, $\quad y = -\sin t$
6. $x = \sin t$, $\quad y = \cos t$
7. $x = \cos(t^2)$, $\quad y = \sin(t^2)$
8. $x = \cos(t^3 - t)$, $\quad y = \sin(t^3 - t)$
9. $x = \cos(\ln t)$, $\quad y = \sin(\ln t)$
10. $x = \cos(\cos t)$, $\quad y = \sin(\cos t)$

11. Describe the similarities and differences among the motions in the plane given by the following three pairs of parametric equations:
    (a) $x = t$, $\quad y = t^2$ (b) $x = t^2$, $\quad y = t^4$ (c) $x = t^3$, $\quad y = t^6$.

Write a parameterization for each of the curves in the xy-plane in Problems 12–18.

12. A circle of radius 3 centered at the origin and traced out clockwise.
13. A vertical line through the point $(-2, -3)$.
14. A circle of radius 5 centered at the point $(2, 1)$ and traced out counterclockwise.
15. A circle of radius 2 centered at the origin traced clockwise starting from $(-2, 0)$ when $t = 0$.
16. The line through the points $(2, -1)$ and $(1, 3)$.
17. An ellipse centered at the origin and crossing the x-axis at ± 5 and the y-axis at ± 7.
18. An ellipse centered at the origin, crossing the x-axis at ± 3 and the y-axis at ± 7. Start at the point $(-3, 0)$ and trace out the ellipse counterclockwise.
19. As t varies, the following parametric equations trace out a line in the plane

$$x = 2 + 3t, \quad y = 4 + 7t.$$

(a) What part of the line is obtained by restricting t to nonnegative numbers?

(b)   What part of the line is obtained if t is restricted to $-1 \le t \le 0$?

(c)   How should t be restricted to give the part of the line to the left of the y-axis?

20.   Suppose $a, b, c, d, m, n, p, q > 0$. Match each pair of parametric equations below with one of the lines l_1, l_2, l_3, l_4 in Figure 16.14.

I. $\begin{cases} x = a + ct, \\ y = -b + dt. \end{cases}$ II. $\begin{cases} x = m + pt, \\ y = n - qt. \end{cases}$

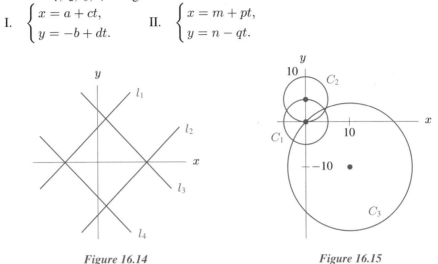

Figure 16.14   Figure 16.15

21.   What can you say about the values of a, b and k if the equations

$$x = a + k \cos t, \quad y = b + k \sin t, \qquad 0 \le t \le 2\pi,$$

trace out each of the circles in Figure 16.15? (a) C_1 (b) C_2 (c) C_3

22.   Describe in words the curve represented by the parametric equations

$$x = 3 + t^3, \quad y = 5 - t^3, \quad z = 7 + 2t^3.$$

Write a parameterization in 3-space for each of the curves in Problems 23–24.

23.   The circle of radius 2 in the xz-plane, centered at the origin.

24.   The circle of radius 3 centered at the point $(0, 0, 2)$ parallel to the xy-plane.

For Problems 25–29, find parametric equations for the given line.

25.   The line through the points $(2, 3, -1)$ and $(5, 2, 0)$.

26.   The line pointing in the direction of the vector $3\vec{i} - 3\vec{j} + \vec{k}$ and through the point $(1, 2, 3)$.

27.   The line parallel to the z-axis passing through the point $(1, 0, 0)$.

28.   The line of intersection of the planes $x - y + z = 3$ and $2x + y - z = 5$.

29.   The line perpendicular to the surface $z = x^2 + y^2$ at the point $(1, 2, 5)$.

30.   Do the lines in Problems 25 and 26 intersect?

31. Is the point $(-3, -4, 2)$ visible from the point $(4, 5, 0)$ if there is an opaque ball of radius 1 centered at the origin?

32. Show that the equations

$$x = 3 + t, \quad y = 2t, \quad z = 1 - t$$

satisfy the equations $x + y + 3z = 6$ and $x - y - z = 2$. What does this tell you about the curve parameterized by these equations?

33. Two particles are traveling through space. At time t the first particle is at the point $(-1 + t, 4 - t, -1 + 2t)$ and the second particle is at $(-7 + 2t, -6 + 2t, -1 + t)$.
    (a) Describe the two paths.
    (b) Do the two particles collide? If so, when and where?
    (c) Do the paths of the two particles cross? If so, where?

34. Imagine a light shining on the helix of Example 9 on page 278 from far down each of the axes. Sketch the shadow cast by the helix on each of the coordinate planes: xy, xz, and yz.

Graph the Lissajous figures in Problems 35–38 using a calculator or computer.

35. $x = \cos 2t, \quad y = \sin 5t$

36. $x = \cos 3t, \quad y = \sin 7t$

37. $x = \cos 2t, \quad y = \sin 4t$

38. $x = \cos 2t, \quad y = \sin \sqrt{3}t$

39. Motion along a straight line is given by a single equation, say, $x = t^3 - t$ where x is distance along the line. It is difficult to see the motion from a plot; it just traces out the x-line, as in Figure 16.16. To visualize the motion, we introduce a y-coordinate and let it slowly increase, giving Figure 16.17. Try the following on a calculator or computer. Let $y = t$. Now plot the parametric equations $x = t^3 - t$, $y = t$ for, say, $-3 \leq t \leq 3$. What does the plot in Figure 16.17 tell you about the particle's motion?

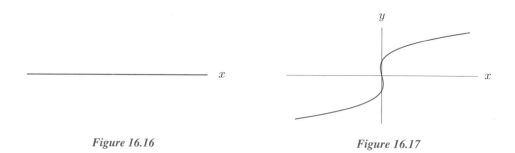

Figure 16.16   Figure 16.17

For Problems 40–42, plot the motion along the x-line by the method of Problem 39. What does the plot tell you about the particle's motion?

40. $x = \cos t, \quad -10 \leq t \leq 10$

41. $x = t^4 - 2t^2 + 3t - 7, \quad -3 \leq t \leq 2$

42. $x = t \ln t, \quad 0.01 \leq t \leq 10$

16.2 MOTION, VELOCITY, AND ACCELERATION

In this section, we write parametric equations using position vectors. This enables us to calculate the velocity and acceleration of a particle moving in 2-space or 3-space.

Using Position Vectors to Write Parameterized Curves as Vector Functions

Recall that a point in the plane with coordinates (x, y) can be represented by the position vector $\vec{r} = x\vec{i} + y\vec{j}$ shown in Figure 16.18. Similarly, in 3-space we write $\vec{r} = x\vec{i} + y\vec{j} + z\vec{k}$. (See Figure 16.19.)

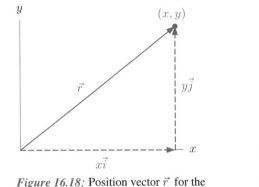

Figure 16.18: Position vector $\vec{r}$ for the point (x, y)

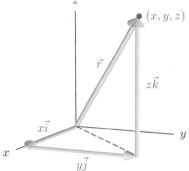

Figure 16.19: Position vector $\vec{r}$ for the point (x, y, z)

We can write the parametric equations $x = f(t)$, $y = g(t)$, $z = h(t)$ as a single vector equation

$$\vec{r}(t) = f(t)\vec{i} + g(t)\vec{j} + h(t)\vec{k}$$

called *parameterization*. As the parameter t varies, the point with position vector $\vec{r}(t)$ traces out a curve in 3-space. For example, the circular motion

$$x = \cos t, y = \sin t \quad \text{can be written as} \quad \vec{r} = (\cos t)\vec{i} + (\sin t)\vec{j}$$

and the helix

$$x = \cos t, y = \sin t, z = t \quad \text{can be written as} \quad \vec{r} = (\cos t)\vec{i} + (\sin t)\vec{j} + t\vec{k}.$$

See Figure 16.20.

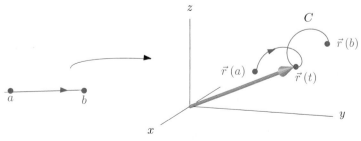

Figure 16.20: The parameterization sends the interval, $a \leq t \leq b$, to the curve, C, in 3-space

Example 1   Give a parameterization for the circle of radius $\frac{1}{2}$ centered at the point $(-1, 2)$.

Solution   The circle of radius 1 centered at the origin is parameterized by the vector-valued function

$$\vec{r_1}(t) = \cos t\vec{i} + \sin t\vec{j}, \quad 0 \le t \le 2\pi.$$

The point $(-1, 2)$ has the position vector $\vec{r}_0 = -\vec{i} + 2\vec{j}$. The position vector, $\vec{r}(t)$, of a point on the circle of radius $\frac{1}{2}$ centered at $(-1, 2)$ is found by adding $\frac{1}{2}\vec{r_1}$ to $\vec{r}_0$. (See Figures 16.21 and 16.22.) Thus,

$$\vec{r}(t) = \vec{r}_0 + \tfrac{1}{2}\vec{r}_1(t) = -\vec{i} + 2\vec{j} + \tfrac{1}{2}(\cos t\vec{i} + \sin t\vec{j}) = (-1 + \tfrac{1}{2}\cos t)\vec{i} + (2 + \tfrac{1}{2}\sin t)\vec{j},$$

or, equivalently,

$$x = -1 + \tfrac{1}{2}\cos t, \quad y = 2 + \tfrac{1}{2}\sin t, \qquad 0 \le t \le 2\pi.$$

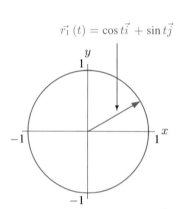

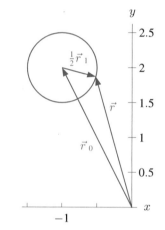

Figure 16.21: The circle $x^2 + y^2 = 1$ parameterized by $\vec{r_1}(t) = \cos t\vec{i} + \sin t\vec{j}$

Figure 16.22: The circle of radius $\frac{1}{2}$ and center $(-1, 2)$ parameterized by $\vec{r}(t) = \vec{r}_0 + \tfrac{1}{2}\vec{r}_1(t)$

Parametric Equation of a Line

Consider a straight line in the direction of a vector $\vec{v}$ passing through the point (x_0, y_0, z_0) with position vector $\vec{r}_0$. We start at $\vec{r}_0$ and move up and down the line, adding different multiples of $\vec{v}$ to $\vec{r}_0$. (See Figure 16.23.)

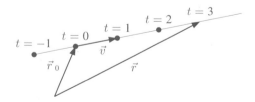

Figure 16.23: The line $\vec{r}(t) = \vec{r}_0 + t\vec{v}$

In this way, every point on the line can be written as $\vec{r}_0 + t\vec{v}$, which yields the following:

Parametric Equation of a Line

The line through the point with position vector $\vec{r_0} = x_0\vec{i} + y_0\vec{j} + z_0\vec{k}$ in the direction of the vector $\vec{v} = a\vec{i} + b\vec{j} + c\vec{k}$ has parametric equation

$$\vec{r}(t) = \vec{r}_0 + t\vec{v}.$$

Example 2   Find the parametric equation for

(a)   The line passing through the points $(2, -1, 3)$ and $(-1, 5, 4)$.

(b)   The line segment from $(2, -1, 3)$ to $(-1, 5, 4)$.

Solution   (a)   The line passes through $(2, -1, 3)$ and is parallel to the displacement vector $\vec{v} = -3\vec{i} + 6\vec{j} + \vec{k}$ from $(2, -1, 3)$ to $(-1, 5, 4)$. Thus the parametric equation is

$$\vec{r}(t) = 2\vec{i} - \vec{j} + 3\vec{k} + t(-3\vec{i} + 6\vec{j} + \vec{k}).$$

(b)   In the parameterization in part (a), $t = 0$ corresponds to the point $(2, -1, 3)$ and $t = 1$ corresponds to the point $(-1, 5, 4)$. So the parameterization of the segment is

$$\vec{r}(t) = 2\vec{i} - \vec{j} + 3\vec{k} + t(-3\vec{i} + 6\vec{j} + \vec{k}), \qquad 0 \le t \le 1.$$

The Velocity Vector

The velocity of a moving particle can be represented by a vector with the following properties:

The **velocity vector** of a moving object is a vector $\vec{v}$ such that:
- The magnitude of $\vec{v}$ is the speed of the object.
- The direction of $\vec{v}$ is the direction of motion.

Thus the speed of the object is $\|\vec{v}\|$ and the velocity vector is tangent to the object's path.

Example 3   A child is sitting on a ferris wheel of diameter 10 meters, making one revolution every 2 minutes. Find the speed of the child and draw velocity vectors at two different times.

Solution   The child moves at a constant speed around a circle of radius 5 meters, completing one revolution every 2 minutes. One revolution around a circle of radius 5 is a distance of 10π, so the child's speed is $10\pi/2 = 5\pi \approx 15.7$ m/min. Hence, the magnitude of the velocity vector is 15.7 m/min. The direction of motion is tangent to the circle, and hence perpendicular to the radius at that point. Figure 16.24 shows the direction of the vector at two different times.

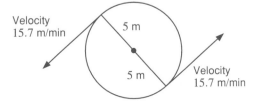

Figure 16.24: Velocity vectors of a child on a ferris wheel (Note that vectors would be in opposite direction if viewed from the other side.)

Computing the Velocity

We find the velocity, as in one-variable calculus, by taking a limit. If the position vector of the particle is $\vec{r}(t)$ at time t, then the displacement vector between its positions at times t and $t + \Delta t$ is $\Delta \vec{r} = \vec{r}(t + \Delta t) - \vec{r}(t)$. (See Figure 16.25.) Over this interval,

$$\text{Average velocity} = \frac{\Delta \vec{r}}{\Delta t}.$$

In the limit as Δt goes to zero we have the instantaneous velocity at time t:

> The **velocity vector**, $\vec{v}(t)$, of a moving object with position vector $\vec{r}(t)$ at time t is
>
> $$\vec{v}(t) = \lim_{\Delta t \to 0} \frac{\Delta \vec{r}}{\Delta t} = \lim_{\Delta t \to 0} \frac{\vec{r}(t + \Delta t) - \vec{r}(t)}{\Delta t},$$
>
> whenever the limit exists. We use the notation $\vec{v} = \dfrac{d\vec{r}}{dt} = \vec{r}'(t)$.

Notice that the direction of the velocity vector $\vec{r}'(t)$ in Figure 16.25 is approximated by the direction of the vector $\Delta \vec{r}$ and that the approximation gets better as $\Delta t \to 0$.

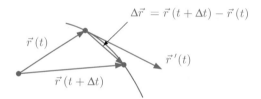

Figure 16.25: The change, $\Delta \vec{r}$, in the position vector for a particle moving on a curve and the velocity vector
$$\vec{v} = \vec{r}'(t)$$

The Components of the Velocity Vector

If we represent a curve parametrically by $x = f(t), y = g(t), z = h(t)$, then we can write its position vector as: $\vec{r}(t) = f(t)\vec{i} + g(t)\vec{j} + h(t)\vec{k}$. Now we can compute the velocity vector:

$$\begin{aligned}
\vec{v}(t) &= \lim_{\Delta t \to 0} \frac{\vec{r}(t + \Delta t) - \vec{r}(t)}{\Delta t} \\
&= \lim_{\Delta t \to 0} \frac{(f(t + \Delta t)\vec{i} + g(t + \Delta t)\vec{j} + h(t + \Delta t)\vec{k}) - (f(t)\vec{i} + g(t)\vec{j} + h(t)\vec{k})}{\Delta t} \\
&= \lim_{\Delta t \to 0} \left(\frac{f(t + \Delta t) - f(t)}{\Delta t}\vec{i} + \frac{g(t + \Delta t) - g(t)}{\Delta t}\vec{j} + \frac{h(t + \Delta t) - h(t)}{\Delta t}\vec{k} \right) \\
&= f'(t)\vec{i} + g'(t)\vec{j} + h'(t)\vec{k} \\
&= \frac{dx}{dt}\vec{i} + \frac{dy}{dt}\vec{j} + \frac{dz}{dt}\vec{k}.
\end{aligned}$$

Thus we have the following result:

The **components of the velocity vector** of a particle moving in space with position vector $\vec{r}(t) = f(t)\vec{i} + g(t)\vec{j} + h(t)\vec{k}$ at time t are given by

$$\vec{v}(t) = f'(t)\vec{i} + g'(t)\vec{j} + h'(t)\vec{k} = \frac{dx}{dt}\vec{i} + \frac{dy}{dt}\vec{j} + \frac{dz}{dt}\vec{k}.$$

Example 4   Find the components of the velocity vector for the child on the ferris wheel in Example 3 using a coordinate system which has its origin at the center of the ferris wheel and which makes the rotation counterclockwise.

Solution   The ferris wheel has radius 5 meters and completes 1 revolution counterclockwise every 2 minutes. The motion is parameterized by an equation of the form

$$\vec{r}(t) = 5\cos(\omega t)\vec{i} + 5\sin(\omega t)\vec{j},$$

where ω is chosen to make the period 2 minutes. Since the period of $\cos(\omega t)$ and $\sin(\omega t)$ is $2\pi/\omega$, we must have

$$\frac{2\pi}{\omega} = 2, \quad \text{so} \quad \omega = \pi.$$

Thus, the motion is described by the equation

$$\vec{r}(t) = 5\cos(\pi t)\vec{i} + 5\sin(\pi t)\vec{j},$$

where t is in minutes. The velocity is given by

$$\vec{v} = \frac{dx}{dt}\vec{i} + \frac{dy}{dt}\vec{j} = -5\pi\sin(\pi t)\vec{i} + 5\pi\cos(\pi t)\vec{j}.$$

To check, we calculate the magnitude of $\vec{v}$,

$$\|\vec{v}\| = \sqrt{(-5\pi)^2 \sin^2(\pi t) + (5\pi)^2 \cos^2(\pi t)} = 5\pi\sqrt{\sin^2(\pi t) + \cos^2(\pi t)} = 5\pi \approx 15.7,$$

which agrees with the speed we calculated in Example 3. To see that the direction is correct, we must show that the vector $\vec{v}$ at any time t is perpendicular to the position vector of the particle at time t. To do this, we compute the dot product of $\vec{v}$ and $\vec{r}$:

$$\vec{v} \cdot \vec{r} = (-5\pi\sin(\pi t)\vec{i} + 5\pi\cos(\pi t)\vec{j}) \cdot (5\cos(\pi t)\vec{i} + 5\sin(\pi t)\vec{j})$$
$$= -25\pi\,\sin(\pi t)\,\cos(\pi t) + 25\pi\cos(\pi t)\,\sin(\pi t) = 0.$$

So the velocity vector, $\vec{v}$, is perpendicular to $\vec{r}$ and hence tangent to the circle. (See Figure 16.26.)

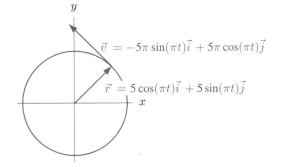

Figure 16.26: Velocity and radius vector of motion around a circle

Velocity Vectors and Tangent Lines

Since the velocity vector is tangent to the path of motion, it can be used to find parametric equations for the tangent line, if there is one.

Example 5 Find the tangent line at the point $(1, 1, 2)$ to the curve defined by the parametric equation

$$\vec{r}(t) = t^2\vec{i} + t^3\vec{j} + 2t\vec{k}.$$

Solution At time $t = 1$ the particle is at the point $(1, 1, 2)$ with position vector $\vec{r}_0 = \vec{i} + \vec{j} + 2\vec{k}$. The velocity vector at time t is $\vec{r}'(t) = 2t\vec{i} + 3t^2\vec{j} + 2\vec{k}$, so at time $t = 1$ the velocity is $\vec{v} = \vec{r}'(1) = 2\vec{i} + 3\vec{j} + 2\vec{k}$. The tangent line passes through $(1, 1, 2)$ in the direction of $\vec{v}$, so it has the parametric equation

$$\vec{r}(t) = \vec{r}_0 + t\vec{v} = (\vec{i} + \vec{j} + 2\vec{k}) + t(2\vec{i} + 3\vec{j} + 2\vec{k}).$$

The Acceleration Vector

Just as the velocity of a particle moving in 2-space or 3-space is a vector quantity, so is the rate of change of the velocity of the particle, namely its acceleration. Figure 16.27 shows a particle at time t with velocity vector $\vec{v}(t)$ and then a little later at time $t + \Delta t$. The vector $\Delta\vec{v} = \vec{v}(t + \Delta t) - \vec{v}(t)$ is the change in velocity and points approximately in the direction of the acceleration. So,

$$\text{Average acceleration} = \frac{\Delta\vec{v}}{\Delta t}.$$

In the limit as $\Delta t \to 0$, we have the instantaneous acceleration at time t:

The **acceleration vector** of an object moving with velocity $\vec{v}(t)$ at time t is

$$\vec{a}(t) = \lim_{\Delta t \to 0} \frac{\Delta\vec{v}}{\Delta t} = \lim_{\Delta t \to 0} \frac{\vec{v}(t + \Delta t) - \vec{v}(t)}{\Delta t},$$

if the limit exists. We use the notation $\vec{a} = \dfrac{d\vec{v}}{dt} = \dfrac{d^2\vec{r}}{dt^2} = \vec{r}''(t)$.

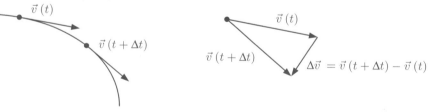

Figure 16.27: Computing the difference between two velocity vectors

Components of the Acceleration Vector

If we represent a curve in space parametrically by $x = f(t)$, $y = g(t)$, $z = h(t)$, we can express the acceleration in components. The velocity vector $\vec{v}(t)$ is given by

$$\vec{v}(t) = f'(t)\vec{i} + g'(t)\vec{j} + h'(t)\vec{k}.$$

From the definition of the acceleration vector, we have

$$\vec{a}\,(t) = \lim_{\Delta t \to 0} \frac{\vec{v}\,(t + \Delta t) - \vec{v}\,(t)}{\Delta t} = \frac{d\vec{v}}{dt}.$$

Using the same method to compute $d\vec{v}/dt$ as we used to compute $d\vec{r}/dt$ on page 287, we obtain

---

The **components of the acceleration vector**, $\vec{a}\,(t)$, at time t of a particle moving in space with position vector $\vec{r}\,(t) = f(t)\vec{i} + g(t)\vec{j} + h(t)\vec{k}$ at time t are given by

$$\vec{a}\,(t) = f''(t)\vec{i} + g''(t)\vec{j} + h''(t)\vec{k} = \frac{d^2x}{dt^2}\vec{i} + \frac{d^2y}{dt^2}\vec{j} + \frac{d^2z}{dt^2}\vec{k}.$$

Motion In a Circle and Along a Line

Example 6   Find the acceleration vector for the child on the ferris wheel in Examples 3 and 4.

Solution   The child's position vector is given by $\vec{r}\,(t) = 5\cos(\pi t)\vec{i} + 5\sin(\pi t)\vec{j}$. In Example 4 we saw that the velocity vector is

$$\vec{v}\,(t) = \frac{dx}{dt}\vec{i} + \frac{dy}{dt}\vec{j} = -5\pi \sin(\pi t)\vec{i} + 5\pi \cos(\pi t)\vec{j}.$$

Thus, the acceleration vector is

$$\begin{aligned}
\vec{a}\,(t) = \frac{d^2x}{dt^2}\vec{i} + \frac{d^2y}{dt^2}\vec{j} &= -(5\pi) \cdot \pi \cos(\pi t)\vec{i} - (5\pi) \cdot \pi \sin(\pi t)\vec{j} \\
&= -5\pi^2 \cos(\pi t)\vec{i} - 5\pi^2 \sin(\pi t)\vec{j}.
\end{aligned}$$

Notice that $\vec{a}\,(t) = -\pi^2 \vec{r}\,(t)$. Thus, the acceleration vector is a multiple of $\vec{r}\,(t)$ and points toward the origin.

The motion of the child on the ferris wheel is an example of uniform circular motion, whose properties follow. (See Problem 26.)

Uniform Circular Motion

A particle whose motion is described by

$$\vec{r}\,(t) = R\cos(\omega t)\vec{i} + R\sin(\omega t)\vec{j}$$

- Moves in a circle of radius R with period $2\pi/\omega$.
- Velocity, $\vec{v}$, is tangent to the circle and speed is constant $\|\vec{v}\| = \omega R$.
- Acceleration, $\vec{a}$, points toward the center of the circle with $\|\vec{a}\| = \|\vec{v}\|^2/R$.

In uniform circular motion, the acceleration vector reflects the fact that the velocity vector does not change in magnitude, only in direction. We now look at straight line motion where the velocity vector always has the same direction but the magnitude changes. We expect that the acceleration vector will point in the same direction as the velocity vector if the speed is increasing and in the opposite direction to the velocity vector if the speed is decreasing.

Example 7   Consider the motion given by the vector equation

$$\vec{r}(t) = 2\vec{i} + 6\vec{j} + (t^3 + t)(4\vec{i} + 3\vec{j} + \vec{k}).$$

Show that this is straight line motion in the direction of the vector $4\vec{i} + 3\vec{j} + \vec{k}$ and relate the acceleration vector to the velocity vector.

Solution   The velocity vector is

$$\vec{v} = (3t^2 + 1)(4\vec{i} + 3\vec{j} + \vec{k}).$$

Since $(3t^2 + 1)$ is a positive scalar, the velocity vector $\vec{v}$ always points in the direction of the vector $4\vec{i} + 3\vec{j} + \vec{k}$. In addition,

$$\text{Speed} = \|\vec{v}\| = (3t^2 + 1)\sqrt{4^2 + 3^2 + 1^2} = \sqrt{26}(3t^2 + 1).$$

Notice that the speed is decreasing until $t = 0$, then starts increasing. The acceleration vector is

$$\vec{a} = 6t(4\vec{i} + 3\vec{j} + \vec{k}).$$

For $t > 0$, the acceleration vector points in the same direction as $4\vec{i} + 3\vec{j} + \vec{k}$, which is the same direction as $\vec{v}$. This makes sense because the object is speeding up. For $t < 0$, the acceleration vector $6t(4\vec{i} + 3\vec{j} + \vec{k})$ points in the opposite direction to $\vec{v}$ because the object is slowing down.

The Length of a Curve

The speed of a particle is the magnitude of its velocity vector:

$$\text{Speed} = \|\vec{v}\| = \sqrt{\left(\frac{dx}{dt}\right)^2 + \left(\frac{dy}{dt}\right)^2 + \left(\frac{dz}{dt}\right)^2}.$$

As in one dimension, we can find the distance traveled by a particle along a curve by integrating its speed. Thus,

$$\text{Distance traveled} = \int_a^b \|\vec{v}(t)\|\, dt.$$

If the particle never stops or reverses its direction as it moves along the curve, the distance it travels will be the same as the length of the curve. This suggests the following formula, which is justified in Problem 33:

If the curve C is given parametrically for $a \le t \le b$ by smooth functions and if the velocity vector $\vec{v}$ is not $\vec{0}$ for $a < t < b$, then

$$\text{Length of } C = \int_a^b \|\vec{v}\|\, dt.$$

Example 8   Find the circumference of the ellipse given by the parametric equations

$$x = 2\cos t, \quad y = \sin t, \quad 0 \le t \le 2\pi.$$

Solution   The circumference of this curve is given by an integral which must be calculated numerically:

$$\text{Circumference} = \int_0^{2\pi} \sqrt{\left(\frac{dx}{dt}\right)^2 + \left(\frac{dy}{dt}\right)^2}\, dt = \int_0^{2\pi} \sqrt{(-2\sin t)^2 + (\cos t)^2}\, dt$$

$$= \int_0^{2\pi} \sqrt{4\sin^2 t + \cos^2 t}\, dt = 9.69.$$

Since the ellipse is inscribed in a circle of radius 2 and circumscribes a circle of radius 1, we would expect the length of the ellipse to be between $2\pi(2) \approx 12.57$ and $2\pi(1) \approx 6.28$, so the value of 9.69 is reasonable.

Problems for Section 16.2

1. (a) Explain how you know that the following two pairs of equations   *Graph it.*
$\vec{r} = (2+t)\vec{i} + (4+3t)\vec{j}, \vec{r} = (1-2t)\vec{i} + (1-6t)\vec{j}$ parameterize the same line.
(b) What are the slope and y intercept of this line?

2. The equation $\vec{r} = 10\vec{k} + t(\vec{i} + 2\vec{j} + 3\vec{k})$ parameterizes a line.

(a) Suppose we restrict ourselves to $t < 0$. What part of the line do we get?
(b) Suppose we restrict ourselves to $0 \le t \le 1$. What part of the line do we get?

3. (a) Explain why the line of intersection of two planes must be parallel to the cross product of a normal vector to the first plane and a normal vector to the second.
(b) Find a vector parallel to the line of intersection of the two planes $x + 2y - 3z = 7$ and $3x - y + z = 0$.
(c) Find parametric equations for the line in part (b).

4. Using time increments of 0.01, give a table of values near $t = 1$ for the position vector of the circular motion
$$\vec{r}(t) = (\cos t)\vec{i} + (\sin t)\vec{j}.$$
Use the table to approximate the velocity vector, $\vec{v}$, at time $t = 1$. Show that $\vec{v}$ is perpendicular to the radius from the origin at $t = 1$.

5. (a) Sketch the parameterized curve $x = t\cos t$, $y = t\sin t$ for $0 \le t \le 4\pi$.
(b) Use difference quotients to approximate the velocity vectors $\vec{v}(t)$ for $t = 2, 4$, and 6.
(c) Compute the velocity vectors $\vec{v}(t)$ for $t = 2, 4$, and 6, exactly and sketch them on the graph of the curve.

For Problems 6–9, find the velocity vector $\vec{v}(t)$ for the given motion of a particle. Also find the speed $\|\vec{v}(t)\|$ and any times when the particle comes to a stop.

6. $x = t^2, \quad y = t^3$

7. $x = \cos(t^2), \quad y = \sin(t^2)$

8. $x = \cos 2t, \quad y = \sin t$

9. $x = t^2 - 2t, y = t^3 - 3t, z = 3t^4 - 4t^3$

10. Find parametric equations for the tangent line at $t = 2$ for Problem 6.

For Problems 11–14, find the velocity and acceleration vectors for the given motions.

11. $x = 3\cos t, y = 4\sin t$

12. $x = t, y = t^3 - t$

13. $x = 2 + 3t, y = 4 + t, z = 1 - t$

14. $x = 3\cos(t^2), y = 3\sin(t^2), z = t^2$

Find the length of the curves in Problems 15–17.

15. $x = 3 + 5t, y = 1 + 4t, z = 3 - t$ for $1 \le t \le 2$. Explain your answer.

16. $x = \cos(e^t), y = \sin(e^t)$ for $0 \le t \le 1$. Explain why your answer is reasonable.

17. $x = \cos 3t, y = \sin 5t$ for $0 \le t \le 2\pi$.

18. A particle that passes through the point $P = (5, 4, -2)$ at time $t = 4$ is moving with constant velocity $\vec{v} = 2\vec{i} - 3\vec{j} + \vec{k}$. Find the parametric equations for its motion.

19. A particle that passes through the point $P = (5, 4, 3)$ at time $t = 7$ is moving with constant velocity $\vec{v} = 3\vec{i} + \vec{j} + 2\vec{k}$. Find the equations for its position at time t.

20. An object moving with constant velocity in 3-space (with coordinates in meters) passes through the point $(1, 1, 1)$, and then passes through the point $(2, -1, 3)$ five seconds later. What is its velocity vector? What is its acceleration vector?

21. Table 16.4 gives x and y coordinates of a particle in the plane at time t. Assuming the path is smooth, estimate the following quantities:

    (a) The velocity vector and speed at time $t = 2$.
    (b) Any times when the particle is moving parallel to the y-axis.
    (c) Any times when the particle has come to a stop.

    TABLE 16.4

t	0	0.5	1.0	1.5	2.0	2.5	3.0	3.5	4.0
x	1	4	6	7	6	3	2	3	5
y	3	2	3	5	8	10	11	10	9

22. Consider the motion of the particle given by the parametric equations

    $$x = t^3 - 3t, \quad y = t^2 - 2t,$$

    where the y-axis is vertical and the x-axis is horizontal.
    (a) Does the particle ever come to a stop? If so, when and where?
    (b) Is the particle ever moving straight up or down? If so, when and where?
    (c) Is the particle ever moving straight horizontally right or left? If so, when and where?

23. Suppose $\vec{r}(t) = \cos t\,\vec{i} + \sin t\,\vec{j} + 2t\,\vec{k}$ represents the position of a particle on a helix, where z is the height of the particle above the ground.

    (a) Is the particle ever moving downwards? When?
    (b) When does the particle reach a point 10 units above the ground?
    (c) What is the velocity of the particle when it is 10 units above the ground?
    (d) Suppose the particle leaves the helix and moves along the tangent line to the spiral at this point. Find parametric equations for this tangent line.

24. Mr. Skywalker is traveling along the curve given by

    $$\vec{r}(t) = -2e^{3t}\vec{i} + 5\cos t\vec{j} - 3\sin(2t)\vec{k}.$$

    If the power thrusters are turned off, his ship flies off on a tangent line to $\vec{r}(t)$. He is almost out of power when he notices that a station on Xardon is open at the point with coordinates $(1.5, 5, 3.5)$. Quickly calculating his position, he turns off the thrusters at $t = 0$. Does he make it to the Xardon station? Explain.

25. Determine the position vector $\vec{r}(t)$ for a rocket which is launched from the origin at time $t = 0$ seconds, reaches its highest point of $(x, y, z) = (1000, 3000, 10{,}000)$, where x, y, z are measured in meters, and after the launch is subject only to the acceleration due to gravity, 9.8 m/sec^2.

26. The motion of a particle is given by $\vec{r}(t) = R\cos(\omega t)\vec{i} + R\sin(\omega t)\vec{j}$, with $R > 0$, $\omega > 0$.

    (a) Show that the particle moves on a circle and find the radius, direction, and period.
    (b) Determine the velocity vector of the particle and its direction and speed.
    (c) What are the direction and magnitude of the acceleration vector of the particle?

27. A stone is swung around on a string at a constant speed with period 2π seconds in a horizontal circle centered at the point $(0, 0, 8)$. When $t = 0$, the stone is at the point $(0, 5, 8)$; it travels clockwise when viewed from above. When the stone is at the point $(5, 0, 8)$, the string breaks and it moves under gravity.

    (a) Parameterize the stone's circular trajectory.
    (b) Find the velocity and acceleration of the stone at the moment before the string breaks.
    (c) Write, but do not solve, the differential equations (with initial conditions) satisfied by the coordinates x, y, z giving the position of the stone after it has left the circle.

28. Emily is standing on the outer edge of a merry-go-round, 10 meters from the center. The merry-go-round completes one full revolution every 20 seconds. As Emily passes over a point P on the ground, she drops a ball from 3 meters above the ground.

    (a) How fast is Emily going?
    (b) How far from P does the ball hit the ground? (The acceleration due to gravity is 9.8 m/sec^2.)
    (c) How far from Emily does the ball hit the ground?

29. A lighthouse L is located on an island in the middle of a lake as shown in Figure 16.28. Consider the motion of the point where the light beam from L hits the shore of the lake.

    (a) Suppose the beam rotates counterclockwise about L at a constant angular velocity. At which of A, B, C, D, or E is the speed of the point greatest? At which point is it smallest?
    (b) Repeat part (a), supposing the beam rotates counterclockwise so that it sweeps out equal areas of the lake in equal times.
    (c) What happens if you place the lighthouse at different points in the lake? Can the speed of the point on the shore ever be infinite for part (a)?
    (d) Suppose now that the lake is rectangular instead. What happens to the velocity vector at the corners? For part (b) show that the speed is constant along each side (possibly a different constant on each side).

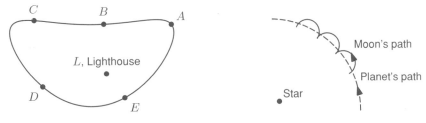

Figure 16.28: The lighthouse in the lake   Figure 16.29

30. A hypothetical moon orbits a planet which in turn orbits a star. Suppose that the orbits are circular and that the moon orbits the planet 12 times in the time it takes for the planet to orbit the star once. In this problem we will investigate whether the moon could come to a stop at some instant. (See Figure 16.29.)

    (a) Suppose the radius of the moon's orbit around the planet is 1 unit and the radius of the planet's orbit around the star is R units. Explain why the motion of the moon relative to the star can be described by the parametric equations

    $$x = R\cos t + \cos(12t), \quad y = R\sin t + \sin(12t).$$

    (b) Find a value for R and t such that the moon stops relative to the star at time t.
    (c) On a graphing calculator, plot the path of the moon for the value of R you obtained in part (b). Experiment with other values for R.

31. Suppose $F(x, y) = 1/(x^2 + y^2 + 1)$ gives the temperature at the point (x, y) in the plane. A ladybug moves along a parabola according to the parametric equations

$$x = t, \quad y = t^2.$$

Find the rate of change in the temperature of the ladybug at time t.

32. This problem generalizes the result of Problem 31. Suppose $F(x, y)$ gives the temperature at any point (x, y) in the plane and that a ladybug moves in the plane with position vector at time t given by $\vec{r}(t) = x(t)\vec{i} + y(t)\vec{j}$ and velocity vector $\vec{r}'(t)$. Use the chain rule to show that

Rate of change in the temperature of the bug at time t = grad $F(\vec{r}(t)) \cdot \vec{r}'(t)$.

33. In this problem we justify the formula for the length of a curve given on page 290. Suppose the curve C is given by smooth parametric equations $x = x(t)$, $y = y(t)$, $z = z(t)$ for $a \leq t \leq b$. By dividing the parameter interval $a \leq t \leq b$ at points $t_1, \ldots, t_{n-1}$ into small segments of length $\Delta t = t_{i+1} - t_i$, we get a corresponding division of the curve C into small pieces. See Figure 16.30, where the points $P_i = (x(t_i), y(t_i), z(t_i))$ on the curve C correspond to parameter values $t = t_i$. Let C_i be the portion of the curve C between P_i and P_{i+1}.

(a) Use local linearity to show that

$$\text{Length of } C_i \approx \sqrt{x'(t_i)^2 + y'(t_i)^2 + z'(t_i)^2} \, \Delta t.$$

(b) Use part (a) and a Riemann sum to explain why

$$\text{Length of } C = \int_a^b \sqrt{x'(t)^2 + y'(t)^2 + z'(t)^2} \, dt.$$

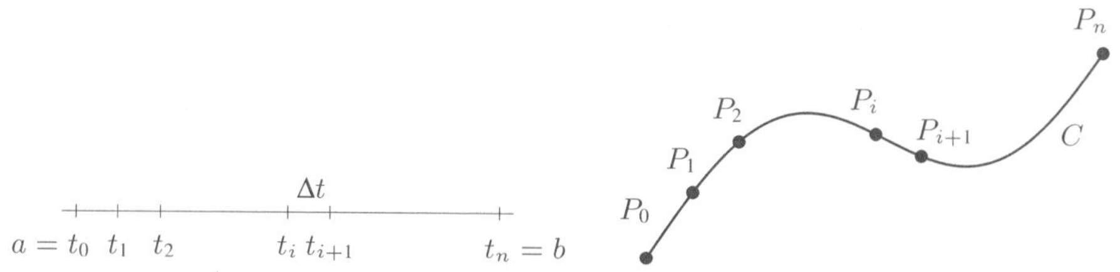

Figure 16.30: A subdivision of the parameter interval and the corresponding subdivision of the curve C

16.3 PARAMETERIZED SURFACES

How Do We Parameterize a Surface?

In Section 16.1 we parameterized a circle in 2-space using the equations

$$x = \cos t, \quad y = \sin t.$$

In 3-space, the same circle in the xy-plane has parametric equations

$$x = \cos t, \quad y = \sin t, \quad z = 0.$$

We add the equation $z = 0$ to specify that the circle is in the xy-plane. If we wanted a circle in the plane $z = 3$, we would use the equations

$$x = \cos t, \quad y = \sin t, \quad z = 3.$$

Suppose now we let z vary freely, as well as t. We get circles in every horizontal plane, forming the cylinder in Figure 16.31. Thus, we need two parameters, t and z, to parameterize the cylinder.

This is true in general. A curve, though it may live in two or three dimensions, is itself one-dimensional; if we move along it we can only move backwards and forwards in one direction. Thus, it only requires one parameter to trace out a curve.

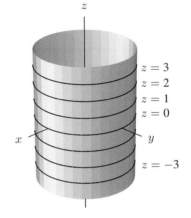

Figure 16.31: The cylinder $x = \cos t$, $y = \sin t$, $z = z$

A surface is 2-dimensional; at any given point there are two independent directions we can move in. For example, on the cylinder we can move vertically, or we can circle around the z-axis horizontally. So we need *two* parameters to describe it. We can think of the parameters as map coordinates, like longitude and latitude on the surface of the earth.

In the case of the cylinder our parameters are t and z, so

$$x = \cos t, \quad y = \sin t, \quad z = z, \quad 0 \le t < 2\pi, \quad -\infty < z < \infty.$$

The last equation, $z = z$, looks strange, but it reminds us that we are in three dimensions, not two, and that the z-coordinate on our surface is allowed to vary freely. In general, we express the coordinates, (x, y, z) of a point on a surface S in terms of two parameters, s and t:

$$x = f_1(s, t), \quad y = f_2(s, t), \quad z = f_3(s, t).$$

As the values of s and t vary, the corresponding point (x, y, z) sweeps out the surface, S. (See Figure 16.32.) The function which sends the point (s, t) to the point (x, y, z) is called the *parameterization of the surface.*

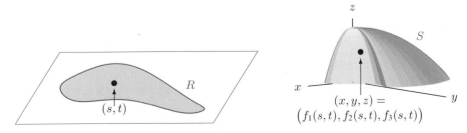

Figure 16.32: The parameterization sends each point (s, t) in the parameter region, R, to a point $(x, y, z) = (f_1(s, t), f_2(s, t), f_3(s, t))$ in the surface, S

Using Position Vectors

We can use the position vector $\vec{r} = x\vec{i} + y\vec{j} + z\vec{k}$ to combine the three parametric equations for a surface into a single vector equation. For example, the parameterization of the cylinder $x = \cos t, y = \sin t, z = z$ can be written as

$$\vec{r}(t,z) = \cos t\vec{i} + \sin t\vec{j} + z\vec{k} \qquad 0 \le t < 2\pi, \quad -\infty < z < \infty.$$

For a general parameterized surface S, we write

$$\vec{r}(s,t) = f_1(s,t)\vec{i} + f_2(s,t)\vec{j} + f_3(s,t)\vec{k}.$$

Parameterizing a Surface of the Form $z = f(x, y)$

The graph of a function $z = f(x, y)$ can be given parametrically simply by letting the parameters s and t be x and y:

$$x = s, \quad y = t, \quad z = f(s, t).$$

Example 1   Give a parametric description of the lower hemisphere of the sphere $x^2 + y^2 + z^2 = 1$.

Solution   The surface is the graph of the function $z = -\sqrt{1 - x^2 - y^2}$ over the region $x^2 + y^2 \le 1$ in the plane. Then parametric equations are $x = s$, $y = t$, $z = -\sqrt{1 - s^2 - t^2}$, where the parameters s and t vary inside the unit circle.

In practice we often think of x and y as parameters rather than introduce new variables s and t. Thus, we may write $x = x, y = y, z = f(x, y)$.

Parameterizing Planes

Consider a plane containing two nonparallel vectors $\vec{v}_1$ and $\vec{v}_2$ and a point P_0 with position vector $\vec{r}_0$. We can get to any point on the plane by starting at P_0 and moving parallel to $\vec{v}_1$ or $\vec{v}_2$, adding multiples of them to $\vec{r}_0$. (See Figure 16.33.)

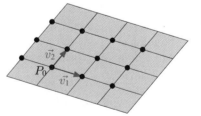

Figure 16.33: The plane $\vec{r}(s,t) = \vec{r}_0 + s\vec{v}_1 + t\vec{v}_2$ and some points corresponding to various choices of s and t

Since $s\vec{v}_1$ is parallel to $\vec{v}_1$ and $t\vec{v}_2$ is parallel to $\vec{v}_2$, we have the following result:

Parametric Equations for a Plane

The plane through the point with position vector $\vec{r}_0$ and containing the two nonparallel vectors $\vec{v}_1$ and $\vec{v}_2$ has parametric equation

$$\vec{r}(s,t) = \vec{r}_0 + s\vec{v}_1 + t\vec{v}_2.$$

If $\vec{r}_0 = x_0\vec{i} + y_0\vec{j} + z_0\vec{k}$, and $\vec{v}_1 = a_1\vec{i} + a_2\vec{j} + a_3\vec{k}$, and $\vec{v}_2 = b_1\vec{i} + b_2\vec{j} + b_3\vec{k}$, then the parametric equations of the plane can be written in the form

$$x = x_0 + sa_1 + tb_1, \quad y = y_0 + sa_2 + tb_2, \quad z = z_0 + sa_3 + tb_3.$$

Notice that the parameterization of the plane expresses the coordinates x, y, and z as linear functions of the parameters s and t.

Example 2   Write a parametric equation for the plane through the point $(2, -1, 3)$ and containing the vectors $\vec{v}_1 = 2\vec{i} + 3\vec{j} - \vec{k}$ and $\vec{v}_2 = \vec{i} - 4\vec{j} + 5\vec{k}$.

Solution   The parametric equation is

$$\vec{r}(s,t) = \vec{r}_0 + s\vec{v}_1 + t\vec{v}_2 = 2\vec{i} - \vec{j} + 3\vec{k} + s(2\vec{i} + 3\vec{j} - \vec{k}) + t(\vec{i} - 4\vec{j} + 5\vec{k})$$
$$= (2 + 2s + t)\vec{i} + (-1 + 3s - 4t)\vec{j} + (3 - s + 5t)\vec{k},$$

or equivalently,

$$x = 2 + 2s + t, \quad y = -1 + 3s - 4t, \quad z = 3 - s + 5t.$$

Parameterizations Using Spherical Coordinates

Recall the spherical coordinates ρ, ϕ, and θ introduced on page 251 of Chapter 15. On a sphere of radius $\rho = a$ we can use ϕ and θ as coordinates, similar to latitude and longitude on the surface of the earth. (See Figure 16.34.) The latitude, however, is measured from the equator, whereas ϕ is measured from the north pole. If the positive x-axis passes through the Greenwich meridian, the longitude and θ are equal for $0 \leq \theta \leq \pi$.

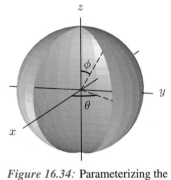

Figure 16.34: Parameterizing the
sphere by ϕ and θ

Example 3   You are at a point on a sphere with $\phi = 3\pi/4$. Are you in the northern or southern hemisphere? If ϕ decreases, do you move closer to or farther from the equator?

Solution   The equator has $\phi = \pi/2$. Since $3\pi/4 > \pi/2$, you are in the southern hemisphere. If ϕ decreases, you move closer to the equator.

Example 4   On a sphere, you are standing at a point with coordinates θ_0 and ϕ_0. Your *antipodal* point is the point on the other side of the sphere on a line through you and the center. What are the θ, ϕ coordinates of your antipodal point?

Solution   Figure 16.35 shows that the coordinates are $\theta = \pi + \theta_0$ if $\theta_0 < \pi$ or $\theta = \theta_0 - \pi$ if $\pi \leq \theta_0 \leq 2\pi$, and $\phi = \pi - \phi_0$. Notice that if you are on the equator, then so is your antipodal point.

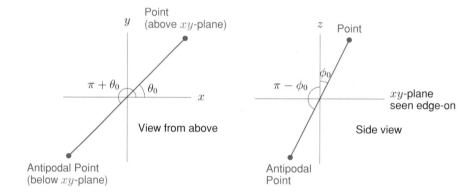

Figure 16.35: Two views of the xyz-coordinate system showing coordinates of antipodal points

Parameterizing a Sphere Using Spherical Coordinates

The sphere with radius 1 centered at the origin is parameterized by

$$x = \sin \phi \cos \theta, \qquad y = \sin \phi \sin \theta, \qquad z = \cos \phi.$$

Where $0 \leq \theta \leq 2\pi$ and $0 \leq \phi \leq \pi$. (See Figure 16.36.)

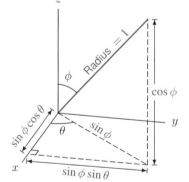

Figure 16.36: The relationship between x, y, z and ϕ, θ on a sphere of radius 1

We can also write these equations in vector form:

$$\vec{r}(\theta, \phi) = \sin \phi \cos \theta \, \vec{i} + \sin \phi \sin \theta \, \vec{j} + \cos \phi \, \vec{k}.$$

Since $x^2 + y^2 + z^2 = \sin^2 \phi(\cos^2 \theta + \sin^2 \theta) + \cos^2 \phi = \sin^2 \phi + \cos^2 \phi = 1$ this verifies that the point with position vector $\vec{r}(\theta, \phi)$ does lie on the sphere of radius 1. Notice that the z-coordinate depends only on the parameter ϕ. Geometrically, this means that all points on the same latitude have the same z-coordinate.

Example 5   Find parametric equations for the following spheres:
  (a)   Center at the origin and radius 2.
  (b)   Center at the point $(2, -1, 3)$ and radius 2.

Solution   (a)   We must scale the distance from the origin by 2. Thus, we have

$$x = 2 \sin \phi \cos \theta, \qquad y = 2 \sin \phi \sin \theta, \qquad z = 2 \cos \phi,$$

where $0 \le \theta \le 2\pi$ and $0 \le \phi \le \pi$. In vector form, this is written

$$\vec{r}(\theta, \phi) = 2 \sin \phi \cos \theta \vec{i} + 2 \sin \phi \sin \theta \vec{j} + 2 \cos \phi \vec{k}.$$

  (b)   To shift the center of the sphere from the origin to the point $(2, -1, 3)$, we add the vector parameterization we found in part (a) to the position vector of $(2, -1, 3)$. (See Figure 16.37.) This gives

$$\vec{r}(\theta, \phi) = 2\vec{i} - \vec{j} + 3\vec{k} + (2 \sin \phi \cos \theta \vec{i} + 2 \sin \phi \sin \theta \vec{j} + 2 \cos \phi \vec{k})$$
$$= (2 + 2 \sin \phi \cos \theta)\vec{i} + (-1 + 2 \sin \phi \sin \theta)\vec{j} + (3 + 2 \cos \phi)\vec{k},$$

where $0 \le \theta \le 2\pi$ and $0 \le \phi \le \pi$. Alternatively

$$x = 2 + 2 \sin \phi \cos \theta, \qquad y = -1 + 2 \sin \phi \sin \theta, \qquad z = 3 + 2 \cos \phi.$$

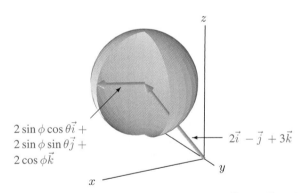

Figure 16.37: Sphere with center at the point $(2, -1, 3)$ and radius 2

Note that the same point can have more than one value for θ or ϕ. For example, points with $\theta = 0$ also have $\theta = 2\pi$, unless we restrict θ to the range $0 \le \theta < 2\pi$. Also, the north pole, at $\phi = 0$, and the south pole, at $\phi = \pi$, can have any value of θ.

Parameterizing Surfaces of Revolution

Many surfaces have an axis of rotational symmetry and circular cross-sections perpendicular to that axis. These surfaces are referred to as *surfaces of revolution*.

Example 6   Find a parameterization of the cone whose base is the circle $x^2 + y^2 = a^2$ in the xy-plane and whose vertex is at a height h above the xy-plane. (See Figure 16.38.)

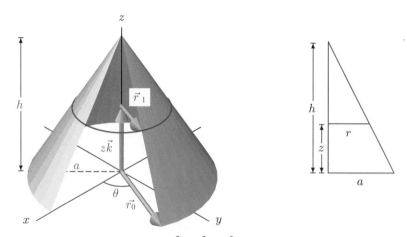

Figure 16.38: The cone whose base is the circle $x^2 + y^2 = a^2$ in the xy-plane and whose vertex is at the point $(0, 0, h)$ and the vertical cross-section through the cone

Solution   We use cylindrical coordinates, r, θ, z. (See Figure 16.38.) In the xy-plane, the radius vector, $\vec{r}_0$, from the z-axis to a point on the cone in the xy-plane is

$$\vec{r}_0 = a \cos \theta \vec{i} + a \sin \theta \vec{j}.$$

Above the xy-plane, the radius of the circular cross section, r, decreases linearly from $r = a$ when $z = 0$ to $r = 0$ when $z = h$. From the similar triangles in Figure 16.38,

$$\frac{a}{h} = \frac{r}{h - z}.$$

Solving for r, we have

$$r = \left(1 - \frac{z}{h}\right) a.$$

The horizontal radius vector, $\vec{r}_1$, at height z has components similar to $\vec{r}_0$, but with a replaced by r:

$$\vec{r}_1 = r \cos \theta \vec{i} + r \sin \theta \vec{j} = \left(1 - \frac{z}{h}\right) a \cos \theta \vec{i} + \left(1 - \frac{z}{h}\right) a \sin \theta \vec{j}.$$

As θ goes from 0 to 2π, the vector $\vec{r}_1$ traces out the horizontal circle in Figure 16.38. We get the position vector, $\vec{r}$, of a point on the cone by adding the vector $z\vec{k}$, so

$$\vec{r} = \vec{r}_1 + z\vec{k} = a\left(1 - \frac{z}{h}\right)\cos\theta\vec{i} + a\left(1 - \frac{z}{h}\right)\sin\theta\vec{j} + z\vec{k}, \quad \text{for } 0 \leq z \leq h \text{ and } 0 \leq \theta \leq 2\pi.$$

These equations can be written as

$$x = \left(1 - \frac{z}{h}\right) a \cos\theta, \quad y = \left(1 - \frac{z}{h}\right) a \sin\theta, \quad z = z.$$

The parameters are θ and z.

Example 7   Consider the bell of a trumpet. A model for the radius $z = f(x)$ of the horn (in cm) at a distance x cm from the large open end is given by the function

$$f(x) = \frac{6}{(x+1)^{0.7}}.$$

The bell is obtained by rotating the graph of f about the x-axis. Find a parameterization for the first 24 cm of the bell. (See Figure 16.39.)

Figure 16.39: The bell of a trumpet obtained by rotating the graph of $z = f(x)$ about the x-axis

Solution   At distance x from the large open end of the horn, the cross section parallel to the yz-plane is a circle of radius $f(x)$, with center on the x-axis. Such a circle can be parameterized by $y = f(x)\cos\theta$, $z = f(x)\sin\theta$. Thus we have the parameterization

$$x = x, \quad y = \left(\frac{6}{(x+1)^{0.7}}\right)\cos\theta, \quad z = \left(\frac{6}{(x+1)^{0.7}}\right)\sin\theta, \quad 0 \le x \le 24, \quad 0 \le \theta \le 2\pi.$$

The parameters are x and θ.

Parameter Curves

On a parameterized surface, the curve obtained by setting one of the parameters equal to a constant and letting the other vary is called a *parameter curve*. If the surface is parameterized by

$$\vec{r}(s,t) = f_1(s,t)\vec{i} + f_2(s,t)\vec{j} + f_3(s,t)\vec{k},$$

there are two families of parameter curves on the surface, one family with t constant and the other with s constant.

Example 8   Consider the vertical cylinder

$$x = \cos t, \quad y = \sin t, \quad z = z.$$

(a)   Describe the two parameter curves through the point $(0, 1, 1)$.
(b)   Describe the family of parameter curves with t constant and the family with z constant.

Solution   (a)   Since the point $(0, 1, 1)$ corresponds to the parameter values $t = \pi/2$ and $z = 1$, there are two parameter curves, one with $t = \pi/2$ and the other with $z = 1$. The parameter curve with $t = \pi/2$ has the parametric equations

$$x = \cos\left(\frac{\pi}{2}\right) = 0, \quad y = \sin\left(\frac{\pi}{2}\right) = 1, \quad z = z,$$

with parameter z. This is a line through the point $(0, 1, 1)$ parallel to the z-axis.
        The parameter curve with $z = 1$ has the parametric equations

$$x = \cos t, \quad y = \sin t, \quad z = 1,$$

with parameter t. This is a unit circle parallel to and one unit above the xy-plane centered on the z-axis.

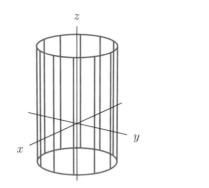

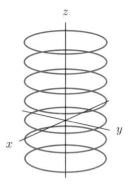

Figure 16.40: The family of parameter
curves with $t = t_0$ for the cylinder
$x = \cos t, y = \sin t, z = z$

Figure 16.41: The family of parameter
curves with $z = z_0$ for the cylinder
$x = \cos t, y = \sin t, z = z$

(b)   First, fix $t = t_0$ for t and let z vary. The curves parameterized by z have equation

$$x = \cos t_0, \quad y = \sin t_0, \quad z = z.$$

These are vertical lines on the cylinder parallel to the z-axis. (See Figure 16.40.)

The other family is obtained by fixing $z = z_0$ and varying t. Curves in this family are
parameterized by t and have equation

$$x = \cos t, \quad y = \sin t, \quad z = z_0.$$

They are circles of radius 1 parallel to the xy-plane centered on the z-axis. (See Figure 16.41.)

Example 9   Describe the families of parameter curves with $\theta = \theta_0$ and $\phi = \phi_0$ for the sphere

$$x = \sin \phi \cos \theta, \qquad y = \sin \phi \sin \theta, \qquad z = \cos \phi,$$

where $0 \leq \theta \leq 2\pi, 0 \leq \phi \leq \pi$.

Solution   Since ϕ measures latitude, the family with ϕ constant consists of the circles of constant latitude. (See
Figure 16.42.) Similarly, the family with θ constant consists of the meridians (semicircles) running
between the north and south poles. (See Figure 16.43.)

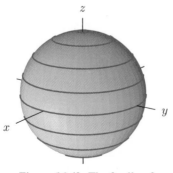

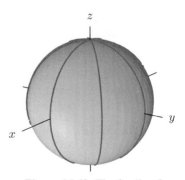

Figure 16.42: The family of
parameter curves with $\phi = \phi_0$ for
the sphere parameterized by (θ, ϕ)

Figure 16.43: The family of
parameter curves with $\theta = \theta_0$ for
the sphere parameterized by (θ, ϕ)

We have seen parameter curves before in pages 15-17 of Section 11.3: The cross-sections with $x = a$ and $y = b$ on a surface $z = f(x, y)$ are examples of parameter curves. So are the grid lines on a computer sketch of a surface. The small regions shaped like parallelograms surrounded by nearby pairs of parameter curves are called *parameter rectangles*. See Figure 16.44.

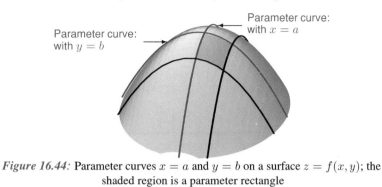

Parameter curve: with $x = a$

Parameter curve: with $y = b$

Figure 16.44: Parameter curves $x = a$ and $y = b$ on a surface $z = f(x, y)$; the shaded region is a parameter rectangle

Problems for Section 16.3

Describe in words the objects parameterized by the equations in Problems 1–8.

1. $x = r \cos \theta \quad 0 \le r \le 5$
   $y = r \sin \theta \quad 0 \le \theta \le 2\pi$
   $z = 7$

2. $x = 5 \cos \theta \quad 0 \le \theta \le 2\pi$
   $y = 5 \sin \theta$
   $z = 7$

3. $x = 5 \cos \theta \quad 0 \le \theta \le 2\pi$
   $y = 5 \sin \theta \quad 0 \le z \le 7$
   $z = z$

4. $x = 5 \cos \theta \quad 0 \le \theta \le 2\pi$
   $y = 5 \sin \theta$
   $z = 5\theta$

5. $x = r \cos \theta \quad 0 \le r \le 5$
   $y = r \sin \theta \quad 0 \le \theta \le 2\pi$
   $z = r$

6. $x = 2z \cos \theta \quad 0 \le z \le 7$
   $y = 2z \sin \theta \quad 0 \le \theta \le 2\pi$
   $z = z$

7. $x = 3 \cos \theta \quad 0 \le \theta \le 2\pi$
   $y = 2 \sin \theta \quad 0 \le z \le 7$
   $z = z$

8. $x = x \quad -5 \le x \le 5$
   $y = x^2 \quad 0 \le z \le 7$
   $z = z$

9. Find a parameterization of a circular cylinder of radius a whose axis is along the z-axis, from $z = 0$ to a height $z = h$. See Figure 16.45.

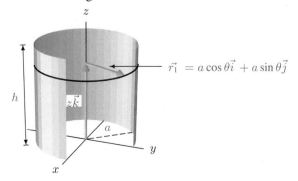

$\vec{r_1} = a \cos \theta \vec{i} + a \sin \theta \vec{j}$

Figure 16.45

10. A city is described parametrically by the equation

$$\vec{r} = (x_0\vec{i} + y_0\vec{j} + z_0\vec{k}) + s\vec{v_1} + t\vec{v_2}$$

where $\vec{v}_1 = 2\vec{i} - 3\vec{j} + 2\vec{k}$ and $\vec{v}_2 = \vec{i} + 4\vec{j} + 5\vec{k}$. A city block is a rectangle determined by $\vec{v}_1$ and $\vec{v}_2$. East is in the direction of $\vec{v}_1$ and north is in the direction of $\vec{v}_2$. Starting at the point (x_0, y_0, z_0), you walk 5 blocks east, 4 blocks north, 1 block west and 2 blocks south. What are the parameters of the point where you end up? What are your x, y and z coordinates at that point?

11. Find a parameterization for the plane through $(1, 3, 4)$ and orthogonal to $\vec{n} = 2\vec{i} + \vec{j} - \vec{k}$.

12. Does the plane $\vec{r}(s, t) = (2 + s)\vec{i} + (3 + s + t)\vec{j} + 4t\vec{k}$ contain the following points?
    (a) $(4, 8, 12)$ (b) $(1, 2, 3)$

13. Are the following two planes parallel?

$$x = 2 + s + t, \quad y = 4 + s - t, \quad z = 1 + 2s, \quad \text{and}$$

$$x = 2 + s + 2t, \quad y = t, \quad z = s - t.$$

14. You are at a point on the earth with longitude 80° West of Greenwich, England, and latitude 40° North of the equator.

    (a) If your latitude decreases have you moved nearer to or farther from the equator?
    (b) If your latitude decreases, have you moved nearer to or farther from the north pole?
    (c) If your longitude increases (say, to 90° West), have you moved nearer to or farther from Greenwich?

15. Describe in words the curve $\phi = \pi/4$ on the surface of the globe.

16. Describe in words the curve $\theta = \pi/4$ on the surface of the globe.

17. Find parametric equations for the sphere centered at the origin and with radius 5.

18. Find parametric equations for the sphere centered at the point $(2, -1, 3)$ and with radius 5.

19. Find parametric equations for the sphere $(x - a)^2 + (y - b)^2 + (z - c)^2 = d^2$.

20. Adapt the parameterization for the sphere to find a parameterization for the ellipsoid

$$\frac{x^2}{a^2} + \frac{y^2}{b^2} + \frac{z^2}{c^2} = 1.$$

21. Suppose you are standing at a point on the equator of a sphere, parameterized by spherical coordinates θ_0, and ϕ_0. If you go halfway around the equator and halfway up toward the north pole along a longitude, what are your new θ and ϕ coordinates?

22. If the sphere is parameterized using the spherical coordinates θ and ϕ, describe in words the part of the sphere given by the following restrictions:

    (a) $0 \le \theta < 2\pi, \quad 0 \le \phi \le \pi/2$ (b) $\pi \le \theta < 2\pi, \quad 0 \le \phi \le \pi$
    (c) $\pi/4 \le \theta < \pi/3, \quad 0 \le \phi \le \pi$ (d) $0 \le \theta \le \pi, \quad \pi/4 \le \phi < \pi/3$

23. Find parametric equations for the cone $x^2 + y^2 = z^2$.

24. Parameterize the cone in Example 6 on page 300 in terms of r and θ.

25. Parameterize a cone of height h and maximum radius a with vertex at the origin and opening upward. Do this in two ways, giving the range of values for each parameter in each case:
    (a) Use r and θ.   (b) Use z and θ.

26. Parameterize the paraboloid $z = x^2 + y^2$ using cylindrical coordinates.

27. Parameterize a vase formed by rotating the curve $z = 10\sqrt{x-1}$, $1 \le x \le 2$, around the z-axis. Sketch the vase.

For Problems 28–31

(a) Write an equation in x, y, z and identify the parametric surface.

(b) Draw a picture of the surface.

28.
$$x = 2s \qquad 0 \le s \le 1$$
$$y = s + t \qquad 0 \le t \le 1$$
$$z = 1 + s - t$$

29.
$$x = s + t \qquad 0 \le s \le 1$$
$$y = s - t \qquad 0 \le t \le 1$$
$$z = s^2 + t^2$$

30.
$$x = 3 \sin s \qquad 0 \le s \le \pi$$
$$y = 3 \cos s \qquad 0 \le t \le 1$$
$$z = t + 1$$

31.
$$x = s \qquad s^2 + t^2 \le 1$$
$$y = t \qquad s, t \ge 0$$
$$z = \sqrt{1 - s^2 - t^2}$$

32. (a) Describe the surface given parametrically by the equations

$$x = \cos(s - t), \quad y = \sin(s - t), \quad z = s + t.$$

(b) Describe the two families of parameter curves on the surface.

33. Give a parameterization of the circle of radius a centered at the point (x_0, y_0, z_0) and in the plane parallel to two given unit vectors $\vec{u}$ and $\vec{v}$ such that $\vec{u} \cdot \vec{v} = 0$.

34. A torus (doughnut) is constructed by rotating a small circle of radius a in a large circle of radius b about the origin. The small circle is in a (rotating) vertical plane through the origin and the large circle is in the xy-plane. (See Figure 16.46.) Parameterize the torus as follows.

(a) Parameterize the large circle.

(b) For a typical point on the large circle, find two unit vectors which are perpendicular to one another and in the plane of the small circle at that point. Use these vectors to parameterize the small circle relative to its center.

(c) Combine your answers to parts (b) and (c) to parameterize the torus.

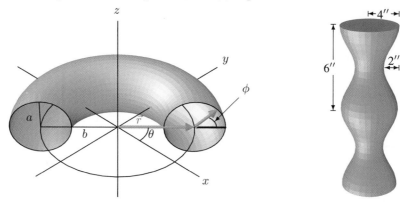

Figure 16.46 *Figure 16.47*

35. A decorative oak post is $48''$ long and is turned on a lathe so that its profile is sinusoidal as shown in Figure 16.47.

(a) Describe the surface of the post parametrically using cylindrical coordinates.

(b) Find the volume of the post.

16.4 THE IMPLICIT FUNCTION THEOREM

In this section we explain the Implicit Function Theorem and how it can be used to find linear approximations to small pieces of smooth curves and surfaces.

Implicit, Explicit, and Parametric Representations of Curves in 2-Space

The circle of radius 1 centered at the origin can be represented implicitly by the equation

$$x^2 + y^2 = 1,$$

or explicitly by the equations

$$y = \sqrt{1 - x^2} \quad \text{and} \quad y = -\sqrt{1 - x^2},$$

or parametrically by the equations

$$x = \cos t, \ y = \sin t, \quad 0 \le t \le 2\pi.$$

In general

- An **implicit** representation of a curve in the xy-plane is given by a single equation in x and y, of the form $f(x, y) = 0$.
- An **explicit** representation of a curve in the xy-plane is given by equations expressing y in terms of x or x in terms of y of the form $y = g(x)$ or $x = h(y)$.
- A **parametric** representation of a curve in the xy-plane is given by a pair of equations expressing x and y in terms of a third variable, often denoted by t.

There can be many different implicit or parametric representations of a given curve.

Example 1 Give implicit, explicit, and parametric representations of the line passing through the points $(3, 0)$ and $(0, 5)$.

Solution An implicit representation is $x/3 + y/5 - 1 = 0$. (You should check that the x-intercept is 3 and the y-intercept is 5.) An explicit representation is $y = 5 - (5/3)x$. A parametric representation is $x = 3t, \ y = 5 - 5t$.

Converting Parametric Representations to Implicit or Explicit

Example 2 Give implicit and explicit representations of the curve having the parametric representation

$$x = 3 + 5\sin t, \quad y = 1 + 2\cos t, \quad 0 \le t \le 2\pi.$$

Solution We need to eliminate the parameter t. Solving for $\sin t$ and $\cos t$, we get $\sin t = (x - 3)/5$, $\cos t = (y - 1)/2$. Since $\sin^2 t + \cos^2 t = 1$, we have

$$\left(\frac{x - 3}{5}\right)^2 + \left(\frac{y - 1}{2}\right)^2 = 1.$$

This is an implicit representation for an ellipse centered at the point $(3, 1)$. To get an explicit representation, we solve for y in terms of x:

$$\left(\frac{y-1}{2}\right)^2 = 1 - \left(\frac{x-3}{5}\right)^2, \qquad \text{so} \qquad \frac{y-1}{2} = \pm\sqrt{1 - \left(\frac{x-3}{5}\right)^2},$$

and thus,

$$y = 1 \pm 2\sqrt{1 - \frac{(x-3)^2}{25}}.$$

We do not get a single explicit representation for the whole ellipse; rather, we get one for the upper half (the positive square root) and one for the lower half (the negative square root).

Explicit and parametric equations are easier to plot than implicit equations. To sketch $y = f(x)$, we evaluate $f(x)$ for various values of x and plot points. Similarly, to sketch a curve given parametrically, we evaluate x and y for various values of t and plot points. For an implicit representation, however, we substitute a value for x, but then we must solve the implicit equation for y; there may be many or no solutions. Moreover, it may be impossible to solve the equation for y algebraically.

Using Linearization to Construct a Local Approximation

Even when we cannot solve an implicit equation explicitly for y in terms of x, we can often replace the equation by a linear approximation valid near a point. This approximation can usually be solved for y. Thus, near a particular point, an implicit equation usually does define an explicit linear approximation.

Example 3   The point $(1, 1)$ is on the curve $y^3 + x^2y + x^3 = 3$. Find an explicit linear equation that gives a good approximation for the piece of the curve near the point $(1, 1)$.

Solution   The linear approximation of the function $f(x, y) = y^3 + x^2y + x^3$ near the point $(1, 1)$ is

$$f(x, y) \approx f(1, 1) + f_x(1, 1)(x - 1) + f_y(1, 1)(y - 1) = 3 + 5(x - 1) + 4(y - 1).$$

The curve has equation $f(x, y) = 3$, so near the point $(1, 1)$ the curve is closely approximated by

$$3 + 5(x - 1) + 4(y - 1) = 3.$$

Solving for y gives the explicit linear equation

$$y = 1 - \frac{5}{4}(x - 1).$$

This is the equation for the tangent line to the curve at the point $(1, 1)$. (See Figure 16.48.)

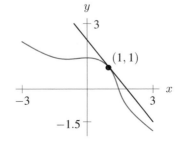

Figure 16.48: The curve $y^3 + x^2y + x^3 = 3$ is well-approximated by its tangent line near the point $(1, 1)$

Implicit, Explicit, and Parametric Representations of Surfaces in 3-Space

Implicit and explicit equations in 3 variables describe surfaces in 3-space rather than curves. For example, the sphere of radius 1 centered at the origin can be represented implicitly by the equation

$$x^2 + y^2 + z^2 = 1,$$

or explicitly by the equations

$$z = \sqrt{1 - x^2 - y^2} \quad \text{and} \quad z = -\sqrt{1 - x^2 - y^2},$$

or parametrically by

$$x = \sin\phi\cos\theta, \quad y = \sin\phi\sin\theta, \quad z = \cos\phi, \quad 0 \leq \theta \leq 2\pi, \quad 0 \leq \phi \leq \pi.$$

Curves in 3-space cannot be represented explicitly or implicitly because a single equation in 3-variables usually describes a surface. However, as we have seen, curves in 3-space can be given parametrically. For example, a helix on the cylinder of radius 1 centered on the z-axis may be described parametrically by

$$x = \cos t, \quad y = \sin t, \quad z = t, \quad -\infty < t < \infty.$$

Notice the difference between the parametric representations of curves and surfaces: Curves require one parameter and surfaces require two. For this reason, we say that curves are 1-dimensional objects and surfaces are 2-dimensional.

Getting an Explicit Function from an Implicit Equation

Notice that although a circle and a sphere are represented by implicit equations, they both give rise to explicit functions, each of which corresponds to part of the graph. For example the sphere

$$x^2 + y^2 + z^2 = 1$$

corresponds to the explicit functions for the top and bottom of the sphere:

$$z = f_1(x, y) = \sqrt{1 - x^2 - y^2} \quad \text{and} \quad z = f_2(x, y) = -\sqrt{1 - x^2 - y^2}.$$

We now look at a more complicated example.

Example 4   Show that the implicit equation $z^3 - 7yz + 6e^x = 0$ does not define z as a function of x and y.

Solution   As an example, let's try to calculate the value of z corresponding to $x = 0, y = 1$. We substitute $x = 0$ and $y = 1$ in the equation and solve for z. Since

$$z^3 - 7z + 6 = (z - 2)(z - 1)(z + 3) = 0,$$

we have three solutions: $z = 2, z = 1, z = -3$. Thus, z is not a function of x and y.

This example shows that if $z^3 - 7yz + 6e^x = 0$, then we cannot expect to write z as an explicit function of x and y. However we may still be able to write z as an explicit function of x and y on part of the graph.

The graph of the equation $z^3 - 7yz + 6e^x = 0$ is a surface which contains the three points $(0, 1, 2)$, $(0, 1, 1)$ and $(0, 1, -3)$. We hope we can find functions

$$z = f_1(x, y), \quad z = f_2(x, y), \quad z = f_3(x, y)$$

such that f_1 gives points on the surface near $(0, 1, 2)$ and f_2 gives points near $(0, 1, 1)$ and f_3 gives points near $(0, 1, -3)$.

What are the formulas for f_1, f_2, and f_3? Since the equation $z^3 - 7yz + 6e^x = 0$ cannot easily be solved for z, we cannot easily find explicit formulas for f_1, f_2 and f_3. However this does *not* mean that there are no such functions. We can still evaluate f_1, f_2, f_3.

Example 5 Suppose f_1, f_2, f_3 are the functions just defined by $z^3 - 7yz + 6e^x = 0$.
  (a) Find $f_1(0, 1)$, $f_2(0, 1)$, $f_3(0, 1)$.
  (b) Find $f_1(0.02, 1.01)$, $f_2(0.02, 1.01)$, $f_3(0.02, 1.01)$.

Solution (a) Since f_1 gives z-values near the point $(0, 1, 2)$ we have

$$f_1(0, 1) = 2.$$

Similarly

$$f_2(0, 1) = 1 \quad \text{and} \quad f_3(0, 1) = -3.$$

  (b) To calculate $f_1(0.02, 1.01)$ we substitute $x = 0.02$ and $y = 1.01$ in the implicit equation

$$z^3 - 7.07z + 6e^{0.02} = 0.$$

Solving numerically for z, we again get three solutions, $z = 2.0038, z = 1.0127, z = -3.0165$. Thus, we expect

$$f_1(0.02, 1.01) = 2.0038, \quad f_2(0.02, 1.01) = 1.0127, \quad f_3(0.02, 1.01) = -3.0165.$$

This example suggests that the functions f_1, f_2, and f_3 are well-defined and we can evaluate them for x near 0 and y near 1.

Finding an Explicit Linear Approximation

Although we cannot find explicit formulas for f_1, f_2 and f_3, we can find explicit linear functions which approximate each one near the corresponding point on the surface. To do this, we make a linear approximation to the original implicit equation $z^3 - 7zy + 6e^x = 0$ and solve for z.

Example 6 (a) Find an explicit linear function, l, which approximates the function f_1 near the point $(0, 1, 2)$.
  (b) Compare the value given by this approximation to the values of $f_1(0, 1)$ and $f_1(0.02, 1.01)$.

Solution (a) To find an explicit function valid near $(0, 1, 2)$, we use a linear approximation and solve for z. Suppose $F(x, y, z) = z^3 - 7yz + 6e^x$. Then

$$F_x(x, y, z) = 6e^x, \quad F_y(x, y, z) = -7z, \quad F_z(x, y, z) = 3z^2 - 7y,$$
$$F_x(0, 1, 2) = 6, \quad F_y(0, 1, 2) = -14, \quad F_z(0, 1, 2) = 5.$$

The linear approximation to F near $(0, 1, 2)$ is

$$F(x, y, z) \approx F(0, 1, 2) + F_x(0, 1, 2)(x - 0) + F_y(0, 1, 2)(y - 1) + F_z(0, 1, 2)(z - 2).$$

Since $F(0, 1, 2) = 0$, we have

$$F(x, y, z) \approx 0 + 6x - 14(y - 1) + 5(z - 2) \quad \text{for } (x, y, z) \text{ near } (0, 1, 2).$$

Since the surface is given by $F(x, y, z) = 0$, we have

$$0 \approx 6x - 14(y - 1) + 5(z - 2).$$

Solving for z tells us that for (x, y, z) near $(0, 1, 2)$ we have

$$z \approx -0.8 - 1.2x + 2.8y.$$

If we define the explicit function, l, by

$$l(x, y) = -0.8 - 1.2x + 2.8y$$

then the function $z = l(x, y)$ is a good approximation to the function $z = f_1(x, y)$ for (x, y) near $(0, 1)$.

(b)   We have $l(0,1) = 2$ and $l(0.02, 1.01) = 2.004$, whereas $f_1(0,1) = 2$ and $f_1(0.02, 1.01) = 2.0038$. Thus, the values of l and f_1 agree exactly at $(0,1)$ and are close for (x,y) near $(0,1)$.

Example 7   Try to find a linear function, $l(x,y)$, such that $z = l(x,y)$ approximates solutions to the equation

$$x^2 + y^2 + z^2 = 25$$

near the solution $(x,y,z) = (3,4,0)$.

Solution   Consider the equivalent equation $F(x,y,z) = x^2 + y^2 + z^2 - 25 = 0$. Since $F_x = 2x, F_y = 2y$ and $F_z = 2z$, we have

$$F_x(3,4,0) = 6, \quad F_y(3,4,0) = 8, \quad F_z(3,4,0) = 0.$$

The local linearization of F for (x,y,z) near $(3,4,0)$ is given by

$$F(x,y,z) \approx 6(x-3) + 8(y-4) + 0(z-0).$$

So, the linearization of the equation $F(x,y,z) = 0$ near $(3,4,0)$ is

$$6(x-3) + 8(y-4) = 0,$$

which cannot be solved for z, because z does not appear in this equation. Therefore, this method does not give an approximation for z as a function of (x,y) near the point $(x,y,z) = (3,4,0)$.

The solution to Example 7 shows that although we cannot solve for z near $(3,4,0)$, we could solve for y, say, and express $y = l_1(x,z)$ near that point. This tells us that although we cannot use x and y to parameterize the sphere near $(3,4,0)$, we can use x and z.

Figure 16.49 shows why the point $(3,4,0)$ causes trouble. The equation $x^2 + y^2 + z^2 = 25$ is a sphere of radius 5 centered at the origin. Do values of (x,y) near $(3,4)$ determine unique values of z near 0 such that (x,y,z) lies on the sphere? The answer is no, for two reasons. For points such as $(x,y) = (3.01, 4.01)$ where $x^2 + y^2 > 25$, there is no z at all such that (x,y,z) is on the sphere. For points such as $(2.99, 3.99)$ where $x^2 + y^2$ is slightly less than 25, there are two z-values near 0 that satisfy the equation, namely $z = \sqrt{25 - x^2 - y^2}$ and $z = -\sqrt{25 - x^2 - y^2}$. (See Figure 16.49.) Since the equation $x^2 + y^2 + z^2 = 25$ does not determine unique z-values for all (x,y) near $(3,4)$, there is no function to approximate near $(3,4)$.

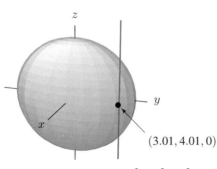

Figure 16.49: The sphere $x^2 + y^2 + z^2 = 25$ and a vertical line through the point $(3.01, 4.01, 0)$

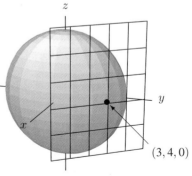
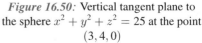

Figure 16.50: Vertical tangent plane to the sphere $x^2 + y^2 + z^2 = 25$ at the point $(3,4,0)$

The tangent plane to the sphere at $(3,4,0)$ is the linearization $6(x-3) + 8(y-4) = 0$. The absence of z in this equation indicates that the tangent plane is vertical and does not determine z as a function of (x, y). (See Figure 16.50.) The fact that z is missing corresponds to the fact that $F_z(3,4,0) = 0$.

We summarize the previous examples in the following theorem, whose proof can be found in more advanced texts.

The Implicit Function Theorem

Suppose $F(x, y, z)$ is a smooth function and (a, b, c) is a point such that
- $F(a, b, c) = 0$
- $F_z(a, b, c) \neq 0$

Then there is a smooth function $z = f(x, y)$ such that, for (x, y) near (a, b),

$$F(x, y, f(x, y)) = 0.$$

The linear approximation, $l(x, y)$, to $f(x, y)$ at the point (a, b) is obtained by solving for z in $L(x, y, z) = 0$, where $L(x, y, z)$ is the linear approximation to $F(x, y, z)$ at the point (a, b, c).

Being able to solve for z and write $z = f(x, y)$ tells us that, near (a, b, c), the surface $F(x, y, z) = 0$ can be parameterized by x and y:

$$x = s, \quad y = t, \quad z = f(s, t).$$

This parameterization is not always global and as in Example 7, we may have to use x and z or y and z instead of x and y.

Problems for Section 16.4

What curves do the parametric equations in Problems 1–3 trace out? Find an implicit or explicit equation for each curve.

1. $x = 2 + \cos t$, $y = 2 - \sin t$
2. $x = 2 + \cos t$, $y = 2 - \cos t$
3. $x = 2 + \cos t$, $y = \cos^2 t$

State whether the equations in Problems 4–6 represent a curve parametrically, implicitly, or explicitly. Give the two other types of representations for the same curve.

4. $xy = 1$ for $x > 0$
5. $x^2 - 2x + y^2 = 0$ for $y < 0$
6. $x = e^t$, $y = e^{2t}$ for all t

7. Find an equation for the line tangent to the curve $xe^y + 2ye^x = 0$ at the point $(0, 0)$.
8. The equation $x \cos y + e^x + y = 1$ cannot be solved explicitly for y in terms of x.
   (a) Find a linear equation that has approximately the same solutions near $(0, 0)$, and solve it for y in terms of x.
   (b) What is the geometric meaning of the linear equation you found in part (a)?

9. Let f_1 be the function in Example 5. Make a table of values for f_1 for
   $$x = -0.02, \ -0.01, \ 0.00, \ 0.01, \ 0.02, \quad y = 0.98, \ 0.99, \ 1.00, \ 1.01, \ 1.02.$$

10. Compare the values of the local linear approximation, l, of Example 6 to the values of f_1 computed in Problem 9.

11. Let $z = f_2(x, y)$ be the function, defined for (x, y) near $(0, 1)$, such that z is near 1 and $z^3 - 7yz + 6e^x = 0$.

    (a) Evaluate $f_2(0.01, 0.98)$.
    (b) Find a linear approximation for $f_2(x, y)$ for (x, y) near $(0, 1)$ and use it to approximate $f_2(0.01, 0.98)$.
    (c) Find $\partial f_2/\partial x$ at $(0, 1)$ and $\partial f_2/\partial y$ at $(0, 1)$.

12. Let $z = f_3(x, y)$ be the function, defined for (x, y) near $(0, 1)$, such that z is near -3 and $z^3 - 7yz + 6e^x = 0$.

    (a) Evaluate $f_3(0.01, 0.98)$.
    (b) Find a linear approximation for $f_3(x, y)$ for (x, y) near $(0, 1)$ and use it to approximate $f_3(0.01, 0.98)$.
    (c) Find $\partial f_3/\partial x$ at $(0, 1)$ and $\partial f_3/\partial y$ at $(0, 1)$.

13. At the point $(3, 5, 7)$ a certain continuously differentiable function $f(x, y, z)$ has local linearization $L(x, y, z) = 2(x - 3) + 4(y - 5) + 5(z - 7)$.

    (a) What can you say about the graph of the equation $f(x, y, z) = 0$?
    (b) What can you say about the solutions of the equation $f(x, y, z) = 0$?

14. Find an equation for the tangent plane to the surface $z^2 + x^2 - y = 0$ at the point $(1, 1, 0)$ using a local linearization.

15. Suppose the satisfaction that a person experiences as a result of consuming a quantity x_1 of one item and a quantity x_2 of another item is given as a function of x_1 and x_2 by

    $$S = f(x_1, x_2) = a \ln x_1 + (1 - a) \ln x_2,$$

    where a is a constant, $0 < a < 1$. The prices of the two items are p_1 and p_2 respectively, and the budget is b.

    (a) Express the maximum satisfaction that can be achieved as a function of p_1, p_2, and b, that is, $S = g(p_1, p_2, b)$.
    (b) Find a function giving the amount of money that must be spent to achieve a particular level of satisfaction, c, as a functions of p_1, p_2, and c, that is $b = h(p_1, p_2, c)$.
    (c) Explain why part (b) is an example of the implicit function theorem.

16. The maximum *utility* (satisfaction) that a person can achieve as a result of consuming x_1 units of one item and x_2 units of another item is a function of the prices p_1 and p_2 of the two items and the budget, m. We write

    $$u = f(p_1, p_2, m).$$

    (a) We assume u is an increasing function of m. What partial derivative is this telling us about? What does this assumption mean in economic terms?
    (b) Use the implicit function theorem to show that there is a function $m = g(p_1, p_2, u)$ satisfying

    $$u = f(p_1, p_2, g(p_1, p_2, u)).$$

    (c) Explain why g is called the *expenditure function*. What does it tell us in economic terms?

16.5 NOTES ON NEWTON, KEPLER, AND PLANETARY MOTION

Each night the dome of stars rotates slowly overhead. At first the stars seem fixed in relation to each other, but observation over many nights reveals some stars moving relative to the rest. These wanderers are not stars, but planets. Five of them are visible to the naked eye: Mercury, Venus, Mars, Jupiter, and Saturn, all named after ancient Roman gods. Their erratic paths have caused people to endow them with supernatural powers; for example, astrology is based largely on the position of the planets with respect to the fixed stars. Newton and Kepler's mathematical explanations of planetary motion were two of the great breakthroughs in the history of science.

First Steps: Eratosthenes and Copernicus

The first step toward the explanation of planetary motion was the realization that

> The earth is round, a sphere of radius about 4000 miles.

This was known to some in ancient Greece. Eratosthenes (276–197 BC) made a reasonable estimate of the earth's radius by observing the angle of the sun at noon on June 21 at two different locations (See Problem 1). Another important step was the realization that

> The earth rotates once each day on a central axis passing through the north and south poles.

The Greek philosopher Aristotle (384–322 BC) thought, on the contrary, that the earth remained fixed while other heavenly bodies moved around it. Indeed, the very idea of the earth spinning may seem preposterous; wouldn't we be spun right off the planet? In fact, the acceleration at the equator caused by the earth's rotation is less than 1% the acceleration due to gravity, and is about the same as the acceleration at the edge of a merry-go-round with 25 foot radius rotating once every $1\frac{1}{2}$ minutes (See Problem 2.)

During the Renaissance, Nicolaus Copernicus (1473–1543) proposed the more modern point of view, that the apparent motion of the stars each night is caused by the earth's rotation. Copernicus also contradicted previous theories by placing the sun at the center of the solar system with the earth and other planets revolving around it:

> The earth and the planets orbit around the sun. The earth completes one revolution around the sun each $365\frac{1}{4}$ days (approximately). The moon orbits around the earth completing one revolution in about $27\frac{1}{3}$ days.

In fact Aristarchus of Samos (310–230 BC) had already proposed such a theory, saying that the motion of the earth and planets around the sun explains the apparent motions of the planets.

Kepler's Laws for Planetary Motion

We now know that the planets do not orbit in circles with the sun at the center, nor does the moon orbit in a circle with the earth at the center. For example, the moon's distance from the earth varies from 220,000 to 260,000 miles. In the last half of the 16th century the Danish astronomer Tycho Brahe (1546–1601) made measurements of the positions of the planets. Johann Kepler (1571–1630) studied this data for years and, after some false starts, arrived at three laws for planetary motion:

Kepler's Laws

I. The orbit of each planet is an ellipse with the sun at one focus. In particular, the orbit lies in a plane containing the sun.

II. As a planet orbits around the sun, the line segment from the sun to the planet sweeps out equal areas in equal times.

III. The ratio p^2/d^3 is the same for every planet orbiting around the sun, where p is the period of the orbit (time to complete one revolution) and d is the mean distance of the orbit (average of the shortest and farthest distances from the sun).

An ellipse is a closed curve in the plane such that the sum of the distances from any point on the curve to two fixed points, called the foci of the ellipse, is constant. If the two foci are located at $(0, -b)$ and $(0, b)$ on the y-axis, and if the constant sum of distance is $2d$, then it can be shown that the equation of the ellipse is

$$\frac{x^2}{c^2} + \frac{y^2}{d^2} = 1,$$

where d is the mean distance and $c^2 = d^2 - b^2$. (See Problem 3.)

Kepler's Second Law says that the line from the planet to the sun always sweeps out the same area in one unit of time. This implies that the speed of the planet is not constant, because the planet must move faster when it is near the sun (See Figure 16.51.)

The Third Law says that p^2/d^3 is the same for all planets. In particular, this means if we know the period p for a planet in earth years, then we know the ratio of its mean distance to the earth's mean distance. Newton later showed that the constant value of p^2/d^3 depends on the mass of the object about which the planets are orbiting.

Kepler's Laws, impressive as they are, were purely descriptive; they didn't explain the motion of the planets. Newton's great achievement was to find an underlying cause for them.

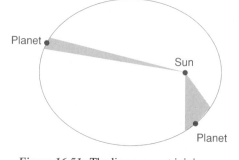

Figure 16.51: The line segment joining a planet to the sun sweeps out equal areas in equal time

Newton's First and Second Laws of Motion and Universal Gravitation

In 1687, Isaac Newton (1642–1727) published the *Principia Mathematica*.[1] In *Principia*, Newton developed a theory of motion based on the concept of force. He began by observing that curving motion is an indication of acceleration, and set about finding a specific law of acceleration that would explain Kepler's Laws. Newton defined the "motion" of a body, which we would call the momentum, to be the product of its mass, m, and velocity, $\vec{v}$.

Newton's First Law of Motion

Every body continues in a state of rest, or of constant motion [momentum] in a straight line, unless it is compelled to change that state by forces acting on it.

In modern language, the First Law says that $m\vec{v}$ is a constant vector if no force is acting. In particular, a planet can move in an ellipse only if there is a force acting on it.

Newton's Second Law of Motion

The rate of change of motion [momentum] of a body is proportional to the force acting on it; and is in the direction of the straight line in which that force acts.

Newton recognized that the rate of change of momentum was a vector quantity. His second law gives the direction of the rate of change: In modern notation, if m is mass and $\vec{v}$ is velocity, then

$$\frac{d}{dt}(m\vec{v}) = m\frac{d\vec{v}}{dt} = m\vec{a}.$$

Writing $\vec{F}$ for the force acting on the body, we get the modern version of Newton's Second Law, $\vec{F} = m\vec{a}$. (The units are chosen so that the constant of proportionality is 1.)

In order to complete his explanation of Kepler's Laws, Newton needed a law giving the gravitational force of the sun on a planet.

The Universal Law of Gravitation

Two objects of mass M and m are attracted to each other by a force, F, proportional to the product of their masses and the inverse square of the distance r between them:

$$F = \frac{GMm}{r^2},$$

where G is a universal constant.

Although Newton's work contains many of the fundamental ideas of calculus, his reasoning used similar triangles and geometry.[2] In the remainder of this section, we will explain Newton's approach with modern proofs using derivatives, vectors, and cross products.

---

[1] The full Latin title was *Philosophiae Naturalis Principia Mathematica*, which means *Mathematical Principles of Natural Philosophy*.

[2] See Tristan Needham, *Newton and the Transmutation of Force*, The American Mathematical Monthly, vol. 100, 1993, pp. 119–137, for an account of Newton's approach.

Newton's Explanation of Kepler's Second Law

Newton's Second Law says that the acceleration vector of a planet points in the direction of the gravitational force acting on it, and the Law of Gravitation says that the gravitational force vector points toward the sun. Thus, the acceleration vector of a planet always points toward the sun. We define *centripetal motion about the fixed point A* to be motion in which the acceleration vector always points toward A (or directly away from A). Newton proved that centripetal motion is equivalent to motion in a plane obeying Kepler's Second Law.

Newton's Theorem: Kepler's Laws and Centripetal Motion

Suppose an object moves in a plane containing the point A in such a way that the line segment from A to the object sweeps out equal areas in equal times. Then the motion is centripetal about A. Conversely, if the motion is centripetal about A, then the object moves in a plane containing A and the line segment from A to the object sweeps out equal areas in equal times.

Newton gave a geometric proof of this theorem. We give a modern proof. Consider the vector $\vec{r} \times \vec{v}$, where $\vec{r}$ is the vector from A to the object and $\vec{v}$ is the velocity vector of the object. (See Figure 16.52.) We show that $\vec{r} \times \vec{v}$ represents the rate at at which area is being swept out by the vector $\vec{r}$.

Suppose in time Δt the vector $\vec{r}$ becomes $\vec{r} + \Delta\vec{r}$. Figure 16.53 shows that the area swept out by the vector $\vec{r}$ during the time interval Δt is approximately triangular and is given by

$$\Delta\vec{A} \approx \frac{1}{2}\vec{r} \times (\vec{r} + \Delta\vec{r}) = \frac{1}{2}\vec{r} \times \vec{r} + \frac{1}{2}\vec{r} \times \Delta\vec{r} = \frac{1}{2}\vec{r} \times \Delta\vec{r}$$

since $\vec{r} \times \vec{r} = \vec{0}$. Dividing by Δt, we have

$$\frac{\Delta\vec{A}}{\Delta t} \approx \frac{1}{2}\vec{r} \times \frac{\Delta\vec{r}}{\Delta t}.$$

Letting $\Delta t \to 0$ gives

$$\frac{d\vec{A}}{dt} = \frac{1}{2}\vec{r} \times \vec{v}.$$

Thus, we see that the area of the triangle determined by $\vec{r}$ and $\vec{v}$ gives the rate at which area is being swept out as the object moves around its orbit.

The direction of $\vec{r} \times \vec{v}$ is perpendicular to the plane containing $\vec{r}$ and $\vec{v}$. Thus, if the area is being swept out at a constant rate and if $\vec{r}$ and $\vec{v}$ always lie in the same plane as the motion, then $\vec{r} \times \vec{v}$ is a constant vector. To see whether $\vec{r} \times \vec{v}$ is constant, we take the derivative of $\vec{r} \times \vec{v}$. Using the product rule, the fact that $\vec{v} \times \vec{v} = \vec{0}$, and writing the acceleration $\vec{a} = d\vec{v}/dt$, we get

$$\frac{d}{dt}(\vec{r} \times \vec{v}) = \frac{d\vec{r}}{dt} \times \vec{v} + \vec{r} \times \frac{d\vec{v}}{dt} = \vec{v} \times \vec{v} + \vec{r} \times \vec{a} = \vec{r} \times \vec{a}.$$

First, let us assume the motion is in a plane containing A and the line segment sweeps out the equal areas in equal times. Then $\vec{r} \times \vec{v}$ is constant so we must have $\vec{r} \times \vec{a} = \vec{0}$. This means that $\vec{a}$ must be parallel to $\vec{r}$. But this means that $\vec{a}$ always points toward (or away from) A, so the motion is centripetal.

Conversely, suppose the motion is centripetal. Then $\vec{r}$ and $\vec{a}$ are parallel, so $\vec{r} \times \vec{a} = \vec{0}$, which implies that $\vec{r} \times \vec{v}$ is constant. This tells us that equal areas are swept out in equal times and that $\vec{r}$ and $\vec{v}$ always lie in the same plane (the plane perpendicular to $\vec{r} \times \vec{v}$).

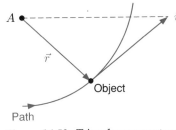

Figure 16.52: Triangle represents rate at which area is swept out by vector $\vec{r}$ when body is moving with velocity $\vec{v}$

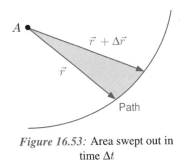

Figure 16.53: Area swept out in time Δt

Newton's Explanation of Kepler's First Law

The equivalence between Kepler's Second Law and centripetal motion tells us that a planet's acceleration vector always points towards the sun. But what about its magnitude? Newton showed that the magnitude of the acceleration could be derived from Kepler's First Law, which says that the planet moves in an ellipse, centripetally about one of its foci. His result was the following:

Newton's Theorem: Motion about the Focus of the Ellipse

Suppose an object is moving in an ellipse centripetally about a focus B of the ellipse. Suppose the mean distance of the ellipse is d and the period of the motion is p. If r is the distance from the object to B, then the magnitude of the acceleration is given by

$$a = \frac{k}{r^2}, \qquad \text{where } k = \frac{4\pi^2 d^3}{p^2}.$$

Thus, a is proportional to the inverse square of r.

The Gravitational Force

So far, we have concentrated on acceleration rather than force. Newton realized there must be a force bending the path of the moon as it circled the earth. He also knew that a falling body, for example an apple from a tree, accelerated toward the earth. His insight was that the force pulling the apple to the earth was the same force that accelerated the moon in its orbit. The magnitude of the force of gravity is given by the Universal Law of Gravitation,

$$F = \frac{GMm}{r^2},$$

where G is a constant. For a planet of mass m orbiting the sun of mass M, we have

$$F = ma = \frac{GMm}{r^2}, \qquad \text{so} \quad a = \frac{GM}{r^2}.$$

Notice that Newton's formulation of the law of gravity explains the fact that Kepler's laws do not involve the masses of the individual planets. Since $F = ma$ and $F = GMm/r^2$, the mass, m, cancels out. This is the explanation for Galileo's striking discovery that the acceleration of a body in free fall does not depend on the mass of the body.

By Newton's theorem, the constant of proportionality is $4\pi^2 d^3/p^2$, so we have

$$GM = \frac{4\pi^2 d^3}{p^2}.$$

This relationship tells us that if we know the gravitational constant G, we can compute the mass M. These calculations assume that M is much larger than m, so that the mass M can be regarded as stationary. This applies to planets orbiting the sun, to the moon orbiting the earth, or to a moon orbiting Jupiter. Since laboratory experiments on earth give us the value of G, Newton's law opened the way to the calculation of the mass of the sun, the earth, and Jupiter.

The work of Newton inaugurated a new era in science in which the use of differential equations brought spectacular advances throughout physics and astronomy.

Problems for Section 16.5

1. In the the 3rd century BC, Eratosthenes estimated the circumference of the earth by the following method. He knew that near Syene, Egypt, on the longest day of the year, the sun could be seen reflected at the bottom of a deep well and so was directly overhead. On the same day, at Alexandria, Egypt, he observed that the sun passed about 1/50 of a full circle south of the zenith (that is, south of directly overhead). By talking to the camel drivers, Eratosthenes also learned that the north-south distance between Alexandria and Syene was about 5,000 stadia (a *stadium* was a Greek unit of length which is thought to be about 185 meters). Use this information to estimate the circumference of the earth.

2. For a point on the equator, compute the magnitude of the acceleration caused by the earth's rotation. Use feet per second per second as the units. The radius of the earth is 4000 miles and (as you know) the period of its rotation is 24 hours. Compare your answer to the value of the acceleration due to gravity, $g = 32$ ft/sec^2. Suppose a point at the edge of a merry-go-round of radius 25 feet has the same acceleration as the point on the equator. What is the speed of the point on the merry-go-round? What is the period of the merry-go-round?

3. Suppose an ellipse has foci at $(0, b)$ and $(0, -b)$ in the xy-plane and that the mean distance to the focus at $(0, b)$ is d. Show that the constant sum of the distances from any point on the ellipse to the two foci is $2d$. Then show that the equation for the ellipse is

$$\frac{x^2}{c^2} + \frac{y^2}{d^2} = 1$$

where d is the mean distance and $c^2 = d^2 - b^2$.

4. Suppose a particle moves in the xy-plane so that its acceleration vector $\vec{a}$ always points to the origin and has magnitude proportional to the distance to the origin. Choose the x-axis so that the closest point to the origin on the particle's path is $(a, 0)$. Explain why at that point the velocity vector is perpendicular to the x-axis. Show that with the given x and y coordinates, if we define time $t = 0$ to be the instant when the particle is at $(a, 0)$, then the particle satisfies the differential equations

$$\frac{d^2x}{dt^2} = -kx, \qquad \frac{d^2y}{dt^2} = -ky, \qquad k > 0,$$

with initial conditions $x(0) = a$, $x'(0) = 0$, and $y(0) = 0$, $y'(0) = c$. Here c is the velocity in the y-direction at time $t = 0$. Now show that if $b = c/\sqrt{k}$, the solution to these differential equations is

$$x = a \cos \sqrt{k}t, \qquad y = b \sin \sqrt{k}t.$$

5. Experiment on a computer or calculator with the orbits resulting from centripetal acceleration. You will need a program that plots trajectories (solutions) for systems of differential equations. For example, to look at orbits with a centripetal acceleration of k/r, you need to solve a system of differential equations with four variables, the position variables x and y and the velocity variables $u = dx/dt$ and $v = dy/dt$, which satisfy the system

$$\frac{dx}{dt} = u, \quad \frac{dy}{dt} = v, \quad \frac{du}{dt} = \frac{-kx}{x^2 + y^2}, \quad \frac{dv}{dt} = \frac{-ky}{x^2 + y^2}.$$

Check that these equations imply that the acceleration vector $(d^2x/dt^2)\vec{i} + (d^2y/dt^2)\vec{j} = (du/dt)\vec{i} + (dv/dt)\vec{j}$ has the correct direction and magnitude. Then use the computer to plot the x and y variables starting from some initial values for x, y, u, and v. Try other laws: k/r^3, kr^2. Are orbits always closed?

6. A hyperbola is a curve such that the *difference* of the distances from any point on the curve to two fixed points (called the foci) is constant. The equation for a hyperbola centered at the origin is

$$-\frac{x^2}{c^2} + \frac{y^2}{d^2} = 1$$

where $2d$ is the constant difference in distances from foci at $(0, b)$ and $(0, -b)$ and $c^2 = b^2 - d^2$. Show that the motion parameterized by

$$x = \frac{c}{2}(e^{kt} - e^{-kt}), \quad y = \frac{d}{2}(e^{kt} + e^{-kt})$$

lies on the hyperbola $-x^2/c^2 + y^2/d^2 = 1$ and also that

$$\frac{d^2x}{dt^2} = k^2 x, \quad \frac{d^2y}{dt^2} = k^2 y.$$

Thus, the given motion has its acceleration pointing away from the origin with magnitude proportional to the distance from the origin.

REVIEW PROBLEMS FOR CHAPTER SIXTEEN

Write a parameterization for each of the curves in Problems 1–10.

1. The horizontal line through the point $(0, 5)$.

2. The circle of radius 2 centered at the origin starting at the point $(0, 2)$ when $t = 0$.

3. The circle of radius 4 centered at the point $(4, 4)$ starting on the x-axis when $t = 0$.

4. The circle of radius 1 in the xy-plane centered at the origin, traversed counterclockwise when viewed from above.

5. The line through the points $(2, -1, 4)$ and $(1, 2, 5)$.

6. The line through the point $(1, 3, 2)$ perpendicular to the xz-plane.

7. The line through the point $(1, 1, 1)$ perpendicular to the plane $2x - 3y + 5z = 4$.

8. The circle of radius 2 parallel to the xy-plane, centered at the point $(0, 0, 1)$, and traversed counterclockwise when viewed from below.

9. The circle of radius 3 parallel to the xz-plane, centered at the point $(0, 5, 0)$, and traversed counterclockwise when viewed from $(0, 10, 0)$.

10. The circle of radius 2 centered at $(0, 1, 0)$ lying in the plane $x + z = 0$.

11. Consider the parametric equations below for $0 \leq t \leq \pi$.

(I) $\vec{r} = \cos(2t)\vec{i} + \sin(2t)\vec{j}$

(II) $\vec{r} = 2\cos t\vec{i} + 2\sin t\vec{j}$

(III) $\vec{r} = \cos(t/2)\vec{i} + \sin(t/2)\vec{j}$

(IV) $\vec{r} = 2\cos t\vec{i} - 2\sin t\vec{j}$

(a) Match the equations above with four of the curves C_1, C_2, C_3, C_4, C_5 and C_6 in Figure 16.54. (Each curve is part of a circle.)

(b) Give parametric equations for the curves which have not been matched, again assuming $0 \leq t \leq \pi$.

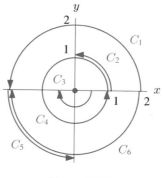

Figure 16.54

12. On a graphing calculator or a computer, plot $x = 2t/(t^2 + 1)$, $y = (t^2 - 1)/(t^2 + 1)$, first for $-50 \leq t \leq 50$ then for $-5 \leq t \leq 5$. Explain what you see. Is the curve really a circle?

13. Let $f(x, y) = \dfrac{x^2 - y^2}{x^2 + y^2}$.

(a) In which direction should you move from the point $(1, 1)$ to obtain the maximum rate of increase of f?

(b) Find a direction in which the directional derivative at the point $(1, 1)$ is equal to zero.

(c) Suppose you move along the curve $x = e^{2t}$, $y = 2t^3 + 6t + 1$. What is df/dt at $t = 0$?

14. Find the parametric equation for the line of intersection of the planes $z = 4 + 2x + 5y$ and $z = 3 + x + 3y$.

15. Find parametric equations of the line passing through the points $(1, 2, 3), (3, 5, 7)$ and calculate the shortest distance from the line to the origin.

16. Are the lines $x = 3 + 2t$, $y = 5 - t$, $z = 7 + 3t$ and $x = 3 + t$, $y = 5 + 2t$, $z = 7 + 2t$ parallel?

17. Plot the Lissajous figure given by $x = \cos 2t$, $y = \sin t$ using a graphing calculator or computer. Explain why it looks like part of a parabola. [Hint: Use a double angle identity from trigonometry.]

18. Suppose that a planet P in the xy-plane orbits the star S counterclockwise in a circle of radius 10 units, completing one orbit in 2π units of time. Suppose in addition a moon M orbits the planet P counterclockwise in a circle of radius 3 units, completing one orbit in $2\pi/8$ units of time. The star S is fixed at the origin $x = 0$, $y = 0$, and at time $t = 0$ the planet P is at the point $(10, 0)$ and the moon M is at the point $(13, 0)$.

(a) Find parametric equations for the x- and y-coordinates of the planet at time t.

(b) Find parametric equations for the x- and y-coordinates of the moon at time t. [Hint: For the moon's position at time t, take a vector from the sun to the planet at time t and add a vector from the planet to the moon].

(c) Plot the path of the planet using a graphing calculator or computer.

(d) Experiment with different radii and speeds for the moon's orbit around the planet.

19. A particle travels along a line, with position at time t given by

$$\vec{r}(t) = (2 + 5t)\vec{i} + (3 + t)\vec{j} + 2t\vec{k}.$$

(a) Where is the particle when $t = 0$?

(b) At what time does the particle reach the point $(12, 5, 4)$?

(c) Does the particle ever reach the point $(12, 4, 4)$? Why or why not?

20. An ant, starting at the origin, moves at 2 units/sec along the x-axis to the point $(1, 0)$. The ant then moves counterclockwise along the unit circle to $(0, 1)$ at a speed of $3\pi/2$ units/sec, then straight down to the origin at a speed of 2 units/sec along the y-axis.

(a) Express the ant's coordinates as a function of time, t, in secs.

(b) Express the reverse path as a function of time.

21. A basketball player shoots the ball from 6 feet above the ground towards a basket that is 10 feet above the ground and 15 feet away horizontally.

(a) Suppose she shoots the ball at an angle of A degrees above the horizontal $(0 < A < \pi/2)$ with an initial speed V. Give the x- and y-coordinates of the position of the basketball at time t. Assume the x-coordinate of the basket is 0 and that the x-coordinate of the shooter is -15. [Hint: There is an acceleration of -32 ft/sec^2 in the y-direction; there is no acceleration in the x-direction. Ignore air resistance.]

(b) Using the parametric equations you obtained in part (a), experiment with different values for V and A, plotting the path of the ball on a graphing calculator or computer to see how close the ball comes to the basket. (The tick marks on the y-axis can be used to locate the basket.)

(c) Find the angle A that minimizes the velocity needed for the ball to reach the basket. (This is a lengthy computation. First find an equation in V and A that holds if the path of the ball passes through the point 15 feet from the shooter and 10 feet above the ground. Then minimize V.)

22. A cheerleader has a 0.4 m long baton with a light on one end. She throws the baton in such a way that its center moves along a parabola, and the baton rotates counterclockwise around the center with a constant angular velocity. The baton is initially horizontal and 1.5 m above the ground; its initial velocity is 8 m/sec horizontally and 10 m/sec vertically, and its angular velocity is 2 revolutions per second. Find parametric equations describing the following motions:

(a) The center of the baton relative to the ground.

(b) The end of the baton relative to its center.

(c) The path traced out by the end of the baton relative to the ground.

(d) Sketch a graph of the motion of the end of the baton.

23. A wheel of radius 1 meter rests on the x-axis with its center on the y-axis. There is a spot on the rim at the point $(1, 1)$. See Figure 16.55. At time $t = 0$ the wheel starts rolling on the x-axis in the direction shown at a rate of 1 radian per second.

(a) Find parametric equations describing the motion of the center of the wheel.

(b) Find parametric equations describing the motion of the spot on the rim. Plot its path.

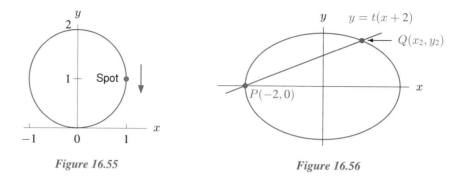

Figure 16.55   Figure 16.56

24. (a) The ellipse $2x^2 + 3y^2 = 8$ intersects the line of slope t through the point $P = (-2, 0)$ in two points, one of which is P. Compute the coordinate of the other point Q. See Figure 16.56.
     (b) Give a parameterization of the ellipse $2x^2 + 3y^2 = 8$ by rational functions.

25. Find parametric equations for the plane $3x + 4y + 5z = 10$.

26. (a) Find a vector normal to the plane $\vec{r} = (3 - 5s + 2t)\vec{i} + (1 + s + 3t)\vec{j} + (s - t)\vec{k}$.
     (b) Find an implicit equation for the plane.

27. You are standing at the point $(-5, 0, 5)$, your friend Jane is standing at the point $(0, -5, -5)$ and her friend Jo is standing at the point $(10, 5, 0)$. Find parametric equations for the plane you are all standing on so that your parameters are $(0, 0)$, and Jo's are $(1, 0)$, and Jane's are $(0, 1)$. Find a parameterization so Jane's and Jo's parameters are switched. Find a parameterization so that your parameters and Jane's are switched.

28. There is a famous way to parameterize a sphere called *stereographic projection*. We work with the sphere $x^2 + y^2 + z^2 = 1$. Draw a line from a point (x, y) in the xy-plane to the north pole $(0, 0, 1)$. This line intersects the sphere in a point (x, y, z). This gives a parameterization of the sphere by points in the plane.
     (a) Which point corresponds to the south pole?
     (b) Which points correspond to the equator?
     (c) Do we get all the points of the sphere by this parameterization?
     (d) Which points correspond to the upper hemisphere?
     (e) Which points correspond to the lower hemisphere?

29. Many brass instruments are approximately Bessel horns, which are surfaces of revolution about the x-axis of

$$z = f(x) = \frac{b}{(x + a)^m}$$

for positive constants $a, b,$ and m. Thus $f(x)$ is the radius of the bell at a distance x from the large open end. Usually m is in the range $0.5 \le m \le 1$.

   Determine a and b and sketch the graph of f for $0 \le x \le 20$ if the radius at $x = 0$ is 15 cm and the radius at $x = 20$ cm is 1 cm for each of the following three cases: $m = 0.5$, $m = 0.7, m = 1$. Why is m called the flare parameter?

30. Give a parameterization of the bell of the horn of Problem 29 with $m = 0.5$.

31. Figure 16.57 is a picture of the parametric surface

$$x = a(s + t), \quad y = b(s - t), \quad z = 4ct^2,$$

for $a = 1$, $b = 1$ and $c = 1$. What happens if you increase a? Increase b? Increase c? How could we flip the surface upside down?

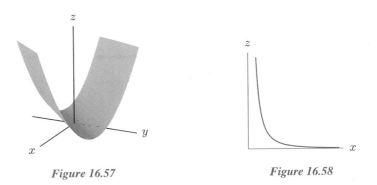

Figure 16.57   *Figure 16.58*

32. Figure 16.58 shows the curve $x^2 z = 1$ in the xz-plane. Obtain a parameterization of the surface obtained by rotating this curve
    (a) About the x-axis, for $x > 0$.   (b) About the z-axis, for $z > 0$.

33. The parametric representation of a line in three dimensions is used to find where the line between two points intersects a given plane. For example, the pictures of curves and surfaces in three-dimensional space in this book are drawn by computer. To do this, the computer first calculates the xyz-coordinates of some points on the curve or surface. For each such point, it then computes the line of sight from that point to the eye of an imaginary viewer and determines where that line intersects an imaginary window (the plane of the computer screen) lying between the point and the viewer's eye. The two-dimensional screen coordinates of that point of intersection are computed so the point can be plotted on the screen. Find formulas for the coordinates of the point of intersection of the plane $Ax + By + Cz = D$ with the line from the point (a, b, c) to a viewer at the point (A, B, C).

34. In Problem 33, the xyz-coordinates are computed for the point where the line of sight from a viewer at (A, B, C) to a point (a, b, c) meets a viewing plane (the screen) $Ax + By + Cz = D$. The computer needs to compute screen coordinates of this point, not the xyz-space coordinates. To do this, take two vectors $\vec{u}$ and $\vec{v}$ at right angles to each other beginning at the screen origin and lying in the plane of the screen. Then any point on the screen can be written as $r\vec{u} + s\vec{v}$ for some numbers r, s; these numbers are the screen coordinates. Choose for the screen origin the point Q where the line of sight from the viewer (A, B, C) to the xyz-origin $(0, 0, 0)$ intersects the viewing plane $Ax + By + Cz = D$. Choose $\vec{u}$ to be a unit vector parallel to the xy-plane pointing to the viewer's right and $\vec{v}$ to be a unit vector at right angles to $\vec{u}$ and pointing up (with positive z-component). The screen coordinates are found by taking the dot product with $\vec{u}$ and $\vec{v}$ of the vector from the screen origin Q to the point of intersection computed in Problem 33.
    (a) Find the xyz-coordinates of the screen origin Q in terms of A, B, C, D.

(b)   Find the vector $\vec{u}$ in terms of A, B, C.

(c)   Find the vector $\vec{v}$ in terms of A, B, C.

(d)   Find the coordinates of the point of intersection computed in Problem 33.

(e)   Find the screen coordinates r and s of the point computed in Problem 33. That is, find $r = \vec{u} \cdot (\vec{P} - \vec{Q})$ and $s = \vec{v} \cdot (\vec{P} - \vec{Q})$. [Hint: Use the fact that $\vec{u} \cdot (A\vec{i} + B\vec{j} + C\vec{k}) = 0$ and $\vec{v} \cdot (A\vec{i} + B\vec{j} + C\vec{k}) = 0$.]

CHAPTER SEVENTEEN

VECTOR FIELDS

Some physical quantities (such as temperature) are best represented by scalars; others (such as velocity) are best represented by vectors. We have looked at functions of many variables whose values are scalars, for example, temperature as a function of position on a weather map. Such functions are called scalar-valued functions.

Some weather maps indicate wind velocity at various points by arrows. Wind velocity is an example of a vector-valued function, since its value at any point is the vector indicating the direction and strength of the wind. Such functions are also called *vector fields*. We have already seen one important example of a vector field, namely the gradient of a scalar-valued function. In this chapter we will look at other examples, such as velocity vector fields describing a fluid flow. We will also look at the path followed by a particle moving with the flow, which is called a *flow line* of the vector field.

17.1 VECTOR FIELDS

Introduction to Vector Fields

A *vector field* is a function that assigns a vector to each point in the plane or in 3-space. One example of a vector field is the gradient of a function $f(x, y)$; at each point (x, y) the vector grad $f(x, y)$ points in the direction of maximum rate of increase of f. In this section we will look at other vector fields representing velocities and forces.

Velocity Vector Fields

Figure 17.1 shows the flow of a part of the Gulf stream, a current in the Atlantic Ocean.[1] It is an example of a *velocity vector field*: each vector shows the velocity of the current at that point. The current is fastest where the velocity vectors are longest in the middle of the stream. Beside the stream are eddies where the water flows round and round in circles.

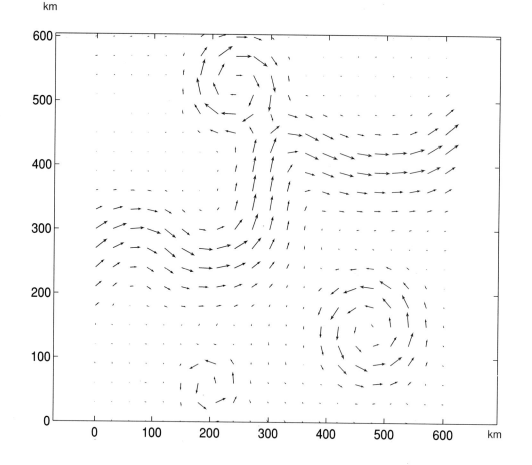

Figure 17.1: The velocity vector field of the Gulf stream

[1] Based on data supplied by Avijit Gangopadhyay of the Jet Propulsion Laboratory

Figure 17.2: The gravitational field of the earth

Force Fields

Another physical quantity represented by a vector is force. When we experience a force, sometimes it results from direct contact with the object that supplies the force (for example, a push). Many forces, however, can be felt at all points in space. For example, the earth exerts a gravitational pull on all other masses. Such forces can be represented by vector fields.

Figure 17.2 shows the gravitational force exerted by the earth on a mass of one kilogram at different points in space. This is a sketch of the vector field in 3-space. You can see that the vectors all point towards the earth (which is not shown in the diagram) and that the vectors further from the earth are smaller in magnitude.

Definition of a Vector Field

Now that you have seen some examples of vector fields we give a more formal definition.

A **vector field** in 2-space is a function $\vec{F}(x, y)$ whose value at a point (x, y) is a 2-dimensional vector. Similarly, a vector field in 3-space is a function $\vec{F}(x, y, z)$ whose values are 3-dimensional vectors.

Notice the arrow over the function, $\vec{F}$, indicating that its value is a vector, not a scalar. We often represent the point (x, y) or (x, y, z) by its position vector $\vec{r}$ and write the vector field as $\vec{F}(\vec{r})$.

Visualizing a Vector Field Given by a Formula

Since a vector field is a function that assigns a vector to each point, a vector field can often be given by a formula.

Example 1   Sketch the vector field in 2-space given by $\vec{F}(x, y) = -y\vec{i} + x\vec{j}$.

Solution   Table 17.1 shows the value of the vector field at a few points. Notice that each value is a vector. To plot the vector field, we plot $\vec{F}(x, y)$ with its tail at (x, y). (See Figure 17.3.)

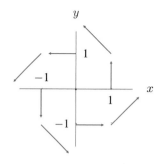

Figure 17.3: The value $\vec{F}(x, y)$
is placed at the point (x, y).

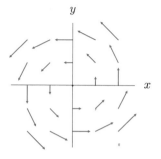

Figure 17.4: The vector field
$\vec{F}(x, y) = -y\vec{i} + x\vec{j}$, vectors
scaled smaller to fit in diagram.

TABLE 17.1 *A few values of*
$\vec{F}(x, y) = -y\vec{i} + x\vec{j}$

		-1	0	1
x	-1	$\vec{i} - \vec{j}$	$-\vec{j}$	$-\vec{i} - \vec{j}$
	0	$\vec{i}$	$\vec{0}$	$-\vec{i}$
	1	$\vec{i} + \vec{j}$	$\vec{j}$	$-\vec{i} + \vec{j}$

Now we look at the formula to get a better sketch. The magnitude of the vector at (x, y) is
$\|\vec{F}(x, y)\| = \|-y\vec{i} + x\vec{j}\| = \sqrt{x^2 + y^2}$, which is the distance from (x, y) to the origin. Therefore, all
the vectors at a fixed distance from the origin (that is, on a circle centered at the origin) have the same
magnitude. The magnitude gets larger as we move further from the origin. What about the direction?
Figure 17.3 suggests that at each point (x, y) the vector $\vec{F}(x, y)$ is perpendicular to the position
vector $\vec{r} = x\vec{i} + y\vec{j}$. We verify this using the dot product: $\vec{r} \cdot \vec{F}(x, y) = (x\vec{i} + y\vec{j}) \cdot (-y\vec{i} + x\vec{j}) = 0$.
This means that the vectors in this vector field are tangent to circles centered at the origin and get
longer as we go out. In Figure 17.4, the vectors have been scaled so that they do not obscure each
other.

Example 2 Draw pictures of the vector fields in 2-space given by (a) $\vec{F}(x, y) = x\vec{j}$ (b) $\vec{G}(x, y) = x\vec{i}$.

Solution (a) The vector $x\vec{j}$ is parallel to the y-direction, pointing up when x is positive and down when x
is negative. Also, the larger $|x|$ is, the longer the vector. The vectors in the field are constant
along vertical lines since the vector field does not depend on y. (See Figure 17.5.)

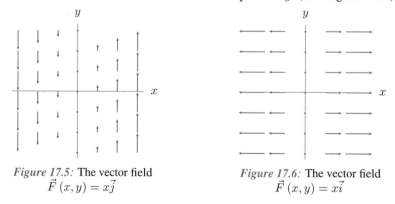

Figure 17.5: The vector field
$\vec{F}(x, y) = x\vec{j}$

Figure 17.6: The vector field
$\vec{F}(x, y) = x\vec{i}$

(b)   This is similar to the previous example, except that the vector $x\vec{i}$ is parallel to the x-direction, pointing to the right when x is positive and to the left when x is negative. Again, the larger $|x|$ is the longer the vector, and the vectors are constant along vertical lines, since the vector field does not depend on y. (See Figure 17.6.)

Example 3   Describe the vector field in 3-space given by $\vec{F}(\vec{r}) = \vec{r}$, where $\vec{r} = x\vec{i} + y\vec{j} + z\vec{k}$.

Solution   The notation $\vec{F}(\vec{r}) = \vec{r}$ means that the value of $\vec{F}$ at the point (x, y, z) with position vector $\vec{r}$ is the vector $\vec{r}$ with its tail at (x, y, z). Thus, the vector field points outward. See Figure 17.7. Note that the lengths of the vectors have been scaled down so as to fit into the diagram. This vector field can also be written as $\vec{F}(x, y, z) = x\vec{i} + y\vec{j} + z\vec{k}$. You can see the notation using $\vec{r}$ is more concise.

Figure 17.7: The vector field
$\vec{F}(\vec{r}) = \vec{r}$

Finding a Formula for a Vector Field

Example 4   Newton's Law of Gravitation states that the magnitude of the gravitational force exerted by an object of mass M on an object of mass m is proportional to M and m and inversely proportional to the square of the distance between them. The direction of the force is from m to M along the line connecting them. (See Figure 17.8.) Find a formula for the vector field $\vec{F}(\vec{r})$ that represents the gravitational force, assuming M is located at the origin and m is located at the point with position vector $\vec{r}$.

$M \bullet$

Figure 17.8: Force exerted
on mass m by mass M

Solution   Since the mass m is located at $\vec{r}$, Newton's law says that the magnitude of the force is given by

$$\|\vec{F}(\vec{r})\| = \frac{GMm}{\|\vec{r}\|^2},$$

where G is called the universal gravitational constant. A unit vector in the direction of the force is $-\vec{r}/\|\vec{r}\|$, where the negative sign indicates that the direction of force is towards the origin (gravity is attractive). By taking the product of the magnitude of the force and a unit vector in the direction of the force we obtain an expression for the force vector field:

$$\vec{F}(\vec{r}) = \frac{GMm}{\|\vec{r}\|^2}\left(-\frac{\vec{r}}{\|\vec{r}\|}\right) = \frac{-GMm\vec{r}}{\|\vec{r}\|^3}.$$

We have already seen a picture of this vector field in Figure 17.2.

Gradient Vector Fields

The gradient of a scalar function f is a function that assigns a vector to each point, and is therefore a vector field. It is called the *gradient field* of f. Many vector fields in physics are gradient fields.

Example 5   Sketch the gradient field of the functions in Figures 17.9–17.11.

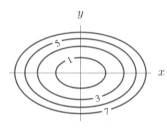

Figure 17.9: The contour map of
$f(x, y) = x^2 + 2y^2$

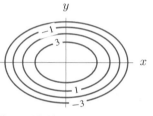

Figure 17.10: The contour map
of $g(x, y) = 5 - x^2 - 2y^2$

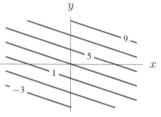

Figure 17.11: The contour map
of $h(x, y) = x + 2y + 3$

Solution   See Figures 17.12–17.14. For a function $f(x, y)$, the gradient vector of f at a point is perpendicular to the contours in the direction of increasing f and its magnitude is the rate of change in that direction. The rate of change is large when the contours are close together and small when they are far apart. Notice that in Figure 17.12 the vectors all point outward, away from the local minimum of f, and in Figure 17.13 the vectors of grad g all point inward, toward the local maximum of g. Since h is a linear function, its gradient is constant, so grad h in Figure 17.14 is a constant vector field.

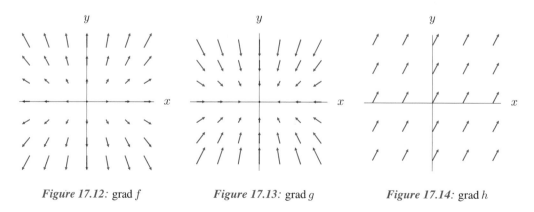

Figure 17.12: grad f *Figure 17.13:* grad g *Figure 17.14:* grad h

Problems for Section 17.1

1. Each vector field in Figures (I)–(IV) represents the force on a particle at different points in space as a result of another particle at the origin. Match up the vector fields with the descriptions below.

   (a) A repulsive force whose magnitude decreases as distance increases, such as between electric charges of the same sign.
   (b) A repulsive force whose magnitude increases as distance increases.
   (c) An attractive force whose magnitude decreases as distance increases, such as gravity.
   (d) An attractive force whose magnitude increases as distance increases.

(I)
(c)

(II)
(d)

(III)
(b)

(IV)
(a)

Sketch the vector fields in Problems 2–7.

2. $\vec{F}(x, y) = 2\vec{i} + 3\vec{j}$ 3. $\vec{F}(x, y) = y\vec{i}$ 4. $\vec{F}(x, y) = 2x\vec{i} + x\vec{j}$

5. $\vec{F}(\vec{r}) = 2\vec{r}$ 6. $\vec{F}(\vec{r}) = \dfrac{\vec{r}}{\|\vec{r}\|}$ 7. $\vec{F}(x, y) = (x + y)\vec{i} + (x - y)\vec{j}$

For Problems 8–13, find formulas for the vector fields. (There are many possible answers.)

8.

9.

10.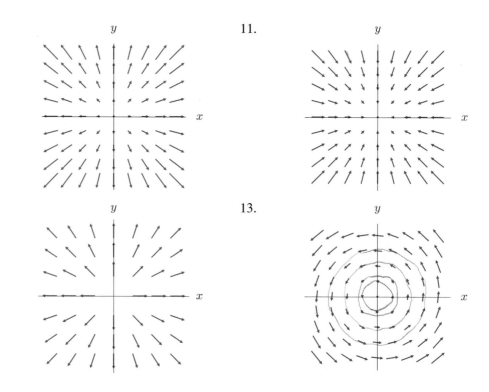

11.

12.

13.

14. Figures 17.15 and 17.16 show the gradient of the functions $z = f(x, y)$ and $z = g(x, y)$.

   (a) For each function, draw a rough sketch of the level curves, showing possible z-values.

   (b) The xz-plane cuts each of the surfaces $z = f(x, y)$ and $z = g(x, y)$ in a curve. Sketch each of these curves, making clear how they are similar and how they are different from one another.

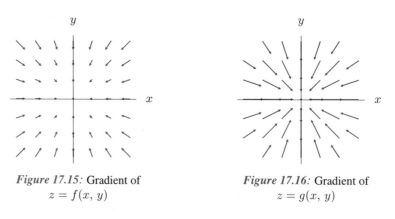

Figure 17.15: Gradient of $z = f(x, y)$

Figure 17.16: Gradient of $z = g(x, y)$

In Problems 15–17, use a computer to print out vector fields with the given properties. Show on your printout the formula used to generate it. (There are many possible answers.)

15. All vectors are parallel to the x-axis; all vectors on a vertical line have the same magnitude.

16. All vectors point towards the origin and have constant length.

17. All vectors are of unit length and perpendicular to the position vector at that point.

18. Imagine a wide, steadily flowing river in the middle of which there is a fountain that spouts water horizontally in all directions.

    (a) Suppose that the river flows in the $\vec{i}$-direction in the xy-plane and that the fountain is at the origin. Explain why the expression
    $$\vec{v} = A\vec{i} + K(x^2 + y^2)^{-1}(x\vec{i} + y\vec{j}), \quad A > 0, K > 0$$
    could represent the velocity field for the combined flow of the river and the fountain.

    (b) What is the significance of the constants A and K?

    (c) Using a computer, sketch the vector field $\vec{v}$ for $K = 1$ and $A = 1$ and $A = 2$, and for $A = 0.2, K = 2$.

17.2 THE FLOW OF A VECTOR FIELD

When an iceberg is spotted in the North Atlantic, it is important to be able to predict where the iceberg is likely to be a day or a week later. To do this, one needs to know the velocity vector field of the ocean currents, that is, how fast and in what direction the water is moving at each point.

In this section we use differential equations to find the path of an object in a fluid flow. This path is called a *flow line*. Figure 17.17 shows several flow lines for the Gulf stream velocity vector field in Figure 17.1 on page 326. The arrows on each flow line indicate the direction of flow along it.

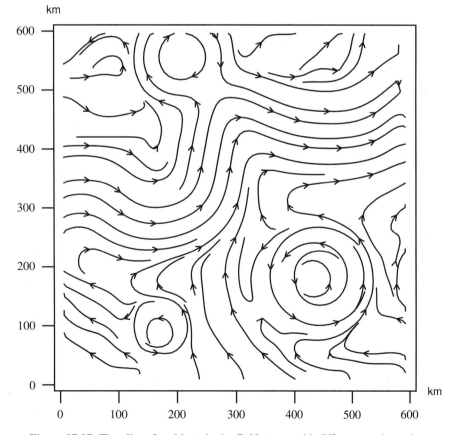

Figure 17.17: Flow lines for objects in the Gulf stream with different starting points

How Do We Find a Flow Line?

Suppose that $\vec{F}$ is the velocity vector field of water on the surface of a creek and imagine a seed being carried along by the current. We want to know the position vector $\vec{r}(t)$ of the seed at time t. We know

$$\begin{array}{cc} \text{Velocity of seed} & \text{Velocity of current at seed's position} \\ \text{at time } t & = \qquad \text{at time } t \end{array}$$

that is,

$$\vec{r}\,'(t) = \vec{F}\,(\vec{r}\,(t)).$$

For an arbitrary vector field we make the following definition:

A **flow line** of a vector field $\vec{v} = \vec{F}\,(\vec{r}\,)$ is a path $\vec{r}\,(t)$ whose velocity vector equals $\vec{v}$. Thus

$$\vec{r}\,'(t) = \vec{v} = \vec{F}\,(\vec{r}\,(t)).$$

The **flow** of a vector field is the family of all of its flow lines.

A flow line is also called an *integral curve* or a *streamline*. We define flow lines for any vector field, as it turns out to be useful to study the flow of fields (for example, electric and magnetic) that are not velocity fields.

Resolving $\vec{F}$ and $\vec{r}$ into components, $\vec{F} = F_1\vec{i} + F_2\vec{j}$ and $\vec{r}\,(t) = x(t)\vec{i} + y(t)\vec{j}$, the definition of a flow line tells us that $x(t)$ and $y(t)$ satisfy the system of differential equations

$$x'(t) = F_1(x(t), y(t)) \quad \text{and} \quad y'(t) = F_2(x(t), y(t)).$$

Solving these differential equations gives a parameterization of the flow line.

Example 1 Find the flow line of the constant velocity field $\vec{v} = 3\vec{i} + 4\vec{j}$ cm/sec that passes through the point $(1, 2)$ at time $t = 0$.

Solution Let $\vec{r}\,(t) = x(t)\vec{i} + y(t)\vec{j}$ be the position in cm of a particle at time t, where t is in seconds. We have

$$x'(t) = 3 \quad \text{and} \quad y'(t) = 4.$$

Thus,

$$x(t) = 3t + x_0 \quad \text{and} \quad y(t) = 4t + y_0.$$

Since the path passes the point $(1, 2)$ at $t = 0$, we have $x_0 = 1$ and $y_0 = 2$ and so

$$x(t) = 3t + 1 \qquad \text{and} \qquad y(t) = 4t + 2.$$

Thus, the path is the line given parametrically by

$$\vec{r}\,(t) = (3t + 1)\vec{i} + (4t + 2)\vec{j}.$$

(See Figure 17.18.) To find an explicit equation for the path, eliminate t between these expressions to get

$$\frac{x-1}{3} = \frac{y-2}{4} \quad \text{or} \quad y = \frac{4}{3}x + \frac{2}{3}.$$

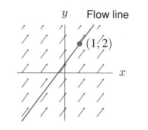

Figure 17.18: Vector field $\vec{F} = 3\vec{i} + 4\vec{j}$ with the flow line through $(1, 2)$

Example 2   The velocity of a flow at the point (x, y) is $\vec{F}(x, y) = \vec{i} + x\vec{j}$. Find the path of motion of an object in the flow that is at the point $(-2, 2)$ at time $t = 0$.

Solution   Figure 17.19 shows a sketch of this field. Since $\vec{r}\,'(t) = \vec{F}(\vec{r}(t))$, we are looking for the flow line that satisfies the system of differential equations

$$x'(t) = 1, \quad y'(t) = x(t).$$

Solving for $x(t)$ first, we get $x(t) = t + x_0$, where x_0 is a constant of integration. Thus, $y'(t) = t + x_0$, so $y(t) = \frac{1}{2}t^2 + x_0 t + y_0$, where y_0 is also a constant of integration. Since $x(0) = x_0 = -2$ and $y(0) = y_0 = 2$, the path of motion is given by

$$x(t) = t - 2, \quad y(t) = \frac{1}{2}t^2 - 2t + 2,$$

or, equivalently,

$$\vec{r}(t) = (t-2)\vec{i} + \left(\frac{1}{2}t^2 - 2t + 2\right)\vec{j}.$$

The graph of this flow line in Figure 17.20 looks like a parabola. We check this by seeing that an explicit equation for the path is $y = \frac{1}{2}x^2$.

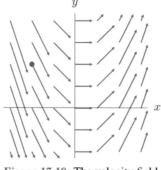

Figure 17.19: The velocity field $\vec{v} = \vec{i} + x\vec{j}$

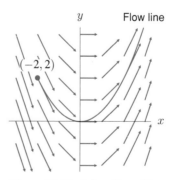

Figure 17.20: A flow line of the velocity field $\vec{v} = \vec{i} + x\vec{j}$

Example 3   Determine the flow of the vector field $\vec{v} = -y\vec{i} + x\vec{j}$.

Solution   Figure 17.21 suggests that the flow consists of concentric counterclockwise circles, centered at the origin. The system of differential equations for the flow is

$$x'(t) = -y(t) \qquad y'(t) = x(t).$$

The equations $(x(t), y(t)) = (a \cos t, a \sin t)$ parameterize a family of counterclockwise circles of radius a, centered at the origin. We check that this family satisfies the system of differential equations:

$$x'(t) = -a \sin t = -y(t) \quad \text{and} \quad y'(t) = a \cos t = x(t).$$

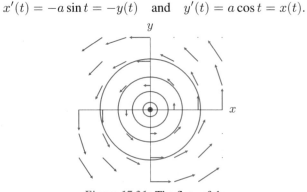

Figure 17.21: The flow of the
vector field $\vec{v} = -y\vec{i} + x\vec{j}$

Finding Flow Lines Numerically

Often it is not possible to find formulas for the flow lines of a vector field. However, we can approximate them numerically by Euler's method for solving differential equations. Since the flow lines $\vec{r}(t) = x(t)\vec{i} + y(t)\vec{j}$ of a vector field $\vec{v} = \vec{F}(x, y)$ satisfy the differential equation $\vec{r}'(t) = \vec{F}(\vec{r}(t))$, we have

$$\vec{r}(t + \Delta t) \approx \vec{r}(t) + (\Delta t)\vec{r}'(t)$$
$$= \vec{r}(t) + (\Delta t)\vec{F}(\vec{r}(t)) \quad \text{for } \Delta t \text{ near } 0.$$

To approximate a flow line, we start at a point $\vec{r}_0 = \vec{r}(0)$ and estimate the position $\vec{r}_1$ of a particle at time Δt later:

$$\vec{r}_1 = \vec{r}(\Delta t) \approx \vec{r}(0) + (\Delta t)\vec{F}(\vec{r}(0))$$
$$= \vec{r}_0 + (\Delta t)\vec{F}(\vec{r}_0).$$

We then repeat the same procedure starting at $\vec{r}_1$, and so on. The general formula for getting from one point to the next is

$$\vec{r}_{n+1} = \vec{r}_n + (\Delta t)\vec{F}(\vec{r}_n).$$

The points with position vectors $\vec{r}_0, \vec{r}_1, \ldots$ trace out the path, as shown in the next example.

Example 4   Use Euler's method to approximate the flow line through $(1, 2)$ for the vector field $\vec{v} = y^2\vec{i} + 2x^2\vec{j}$.

Solution   The flow is determined by the differential equations $\vec{r}'(t) = \vec{v}$, or equivalently

$$x'(t) = y^2, \qquad y'(t) = 2x^2.$$

We use Euler's method with $\Delta t = 0.02$, giving

$$\vec{r}_{n+1} = \vec{r}_n + 0.02\,\vec{v}\,(x_n, y_n)$$
$$= x_n\vec{i} + y_n\vec{j} + 0.02(y_n^2\vec{i} + 2x_n^2\vec{j}),$$

or equivalently,

$$x_{n+1} = x_n + 0.02y_n{}^2, \qquad y_{n+1} = y_n + 0.02 \cdot 2x_n{}^2.$$

When $t = 0$, we have $(x_0, y_0) = (1, 2)$. Then

$$x_1 = x_0 + 0.02 \cdot y_0{}^2 = 1 + 0.02 \cdot 2^2 = 1.08,$$
$$y_1 = y_0 + 0.02 \cdot 2x_0^2 = 2 + 0.02 \cdot 2 \cdot 1^2 = 2.04.$$

So after one step $x(0.02) \approx 1.08$ and $y(0.02) \approx 2.04$. Similarly, $x(0.04) = x(2\Delta t) \approx 1.16$, $y(0.04) = y(2\Delta t) \approx 2.08$ and so on. Further values along the flow line are given in Table 17.2 and plotted in Figure 17.22.

TABLE 17.2 *The approximated flow line for the vector field $\vec{v} = y^2\vec{i} + 2x^2\vec{j}$ starting at the point $(1, 2)$*

t	0	.02	.04	.06	.08	.1	.12	.14	.16	.18
x	1	1.08	1.16	1.25	1.34	1.44	1.54	1.65	1.77	1.90
y	2	2.04	2.08	2.14	2.20	2.28	2.36	2.45	2.56	2.69

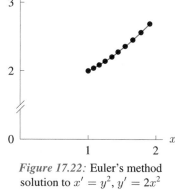

Figure 17.22: Euler's method solution to $x' = y^2$, $y' = 2x^2$

Problems for Section 17.2

For Problems 1–3, sketch the vector field and its flow.

1. $\vec{v} = 3\vec{i}$
2. $\vec{v} = 2\vec{j}$
3. $\vec{v} = 3\vec{i} - 2\vec{j}$

For Problems 4–7, sketch the vector field and the flow. Then find the system of differential equations associated with the vector field and verify that the flow satisfies the system.

4. $\vec{v} = y\vec{i} + x\vec{j}$; $x(t) = a(e^t + e^{-t})$, $y(t) = a(e^t - e^{-t})$.
5. $\vec{v} = y\vec{i} - x\vec{j}$; $x(t) = a\sin t$, $y(t) = a\cos t$.
6. $\vec{v} = x\vec{i} + y\vec{j}$; $x(t) = ae^t$, $y(t) = be^t$.
7. $\vec{v} = x\vec{i} - y\vec{j}$; $x(t) = ae^t$, $y(t) = be^{-t}$.
8. Use a computer or calculator with Euler's method to approximate the flow line through $(1, 2)$ for the vector field $\vec{v} = y^2\vec{i} + 2x^2\vec{j}$ using 5 steps with a time interval $\Delta t = 0.1$.

9. Match the following vector fields with their flow lines. Put arrows on the flow lines indicating the direction of flow.

(a) $y\vec{i} + x\vec{j}$
(b) $-y\vec{i} + x\vec{j}$
(c) $x\vec{i} + y\vec{j}$
(d) $-y\vec{i} + (x + y/10)\vec{j}$
(e) $-y\vec{i} + (x - y/10)\vec{j}$
(f) $(x - y)\vec{i} + (x - y)\vec{j}$

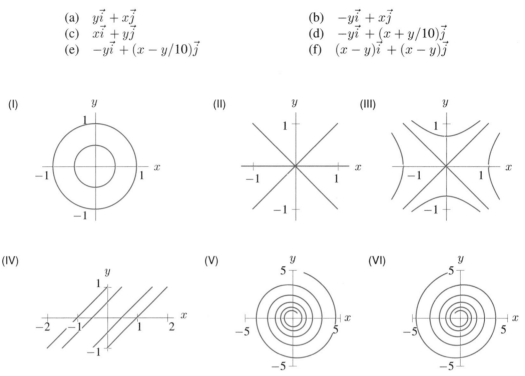

10. We have a fixed set of axes and a solid metal ball whose center is at the origin. The ball rotates once every 24 hours around the z-axis. The direction of rotation is counterclockwise when viewed from above. Consider the point (x, y, z) in our coordinate system lying within the ball. Let $\vec{v}(x, y, z)$ be the velocity vector of the particle of metal at this point. Assume x, y, z are in meters and time is in hours.

(a) Find a formula for the vector field $\vec{v}$. Give units for your answer.
(b) Describe in words the flow lines of $\vec{v}$.

REVIEW PROBLEMS FOR CHAPTER SEVENTEEN

1. (a) What is meant by a vector field?
    (b) Suppose $\vec{a} = a_1\vec{i} + a_2\vec{j} + a_3\vec{k}$ is a constant vector. Which of the following are vector fields? Explain.

    (i) $\vec{r} + \vec{a}$
    (ii) $\vec{r} \cdot \vec{a}$
    (iii) $x^2\vec{i} + y^2\vec{j} + z^2\vec{k}$
    (iv) $x^2 + y^2 + z^2$

Sketch the vector fields in Problems 2–4.

2. $\vec{F} = \left(\dfrac{y}{\sqrt{x^2 + y^2}}\right)\vec{i} - \left(\dfrac{x}{\sqrt{x^2 + y^2}}\right)\vec{j}$ 3. $\vec{F} = \left(\dfrac{y}{x^2 + y^2}\right)\vec{i} - \left(\dfrac{x}{x^2 + y^2}\right)\vec{j}$

4. $\vec{F} = y\vec{i} - x\vec{j}$

5. If $\vec{F} = \vec{r}/\|\vec{r}\|^3$, find the following quantities in terms of x, y, z, or t.

    (a) $\|\vec{F}\|$

    (b) $\vec{F} \cdot \vec{r}$

    (c) A unit vector parallel to $\vec{F}$ and pointing in the same direction.

    (d) A unit vector parallel to $\vec{F}$ and pointing in the opposite direction.

    (e) $\vec{F}$ if $\vec{r} = \cos t\vec{i} + \sin t\vec{j} + \vec{k}$

    (f) $\vec{F} \cdot \vec{r}$ if $\vec{r} = \cos t\vec{i} + \sin t\vec{j} + \vec{k}$

For Problems 6–9, find the region of the Gulf stream velocity field in figure 17.23 represented by the given table of velocity vectors (in cm/sec).

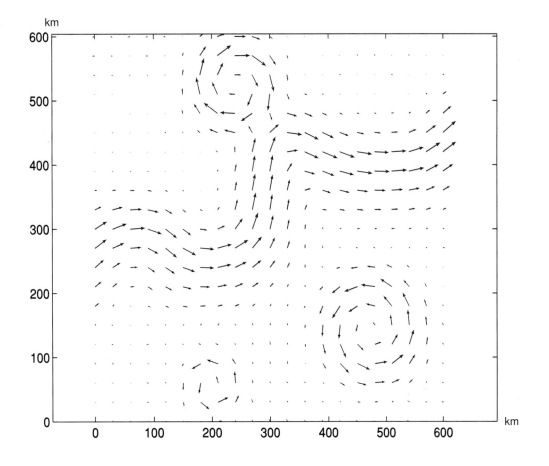

Figure 17.23: The velocity field of the Gulf stream

6.

$35\vec{i} + 131\vec{j}$	$48\vec{i} + 92\vec{j}$	$47\vec{i} + \vec{j}$
$-32\vec{i} + 132\vec{j}$	$-44\vec{i} + 92\vec{j}$	$-42\vec{i} + \vec{j}$
$-51\vec{i} + 73\vec{j}$	$-119\vec{i} + 84\vec{j}$	$-128\vec{i} + 6\vec{j}$

7.

$10\vec{i} - 3\vec{j}$	$11\vec{i} + 16\vec{j}$	$20\vec{i} + 75\vec{j}$
$53\vec{i} - 7\vec{j}$	$58\vec{i} + 23\vec{j}$	$64\vec{i} + 80\vec{j}$
$119\vec{i} - 8\vec{j}$	$121\vec{i} + 31\vec{j}$	$114\vec{i} + 66\vec{j}$

8.

$97\vec{i} - 41\vec{j}$	$72\vec{i} - 24\vec{j}$	$54\vec{i} - 10\vec{j}$
$134\vec{i} - 49\vec{j}$	$131\vec{i} - 44\vec{j}$	$129\vec{i} - 18\vec{j}$
$103\vec{i} - 36\vec{j}$	$122\vec{i} - 30\vec{j}$	$131\vec{i} - 17\vec{j}$

9.

$-95\vec{i} - 60\vec{j}$	$18\vec{i} - 48\vec{j}$	$82\vec{i} - 22\vec{j}$
$-29\vec{i} + 48\vec{j}$	$76\vec{i} + 63\vec{j}$	$128\vec{i} - 16\vec{j}$
$26\vec{i} + 105\vec{j}$	$49\vec{i} + 119\vec{j}$	$88\vec{i} + 13\vec{j}$

10. Each of the following vector fields represents an ocean current. Sketch the vector field, and sketch the path of an iceberg in this current. Determine the location of an iceberg at time $t = 7$ if it is at the point $(1, 3)$ at time $t = 0$.

  (a) The current everywhere is $\vec{i}$.
  (b) The current at (x, y) is $2x\vec{i} + y\vec{j}$.
  (c) The current at (x, y) is $-y\vec{i} + x\vec{j}$.

Suppose $q_1, \ldots, q_n$ are electric charges at points with position vectors $\vec{r}_1, \ldots, \vec{r}_n$. Problems 11–12 use Coulomb's Law which states that, at the point with position vector $\vec{r}$, the resulting electric field $\vec{E}$ is given by

$$\vec{E}(\vec{r}) = \sum_{i=1}^{n} q_i \frac{(\vec{r} - \vec{r}_i)}{\|\vec{r} - \vec{r}_i\|^3}.$$

11. A charge configuration with just two charges q_1 and q_2 in 3-space is called an *electric dipole*. Suppose $\vec{r}_1 = \vec{i}$ and $\vec{r}_2 = -\vec{i}$.

  (a) If $q_1 = q$ and $q_2 = -q$, use a computer to sketch the vector field $\vec{E}$ in the xy-plane produced by these two opposite charges.
  (b) If $q_1 = q_2 = q$, sketch the vector field $\vec{E}$ in the xy-plane produced by these two like charges.

12. An *ideal electric dipole* can be thought of as an infinitesimal dipole; its magnitude and direction are given by its dipole moment vector $\vec{p}$. The resulting electric field $\vec{D}$ at the point with position vector $\vec{r}$ is given by

$$\vec{D}(\vec{r}) = 3\frac{(\vec{r} \cdot \vec{p})\vec{r}}{\|\vec{r}\|^5} - \frac{\vec{p}}{\|\vec{r}\|^3}.$$

Assume $\vec{p} = p\vec{i}$, so that the dipole points in the $\vec{i}$ direction and has magnitude p.

  (a) Use a computer to plot the vector field $\vec{D}$ in the xy-plane for three different values of p.
  (b) The field $\vec{D}$ is an approximation to the electric field $\vec{E}$ produced by two opposite charges, q at $\vec{r}_2$ and $-q$ at $\vec{r}_1$, when the distance $\|\vec{r}_2 - \vec{r}_1\|$ is small. The dipole moment of this configuration of charges is defined to be $\vec{p} = q(\vec{r}_2 - \vec{r}_1)$. Suppose $\vec{r}_2 = (\ell/2)\vec{i}$ and $\vec{r}_1 = -(\ell/2)\vec{i}$, so that $\vec{p} = q\ell\vec{i}$.

  (i) Plot the vector field $\vec{E}$ using the same values of $p = q\ell$ you used to plot $\vec{D}$.
  (ii) Where is the vector field $\vec{D}$ a good approximation to $\vec{E}$? Where is it a poor approximation?
  (iii) The magnitude of each term in the expression for $\vec{E}$ decays like $1/\|\vec{r}\|^2$, while the magnitude of $\vec{D}$ decays like $1/\|\vec{r}\|^3$. If the vector field $\vec{D}$ is supposed to be a good approximation to $\vec{E}$ when the distance $\|\vec{r}\|$ from the origin is large, suggest a reason for this apparent discrepancy.

CHAPTER EIGHTEEN

LINE INTEGRALS

When a constant force, $\vec{F}$, acts on an object as it moves through a displacement, $\vec{d}$, the work done by the force is the dot product, $\vec{F} \cdot \vec{d}$. What if the object moves along a curved path through a variable force field? In this chapter we define a *line integral* that calculates the work in this situation. We also define the *circulation* which is used to measure the strength of eddies in a fluid flow.

Line integrals provide us with an analogue of the Fundamental Theorem of Calculus, which tells us how to recover a function from its derivative. The line integral analogue of the Fundamental Theorem shows how a line integral can be used to recover a multivariable function from its gradient field.

In contrast with the one-variable situation, not all vector fields are gradient fields. The line integral can be used to distinguish those that are, using the concept of a *path-independent* (or *conservative*) *vector* field. We study path-independent fields, which are central in physics, and end with Green's Theorem.

18.1 THE IDEA OF A LINE INTEGRAL

Imagine that you are rowing in a river with a noticeable current. At times you may be working against the current and at other times you may be moving with it. At the end you have a sense of whether, overall, you were helped or hindered by the current. The line integral, defined in this section, measures the extent to which a curve in a vector field is, overall, going with the vector field or against it.

Orientation of a Curve

A curve can be traced out in two directions, as shown in Figure 18.1. We need to choose one direction before we can define a line integral.

> A curve is said to be **oriented** if we have chosen a direction of travel on it.

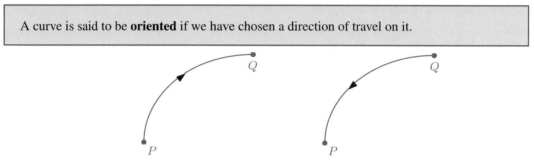

Figure 18.1: A curve with two different orientations represented by arrowheads

Definition of the Line Integral

Consider a vector field $\vec{F}$ and an oriented curve C. We begin by dividing C into n small, almost straight pieces along which $\vec{F}$ is approximately constant. Each piece can be represented by a displacement vector $\Delta\vec{r}_i = \vec{r}_{i+1} - \vec{r}_i$ and the value of $\vec{F}$ at each point of this small piece of C is approximately $\vec{F}(\vec{r}_i)$. See Figures 18.2 and 18.3.

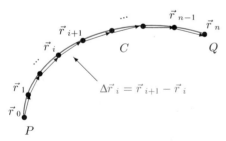

Figure 18.2: The curve C, oriented from P to Q, approximated by straight line segments represented by displacement vectors
$\Delta\vec{r}_i = \vec{r}_{i+1} - \vec{r}_i$

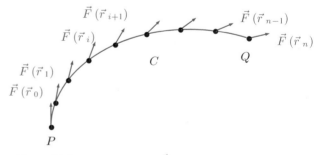

Figure 18.3: The vector field $\vec{F}$ evaluated at the points with position vector $\vec{r}_i$ on the curve C oriented from P to Q

For each point with position vector $\vec{r}_i$ on C, we form the dot product $\vec{F}(\vec{r}_i) \cdot \Delta\vec{r}_i$. Summing over all such pieces, we get a Riemann sum:

$$\sum_{i=0}^{n-1} \vec{F}(\vec{r}_i) \cdot \Delta\vec{r}_i.$$

We define the line integral, written $\int_C \vec{F} \cdot d\vec{r}$, by taking the limit as $\|\Delta \vec{r}_i\| \to 0$. Provided the limit exists, we say

> The **line integral** of a vector field $\vec{F}$ along an oriented curve C is
>
> $$\int_C \vec{F} \cdot d\vec{r} = \lim_{\|\Delta \vec{r}_i\| \to 0} \sum_{i=0}^{n-1} \vec{F}(\vec{r}_i) \cdot \Delta \vec{r}_i.$$

How Does the Limit Defining a Line Integral Work?

The limit in the definition of a line integral exists if $\vec{F}$ is continuous on the curve C and if C is made by joining end to end a finite number of smooth curves, that is, curves which can be parameterized by smooth functions. We can use the parameterization to subdivide a smooth curve, by subdividing the parameter interval in the same way as for ordinary one-variable integrals. The parameterization must go from one end of the curve to the other, in the forward direction, without retracing any portion of the curve. Under these conditions, the line integral is independent of the way in which the subdivisions are made. All the curves we consider in this book are *piecewise smooth* in this sense. Section 18.2 shows how to use a parameterization to compute a line integral.

Example 1   Find the line integral of the constant vector field $\vec{F} = \vec{i} + 2\vec{j}$ along the path from $(1, 1)$ to $(10, 10)$ shown in Figure 18.4.

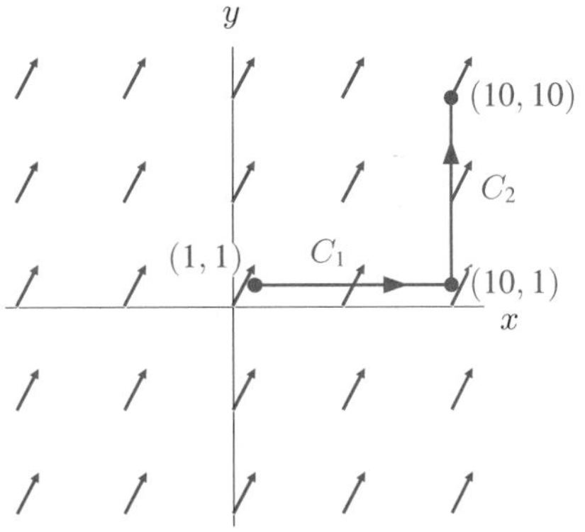

Figure 18.4: The constant vector field $\vec{F} = \vec{i} + 2\vec{j}$ and the path from $(1, 1)$ to $(10, 10)$

Solution   Let C_1 be the horizontal segment of the path going from $(1, 1)$ to $(10, 1)$. When we break this path into pieces, each piece $\Delta \vec{r}$ is horizontal, so $\Delta \vec{r} = \Delta x \vec{i}$ and $\vec{F} \cdot \Delta \vec{r} = (\vec{i} + 2\vec{j}) \cdot \Delta x \vec{i} = \Delta x$. Hence,

$$\int_{C_1} \vec{F} \cdot d\vec{r} = \int_{x=1}^{x=10} dx = 9.$$

Similarly, along the vertical segment C_2, we have $\Delta \vec{r} = \Delta y \vec{j}$ and $\vec{F} \cdot \Delta \vec{r} = (\vec{i} + 2\vec{j}) \cdot \Delta y \vec{j} = 2\Delta y$, so

$$\int_{C_2} \vec{F} \cdot d\vec{r} = \int_{y=1}^{y=10} 2\,dy = 18.$$

Thus,

$$\int_C \vec{F} \cdot d\vec{r} = \int_{C_1} \vec{F} \cdot d\vec{r} + \int_{C_2} \vec{F} \cdot d\vec{r} = 9 + 18 = 27.$$

What Does the Line Integral Tell Us?

Remember that for any two vectors $\vec{u}$ and $\vec{v}$, the dot product $\vec{u} \cdot \vec{v}$ is positive if $\vec{u}$ and $\vec{v}$ point roughly in the same direction (that is, if the angle between them is less than $\pi/2$). The dot product is zero if $\vec{u}$ is perpendicular to $\vec{v}$ and is negative if they point roughly in opposite directions (that is, if the angle between them is greater than $\pi/2$).

The line integral of $\vec{F}$ adds up the dot products of $\vec{F}$ and $\Delta\vec{r}$ along the path. If $\|\vec{F}\|$ is constant, the line integral gives a positive number if $\vec{F}$ is mostly pointing in the same direction as $\Delta\vec{r}$, a negative number if $\vec{F}$ is mostly pointing in the opposite direction. The line integral is zero if $\vec{F}$ is perpendicular to the path at all points or if the positive and negative contributions cancel out. In general, the line integral of a vector field $\vec{F}$ along a curve C measures the extent to which C is going with $\vec{F}$ or against it.

Example 2   The vector field $\vec{F}$ and the oriented curves C_1, C_2, C_3, C_4 are shown in Figure 18.5. The curves C_1 and C_3 are the same length. Which of the line integrals $\int_{C_i} \vec{F} \cdot d\vec{r}$, for $i = 1, 2, 3, 4$, are positive? Which are negative? Arrange these line integrals in ascending order.

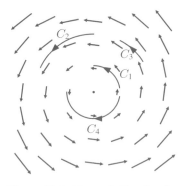

Figure 18.5: Vector field and paths
C_1, C_2, C_3, C_4

Solution   The vector field $\vec{F}$ and the line segments $\Delta\vec{r}$ are approximately parallel and in the same direction for the curves C_1, C_2, and C_3. So the contributions of each term $\vec{F} \cdot \Delta\vec{r}$ are positive for these curves. Thus, $\int_{C_1} \vec{F} \cdot d\vec{r}$, $\int_{C_2} \vec{F} \cdot d\vec{r}$, and $\int_{C_3} \vec{F} \cdot d\vec{r}$ are each positive. For the curve C_4, the vector field and the line segments are in opposite directions, so each term $\vec{F} \cdot \Delta\vec{r}$ is negative, and therefore the integral $\int_{C_4} \vec{F} \cdot d\vec{r}$ is negative.

Since the magnitude of the vector field is smaller along C_1 than along C_3, and these two curves are the same length, we have

$$\int_{C_1} \vec{F} \cdot d\vec{r} < \int_{C_3} \vec{F} \cdot d\vec{r}.$$

In addition, the magnitude of the vector field is the same along C_2 and C_3, but the curve C_2 is longer than the curve C_3. Thus,

$$\int_{C_3} \vec{F} \cdot d\vec{r} < \int_{C_2} \vec{F} \cdot d\vec{r}.$$

Putting these results together with the fact that $\int_{C_4} \vec{F} \cdot d\vec{r}$ is negative, we have

$$\int_{C_4} \vec{F} \cdot d\vec{r} < \int_{C_1} \vec{F} \cdot d\vec{r} < \int_{C_3} \vec{F} \cdot d\vec{r} < \int_{C_2} \vec{F} \cdot d\vec{r}.$$

Interpretations of the Line Integral

Work

Recall from Section 12.3 that if a constant force $\vec{F}$ acts on an object while it moves along a straight line through a displacement $\vec{d}$, the work done by the force on the object is

$$\text{Work done} = \vec{F} \cdot \vec{d}.$$

Now suppose we want to find the work done by gravity on an object moving far above the surface of the earth. Since the force of gravity varies with distance from the earth and the path may not be straight, we can't use the formula $\vec{F} \cdot \vec{d}$. We approximate the path by line segments which are small enough that the force is approximately constant on each one. Suppose the force at a point with position vector $\vec{r}$ is $\vec{F}(\vec{r})$, as in Figures 18.2 and 18.3. Then

$$\begin{array}{c}\text{Work done by force } \vec{F}(\vec{r}_i) \\ \text{over small displacement } \Delta \vec{r}_i\end{array} \approx \quad \vec{F}(\vec{r}_i) \cdot \Delta \vec{r}_i,$$

and so,

$$\begin{array}{c}\text{Total work done by force} \\ \text{along oriented curve } C\end{array} \approx \quad \sum_i \vec{F}(\vec{r}_i) \cdot \Delta \vec{r}_i.$$

Taking the limit as $\|\Delta \vec{r}_i\| \to 0$, we get

$$\begin{array}{c}\text{Work done by force } \vec{F}(\vec{r}) \\ \text{along curve } C\end{array} = \lim_{\|\Delta \vec{r}_i\| \to 0} \sum_i \vec{F}(\vec{r}_i) \cdot \Delta \vec{r}_i = \int_C \vec{F} \cdot d\vec{r}.$$

Example 3   A mass lying on a flat table is attached to a spring whose other end is fastened to the wall. (See Figure 18.6.) The spring is extended 20 cm beyond its rest position and released. If the axes are as shown in Figure 18.6, when the spring is extended by a distance of x, the force exerted by the spring on the mass is given by

$$\vec{F}(x) = -kx\vec{i},$$

where k is a positive constant that depends on the strength of the spring.

Suppose the mass moves back to the rest position. How much work is done by the force exerted by the spring?

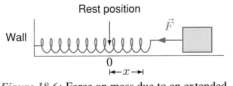

Figure 18.6: Force on mass due to an extended spring

Figure 18.7: Dividing up the interval $0 \le x \le 20$ in order to calculate the work done

Solution   The path from $x = 20$ to $x = 0$ is divided as shown in Figure 18.7, with a typical segment represented by

$$\Delta \vec{r} = \Delta x \vec{i}.$$

Since we are moving from $x = 20$ to $x = 0$, the quantity Δx will be negative. The work done by the force as the mass moves through this segment is approximated by

$$\text{Work done} \approx \vec{F} \cdot \Delta \vec{r} = (-kx\vec{i}) \cdot (\Delta x \vec{i}) = -kx \, \Delta x.$$

Thus, we have

$$\text{Total work done} \approx \sum -kx \, \Delta x.$$

In the limit, as $\|\Delta x\| \to 0$, this sum becomes an ordinary definite integral. Since the path starts at $x = 20$, this is the lower limit of integration; $x = 0$ is the upper limit. Thus, we get

$$\text{Total work done} = \int_{x=20}^{x=0} -kx \, dx = -\frac{kx^2}{2}\bigg|_{20}^{0} = \frac{k(20)^2}{2} = 200k.$$

Note that the work done is positive, since the force acts in the direction of motion.

Example 3 shows how a line integral over a path parallel to the x-axis reduces to a one-variable integral. Section 18.2 shows how to convert *any* line integral into a one-variable integral.

Example 4   A particle with position vector $\vec{r}$ is subject to a force, $\vec{F}$, due to gravity. What is the *sign* of the work done by $\vec{F}$ as the particle moves along the path C_1, a radial line through the center of the earth, starting 8000 km from the center and ending 10,000 km from the center? (See Figure 18.8.)

Solution   We divide the path into small radial segments, $\Delta \vec{r}$, pointing away from the center of the earth and parallel to the gravitational force. The vectors $\vec{F}$ and $\Delta \vec{r}$ point in opposite directions, so each term $\vec{F} \cdot \Delta \vec{r}$ is negative. Adding all these negative quantities and taking the limit results in a negative value for the total work. Thus, the work done by gravity is negative. The negative sign indicates that we would have to do work *against* gravity to move the particle along the path C_1.

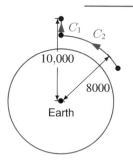

Figure 18.8: The earth

Example 5   Find the sign of the work done by gravity along the curve C_1 in Example 4, but with the opposite orientation.

Solution   Tracing a curve in the opposite direction changes the sign of the line integral because all the segments $\Delta \vec{r}$ change direction, and so every term $\vec{F} \cdot \Delta \vec{r}$ changes sign. Thus, the result will be the negative of the answer found in Example 4. Therefore, the work done by gravity as a particle moves along C_1 toward the center of the earth is positive.

Example 6   Find the work done by gravity as a particle moves along C_2, an arc of a circle 8000 km long at a distance of 8000 km from the center of the earth. (See Figure 18.8.)

Solution   Since C_2 is everywhere perpendicular to the gravitational force, $\vec{F} \cdot \Delta \vec{r} = 0$ for all $\Delta \vec{r}$ along C_2. Thus,

$$\text{Work done} = \int_{C_2} \vec{F} \cdot d\vec{r} = 0,$$

so the work done is zero. This is why satellites can remain in orbit without expending any fuel, once they have attained the correct altitude and velocity.

Circulation

The velocity vector field for the Gulf stream on page 326 shows distinct eddies or regions where the water circulates. We can measure this circulation using a *closed curve*, that is, one that starts and ends at the same point.

> If C is an oriented closed curve, the line integral of a vector field $\vec{F}$ around C is called the **circulation** of $\vec{F}$ around C.

Circulation is a measure of the net tendency of the vector field to point around the curve C. To emphasize that C is closed, the circulation is sometimes denoted $\oint_C \vec{F} \cdot d\vec{r}$, with a small circle on the integral sign.

Example 7   Describe the rotation of the vector fields in Figures 18.9 and 18.10. Find the sign of the circulation of the vector fields around the indicated paths.

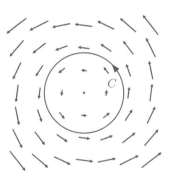

Figure 18.9: A circulating flow

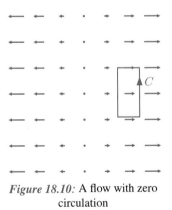

Figure 18.10: A flow with zero circulation

Solution   Consider the vector field in Figure 18.9. If you think of this as representing the velocity of water flowing in a pond, you see that the water is circulating. The line integral around C, measuring the circulation around C, is positive, because the vectors of the field are all pointing in the direction of the path. By way of contrast, look at the vector field in Figure 18.10. Here the line integral around C is zero because the vertical portions of the path are perpendicular to the field and the contributions from the two horizontal portions cancel out. This means that there is no net tendency for the water to circulate around C.

It turns out that the vector field in Figure 18.10 has the property that its circulation around *any* closed path is zero. Water moving according to this vector field has no tendency to circulate around any point and a leaf dropped into the water will not spin. We'll look at such special fields again later when we introduce the notion of the *curl* of a vector field.

Properties of Line Integrals

Line integrals share some basic properties with ordinary one-variable integrals:

For a scalar constant λ, vector fields $\vec{F}$ and $\vec{G}$, and oriented curves C, C_1, and C_2

1. $\displaystyle\int_C \lambda\vec{F} \cdot d\vec{r} = \lambda \int_C \vec{F} \cdot d\vec{r}.$ **2.** $\displaystyle\int_C (\vec{F} + \vec{G}) \cdot d\vec{r} = \int_C \vec{F} \cdot d\vec{r} + \int_C \vec{G} \cdot d\vec{r}.$

3. $\displaystyle\int_{-C} \vec{F} \cdot d\vec{r} = - \int_C \vec{F} \cdot d\vec{r}.$ **4.** $\displaystyle\int_{C_1+C_2} \vec{F} \cdot d\vec{r} = \int_{C_1} \vec{F} \cdot d\vec{r} + \int_{C_2} \vec{F} \cdot d\vec{r}.$

Properties 3 and 4 are concerned with the curve C over which the line integral is taken. If C is an oriented curve, then $-C$ is the same curve traversed in the opposite direction, that is, with the opposite orientation. (See Figure 18.11.) Property 3 holds because if we integrate along $-C$, the vectors $\Delta\vec{r}$ point in the opposite direction and the dot products $\vec{F} \cdot \Delta\vec{r}$ are the negatives of what they were along C.

If C_1 and C_2 are oriented curves with C_1 ending where C_2 begins, we construct a new oriented curve, called $C_1 + C_2$, by joining them together. (See Figure 18.12.) Property 4 is the analogue for line integrals of the property for definite integrals which says that

$$\int_a^b f(x)\, dx = \int_a^c f(x)\, dx + \int_c^b f(x)\, dx.$$

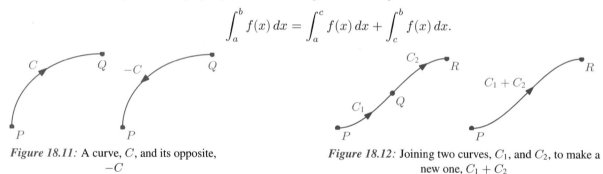

Figure 18.11: A curve, C, and its opposite, $-C$

Figure 18.12: Joining two curves, C_1, and C_2, to make a new one, $C_1 + C_2$

Problems for Section 18.1

In Problems 1–4, say whether you expect the line integral of the pictured vector field over the given curve to be positive, negative, or zero.

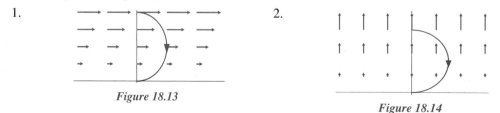

1.

Figure 18.13

2.

Figure 18.14

3.

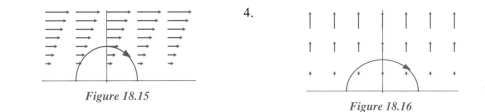

Figure 18.15

4.

Figure 18.16

5. Consider the vector field $\vec{F}$ shown in Figure 18.17, together with the paths C_1, C_2, and C_3. Arrange the line integrals $\int_{C_1} \vec{F} \cdot d\vec{r}$, $\int_{C_2} \vec{F} \cdot d\vec{r}$ and $\int_{C_3} \vec{F} \cdot d\vec{r}$ in ascending order.

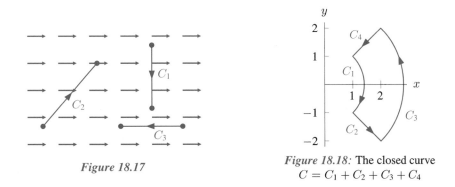

Figure 18.17

Figure 18.18: The closed curve
$C = C_1 + C_2 + C_3 + C_4$

For Problems 6–10, say whether you expect the given vector field to have positive, negative, or zero circulation around the curve C in Figure 18.18. The segments C_1 and C_3 are circular arcs centered at the origin; C_2 and C_4 are radial line segments. You may find it helpful to sketch the vector field.

6. $\vec{F}(x, y) = x\vec{i} + y\vec{j}$

7. $\vec{F}(x, y) = -y\vec{i} + x\vec{j}$

8. $\vec{F}(x, y) = y\vec{i} - x\vec{j}$

9. $\vec{F}(x, y) = x^2\vec{i}$

10. $\vec{F}(x, y) = -\dfrac{y}{x^2 + y^2}\vec{i} + \dfrac{x}{x^2 + y^2}\vec{j}$

In Problems 11–16, calculate the line integral along the line between the given points.

11. $\vec{F} = x\vec{j}$,   from $(1, 0)$ to $(3, 0)$

12. $\vec{F} = x\vec{j}$,   from $(2, 0)$ to $(2, 5)$

13. $\vec{F} = x\vec{i}$,   from $(2, 0)$ to $(6, 0)$

14. $\vec{F} = x\vec{i} + y\vec{j}$,   from $(2, 0)$ to $(6, 0)$

15. $\vec{F} = \vec{r}$,   from $(2, 2)$ to $(6, 6)$

16. $\vec{F} = 3\vec{i} + 4\vec{j}$,   from $(0, 6)$ to $(0, 13)$

17. Draw an oriented curve C and a vector field $\vec{F}$ along C that is not always perpendicular to C, but for which $\int_C \vec{F} \cdot d\vec{r} = 0$.

18. Given the force field $\vec{F}(x, y) = y\vec{i} + x^2\vec{j}$ and the right-angle curve, C, from the points $(0, -1)$ to $(4, -1)$ to $(4, 3)$ shown in Figure 18.19:

   (a) Evaluate $\vec{F}$ at the points $(0, -1)$, $(1, -1)$, $(2, -1)$, $(3, -1)$, $(4, -1)$, $(4, 0)$, $(4, 1)$, $(4, 2)$, $(4, 3)$.
   (b) Make a sketch showing the force field along C.
   (c) Estimate the work done by the indicated force field on an object traversing the curve C.

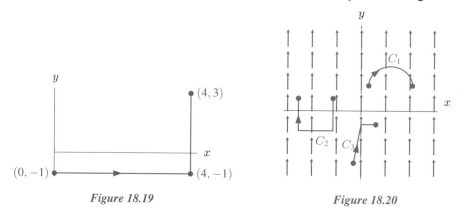

Figure 18.19   Figure 18.20

19. If $\vec{F}$ is the constant force field $\vec{j}$, consider the work done by the field on particles traveling on paths C_1, C_2, and C_3 of Figure 18.20. On which of these paths will zero work be done? Explain.

For Problems 20–23, use a computer to calculate the following integrals.
(a) The line integral of $\vec{F}$ around several closed curves. What do you get?
(b) The line integral of $\vec{F}$ along three curves, each of which starts at the origin and ends at the point $(\frac{1}{2}, \frac{1}{2})$. What do you notice?

20. $\vec{F} = x\vec{i} + y\vec{j}$ 21. $\vec{F} = -y\vec{i} + x\vec{j}$ 22. $\vec{F} = \vec{i} + y\vec{j}$ 23. $\vec{F} = \vec{i} + x\vec{j}$

24. As a result of your answers to Problems 20–23, you should have noticed that the following statement is true: Whenever the line integral of a vector field around any closed curve is zero, the line integral along a curve with fixed endpoints has a constant value (that is, the line integral is independent of the path the curve takes between the endpoints). Explain why this is so.

25. As a result of your answers to Problems 20–23, you should have noticed that the converse to the statement in Problem 24 is also true: Whenever the line integral of a vector field depends only on endpoints and not on paths, the circulation is always zero. Explain why this is so.

In Problems 26–27, use the fact that the force of gravity on a particle of mass m at the point with position vector $\vec{r}$ is given by

$$\vec{F} = -\frac{GMm\vec{r}}{r^3}$$

where $r = \|\vec{r}\|$, and G is the gravitational constant, and M is the mass of the earth.

26. Calculate the work done by the force of gravity on a particle of mass m as it moves from 8000 km to 10,000 km from the center of the earth.

27. Calculate the work done by the force of gravity on a particle of mass m as it moves from 8000 km from the center of the earth to infinitely far away.

28. The fact that an electric current gives rise to a magnetic field is the basis for some electric motors. Ampère's Law relates the magnetic field $\vec{B}$ to a steady current I. It says

$$\int_C \vec{B} \cdot d\vec{r} = kI$$

where I is the current[1] flowing through a closed curve C and k is a constant. Figure 18.21 shows a rod carrying a current and the magnetic field induced around the rod. If the rod is very long and thin, experiments show that the magnetic field $\vec{B}$ is tangent to every circle that is perpendicular to the rod and has center on the axis of the rod (like C in Figure 18.21). The magnitude of $\vec{B}$ is constant along every such circle. Use Ampère's Law to show that around a circle of radius r, the magnetic field due to a current I has magnitude given by

$$\|\vec{B}\| = \frac{kI}{2\pi r}.$$

(In other words, the strength of the field is inversely proportional to the radial distance from the rod.)

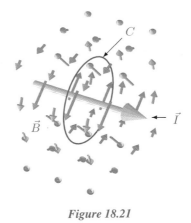

Figure 18.21

18.2 COMPUTING LINE INTEGRALS OVER PARAMETERIZED CURVES

The goal of this section is to show how to use a parameterization of a curve to convert a line integral into an ordinary one-variable integral.

Using a Parameterization to Evaluate a Line Integral

Recall the definition of the line integral,

$$\int_C \vec{F} \cdot d\vec{r} = \lim_{\|\Delta\vec{r}_i\| \to 0} \sum \vec{F}(\vec{r}_i) \cdot \Delta\vec{r}_i,$$

where the $\vec{r}_i$ are the position vectors of points subdividing the curve into short pieces. Now suppose we have a smooth parameterization, $\vec{r}(t)$, of C for $a \leq t \leq b$, so that $\vec{r}(a)$ is the position vector

---

[1]More precisely, I is the net current through any surface that has C as its boundary.

of the beginning of the curve and $\vec{r}(b)$ is the position vector of the end. Then we can divide C into n pieces by dividing the interval $a \leq t \leq b$ into n pieces, each of size $\Delta t = (b - a)/n$. See Figures 18.22 and 18.23.

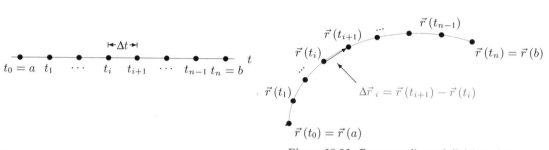

Figure 18.22: Subdivision of the interval $a \leq t \leq b$

Figure 18.23: Corresponding subdivision of the parameterized path C

At each point $\vec{r}_i = \vec{r}(t_i)$ we want to compute

$$\vec{F}(\vec{r}_i) \cdot \Delta \vec{r}_i.$$

Since $t_{i+1} = t_i + \Delta t$, the displacement vectors $\Delta \vec{r}_i$ are given by

$$
\begin{aligned}
\Delta \vec{r}_i &= \vec{r}(t_{i+1}) - \vec{r}(t_i) \\
&= \vec{r}(t_i + \Delta t) - \vec{r}(t_i) \\
&= \frac{\vec{r}(t_i + \Delta t) - \vec{r}(t_i)}{\Delta t} \cdot \Delta t \\
&\approx \vec{r}'(t_i) \Delta t,
\end{aligned}
$$

where we use the facts that Δt is small and that $\vec{r}(t)$ is differentiable to obtain the last approximation. Therefore,

$$\int_C \vec{F} \cdot d\vec{r} \approx \sum \vec{F}(\vec{r}_i) \cdot \Delta \vec{r}_i \approx \sum \vec{F}(\vec{r}(t_i)) \cdot \vec{r}'(t_i) \, \Delta t.$$

Notice that $\vec{F}(\vec{r}(t_i)) \cdot \vec{r}'(t_i)$ is the value at t_i of a one-variable function of t, so this last sum is really a one-variable Riemann sum. In the limit as $\Delta t \to 0$, we get a definite integral:

$$\lim_{\Delta t \to 0} \sum \vec{F}(\vec{r}(t_i)) \cdot \vec{r}'(t_i) \, \Delta t = \int_a^b \vec{F}(\vec{r}(t)) \cdot \vec{r}'(t) \, dt.$$

Thus, we have the following result:

If $\vec{r}(t)$, for $a \leq t \leq b$, is a smooth parameterization of an oriented curve C and $\vec{F}$ is a vector field which is continuous on C, then

$$\int_C \vec{F} \cdot d\vec{r} = \int_a^b \vec{F}(\vec{r}(t)) \cdot \vec{r}'(t) \, dt.$$

In words: To compute the line integral of $\vec{F}$ over C, take the dot product of $\vec{F}$ evaluated on C with the velocity vector, $\vec{r}'(t)$, of the parameterization of C, then integrate along the curve.

Even though we assumed that C is smooth, we can use the same formula to compute line integrals over curves which are only *piecewise smooth*, such as the boundary of a rectangle: If C is piecewise smooth, we apply the formula to each one of the smooth pieces and add the results.

Example 1   Compute $\int_C \vec{F} \cdot d\vec{r}$ where $\vec{F} = (x+y)\vec{i} + y\vec{j}$ and C is the quarter unit circle, oriented counterclockwise as shown in Figure 18.24.

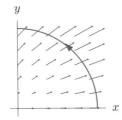

Figure 18.24: The vector field $\vec{F} = (x+y)\vec{i} + y\vec{j}$ and the quarter circle C

Solution   Since all of the vectors in $\vec{F}$ along C point generally in a direction opposite to the orientation of C, we expect our answer to be negative. The first step is to parameterize C by

$$\vec{r}(t) = x(t)\vec{i} + y(t)\vec{j} = \cos t\,\vec{i} + \sin t\,\vec{j}, \quad 0 \leq t \leq \frac{\pi}{2}.$$

Substituting the parameterization into $\vec{F}$ we get $\vec{F}(x(t), y(t)) = (\cos t + \sin t)\vec{i} + \sin t\,\vec{j}$. The vector $\vec{r}\,'(t) = x'(t)\vec{i} + y'(t)\vec{j} = -\sin t\,\vec{i} + \cos t\,\vec{j}$. Then

$$\int_C \vec{F} \cdot d\vec{r} = \int_0^{\pi/2} ((\cos t + \sin t)\vec{i} + \sin t\,\vec{j}) \cdot (-\sin t\,\vec{i} + \cos t\,\vec{j})dt$$

$$= \int_0^{\pi/2} (-\cos t \sin t - \sin^2 t + \sin t \cos t)dt$$

$$= \int_0^{\pi/2} -\sin^2 t\, dt = -\frac{\pi}{4} \approx -0.7854.$$

So the answer is negative, as expected.

Example 2   Consider the vector field $\vec{F} = x\vec{i} + y\vec{j}$.

(a)   Suppose C_1 is the line segment joining $(1, 0)$ to $(0, 2)$ and C_2 is a part of a parabola with its vertex at $(0, 2)$, joining the same points in the same order. (See Figure 18.25.) Verify that

$$\int_{C_1} \vec{F} \cdot d\vec{r} = \int_{C_2} \vec{F} \cdot d\vec{r}.$$

(b)   If C is the triangle shown in Figure 18.26, show that $\int_C \vec{F} \cdot d\vec{r} = 0$.

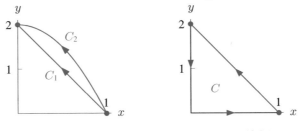

Figure 18.25   *Figure 18.26*

Solution   (a)   We parameterize C_1 by $\vec{r}(t) = (1 - t)\vec{i} + 2t\vec{j}$ with $0 \leq t \leq 1$. Then $\vec{r}'(t) = -\vec{i} + 2\vec{j}$, so

$$\int_{C_1} \vec{F} \cdot d\vec{r} = \int_0^1 \vec{F}(1 - t, 2t) \cdot (-\vec{i} + 2\vec{j})\, dt = \int_0^1 ((1 - t)\vec{i} + 2t\vec{j}) \cdot (-\vec{i} + 2\vec{j})\, dt$$

$$= \int_0^1 (5t - 1)\, dt = \frac{3}{2}.$$

To parameterize C_2, we use the fact that it is part of a parabola with vertex at $(0, 2)$, so its equation is of the form $y = -kx^2 + 2$ for some k. Since the parabola crosses the x-axis at $(1, 0)$, we find that $k = 2$ and $y = -2x^2 + 2$. Therefore, we use the parameterization $\vec{r}(t) = t\vec{i} + (-2t^2 + 2)\vec{j}$ with $0 \leq t \leq 1$, which has $\vec{r}' = \vec{i} - 4t\vec{j}$. This traces out C_2 in reverse, since $t = 0$ gives $(0, 2)$, and $t = 1$ gives $(1, 0)$. Thus, we make $t = 0$ the upper limit of integration and $t = 1$ the lower limit:

$$\int_{C_2} \vec{F} \cdot d\vec{r} = \int_1^0 \vec{F}(t, -2t^2 + 2) \cdot (\vec{i} - 4t\vec{j})\, dt = -\int_0^1 (t\vec{i} + (-2t^2 + 2)\vec{j}) \cdot (\vec{i} - 4t\vec{j})\, dt$$

$$= -\int_0^1 (8t^3 - 7t)\, dt = \frac{3}{2}.$$

So the line integrals along C_1 and C_2 have the same value.

(b)   We break $\int_C \vec{F} \cdot d\vec{r}$ into three pieces, one of which we have already computed (namely, the piece connecting $(1, 0)$ to $(0, 2)$, where the line integral has value $3/2$). The piece running from $(0, 2)$ to $(0, 0)$ can be parameterized by $\vec{r}(t) = (2 - t)\vec{j}$ with $0 \leq t \leq 2$. The piece running from $(0, 0)$ to $(1, 0)$ can be parameterized by $\vec{r}(t) = t\vec{i}$ with $0 \leq t \leq 1$. Then

$$\int_C \vec{F} \cdot d\vec{r} = \frac{3}{2} + \int_0^2 \vec{F}(0, 2 - t) \cdot (-\vec{j})\, dt + \int_0^1 \vec{F}(t, 0) \cdot \vec{i}\, dt$$

$$= \frac{3}{2} + \int_0^2 (2 - t)\vec{j} \cdot (-\vec{j})\, dt + \int_0^1 t\vec{i} \cdot \vec{i}\, dt$$

$$= \frac{3}{2} + \int_0^2 (t - 2)\, dt + \int_0^1 t\, dt = \frac{3}{2} + (-2) + \frac{1}{2} = 0.$$

Example 3   Let C be the closed curve consisting of the upper half-circle of radius 1 and the line forming its diameter along the x-axis, oriented counterclockwise. (See Figure 18.27.) Find $\int_C \vec{F} \cdot d\vec{r}$ where $\vec{F}(x, y) = -y\vec{i} + x\vec{j}$.

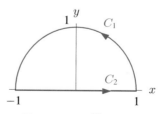

Figure 18.27: The curve
$C = C_1 + C_2$ for Example 3

Solution   We write $C = C_1 + C_2$ where C_1 is the half-circle and C_2 is the line, and compute $\int_{C_1} \vec{F} \cdot d\vec{r}$ and $\int_{C_2} \vec{F} \cdot d\vec{r}$ separately. We parameterize C_1 by $\vec{r}(t) = \cos t\vec{i} + \sin t\vec{j}$, with $0 \le t \le \pi$. Then

$$\int_{C_1} \vec{F} \cdot d\vec{r} = \int_0^\pi (-\sin t\vec{i} + \cos t\vec{j}) \cdot (-\sin t\vec{i} + \cos t\vec{j})\, dt$$

$$= \int_0^\pi (\sin^2 t + \cos^2 t)\, dt = \int_0^\pi 1\, dt = \pi.$$

For C_2, we have $\int_{C_2} \vec{F} \cdot d\vec{r} = 0$, since the vector field $\vec{F}$ has no $\vec{i}$ component along the x-axis (where $y = 0$) and is therefore perpendicular to C_2 at all points.
Finally, we can write

$$\int_C \vec{F} \cdot d\vec{r} = \int_{C_1} \vec{F} \cdot d\vec{r} + \int_{C_2} \vec{F} \cdot d\vec{r} = \pi + 0 = \pi.$$

It is no accident that the result for $\int_{C_1} \vec{F} \cdot d\vec{r}$ is the same as the length of the curve C_1. See Problems 14–15 on page 357.

The next example illustrates the computation of a line integral over a path in 3-space.

Example 4   A particle travels along the helix C given by $\vec{r}(t) = \cos t\vec{i} + \sin t\vec{j} + 2t\vec{k}$ and is subject to a force $\vec{F} = x\vec{i} + z\vec{j} - xy\vec{k}$. Find the total work done on the particle by the force for $0 \le t \le 3\pi$.

Solution   The work done is given by a line integral, which we evaluate using the given parameterization:

$$\text{Work done} = \int_C \vec{F} \cdot d\vec{r} = \int_0^{3\pi} \vec{F}(\vec{r}(t)) \cdot \vec{r}'(t)\, dt$$

$$= \int_0^{3\pi} (\cos t\vec{i} + 2t\vec{j} - \cos t \sin t\vec{k}) \cdot (-\sin t\vec{i} + \cos t\vec{j} + 2\vec{k})\, dt$$

$$= \int_0^{3\pi} (-\cos t \sin t + 2t \cos t - 2\cos t \sin t)\, dt$$

$$= \int_0^{3\pi} (-3\cos t \sin t + 2t \cos t)\, dt = -4.$$

The Notation $\int_C P\, dx + Q\, dy + R\, dz$

There is an alternative notation for line integrals that is quite common. Given functions $P(x, y, z)$ and $Q(x, y, z)$ and $R(x, y, z)$, and an oriented curve C, we consider the vector field $\vec{F} = P\vec{i} + Q\vec{j} + R\vec{k}$. We can write

$$\int_C \vec{F} \cdot d\vec{r} = \int_C P(x, y, z)dx + Q(x, y, z)dy + R(x, y, z)dz,$$

The relation between the two notations can be remembered using $d\vec{r} = dx\vec{i} + dy\vec{j} + dz\vec{k}$.

Example 5   Evaluate $\int_C xy\, dx - y^2\, dy$ where C is the line segment from $(0, 0)$ to $(2, 6)$.

Solution   We parameterize C by $\vec{r}(t) = x(t)\vec{i} + y(t)\vec{j} = t\vec{i} + 3t\vec{j}$, for $0 \le t \le 2$. Thus,

$$\int_C xy\, dx - y^2\, dy = \int_C (xy\vec{i} - y^2\vec{j}) \cdot d\vec{r} = \int_0^2 (3t^2\vec{i} - 9t^2\vec{j}) \cdot (\vec{i} + 3\vec{j})\, dt = \int_0^2 (-24t^2)\, dt = -64.$$

Independence of Parameterization

Since there are many different ways of parameterizing a given oriented curve, you may be wondering what happens to the value of a given line integral if you choose another parameterization. The answer is that the choice of parameterization makes no difference. Since we initially defined the line integral without reference to any particular parameterization, this is exactly as we would expect.

Example 6   Consider the oriented path which is a straight line segment L running from $(0,0)$ to $(1,1)$. Calculate the line integral of the vector field $\vec{F} = (3x - y)\vec{i} + x\vec{j}$ along L using each of the parameterizations

(a) $A(t) = (t, t), \quad 0 \le t \le 1,$

(b) $D(t) = (e^t - 1, e^t - 1), \quad 0 \le t \le \ln 2.$

Solution   The line L has equation $y = x$. Both $A(t)$ and $D(t)$ give a parameterization of L: each has both coordinates equal and each begins at $(0,0)$ and ends at $(1,1)$. Now let's calculate the line integral of the vector field $\vec{F} = (3x - y)\vec{i} + x\vec{j}$ using each parameterization.

(a)   Using $A(t)$, we get

$$\int_L \vec{F} \cdot d\vec{r} = \int_0^1 ((3t - t)\vec{i} + t\vec{j}) \cdot (\vec{i} + \vec{j}) \, dt = \int_0^1 3t \, dt = \left. \frac{3t^2}{2} \right|_0^1 = \frac{3}{2}.$$

(b)   Using $D(t)$, we get

$$\int_L \vec{F} \cdot d\vec{r} = \int_0^{\ln 2} \left((3(e^t - 1) - (e^t - 1)) \vec{i} + (e^t - 1)\vec{j} \right) \cdot (e^t \vec{i} + e^t \vec{j}) \, dt$$

$$= \int_0^{\ln 2} 3(e^{2t} - e^t) \, dt = 3 \left(\frac{e^{2t}}{2} - e^t \right) \Big|_0^{\ln 2} = \frac{3}{2}.$$

The fact that both answers are the same illustrates that the value of a line integral is independent of the parameterization of the path. Problems 17–19 at the end of this section give another way of seeing this.

Problems for Section 18.2 ▬▬▬▬

In Problems 1–10, compute the line integral of the given vector field along the given path.

1. $\vec{F}(x, y) = \ln y \vec{i} + \ln x \vec{j}$ and C is the curve given parametrically by $(2t, t^3)$, for $2 \le t \le 4$.
2. $\vec{F} = x\vec{i} + y\vec{j}$ and C is the line from the origin to the point $(3, 3)$.
3. $\vec{F}(x, y) = x^2 \vec{i} + y^2 \vec{j}$ and C is the line from the point $(1, 2)$ to the point $(3, 4)$.
4. $\vec{F} = 2y\vec{i} - (\sin y)\vec{j}$ counterclockwise around the unit circle C starting at the point $(1, 0)$.
5. $\vec{F}(x, y) = e^x \vec{i} + e^y \vec{j}$ and C is the part of the ellipse $x^2 + 4y^2 = 4$ joining the point $(0, 1)$ to the point $(2, 0)$ in the clockwise direction.
6. $\vec{F}(x, y) = xy\vec{i} + (x - y)\vec{j}$ and C is the triangle joining the points $(1, 0)$, $(0, 1)$ and $(-1, 0)$ in the clockwise direction.
7. $\vec{F} = x\vec{i} + 2zy\vec{j} + x\vec{k}$ and C is given by $\vec{r} = t\vec{i} + t^2 \vec{j} + t^3 \vec{k}$ for $1 \le t \le 2$.
8. $\vec{F} = x^3 \vec{i} + y^2 \vec{j} + z\vec{k}$ and C is the line from the origin to the point $(2, 3, 4)$.
9. $\vec{F} = -y\vec{i} + x\vec{j} + 5\vec{k}$ and C is the helix $x = \cos t, y = \sin t, z = t$, for $0 \le t \le 4\pi$.

10. $\vec{F} = e^y\vec{i} + \ln(x^2 + 1)\vec{j} + \vec{k}$ and C is the circle of radius 2 in the yz-plane centered at the origin and traversed as shown in Figure 18.28.

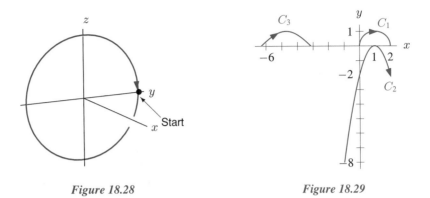

| **Figure 18.28** | **Figure 18.29** |

11. Find parameterizations for the oriented curves shown in Figure 18.29. The curve C_1 is a semicircle of radius 1, centered at the point $(1, 0)$. The curve C_2 is a portion of a parabola, with vertex at the point $(1, 0)$ and y-intercept -2. The curve C_3 is one arc of the sine curve.

12. Suppose C is the line segment from the point $(0, 0)$ to the point $(4, 12)$ and $\vec{F} = xy\vec{i} + x\vec{j}$.

 (a) Is $\int_C \vec{F} \cdot d\vec{r}$ greater than, less than, or equal to zero? Give a geometric explanation.
 (b) A parameterization of C is $(x(t), y(t)) = (t, 3t)$ for $0 \leq t \leq 4$. Use this to compute $\int_C \vec{F} \cdot d\vec{r}$.
 (c) Suppose a particle leaves the point $(0, 0)$, moves along the line towards the point $(4, 12)$, stops before reaching it and backs up, stops again and reverses direction, then completes its journey to the endpoint. All travel takes place along the line segment joining the point $(0, 0)$ to the point $(4, 12)$. If we call this path C', explain why $\int_{C'} \vec{F} \cdot d\vec{r} = \int_C \vec{F} \cdot d\vec{r}$.
 (d) A parameterization for a path like C' is given by

$$(x(t), y(t)) = \left(\frac{1}{3}(t^3 - 6t^2 + 11t), (t^3 - 6t^2 + 11t)\right), \qquad 0 \leq t \leq 4.$$

 Check that this begins at the point $(0, 0)$ and ends at the point $(4, 12)$. Check also that all points of C' lie on the line segment connecting the point $(0, 0)$ to the point $(4, 12)$. What are the values of t at which the particle changes direction?
 (e) Find $\int_{C'} \vec{F} \cdot d\vec{r}$ using the parameterization in part (d). Do you get the same answer as in part (b)?

13. In Example 6 on page 356 we integrated $\vec{F} = (3x - y)\vec{i} + x\vec{j}$ over two parameterizations of the line from $(0, 0)$ to $(1, 1)$, getting $3/2$ each time. Now compute the line integral along two different paths with the same endpoints, and show that the answers are different.

 (a) The path (t, t^2), with $0 \leq t \leq 1$ (b) The path (t^2, t), with $0 \leq t \leq 1$

14. Consider the vector field $\vec{F} = -y\vec{i} + x\vec{j}$. Let C be the unit circle oriented counterclockwise.

 (a) Show that $\vec{F}$ has a constant magnitude of 1 on the circle C.
 (b) Show that $\vec{F}$ is always tangent to the circle C.
 (c) Show that $\int_C \vec{F} \cdot d\vec{r}$ = Length of C.

15. Suppose that along a curve C, a vector field $\vec{F}$ is always tangent to C in the direction of orientation and has constant magnitude $\|\vec{F}\| = m$. Use the definition of the line integral to explain why

$$\int_C \vec{F} \cdot d\vec{r} = m \cdot \text{Length of } C.$$

16. Consider the oriented path which is a straight line segment L running from $(0,0)$ to $(1,1)$. Calculate the line integral of the vector field $\vec{F} = (3x - y)\vec{i} + x\vec{j}$ along L using each of the parameterizations

    (a) $B(t) = (2t, 2t), \quad 0 \leq t \leq 1/2,$ (b) $C(t) = \left(\dfrac{t^2 - 1}{3}, \dfrac{t^2 - 1}{3}\right), \quad 1 \leq t \leq 2,$

In Example 6 on page 356 two parameterizations, $A(t)$, and $D(t)$, are used to convert a line integral into a definite integral. In Problem 16, two other parameterizations, $B(t)$ and $C(t)$, are used on the same line integral. In Problems 17–19 show that two definite integrals corresponding to two of the given parameterizations are equal by finding a substitution which converts one integral to the other. This gives us another way of seeing why changing the parameterization of the curve does not change the value of the line integral.

17. $A(t)$ and $B(t)$ 18. $A(t)$ and $C(t)$ 19. $A(t)$ and $D(t)$

20. A spiral staircase in a building is in the shape of a helix of radius 5 meters. Between two floors of the building, the stairs make one full revolution and climb by 4 meters. A person carries a bag of groceries up two floors. The combined mass of the person and the groceries is 70 kg and the gravitational force is $70g$ downward, where g is the acceleration due to gravity. Calculate the work done by the person against gravity.

18.3 GRADIENT FIELDS AND PATH-INDEPENDENT FIELDS

For a function, f, of one variable, the Fundamental Theorem of Calculus tells us that the definite integral of a rate of change, f', gives the total change in f:

$$\int_a^b f'(t)\, dt = f(b) - f(a).$$

What about functions of two or more variables? The quantity that describes the rate of change is the gradient vector field. If we know the gradient of a function f, can we compute the total change in f between two points? The answer is yes, using a line integral.

Finding the Total Change in f from grad f: The Fundamental Theorem

To find the change in f between two points P and Q, we choose a smooth path C from P to Q, then divide the path into many small pieces. See Figure 18.30. First we estimate the change in f as we move through a displacement $\Delta\vec{r}_i$ from $\vec{r}_i$ to $\vec{r}_{i+1}$. Suppose $\vec{u}$ is a unit vector in the direction of $\Delta\vec{r}_i$. Then the change in f is given by

$$f(\vec{r}_{i+1}) - f(\vec{r}_i) \approx \text{Rate of change of } f \times \text{Distance moved in direction of } \vec{u}$$
$$= f_{\vec{u}}(\vec{r}_i)\|\Delta\vec{r}_i\|$$
$$= \text{grad } f \cdot \vec{u}\, \|\Delta\vec{r}_i\|$$
$$= \text{grad } f \cdot \Delta\vec{r}_i. \qquad \text{since } \Delta\vec{r}_i = \|\Delta\vec{r}_i\|\vec{u}$$

Therefore, summing over all pieces of the path, the total change in f is given by

$$\text{Total change} = f(Q) - f(P) \approx \sum_{i=0}^{n-1} \text{grad } f(\vec{r}_i) \cdot \Delta\vec{r}_i.$$

In the limit as $\|\Delta\vec{r}_i\|$ approaches zero, we obtain the following result:

The Fundamental Theorem of Calculus for Line Integrals

Suppose C is a piecewise smooth oriented path with starting point P and endpoint Q. If f is a function whose gradient is continuous on the path C, then

$$\int_C \text{grad } f \cdot d\vec{r} = f(Q) - f(P).$$

Notice that there are many different paths from P to Q. (See Figure 18.31.) However, the value of the line integral $\int_C \text{grad } f \cdot d\vec{r}$ depends only on the endpoints of C; it does not depend on where C goes in between. [2]

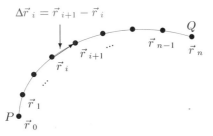

Figure 18.30: Subdivision of the path from P to Q. We estimate the change in f along $\Delta\vec{r}_i$

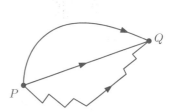

Figure 18.31: There are many different paths from P to Q: all give the same value of $\int_C \text{grad } f \cdot d\vec{r}$

Example 1   Suppose that grad f is everywhere perpendicular to the curve joining P and Q shown in Figure 18.32.
    (a) Explain why you expect the path joining P and Q to be a contour.
    (b) Using a line integral, show that $f(P) = f(Q)$.

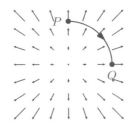

Figure 18.32: The gradient vector field of the function f

Solution   (a) The gradient of f is everywhere perpendicular to the path from P to Q, as you expect along a contour.
    (b) Consider the path from P to Q shown in Figure 18.32 and evaluate the line integral

$$\int_C \text{grad } f \cdot d\vec{r} = f(Q) - f(P).$$

Since grad f is everywhere perpendicular to the path, the line integral is 0. Thus, $f(Q) = f(P)$.

---

[2]Problem 13 on page 382 shows how the Fundamental Theorem for Line Integrals can be derived from the one-variable Fundamental Theorem of Calculus.

Example 2   Consider the vector field $\vec{F} = x\vec{i} + y\vec{j}$. In Example 2 on page 353 we calculated $\int_{C_1} \vec{F} \cdot d\vec{r}$ and $\int_{C_2} \vec{F} \cdot d\vec{r}$ over the oriented curves shown in Figure 18.33 and found they were the same. Find a scalar function f with grad $f = \vec{F}$. Hence, find an easy way to calculate the line integrals, and explain how we could have expected them to be the same.

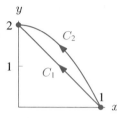

Figure 18.33: Find the line integral of $\vec{F} = x\vec{i} + y\vec{j}$ over the curves C_1 and C_2

Solution   One possibility for f is

$$f(x, y) = \frac{x^2}{2} + \frac{y^2}{2}.$$

You can check that grad $f = x\vec{i} + y\vec{j}$. Now we can use the Fundamental Theorem to compute the line integral. Since $\vec{F}$ = grad f we have

$$\int_{C_1} \vec{F} \cdot d\vec{r} = \int_{C_1} \text{grad } f \cdot d\vec{r} = f(0, 2) - f(1, 0) = \frac{3}{2}.$$

Notice that the calculation looks exactly the same for C_2. Since the value of the integral depends only on the value of f at the endpoints, it is the same no matter what path we choose.

Path-Independent, or Conservative, Vector Fields

In the previous example, the line integral was independent of the path taken between the two (fixed) endpoints. We give vector fields whose line integrals have this property a special name.

> A vector field $\vec{F}$ is said to be **path-independent**, or **conservative**, if for any two points P and Q, the line integral $\int_C \vec{F} \cdot d\vec{r}$ has the same value along any path C from P to Q lying in the domain of $\vec{F}$.

If, on the other hand, the line integral $\int_C \vec{F} \cdot d\vec{r}$ does depend on the path C joining P to Q, then $\vec{F}$ is said to be a *path-dependent* vector field.

Now suppose that $\vec{F}$ is any gradient field, so $\vec{F}$ = grad f. If C is a path from P to Q, the Fundamental Theorem for Line Integrals tells us that

$$\int_C \vec{F} \cdot d\vec{r} = f(Q) - f(P).$$

Since the right-hand side of this equation does not depend on the path, but only on the endpoints of the path, the vector field $\vec{F}$ is path-independent. Thus, we have the following important result:

> If $\vec{F}$ is a gradient vector field, then $\vec{F}$ is path-independent.

Why Do We Care about Path-Independent, or Conservative, Vector Fields?

Many of the fundamental vector fields of nature are path-independent — for example the gravitational field and the electric field of particles at rest. The fact that the gravitational field is path-independent means that the work done by gravity when an object moves depends only on the starting and ending points and not on the path taken. For example the work done by gravity (computed by the line integral) on a bicycle being carried to a sixth floor apartment is the same whether it is carried up the stairs in a zig-zag path or taken straight up in an elevator.

When a vector field is path-independent we can define the *potential energy* of a body. When the body moves to another position, the potential energy changes by an amount equal to the work done by the vector field, which depends only on the starting and ending positions. If the work done had not been path-independent, the potential energy would depend both on the body's current position *and* on how it got there, making it impossible to define a useful potential energy.

Problem 22 on page 366 explains why path-independent force vector fields are also called *conservative* vector fields: When a particle moves under the influence of a conservative vector field, the *total energy* of the particle is *conserved*. It turns out that the force field is obtained from the gradient of the potential energy function.

Path-Independent Fields and Gradient Fields

We have seen that every gradient field is path-independent. What about the converse? That is, given a path-independent vector field $\vec{F}$, can we find a function f such that $\vec{F} = \text{grad } f$? The answer is yes.

How to Construct f from $\vec{F}$

First, notice that there are many different choices for f, since we can add a constant to f without changing grad f. If we pick a fixed starting point P, then by adding or subtracting a constant to f we can ensure that $f(P) = 0$. For any other point Q, we define $f(Q)$ by the formula

$$f(Q) = \int_C \vec{F} \cdot d\vec{r}, \quad \text{where } C \text{ is any path from } P \text{ to } Q.$$

Since $\vec{F}$ is path-independent, it doesn't matter which path we choose from P to Q. On the other hand, if $\vec{F}$ is not path-independent, then different choices might give different values for $f(Q)$, so f would not be a function (a function has to have a single value at each point).

We still have to show that the gradient of the function f really is $\vec{F}$; we do this on page 363. However, by constructing a function f in this manner, we have the following result:

> If $\vec{F}$ is a path-independent vector field, then $\vec{F} = \text{grad } f$ for some f.

Combining the two results, we have

> A vector field $\vec{F}$ is path-independent if and only if $\vec{F}$ is a gradient vector field.

The function f is sufficiently important that it is given a special name:

> If a vector field $\vec{F}$ is of the form $\vec{F} = \text{grad } f$ for some scalar function f, then f is called a **potential function** for the vector field $\vec{F}$.

Warning

Physicists use the convention that a function ϕ is a potential function for a vector field $\vec{F}$ if $\vec{F} = -\operatorname{grad} \phi$. See Problem 21 on page 366.

Example 3 Show that the vector field $\vec{F}(x, y) = y \cos x \vec{i} + \sin x \vec{j}$ is path-independent.

Solution If we can find a potential function f, then $\vec{F}$ must be path-independent. We want $\operatorname{grad} f = \vec{F}$. Since

$$\frac{\partial f}{\partial x} = y \cos x,$$

f must be of the form

$$f(x, y) = y \sin x + g(y) \qquad \text{where } g(y) \text{ is a function of } y \text{ only.}$$

In addition, since $\operatorname{grad} f = \vec{F}$, we must have

$$\frac{\partial f}{\partial y} = \sin x,$$

and differentiating $f(x, y) = y \sin x + g(y)$ gives

$$\frac{\partial f}{\partial y} = \sin x + g'(y).$$

Thus, we must have $g'(y) = 0$, so $g(y) = C$ where C is some constant. Thus,

$$f(x, y) = y \sin x + C$$

is a potential function for $\vec{F}$. Therefore $\vec{F}$ is path-independent.

Example 4 The gravitational force field, $\vec{F}$, of an object of mass M, is given by

$$\vec{F} = -\frac{GM}{r^3} \vec{r}.$$

Show that $\vec{F}$ is a gradient field by finding a potential function for $\vec{F}$.

Solution All the force vectors point in toward the origin. If $\vec{F} = \operatorname{grad} f$, the force vectors must be perpendicular to the level surfaces of f, so the level surfaces of f must be spheres. Also, if $\operatorname{grad} f = \vec{F}$, then $\|\operatorname{grad} f\| = \|\vec{F}\| = GM/r^2$ is the rate of change of f in the direction toward the origin. Now, differentiating with respect to r gives the rate of change in a radially outward direction. Thus, if $w = f(x, y, z)$ we have

$$\frac{dw}{dr} = -\frac{GM}{r^2} = GM\left(-\frac{1}{r^2}\right) = GM\frac{d}{dr}\left(\frac{1}{r}\right).$$

So let's try

$$w = \frac{GM}{r} \quad \text{or} \quad f(x, y, z) = \frac{GM}{\sqrt{x^2 + y^2 + z^2}}.$$

We calculate

$$f_x = \frac{\partial}{\partial x} \frac{GM}{\sqrt{x^2 + y^2 + z^2}} = \frac{-GMx}{(x^2 + y^2 + z^2)^{3/2}},$$

$$f_y = \frac{\partial}{\partial y} \frac{GM}{\sqrt{x^2 + y^2 + z^2}} = \frac{-GMy}{(x^2 + y^2 + z^2)^{3/2}},$$

$$f_z = \frac{\partial}{\partial z} \frac{GM}{\sqrt{x^2 + y^2 + z^2}} = \frac{-GMz}{(x^2 + y^2 + z^2)^{3/2}}.$$

So

$$\text{grad } f = f_x \vec{i} + f_y \vec{j} + f_z \vec{k} = \frac{-GM}{(x^2 + y^2 + z^2)^{3/2}} (x\vec{i} + y\vec{j} + z\vec{k}) = \frac{-GM}{r^3} \vec{r} = \vec{F}.$$

Our computations show that $\vec{F}$ is a gradient field and that $f = GM/r$ is a potential function for $\vec{F}$.

Why Path-Independent Vector Fields are Gradient Fields: Showing grad $f = \overrightarrow{F}$

Suppose $\vec{F}$ is a path-independent vector field. On page 361 we defined the function f, which we hope will satisfy grad $f = \vec{F}$, as follows:

$$f(x_0, y_0) = \int_C \vec{F} \cdot d\vec{r},$$

where C is a path from a fixed starting point P to a point $Q = (x_0, y_0)$. This integral has the same value for any path from P to Q because $\vec{F}$ is path-independent. Now we show why grad $f = \vec{F}$. We consider vector fields in 2-space; the argument in 3-space is essentially the same.

First, we write the line integral in terms of the components $\vec{F}(x, y) = F_1(x, y)\vec{i} + F_2(x, y)\vec{j}$ and the components $d\vec{r} = dx\vec{i} + dy\vec{j}$:

$$f(x_0, y_0) = \int_C F_1(x, y)dx + F_2(x, y)dy.$$

We want to compute the partial derivatives of f, that is, the rate of change of f at (x_0, y_0) parallel to the axes. To do this easily, we choose a path which reaches the point (x_0, y_0) on a horizontal or vertical line segment. Let C' be a path from P which stops short of Q at a fixed point (a, b) and let L_x and L_y be the paths shown in Figure 18.34. Then we can split the line integral into three pieces. Since $d\vec{r} = \vec{j}\, dy$ on L_y and $d\vec{r} = \vec{i}\, dx$ on L_x, we have:

$$f(x_0, y_0) = \int_{C'} \vec{F} \cdot d\vec{r} + \int_{L_y} \vec{F} \cdot d\vec{r} + \int_{L_x} \vec{F} \cdot d\vec{r} = \int_{C'} \vec{F} \cdot d\vec{r} + \int_b^{y_0} F_2(a, y)dy + \int_a^{x_0} F_1(x, y_0)dx.$$

The first two integrals do not involve x_0. Thinking of x_0 as a variable and differentiating with respect to it gives

$$f_{x_0}(x_0, y_0) = \frac{\partial}{\partial x_0} \int_{C'} \vec{F} \cdot d\vec{r} + \frac{\partial}{\partial x_0} \int_b^{y_0} F_2(a, y)dy + \frac{\partial}{\partial x_0} \int_a^{x_0} F_1(x, y_0)dx$$
$$= 0 + 0 + F_1(x_0, y_0) = F_1(x_0, y_0),$$

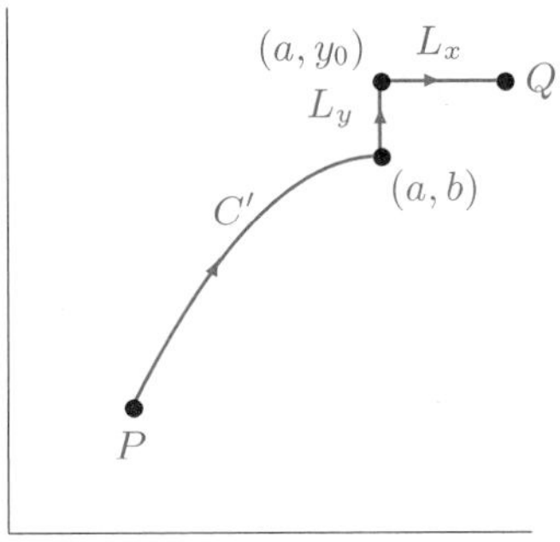

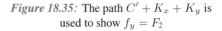

Figure 18.34: The path $C' + L_y + L_x$ is used to show $f_x = F_1$

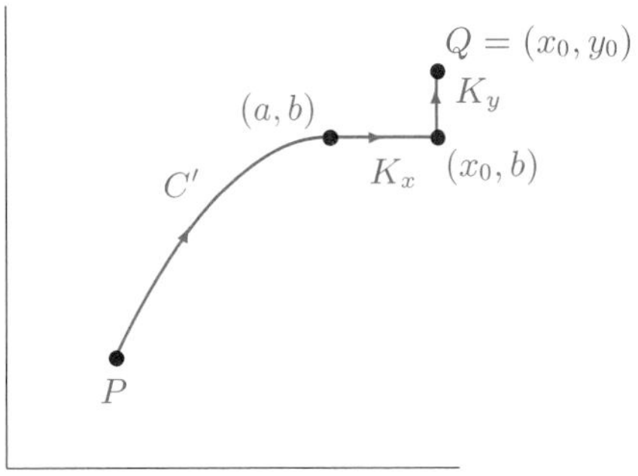

Figure 18.35: The path $C' + K_x + K_y$ is used to show $f_y = F_2$

and thus
$$f_x(x, y) = F_1(x, y).$$
A similar calculation for y using the path from P to Q shown in Figure 18.35 gives
$$f_{y_0}(x_0, y_0) = F_2(x_0, y_0).$$
Therefore, as we claimed
$$\text{grad } f = f_x \vec{i} + f_y \vec{j} = F_1 \vec{i} + F_2 \vec{j} = \vec{F}.$$

Summary

We have studied two apparently different types of vector field: path-independent vector fields and gradient vector fields. It turns out that these are the same. Here is a summary of the definitions and properties of these vector fields:

- **Path-independent vector fields** have the property that for any two points P and Q, the line integral along a path from P to Q is the same no matter what path we choose.

- **Gradient vector fields** are of the form grad f for some scalar function f, called the potential function of the vector field.

- **Gradient vector fields are path-independent** by the Fundamental Theorem of Line Integrals,
$$\int_C \text{grad } f \cdot d\vec{r} = f(Q) - f(P).$$

- **Path-independent vector fields are gradient fields** because we can use a line integral and path independence to construct a potential function.

Problems for Section 18.3

1. The vector field $\vec{F}(x, y) = x\vec{i} + y\vec{j}$ is path-independent. Compute geometrically the line integrals over the three paths A, B, and C shown in Figure 18.36 from $(1, 0)$ to $(0, 1)$ and verify that they are equal. Here A is a portion of a circle, B is a line, and C consists of two line segments meeting at a right angle.

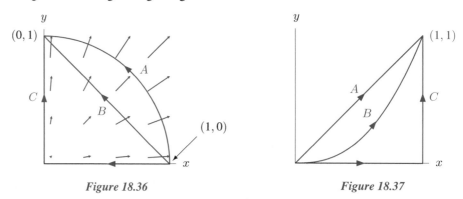

Figure 18.36   *Figure 18.37*

2. The vector field $\vec{F}(x, y) = x\vec{i} + y\vec{j}$ is path-independent. Compute algebraically the line integrals over the three paths A, B, and C shown in Figure 18.37 from $(0, 0)$ to $(1, 1)$ and verify that they are equal. Here A is a line segment, B is part of the graph of $f(x) = x^2$, and C consists of two line segments meeting at a right angle.

For Problems 3–6, decide whether or not the given vector fields could be gradient vector fields. Give a justification for your answer.

3. $\vec{F}(x,y) = x\vec{i}$

4. $\vec{G}(x,y) = (x^2 - y^2)\vec{i} - 2xy\vec{j}$

5. $\vec{F}(x,y,z) = \dfrac{-z}{\sqrt{x^2+z^2}}\vec{i} + \dfrac{y}{\sqrt{x^2+z^2}}\vec{j} + \dfrac{x}{\sqrt{x^2+z^2}}\vec{k}$

6. $\vec{F}(\vec{r}) = \vec{r}/\|\vec{r}\|^3$, where $\vec{r} = x\vec{i} + y\vec{j} + z\vec{k}$

For the vector fields in Problems 7–10, find the line integral along the curve C from the origin along the x-axis to the point $(3,0)$ and then counterclockwise around the circumference of the circle $x^2 + y^2 = 9$ to the point $(3/\sqrt{2}, 3/\sqrt{2})$.

7. $\vec{F} = x\vec{i} + y\vec{j}$

8. $\vec{H} = -y\vec{i} + x\vec{j}$

9. $\vec{F} = y(x+1)^{-1}\vec{i} + \ln(x+1)\vec{j}$

10. $\vec{G} = (ye^{xy} + \cos(x+y))\vec{i} + (xe^{xy} + \cos(x+y))\vec{j}$

11. Suppose that grad $f = 2xe^{x^2}\sin y\,\vec{i} + e^{x^2}\cos y\,\vec{j}$. Find the change in f between $(0,0)$ and $(1, \pi/2)$: (a) By computing a line integral   (b) By computing f.

12. The line integral of $\vec{F} = (x+y)\vec{i} + x\vec{j}$ along each of the following paths is $3/2$:

    (i)   The path (t, t^2), with $0 \le t \le 1$
    (ii) The path (t^2, t), with $0 \le t \le 1$
    (iii) The path (t, t^n), with $n > 0$ and $0 \le t \le 1$

    Verify this
    (a) Using the given parameterization to compute the line integral.
    (b) Using the Fundamental Theorem of Calculus for Line Integrals.

13. Consider the vector field $\vec{F}(x,y) = x\vec{j}$ shown in Figure 18.38.

    (a) Find paths C_1, C_2, and C_3 from P to Q such that

    $$\int_{C_1} \vec{F} \cdot d\vec{r} = 0, \qquad \int_{C_2} \vec{F} \cdot d\vec{r} > 0, \quad \text{and} \quad \int_{C_3} \vec{F} \cdot d\vec{r} < 0.$$

    (b) Is $\vec{F}$ a gradient field?

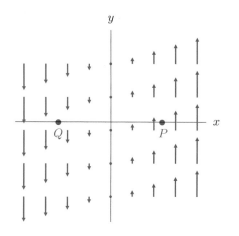

Figure 18.38

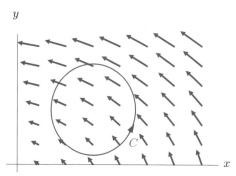

Figure 18.39

14. Consider the vector field $\vec{F}$ graphed in Figure 18.39.

    (a) Is the line integral $\int_C \vec{F} \cdot d\vec{r}$ positive, negative, or zero?
    (b) From your answer to part (a), can you determine whether or not $\vec{F} = \operatorname{grad} f$ for some function f?
    (c) Which of the following formulas best fits this vector field?

    $$\vec{F}_1 = \frac{x}{x^2 + y^2}\vec{i} + \frac{y}{x^2 + y^2}\vec{j}, \quad \vec{F}_2 = -y\vec{i} + x\vec{j}, \quad \vec{F}_3 = \frac{-y}{(x^2 + y^2)^2}\vec{i} + \frac{x}{(x^2 + y^2)^2}\vec{j}.$$

In Problems 15–18, each of the statements is *false*. Explain why or give a counterexample.

15. If $\int_C \vec{F} \cdot d\vec{r} = 0$ for one particular closed path C, then $\vec{F}$ is path-independent.

16. $\int_C \vec{F} \cdot d\vec{r}$ is the total change in $\vec{F}$ along C.

17. If the vector fields $\vec{F}$ and $\vec{G}$ have $\int_C \vec{F} \cdot d\vec{r} = \int_C \vec{G} \cdot d\vec{r}$ for a particular path C, then $\vec{F} = \vec{G}$.

18. If the total change of a function f along a curve C is zero, then C must be a contour of f.

19. Suppose a particle subject to a force $\vec{F}(x, y) = y\vec{i} - x\vec{j}$ moves clockwise along the arc of the unit circle, centered at the origin, that begins at $(-1, 0)$ and ends at $(0, 1)$.

    (a) Find the work done by $\vec{F}$. Explain the sign of your answer.
    (b) Is $\vec{F}$ path-independent? Explain.

20. A particle moves with position vector $\vec{r}(t) = x(t)\vec{i} + y(t)\vec{j} + z(t)\vec{k}$. Let $\vec{v}(t)$ and $\vec{a}(t)$ be its velocity and acceleration vectors. Show that

    $$\frac{1}{2}\frac{d}{dt}\|\vec{v}(t)\|^2 = \vec{a}(t) \cdot \vec{v}(t).$$

21. Let $\vec{F}$ be a path-independent vector field. In physics, the potential function ϕ is usually required to satisfy the equation $\vec{F} = -\nabla\phi$. This problem illustrates the significance of the negative sign.[3]

    (a) Let the xy-plane represent part of the earth's surface with the z-axis pointing away from the earth. (We assume the scale is small enough so that a flat plane is a good approximation to the earth's surface.) Let $\vec{r} = x\vec{i} + y\vec{j} + z\vec{k}$, with $z \geq 0$, and x, y, z in meters, be the position vector of a rock of unit mass. The gravitational potential energy function for the rock is $\phi(x, y, z) = gz$, where $g \approx 9.8$ m/sec^2. Describe in words the level surfaces of ϕ. Does the potential energy increase or decrease with height above the earth?
    (b) What is the relation between the gravitational vector, $\vec{F}$, and the vector $\nabla\phi$? Explain the significance of the negative sign in the equation $\vec{F} = -\nabla\phi$.

22. In this problem we derive the principle of Conservation of Energy. The kinetic energy of a particle of mass m moving with speed v is $(1/2)mv^2$. Suppose the particle has potential energy $f(\vec{r})$ at the position $\vec{r}$ due to a force field $\vec{F} = -\nabla f$. If the particle moves with position vector $\vec{r}(t)$ and velocity $\vec{v}(t)$, then the Conservation of Energy principle says that

$$\text{Total energy} = \text{Kinetic energy} + \text{Potential energy} = \frac{1}{2}m\|\vec{v}(t)\|^2 + f(\vec{r}(t)) = \text{Constant}.$$

---

[3] Adapted from V.I. Arnold, *Mathematical Methods of Classical Mechanics*, 2nd Edition, Graduate Texts in Mathematics, Springer

Let P and Q be two points in space and let C be a path from P to Q parameterized by $\vec{r}(t)$ for $t_0 \leq t \leq t_1$, where $\vec{r}(t_0) = P$ and $\vec{r}(t_1) = Q$.

(a) Using the result of Problem 20 and Newton's law $\vec{F} = m\vec{a}$, show

$$\begin{array}{c} \text{Work done by } \vec{F} \\ \text{as particle moves along } C \end{array} = \text{Kinetic energy at } Q - \text{Kinetic energy at } P.$$

(b) Use the Fundamental Theorem of Calculus for Line Integrals to show that

$$\begin{array}{c} \text{Work done by } \vec{F} \\ \text{as particle moves along } C \end{array} = \text{Potential energy at } P - \text{Potential energy at } Q.$$

(c) Use parts (a) and (b) to show that the total energy at P is the same as at Q.
This problem explains why force vector fields which are *path-independent* are usually called *conservative* (force) vector fields.

18.4 PATH-DEPENDENT VECTOR FIELDS AND GREEN'S THEOREM

Suppose we are given a vector field but are not told whether it is path-independent. How can we tell if it has a potential function, that is, if it is a gradient field?

How to Tell if a Vector Field is Path-Dependent Using Line Integrals

Example 1   Is the vector field, $\vec{F}$, shown in Figure 18.40 path-independent? At any point $\vec{F}$ has magnitude equal to the distance from the origin and direction perpendicular to the line joining the point to the origin.

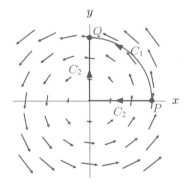

Figure 18.40: Is this vector field
path-independent?

Solution   We choose $P = (1, 0)$ and $Q = (0, 1)$ and two paths between them: C_1, a quarter circle of radius 1, and C_2, formed by parts of the x- and y-axes. (See Figure 18.40.) Along C_1, the line integral $\int_{C_1} \vec{F} \cdot d\vec{r} > 0$, since $\vec{F}$ points in the direction of the curve. Along C_2, however, we have $\int_{C_2} \vec{F} \cdot d\vec{r} = 0$, since $\vec{F}$ is perpendicular to C_2 everywhere. Thus, $\vec{F}$ is not path-independent.

Path-Dependent Fields and Circulation

Notice that the vector field in the previous example has nonzero circulation around the origin. What can we say about the circulation of a general path-independent vector field around a closed curve, C? Suppose C is a *simple* closed curve, that is, a curve which does not cross itself. If P and Q are any two points on the path, then we can think of C (oriented as shown in Figure 18.41) as made up of the path C_1 followed by $-C_2$. Since $\vec{F}$ is path-independent, we know that

$$\int_{C_1} \vec{F} \cdot d\vec{r} = \int_{C_2} \vec{F} \cdot d\vec{r}.$$

Thus, we see that the circulation around C is zero:

$$\int_{C} \vec{F} \cdot d\vec{r} = \int_{C_1} \vec{F} \cdot d\vec{r} + \int_{-C_2} \vec{F} \cdot d\vec{r} = \int_{C_1} \vec{F} \cdot d\vec{r} - \int_{C_2} \vec{F} \cdot d\vec{r} = 0.$$

If the curve C does cross itself, we break it into simple closed curves as shown in Figure 18.42 and apply the same argument to each one.

Now suppose we know that the line integral around any closed curve is zero. For any two points, P and Q, with two paths, C_1 and C_2, between them, create a closed curve, C, as in Figure 18.41. Since the circulation around this closed curve, C, is zero, the line integrals along the two paths, C_1 and C_2, are equal. Thus, $\vec{F}$ is path-independent.

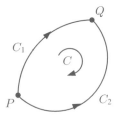

Figure 18.41: A simple closed curve C broken into two pieces, C_1 and C_2

Figure 18.42: A curve C which crosses itself can be broken into simple closed curves

Thus, we have the following result:

A vector field is path-independent if and only if $\displaystyle\int_{C} \vec{F} \cdot d\vec{r} = 0$ for every closed curve C.

Hence, to see if a field is *path-dependent*, we look for a closed path with nonzero circulation. For instance, the vector field in Example 1 has nonzero circulation around a circle around the origin, showing it is path-dependent.

How to Tell If a Vector Field is Path-Dependent Algebraically: The Curl

---

Example 2   Does the vector field $\vec{F} = 2xy\vec{i} + xy\vec{j}$ have a potential function? If so, find it.

Solution   Let's suppose $\vec{F}$ does have a potential function, f, so $\vec{F} = \text{grad } f$. This means that

$$\frac{\partial f}{\partial x} = 2xy \quad \text{and} \quad \frac{\partial f}{\partial y} = xy.$$

Integrating the expression for $\partial f/\partial x$ shows that we must have

$$f(x, y) = x^2 y + C(y) \qquad \text{where } C(y) \text{ is a function of } y.$$

Differentiating this expression for $f(x, y)$ with respect to y and using the fact that $\partial f/\partial y = xy$, we get

$$\frac{\partial f}{\partial y} = x^2 + C'(y) = xy.$$

Thus, we must have

$$C'(y) = xy - x^2.$$

But this expression for $C'(y)$ is impossible because $C'(y)$ is a function of y alone. This argument shows that there is no potential function for the vector field $\vec{F}$.

Is there an easier way to see that a vector field has no potential function, other than by trying to find the potential function and failing? The answer is yes. First we look at a 2-dimensional vector field $\vec{F} = F_1 \vec{i} + F_2 \vec{j}$. If $\vec{F}$ is a gradient field, then there is a potential function f such that

$$\vec{F} = F_1 \vec{i} + F_2 \vec{j} = \frac{\partial f}{\partial x} \vec{i} + \frac{\partial f}{\partial y} \vec{j}.$$

Thus,

$$F_1 = \frac{\partial f}{\partial x} \quad \text{and} \quad F_2 = \frac{\partial f}{\partial y}.$$

Let us assume that f has continuous second partial derivatives. Then, by the equality of mixed partial derivatives

$$\frac{\partial F_1}{\partial y} = \frac{\partial^2 f}{\partial y \partial x} = \frac{\partial^2 f}{\partial x \partial y} = \frac{\partial F_2}{\partial x}$$

Thus we have the following result:

If $\vec{F}(x, y) = F_1 \vec{i} + F_2 \vec{j}$ is a gradient vector field with continuous partial derivatives, then

$$\frac{\partial F_2}{\partial x} - \frac{\partial F_1}{\partial y} = 0.$$

We call $\dfrac{\partial F_2}{\partial x} - \dfrac{\partial F_1}{\partial y}$ the 2-dimensional or scalar **curl** of the vector field $\vec{F}$.

Notice that we now know that if $\vec{F}$ is a gradient field, then its curl is 0. We do not (yet) know whether the converse is true. (That is: If the curl is 0, does $\vec{F}$ have to be a gradient field?) However, the curl already enables us to show that a vector field is *not* a gradient field.

Example 3   Show that $\vec{F} = 2xy\vec{i} + xy\vec{j}$ cannot be a gradient vector field.

Solution   We have $F_1 = 2xy$ and $F_2 = xy$. Since $\partial F_1/\partial y = 2x$ and $\partial F_2/\partial x = y$, in this case

$$\partial F_2/\partial x - \partial F_1/\partial y \neq 0$$

so $\vec{F}$ cannot be a gradient field.

We now have two ways of seeing that a vector field $\vec{F}$ in the plane is path-dependent. We can evaluate $\int_C \vec{F} \cdot d\vec{r}$ for some closed curve and find it is not zero, or we can show that $\partial F_2/\partial x - \partial F_1/\partial y \neq 0$. It's natural to think that

$$\int_C \vec{F} \cdot d\vec{r} \quad \text{and} \quad \frac{\partial F_2}{\partial x} - \frac{\partial F_1}{\partial y}$$

might be related. The relation is called Green's Theorem.

Green's Theorem

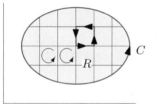

Figure 18.43: Region R bounded by a closed curve C and split into many small regions, ΔR

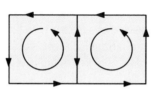

Figure 18.44: Two adjacent small closed curves

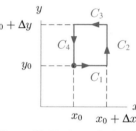

Figure 18.45: A small closed curve ΔC broken into C_1, C_2, C_3, C_4

We obtain the statement of Green's Theorem by splitting a region up into small pieces and looking at the relation between the line integral and the curl on each one.

In 2-space, a simple closed curve C separates the plane into points inside and points outside C. We consider $\int_C \vec{F} \cdot d\vec{r}$, where C is oriented as shown in Figure 18.43. We divide the region R inside C into small pieces, each bounded by a closed curve with the orientation shown. Figure 18.44 shows that if we add the circulation around all of these small closed curves, each common edge of a pair of adjacent curves is counted twice, once in each direction. Hence the integrals along these edges cancel. Thus, the line integrals along all the edges inside the region R cancel, giving

$$\begin{array}{c}\text{Circulation of } \vec{F} \\ \text{around } C\end{array} = \sum_{\Delta C} \begin{array}{c}\text{Circulation of } \vec{F} \\ \text{around small curve, } \Delta C\end{array}$$

Now we estimate the line integral around one of these small closed curves, ΔC. We break ΔC into C_1, C_2, C_3, and C_4, as shown in Figure 18.45. Then we calculate the line integrals along C_1 and C_3, where $\Delta\vec{r}$ is parallel to the x-axis, so

$$\Delta\vec{r} = \Delta x\vec{i}.$$

Thus, on C_1 and C_3,

$$\vec{F} \cdot \Delta\vec{r} = (F_1\vec{i} + F_2\vec{j}) \cdot \Delta x\vec{i} = F_1 \Delta x.$$

However, the function F_1 is evaluated at (x, y_0) on C_1 and at $(x, y_0 + \Delta y)$ on C_3, so

$$
\int_{C_1} \vec{F} \cdot d\vec{r} + \int_{C_3} \vec{F} \cdot d\vec{r} = = \int_{C_1} F_1(x, y_0) \, dx + \int_{C_3} F_1(x, y_0 + \Delta y) \, dx
$$

$$
= \int_{x=x_0}^{x_0+\Delta x} F_1(x, y_0) \, dx + \int_{x_0+\Delta x}^{x=x_0} F_1(x, y_0 + \Delta y) \, dx
$$

$$
= \int_{x=x_0}^{x_0+\Delta x} F_1(x, y_0) \, dx - \int_{x=x_0}^{x_0+\Delta x} F_1(x, y_0 + \Delta y) \, dx
$$

$$
= \int_{x=x_0}^{x_0+\Delta x} (F_1(x, y_0) - F_1(x, y_0 + \Delta y)) \, dx.
$$

Since F_1 is differentiable and Δy is small,

$$
F_1(x, y_0) - F_1(x, y_0 + \Delta y) = -(F_1(x, y_0 + \Delta y) - F_1(x, y_0)) \approx -\frac{\partial F_1}{\partial y}(x, y_0) \, \Delta y,
$$

so we have

$$
\int_{x=x_0}^{x_0+\Delta x} (F_1(x, y_0) - F_1(x, y_0 + \Delta y)) \, dx \approx -\left(\int_{x=x_0}^{x_0+\Delta x} \frac{\partial F_1}{\partial y} \, dx \right) \Delta y.
$$

Assuming $\dfrac{\partial F_1}{\partial y}(x, y_0)$ is approximately constant over the interval $x_0 \leq x \leq x_0 + \Delta x$, we get

$$
-\left(\int_{x=x_0}^{x_0+\Delta x} \frac{\partial F_1}{\partial y} \, dx \right) \Delta y \approx -\frac{\partial F_1}{\partial y}(x_0, y_0) \left(\int_{x=x_0}^{x_0+\Delta x} dx \right) \Delta y = -\frac{\partial F_1}{\partial y}(x_0, y_0) \, \Delta x \, \Delta y.
$$

By a similar argument on C_2 and C_4, we have

$$
\int_{C_2} \vec{F} \cdot d\vec{r} + \int_{C_4} \vec{F} \cdot d\vec{r} \approx \frac{\partial F_2}{\partial x} \Delta x \, \Delta y.
$$

Combining these results for C_1, C_2, C_3, and C_4, we get

$$
\int_{\Delta C} \vec{F} \cdot d\vec{r} = \int_{C_1} \vec{F} \cdot d\vec{r} + \int_{C_2} \vec{F} \cdot d\vec{r} + \int_{C_3} \vec{F} \cdot d\vec{r} + \int_{C_4} \vec{F} \cdot d\vec{r} \approx \frac{\partial F_2}{\partial x} \Delta x \, \Delta y - \frac{\partial F_1}{\partial y} \Delta x \, \Delta y.
$$

Summing over all small regions ΔC gives

$$
\int_C \vec{F} \cdot d\vec{r} \approx \sum_{\Delta C} \int_{\Delta C} \vec{F} \cdot d\vec{r} \approx \sum_{\Delta R} \left(\frac{\partial F_2}{\partial x} - \frac{\partial F_1}{\partial y} \right) \Delta x \, \Delta y.
$$

The last sum is a Riemann sum approximating a double integral; taking the limit as Δx, Δy tend to zero, we get

---

Green's Theorem

Suppose C is a simple closed curve surrounding a region R in the plane and oriented so that the region is on the left as we move around the curve. Suppose $\vec{F} = F_1 \vec{i} + F_2 \vec{j}$ is a smooth vector field defined at every point of the region R and boundary C. Then

$$
\int_C \vec{F} \cdot d\vec{r} = \int_R \left(\frac{\partial F_2}{\partial x} - \frac{\partial F_1}{\partial y} \right) dx \, dy.
$$

Section 18.5 contains a proof of Green's Theorem using the change of variables formula for double integrals.

The Curl Test for Vector Fields in the Plane

We already know that if $\vec{F} = F_1\vec{i} + F_2\vec{j}$ is a gradient field with continuous partial derivatives, then

$$\frac{\partial F_2}{\partial x} - \frac{\partial F_1}{\partial y} = 0.$$

Now we show that the converse is true if the domain of $\vec{F}$ has no holes in it. This means that we assume that

$$\frac{\partial F_2}{\partial x} - \frac{\partial F_1}{\partial y} = 0$$

and show that $\vec{F}$ is path-independent. If C is any oriented closed curve in the domain of $\vec{F}$ and R is the region inside C, then

$$\int_R \left(\frac{\partial F_2}{\partial x} - \frac{\partial F_1}{\partial y} \right) dx\, dy = 0$$

since the integrand is identically 0. Therefore, by Green's Theorem

$$\int_C \vec{F} \cdot d\vec{r} = \int_R \left(\frac{\partial F_2}{\partial x} - \frac{\partial F_1}{\partial y} \right) dx\, dy = 0.$$

Thus, $\vec{F}$ is path-independent and therefore a gradient field. This argument is valid for every closed curve, C, provided the region R is entirely in the domain of $\vec{F}$. Thus we have the following result:

The Curl Test for Vector Fields in 2-Space

Suppose $\vec{F} = F_1\vec{i} + F_2\vec{j}$ is a vector field with continuous partial derivatives, such that
- The domain of $\vec{F}$ has the property that every closed curve in it encircles a region that lies entirely within the domain. In particular, the domain of $\vec{F}$ has no holes.
- $\frac{\partial F_2}{\partial x} - \frac{\partial F_1}{\partial y} = 0.$

Then $\vec{F}$ is path-independent, so $\vec{F}$ is a gradient field and has a potential function.

Why Are Holes in the Domain of the Vector Field Important?

The reason for assuming that the domain of the vector field $\vec{F}$ has no holes is to ensure that the region R inside C is actually contained in the domain of $\vec{F}$. Otherwise, we cannot apply Green's Theorem. The next two examples show that if $\partial F_2/\partial x - \partial F_1/\partial y = 0$ but the domain of $\vec{F}$ contains a hole, then $\vec{F}$ can either be path-independent or path-dependent.

Example 4   Let $\vec{F}$ be the vector field given by $\vec{F}(x, y) = \dfrac{-y\vec{i} + x\vec{j}}{x^2 + y^2}$.

(a)   Calculate $\dfrac{\partial F_2}{\partial x} - \dfrac{\partial F_1}{\partial y}$. Does the curl test imply that $\vec{F}$ is path-independent?

(b)   Calculate $\displaystyle\int_C \vec{F} \cdot d\vec{r}$, where C is the unit circle centered at the origin and oriented counterclockwise. Is $\vec{F}$ a path-independent vector field?

(c)   Explain why the answers to parts (a) and (b) do not contradict Green's Theorem.

Solution   (a)   Taking partial derivatives, we have

$$\frac{\partial F_2}{\partial x} = \frac{\partial}{\partial x}\left(\frac{x}{x^2 + y^2}\right) = \frac{1}{x^2 + y^2} - \frac{x \cdot 2x}{(x^2 + y^2)^2} = \frac{y^2 - x^2}{(x^2 + y^2)^2}.$$

Similarly,

$$\frac{\partial F_1}{\partial y} = \frac{\partial}{\partial y}\left(\frac{-y}{x^2 + y^2}\right) = \frac{-1}{x^2 + y^2} + \frac{y \cdot 2y}{(x^2 + y^2)^2} = \frac{y^2 - x^2}{(x^2 + y^2)^2}.$$

Thus,

$$\frac{\partial F_2}{\partial x} - \frac{\partial F_1}{\partial y} = 0.$$

Since $\vec{F}$ is undefined at the origin, the domain of $\vec{F}$ contains a hole. Therefore, the curl test does not apply.

(b)   On the unit circle, $\vec{F}$ is tangent to the circle and $||\vec{F}|| = 1$. Thus,

$$\int_C \vec{F} \cdot d\vec{r} = ||\vec{F}|| \cdot \text{Length of curve} = 1 \cdot 2\pi = 2\pi.$$

Since the line integral around the closed curve C is nonzero, $\vec{F}$ is not path-independent.

(c)   The domain of $\vec{F}$ is the "punctured plane," as shown in Figure 18.46. Since $\vec{F}$ is not defined at the origin, which is inside C, Green's Theorem does not apply. In this case

$$2\pi = \int_C \vec{F} \cdot d\vec{r} \neq \int_R \left(\frac{\partial F_2}{\partial x} - \frac{\partial F_1}{\partial y}\right) dxdy = 0.$$

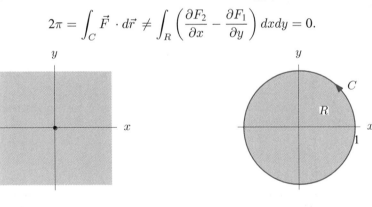

Figure 18.46: The domain of $\vec{F}(x, y) = \frac{-y\vec{i} + x\vec{j}}{x^2 + y^2}$ is the plane minus the origin

Figure 18.47: The region R is *not* contained in the domain of $\vec{F}(x, y) = \frac{-y\vec{i} + x\vec{j}}{x^2 + y^2}$

Although the vector field $\vec{F}$ in the last example was not defined at the origin, this by itself does not prevent the vector field from being path-independent as we see in the following example.

Example 5   Consider the vector field $\vec{F}$ given by $\vec{F}(x, y) = \dfrac{x\vec{i} + y\vec{j}}{x^2 + y^2}$.

(a)   Calculate $\dfrac{\partial F_2}{\partial x} - \dfrac{\partial F_1}{\partial y}$. Does the curl test imply that $\vec{F}$ is path-independent?

(b)   Explain how we know that $\displaystyle\int_C \vec{F} \cdot d\vec{r} = 0$, where C is the unit circle centered at the origin and oriented counterclockwise. Does this imply that $\vec{F}$ is path-independent?

(c)   Check that $f(x, y) = \frac{1}{2}\ln(x^2 + y^2)$ is a potential function for $\vec{F}$. Does this imply that $\vec{F}$ is path-independent?

Solution (a) Taking partial derivatives, we have

$$\frac{\partial F_2}{\partial x} = \frac{\partial}{\partial x}\left(\frac{y}{x^2 + y^2}\right) = \frac{-2xy}{(x^2 + y^2)^2}, \quad \text{and} \quad \frac{\partial F_1}{\partial y} = \frac{\partial}{\partial y}\left(\frac{x}{x^2 + y^2}\right) = \frac{-2xy}{(x^2 + y^2)^2}.$$

Therefore,

$$\frac{\partial F_2}{\partial x} - \frac{\partial F_1}{\partial y} = 0.$$

This does *not* imply that $\vec{F}$ is path-independent: The domain of $\vec{F}$ contains a hole since $\vec{F}$ is undefined at the origin. Thus, the curl test does not apply.

(b) Since $\vec{F}(x, y) = x\vec{i} + y\vec{j} = \vec{r}$ on the unit circle C, the field $\vec{F}$ is everywhere perpendicular to C. Thus

$$\int_C \vec{F} \cdot d\vec{r} = 0.$$

The fact that $\int_C \vec{F} \cdot d\vec{r} = 0$ when C is the unit circle does *not* imply that $\vec{F}$ is path-independent. To be sure that $\vec{F}$ is path-independent, we would have to show that $\int_C \vec{F} \cdot d\vec{r} = 0$ for *every* closed curve C in the domain of $\vec{F}$, not just the unit circle.

(c) To check that grad $f = \vec{F}$, we differentiate f:

$$f_x = \frac{1}{2}\frac{\partial}{\partial x}\ln(x^2 + y^2) = \frac{1}{2}\frac{2x}{x^2 + y^2} = \frac{x}{x^2 + y^2},$$

and

$$f_y = \frac{1}{2}\frac{\partial}{\partial y}\ln(x^2 + y^2) = \frac{1}{2}\frac{2y}{x^2 + y^2} = \frac{y}{x^2 + y^2},$$

so that

$$\text{grad } f = \frac{x\vec{i} + y\vec{j}}{x^2 + y^2} = \vec{F}.$$

Thus, $\vec{F}$ is a gradient field and therefore is path-independent — even though $\vec{F}$ is undefined at the origin.

The Curl Test for Vector Fields in 3-Space

The curl test is a convenient way of deciding whether a 2-dimensional vector field is path-independent. Fortunately, there is an analogous test for 3-dimensional vector fields, although we cannot justify it until Chapter 20.

If $\vec{F}(x, y, z) = F_1\vec{i} + F_2\vec{i} + F_3\vec{k}$ is a vector field on 3-space we define a new vector field, curl $\vec{F}$, on 3-space by

$$\text{curl } \vec{F} = \left(\frac{\partial F_3}{\partial y} - \frac{\partial F_2}{\partial z}\right)\vec{i} + \left(\frac{\partial F_1}{\partial z} - \frac{\partial F_3}{\partial x}\right)\vec{j} + \left(\frac{\partial F_2}{\partial x} - \frac{\partial F_1}{\partial y}\right)\vec{k}.$$

The vector field curl $\vec{F}$ can be used to determine whether the vector field $\vec{F}$ is path-independent.

The Curl Test for Vector Fields in 3-Space

Suppose $\vec{F}$ is a vector field on 3-space with continuous partial derivatives, such that
- The domain of $\vec{F}$ has the property that every closed curve in it can be contracted to a point in a smooth way, staying at all times within the domain.
- curl $\vec{F} = \vec{0}$.

Then $\vec{F}$ is path-independent, so $\vec{F}$ is a gradient field and has a potential function.

For the 2-dimensional curl test, the domain of $\vec{F}$ must have no holes. This meant that if $\vec{F}$ was defined on a closed curve C, then it was also defined at all points inside C. One way to test for holes is to try to "lasso" them with a closed curve. If every closed curve in the domain can be pulled to a point without hitting a hole, that is, without straying outside the domain, then the domain has no holes. In 3-space, we need the same condition to be satisfied; we must be able to pull every closed curve to a point, like a lasso, without straying outside the domain.

Example 6   Decide if the following vector fields are path-independent and whether or not the curl test applies.

(a) $\vec{F} = \dfrac{x\vec{i} + y\vec{j} + z\vec{k}}{(x^2 + y^2 + z^2)^{3/2}}$ (b) $\vec{G} = \dfrac{-y\vec{i} + x\vec{j}}{x^2 + y^2}$

Solution   (a) Suppose $f = -(x^2 + y^2 + z^2)^{-1/2}$. Then $f_x = x(x^2 + y^2 + z^2)^{-3/2}$ and similarly grad $f = \vec{F}$. Thus, $\vec{F}$ is a gradient field and therefore path-independent. Calculations show curl $\vec{F} = \vec{0}$. The domain of $\vec{F}$ is all of 3-space minus the origin, and any closed curve in the domain can be pulled to a point without leaving the domain. Thus, the curl test applies.

(b) Let C be the circle $x^2 + y^2 = 1, z = 0$ traversed counterclockwise when viewed from the positive z-axis. The vector field is everywhere tangent to this curve and of magnitude 1, so

$$\int_C \vec{G} \cdot d\vec{r} = \|\vec{G}\| \cdot \text{Length of curve} = 1 \cdot 2\pi = 2\pi.$$

Since the line integral around this closed curve is nonzero, $\vec{G}$ is path-dependent. Computations show curl $\vec{G} = \vec{0}$. However, the domain of $\vec{G}$ is all of 3-space minus the z-axis, and it does not satisfy the curl test domain criterion. For example, the circle, C, is lassoed around the z-axis, and cannot be pulled to a point without hitting the z-axis. Thus the curl test does not apply.

Problems for Section 18.4

1. Example 1 on page 367 showed that the vector field in Figure 18.48 could not be a gradient field by showing that it is not path-independent. Here is another way to see the same thing. Suppose that the vector field were the gradient of a function f. Draw and label a diagram showing what the contours of f would have to look like, and explain why it would not be possible for f to have a single value at any given point.

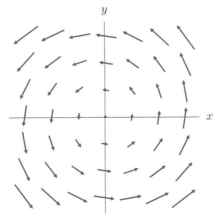

Figure 18.48

2. Repeat Problem 1 for the vector field in Problem 13 on page 365.

3. Find f if grad $f = 2xy\vec{i} + (x^2 + 8y^3)\vec{j}$

4. Find f if grad $f = (yze^{xyz} + z^2\cos(xz^2))\vec{i} + xze^{xyz}\vec{j} + (xye^{xyz} + 2xz\cos(xz^2))\vec{k}$

For Problems 5–6, decide whether the given vector field is the gradient of a function f. If so, find such an f. If not, explain why not.

5. $y\vec{i} + y\vec{j}$

6. $(x^2 + y^2)\vec{i} + 2xy\vec{j}$

Do the vector fields in Problems 7–10 have potential functions, given that $\vec{F} = $ grad f? If so, compute them.

7. $\vec{F} = (2xy^3 + y)\vec{i} + (3x^2y^2 + x)\vec{j}$

8. $\vec{F} = \dfrac{\vec{i}}{x} + \dfrac{\vec{j}}{y} + \dfrac{\vec{k}}{xy}$

9. $\vec{F} = \dfrac{\vec{i}}{x} + \dfrac{\vec{j}}{y} + \dfrac{\vec{k}}{z}$

10. $\vec{F} = 2x\cos(x^2 + z^2)\vec{i} + \sin(x^2 + z^2)\vec{j} + 2z\cos(x^2 + z^2)\vec{k}$

11. Consider the vector field $\vec{F} = y\vec{i}$.

    (a) Sketch $\vec{F}$ and hence decide the sign of the circulation of $\vec{F}$ around the unit circle centered at the origin and traversed counterclockwise.

    (b) Use Green's Theorem to compute the circulation in part (a) exactly.

12. Suppose $\vec{F} = x\vec{j}$. Show that the line integral of $\vec{F}$ around a closed curve in the xy-plane, oriented as in Green's Theorem, measures the area of the region enclosed by the curve.

Use the result of Problem 12 to calculate the area of the region within the parameterized curves in Problems 13–15. In each case, sketch the curve.

13. The ellipse $x^2/a^2 + y^2/b^2 = 1$ parameterized by $x = a\cos t$, $y = b\sin t$, for $0 \le t \le 2\pi$.

14. The hypocycloid $x^{2/3} + y^{2/3} = a^{2/3}$ parameterized by $x = a\cos^3 t$, $y = a\sin^3 t$, $0 \le t \le 2\pi$.

15. The folium of Descartes, $x^3 + y^3 = 3xy$, parameterized by $x = \dfrac{3t}{1+t^3}$, $y = \dfrac{3t^2}{1+t^3}$, for $0 \le t < \infty$.

16. Suppose that R is the region inside the top half of the unit circle, C, centered at the origin and that we want to compute the double integral

$$\int_R (2x - 2y)e^{x^2 + y^2}\, dA$$

(a) Explain why converting this integral to an iterated integral in Cartesian coordinates does not help us compute it.

(b) Since iterated integration fails, we use a numerical method. Show how Green's Theorem can be used to convert the integral to a one-variable integral. Then evaluate the one-variable integral, using ordinary numerical methods.

18.5 PROOF OF GREEN'S THEOREM

In this section we will give a proof of Green's Theorem based on the change of variables formula for double integrals. Assume the vector field $\vec{F}$ is given in components by

$$\vec{F}(x,y) = F_1(x,y)\vec{i} + F_2(x,y)\vec{j}.$$

Proof for Rectangles

We prove Green's Theorem first when R is a rectangular region, as shown in Figure 18.49. The line integral in Green's theorem can be written as

$$\int_C \vec{F} \cdot d\vec{r} = \int_{C_1} \vec{F} \cdot d\vec{r} + \int_{C_2} \vec{F} \cdot d\vec{r} + \int_{C_3} \vec{F} \cdot d\vec{r} + \int_{C_4} \vec{F} \cdot d\vec{r}$$

$$= \int_a^b F_1(x,c)\,dx + \int_c^d F_2(b,y)\,dy - \int_a^b F_1(x,d)\,dx - \int_c^d F_2(a,y)\,dy$$

$$= \int_c^d (F_2(b,y) - F_2(a,y))\,dy + \int_a^b (-F_1(x,d) + F_1(x,c))\,dx.$$

On the other hand, the double integral in Green's theorem can be written as an iterated integral. We evaluate the inside integral using the Fundamental Theorem of Calculus.

$$\int_R \left(\frac{\partial F_2}{\partial x} - \frac{\partial F_1}{\partial y}\right) dx\,dy = \int_R \frac{\partial F_2}{\partial x}\,dx\,dy + \int_R -\frac{\partial F_1}{\partial y}\,dx\,dy$$

$$= \int_c^d \int_a^b \frac{\partial F_2}{\partial x}\,dx\,dy + \int_a^b \int_c^d -\frac{\partial F_1}{\partial y}\,dy\,dx$$

$$= \int_c^d (F_2(b,y) - F_2(a,y))\,dy + \int_a^b (-F_1(x,d) + F_1(x,c))\,dx.$$

Since the line integral and the double integral are equal, we have proved Green's theorem for rectangles.

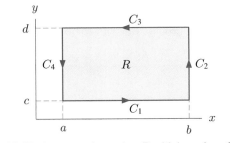

Figure 18.49: A rectangular region R with boundary C broken
into C_1, C_2, C_3, and C_4

Proof for Regions Parameterized by Rectangles

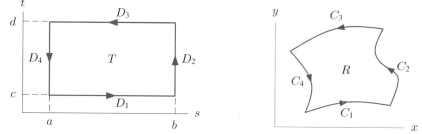

Figure 18.50: A curved region R in the xy-plane corresponding to a rectangular region T in the st-plane

Now we prove Green's Theorem for a region R which can be transformed into a rectangular region. Suppose we have a smooth change of coordinates

$$x = x(s, t), \qquad y = y(s, t).$$

Consider a curved region R in the xy-plane corresponding to a rectangular region T in the st-plane, as in Figure 18.50. We suppose that the change of coordinates is one-to-one on the interior of T.

We prove Green's theorem for R using Green's theorem for T and the change of variables formula for double integrals given on page 267. First we express the line integral around C

$$\int_C \vec{F} \cdot d\vec{r},$$

as a line integral in the st-plane around the rectangle $D = D_1 + D_2 + D_3 + D_4$. In vector notation, the change of coordinates is

$$\vec{r} = \vec{r}(s, t) = x(s, t)\vec{i} + y(s, t)\vec{j}$$

and so

$$\vec{F} \cdot d\vec{r} = \vec{F}(\vec{r}(s, t)) \cdot \frac{\partial \vec{r}}{\partial s} \, ds + \vec{F}(\vec{r}(s, t)) \cdot \frac{\partial \vec{r}}{\partial t} \, dt.$$

We define a vector field $\vec{G}$ on the st-plane with components

$$G_1 = \vec{F} \cdot \frac{\partial \vec{r}}{\partial s} \quad \text{and} \quad G_2 = \vec{F} \cdot \frac{\partial \vec{r}}{\partial t}.$$

Then, if $\vec{u}$ is the position vector of a point in the st-plane, we have $\vec{F} \cdot d\vec{r} = G_1 \, ds + G_2 \, dt = \vec{G} \cdot d\vec{u}$. Problem 5 at the end of this section asks you to show that the formula for line integrals along parameterized paths leads to the following result:

$$\int_C \vec{F} \cdot d\vec{r} = \int_D \vec{G} \cdot d\vec{u}.$$

In addition, using the product rule and chain rule we can show that

$$\frac{\partial G_2}{\partial s} - \frac{\partial G_1}{\partial t} = \left(\frac{\partial F_2}{\partial x} - \frac{\partial F_1}{\partial y} \right) \begin{vmatrix} \frac{\partial x}{\partial s} & \frac{\partial y}{\partial s} \\ \frac{\partial x}{\partial t} & \frac{\partial y}{\partial t} \end{vmatrix}.$$

(See Problem 6 at the end of this section.) Hence, by the change of variables formula for double integrals on page 267,

$$\int_R \left(\frac{\partial F_2}{\partial x} - \frac{\partial F_1}{\partial y} \right) dx \, dy = \int_T \left(\frac{\partial F_2}{\partial x} - \frac{\partial F_1}{\partial y} \right) \begin{vmatrix} \frac{\partial x}{\partial s} & \frac{\partial y}{\partial s} \\ \frac{\partial x}{\partial t} & \frac{\partial y}{\partial t} \end{vmatrix} ds \, dt = \int_T \left(\frac{\partial G_2}{\partial s} - \frac{\partial G_1}{\partial t} \right) ds \, dt.$$

Thus we have shown that

$$\int_C \vec{F} \cdot d\vec{r} = \int_D \vec{G} \cdot d\vec{u}$$

and that

$$\int_R \left(\frac{\partial F_2}{\partial x} - \frac{\partial F_1}{\partial y} \right) dx \, dy = \int_T \left(\frac{\partial G_2}{\partial s} - \frac{\partial G_1}{\partial t} \right) ds \, dt.$$

The integrals on the right are equal, by Green's Theorem for rectangles; hence the integrals on the left are equal as well, which is Green's Theorem for the region R.

Pasting Regions Together

Lastly we show that Green's Theorem holds for a region formed by pasting together regions which can be transformed into rectangles. Figure 18.51 shows two regions R_1 and R_2 that fit together to form a region R. We break the boundary of R into C_1, the part shared with R_1, and C_2, the part shared with R_2. We let C be the part of the the boundary of R_1 which it shares with R_2. So

$$\text{Boundary of } R = C_1 + C_2, \quad \text{Boundary of } R_1 = C_1 + C, \quad \text{Boundary of } R_2 = C_2 + (-C).$$

Note that when the curve C is considered as part of the boundary of R_2, it receives the opposite orientation from the one it receives as the boundary of R_1. Thus

$$\int_{\text{Boundary of } R_1} \vec{F} \cdot d\vec{r} + \int_{\text{Boundary of } R_2} \vec{F} \cdot d\vec{r} = \int_{C_1+C} \vec{F} \cdot d\vec{r} + \int_{C_2+(-C)} \vec{F} \cdot d\vec{r}$$

$$= \int_{C_1} \vec{F} \cdot d\vec{r} + \int_C \vec{F} \cdot d\vec{r} + \int_{C_2} \vec{F} \cdot d\vec{r} - \int_C \vec{F} \cdot d\vec{r}$$

$$= \int_{C_1} \vec{F} \cdot d\vec{r} + \int_{C_2} \vec{F} \cdot d\vec{r}$$

$$= \int_{\text{Boundary of } R} \vec{F} \cdot d\vec{r}.$$

So, applying Green's Theorem for R_1 and R_2, we get

$$\int_R \left(\frac{\partial F_2}{\partial x} - \frac{\partial F_1}{\partial y} \right) dx \, dy = \int_{R_1} \left(\frac{\partial F_2}{\partial x} - \frac{\partial F_1}{\partial y} \right) dx \, dy + \int_{R_2} \left(\frac{\partial F_2}{\partial x} - \frac{\partial F_1}{\partial y} \right) dx \, dy$$

$$= \int_{\text{Boundary of } R_1} \vec{F} \cdot d\vec{r} + \int_{\text{Boundary of } R_2} \vec{F} \cdot d\vec{r}$$

$$= \int_{\text{Boundary of } R} \vec{F} \cdot d\vec{r},$$

which is Green's Theorem for R. Thus, we have proved Green's Theorem for any region formed by pasting together regions that are smoothly parameterized by rectangles.

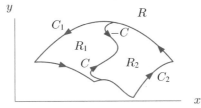

Figure 18.51: Two regions R_1 and R_2
pasted together to form a region R

Example 1   Let R be the annulus (ring) centered at the origin with inner radius 1 and outer radius 2. Using polar coordinates, show that the proof of Green's Theorem applies to R. See Figure 18.52.

Solution   In polar coordinates, $x = r \cos t$ and $y = r \sin t$, the annulus corresponds to the rectangle in the rt-plane $1 \le r \le 2$, $0 \le t \le 2\pi$. The sides $t = 0$ and $t = 2\pi$ are pasted together in the xy-plane along the x-axis; the other two sides become the inner and outer circles of the annulus. Thus R is formed by pasting the ends of a rectangle together.

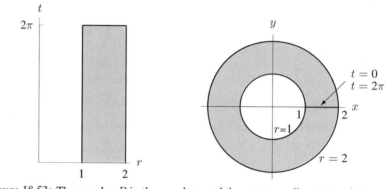

Figure 18.52: The annulus R in the xy-plane and the corresponding rectangle $1 \le r \le 2$,
$0 \le t \le 2\pi$ in the rt-plane

Problems for Section 18.5

1. Let R be the annulus centered at $(-1, 2)$ with inner radius 2 and outer radius 3. Show that R can be parameterized by a rectangle.

2. Let R be the region under the first arc of the graph of the sine function. Show that R can be parameterized by a rectangle.

3. Let $f(x)$ and $g(x)$ be two smooth functions, and suppose that $f(x) \le g(x)$ for $a \le x \le b$. Let R be the region $f(x) \le y \le g(x)$, $a \le x \le b$.

    (a) Sketch an example of such a region.
    (b) For a constant x_0, parameterize the vertical line segment in R where $x = x_0$. Choose your parameterization so that the parameter starts at 0 and ends at 1.
    (c) By putting together the parameterizations in part (b) for different values of x_0, show that R can be parameterized by a rectangle.

4. Let $f(y)$ and $g(y)$ be two smooth functions, and suppose that $f(y) \le g(y)$ for $c \le y \le d$. Let R be the region $f(y) \le x \le g(y)$, $c \le y \le d$.

    (a) Sketch an example of such a region.
    (b) For a constant y_0, parameterize the horizontal line segment in R where $y = y_0$. Choose your parameterization so that the parameter starts at 0 and ends at 1.
    (c) By putting together the parameterizations in part (b) for different values of y_0, show that R can be parameterized by a rectangle.

5. Use the formula for calculating line integrals by parameterization to prove the statement on page 378:

$$\int_C \vec{F} \cdot d\vec{r} = \int_D \vec{G} \cdot d\vec{u} .$$

6. Use the product rule and the chain rule to prove the formula on page 378:

$$\frac{\partial G_2}{\partial s} - \frac{\partial G_1}{\partial t} = \left(\frac{\partial F_2}{\partial x} - \frac{\partial F_1}{\partial y} \right) \begin{vmatrix} \frac{\partial x}{\partial s} & \frac{\partial y}{\partial s} \\ \frac{\partial x}{\partial t} & \frac{\partial y}{\partial t} \end{vmatrix} .$$

REVIEW PROBLEMS FOR CHAPTER EIGHTEEN

For Problems 1–2, consider the vector field $\vec{F}$ shown. Say whether you expect the line integral $\int_C \vec{F} \cdot d\vec{r}$ to be positive, negative, or zero along
(a) A (b) C_1, C_2, C_3, C_4 (c) C, the closed curve consisting of all the C's together.

1.

2.

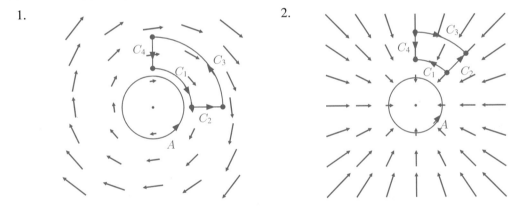

For Problems 3–5, compute $\int_C \vec{F} \cdot d\vec{r}$ for the given $\vec{F}$ and C.

3. $\vec{F} = (x^2 - y)\vec{i} + (y^2 + x)\vec{j}$ and C is the parabola $y = x^2 + 1$ traversed from $(0, 1)$ to $(1, 2)$.

4. $\vec{F} = (3x - 2y)\vec{i} + (y + 2z)\vec{j} - x^2\vec{k}$ and C is the path consisting of the line joining the point $(0, 0, 0)$ to $(1, 1, 1)$.

5. $\vec{F} = (2x - y + 4)\vec{i} + (5y + 3x - 6)\vec{j}$ and C is the triangle with vertices $(0, 0), (3, 0), (3, 2)$ traversed counterclockwise.

Are the statements in Problems 6–9 true or false? Explain why or give a counterexample.

6. $\int_C \vec{F} \cdot d\vec{r}$ is a vector.

7. $\int_C \vec{F} \cdot d\vec{r} = \vec{F}(Q) - \vec{F}(P)$ when P and Q are the endpoints of C.

8. The fact that the line integral of a vector field $\vec{F}$ is zero around the unit circle $x^2 + y^2 = 1$ means that $\vec{F}$ must be a gradient vector field.

9. Suppose C_1 is the unit square joining the points $(0,0)$, $(1,0)$, $(1,1)$, $(0,1)$ oriented clockwise and C_2 is the same square but traversed twice in the opposite direction. If $\int_{C_1} \vec{F} \cdot d\vec{r} = 3$, then $\int_{C_2} \vec{F} \cdot d\vec{r} = -6$.

10. Let $\vec{F} = x\vec{i} + y\vec{j}$, and let C_1 be the line joining the point $(1,0)$ to the point $(0,2)$ and let C_2 be the line joining the point $(0,2)$ to the point $(-1,0)$. Is $\int_{C_1} \vec{F} \cdot d\vec{r} = -\int_{C_2} \vec{F} \cdot d\vec{r}$? Explain.

11. What is the value of $\int_C \vec{F} \cdot d\vec{r}$ if C is an oriented curve that runs from the point $(2,-6)$ to the point $(4,4)$ and if $\vec{F} = 6\vec{i} - 7\vec{j}$?

12. Suppose P and Q both lie on the same contour of f. What can you say about the total change in f from P to Q? Explain your answer in terms of $\int_C \operatorname{grad} f \cdot d\vec{r}$ where C is a portion of the contour that goes from P to Q.

13. In this problem, we see how the Fundamental Theorem for Line Integrals can be derived from the Fundamental Theorem for ordinary definite integrals. Suppose that $(x(t), y(t))$, for $a \le t \le b$, is a parameterization of C, with endpoints $P = (x(a), y(a))$ and $Q = (x(b), y(b))$. The values of f along C are given by the single variable function $h(t) = f(x(t), y(t))$.

    (a) Use the chain rule to show that

$$h'(t) = f_x(x(t), y(t))x'(t) + f_y(x(t), y(t))y'(t).$$

    (b) Use the Fundamental Theorem of Calculus applied to $h(t)$ to show

$$\int_C \operatorname{grad} f \cdot d\vec{r} = f(Q) - f(P).$$

14. Let $\vec{F}(x, y)$ be the path-independent vector field in Figure 18.53. The vector field $\vec{F}$ associates with each point a unit vector pointing radially outward. The curves $C_1, C_2, \ldots, C_7$ have the directions shown. Consider the line integrals $\int_{C_i} \vec{F} \cdot d\vec{r}$, $i = 1, \ldots, 7$. Without computing any integrals

    (a) List all the line integrals which you expect to be zero.
    (b) List all the line integrals which you expect to be negative.
    (c) Arrange the positive line integrals in what you believe to be ascending order.

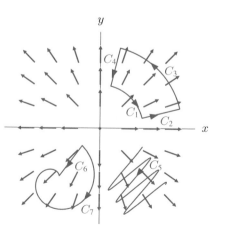

Figure 18.53

15. A *free vortex* circulating about the origin in the xy-plane (or about the z-axis in 3-space) has vector field $\vec{v} = K(x^2 + y^2)^{-1}(-y\vec{i} + x\vec{j})$ where K is a constant. The Rankine model of a tornado hypothesizes an inner core that rotates at constant angular velocity, surrounded by a free vortex. Suppose that the inner core has radius 100 meters and that $\|\vec{v}\| = 3 \cdot 10^5$ meters/hr at a distance of 100 meters from the center.

(a) Assuming that the tornado rotates counterclockwise (viewed from above the xy-plane) and that $\vec{v}$ is continuous, determine ω and K such that

$$\vec{v} = \begin{cases} \omega(-y\vec{i} + x\vec{j}) & \text{if } \sqrt{x^2 + y^2} < 100 \\ K(x^2 + y^2)^{-1}(-y\vec{i} + x\vec{j}) & \text{if } \sqrt{x^2 + y^2} \geq 100. \end{cases}$$

(b) Sketch the vector field $\vec{v}$.
(c) Compute the circulation of $\vec{v}$ about the circle of radius r centered at the origin, traversed counterclockwise.

16. Figure 18.54 shows the tangential velocity as a function of radius for the tornado that hit Dallas on April 2, 1957. Use it and Problem 15 to estimate K and ω for the Rankine model of this tornado.[4]

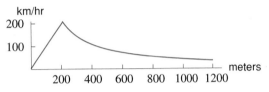

Figure 18.54

17. A *central vector field* is a vector field whose direction is always toward (or away from) a fixed point O (the center) and whose magnitude at a point P is a function only of the distance from P to O. In two dimensions this means that the vector field has constant magnitude on circles centered at O. The gravitational and electrical fields of spherically symmetric sources are both central fields.

(a) Sketch an example of a central vector field.
(b) Suppose that the central field $\vec{F}$ is a gradient field, that is, $\vec{F} = \text{grad } f$. What must be the shape of the contours of f? Sketch some contours for this case.
(c) Is every gradient field a central vector field? Explain.
(d) In Figure 18.55, two paths are shown between the points Q and P. Assuming that the three circles C_1, C_2, and C_3 are centered at O, explain why the work done by a central vector field $\vec{F}$ is the same for either path.
(e) It is in fact true that every central vector field is a gradient field. Use an argument suggested by Figure 18.55 to explain why any central vector field must be path-independent.

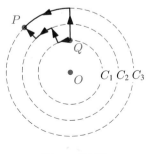

Figure 18.55

---

[4] Adapted from *Encyclopedia Britannica, Macropedia*, Vol. 16, page 477, "Climate and the Weather", Tornadoes and Waterspouts, 1991.

18.   Consider the vector field

$$\vec{F} = (-y^3 + y \sin(xy))\vec{i} + (4x(1 - y^2) + x \sin(xy))\vec{j}$$

defined on the disk D of radius 5 centered at the origin in the plane. Consider the line integral $\int_C \vec{F} \cdot d\vec{r}$, where C is some closed curve contained in D. For which C is the value of this integral the largest? [Hint: Assume C is a closed curve, made up of smooth pieces and never crossing itself, and oriented counterclockwise.]

CHAPTER NINETEEN

FLUX INTEGRALS

In the previous chapter we saw how to integrate vector fields along curves. In this chapter we shall define a new sort of integral, the flux integral, which goes over a surface rather than along a curve. If we view a vector field as representing the velocity of a fluid flow, the flux integral tells us about the rate at which fluid is flowing through the surface. In addition, the flux integral appears in the theory of electricity and magnetism.

19.1   THE IDEA OF A FLUX INTEGRAL

Flow Through a Surface

Imagine water flowing through a fishing net stretched across a stream. Suppose we want to measure the flow rate of water through the net, that is, the volume of fluid that passes through the surface per unit time. (See Figure 19.1.) This flow rate is called the *flux* of the fluid through the surface. We can also compute the flux of vector fields, such as electric and magnetic fields, where no flow is actually taking place.

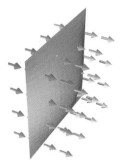

Figure 19.1: Flux measures rate of flow
through a surface

Orientation of a Surface

Before computing the flux of a vector field through a surface, we need to decide which direction of flow through the surface is the positive direction; this is described as choosing an orientation. [1]

> At each point on a smooth surface there are two unit normals, one in each direction. **Choosing an orientation** means picking one of these normals at every point of the surface in a continuous way. The normal vector in the direction of the orientation is denoted by $\vec{n}$. For a closed surface, we usually choose the outward orientation.

We say the flux through a piece of surface is positive if the flow is in the direction of the orientation and negative if it is in the opposite direction. (See Figure 19.2.)

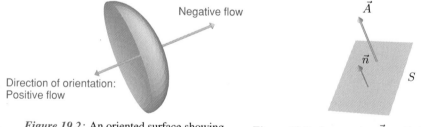

Figure 19.2: An oriented surface showing
directions of positive and negative flow

Figure 19.3: Area vector $\vec{A} = \vec{n}\,A$ of flat surface
with area A and orientation $\vec{n}$

The Area Vector

The flux through a flat surface depends both on the area of the surface and its orientation. Thus, it is useful to represent its area by a vector as shown in Figure 19.3.

---

[1] Although we will not study them, there are a few surfaces for which this cannot be done. See page 392.

> The **area vector** of a flat, oriented surface is a vector $\vec{A}$ such that
> - The magnitude of $\vec{A}$ is the area of the surface.
> - The direction of $\vec{A}$ is the direction of the orientation vector $\vec{n}$.

The Flux of a Constant Vector Field Through a Flat Surface

Suppose the velocity vector field, $\vec{v}$, of a fluid is constant and $\vec{A}$ is the area vector of a flat surface. The flux through this surface is the volume of fluid that flows through in one unit of time. The volume of the skewed box in Figure 19.4 has cross-sectional area $\|\vec{A}\|$ and height $\|\vec{v}\| \cos\theta$, so its volume is $\left(\|\vec{v}\| \cos\theta\right)\|\vec{A}\| = \vec{v} \cdot \vec{A}$. Thus we have the following result:

> If $\vec{v}$ is constant and $\vec{A}$ is the area vector of a flat surface, then
>
> $$\text{Flux through surface} = \vec{v} \cdot \vec{A}.$$

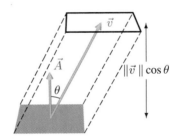

Figure 19.4: Flux of $\vec{v}$ through a surface with area vector $\vec{A}$ is the volume of this skewed box

---

Example 1   Water is flowing down a cylindrical pipe 2 cm in radius with a velocity of 3 cm/sec. Find the flux of the velocity vector field through the ellipse-shaped region shown in Figure 19.5. The normal to the ellipse makes an angle of θ with the direction of flow and the area of the ellipse is $4\pi/(\cos\theta)$ cm^2.

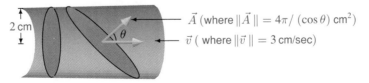

Figure 19.5: Flux through ellipse-shaped region across a cylindrical pipe

Solution   There are two ways to approach this problem. One is to use the formula we just derived which gives

$$\text{Flux through ellipse} = \vec{v} \cdot \vec{A} = \|\vec{v}\|\|\vec{A}\| \cos\theta = 3(\text{Area of ellipse}) \cos\theta$$

$$= 3\left(\frac{4\pi}{\cos\theta}\right) \cos\theta = 12\pi \text{ cm}^3/\text{sec}.$$

The second way is to notice that the flux through the ellipse is equal to the flux through the circle perpendicular to the pipe in Figure 19.5. Since the flux is the rate at which water is flowing down the pipe, we have

$$\text{Flux through circle} = \begin{matrix} \text{Velocity} \\ \text{of water} \end{matrix} \times \begin{matrix} \text{Area of} \\ \text{circle} \end{matrix} = \left(3\,\frac{\text{cm}}{\text{sec}}\right)(\pi 2^2\,\text{cm}^2) = 12\pi\,\text{cm}^3/\text{sec}.$$

When the vector field is not constant or the surface is not flat, we divide the surface up into small, almost flat pieces such that the vector field is approximately constant on each one, as follows.

The Flux Integral

To calculate the flux of a vector field $\vec{F}$ which is not necessarily constant through a curved, oriented surface S, we divide the surface into a patchwork of small, approximately flat pieces (like a wireframe representation of the surface) as shown in Figure 19.6. Suppose a particular patch has area ΔA. We pick an orientation vector $\vec{n}$ at a point on the patch and define the area vector of the patch, $\Delta\vec{A}$, as

$$\Delta\vec{A} = \vec{n}\,\Delta A.$$

(See Figure 19.6.) If the patches are small enough, we can assume that $\vec{F}$ is approximately constant on each piece. Then we know that

$$\text{Flux through patch} \approx \vec{F}\cdot\Delta\vec{A}$$

and so

$$\text{Flux through whole surface} \approx \sum \vec{F}\cdot\Delta\vec{A},$$

where the sum adds the fluxes through all the small pieces. As each patch becomes smaller and $\|\Delta\vec{A}\| \to 0$, the approximation gets better and we get

$$\text{Flux through } S = \lim_{\|\Delta\vec{A}\|\to 0} \sum \vec{F}\cdot\Delta\vec{A}.$$

Thus, we make the following definition:

The **flux integral** of the vector field $\vec{F}$ through the oriented surface S is

$$\int_S \vec{F}\cdot d\vec{A} = \lim_{\|\Delta\vec{A}\|\to 0} \sum \vec{F}\cdot\Delta\vec{A}.$$

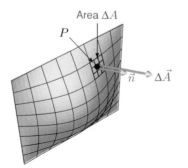

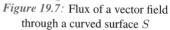

Figure 19.6: Surface S divided into small, almost flat pieces, showing a typical orientation vector $\vec{n}$ and area vector $\Delta\vec{A}$

Figure 19.7: Flux of a vector field through a curved surface S

In computing a flux integral, we have to divide the surface up in a reasonable way, or the limit might not exist. In practice this problem seldom arises; however, one way to avoid it is to use the method for calculating flux integrals introduced in Section 19.3 as the definition of the flux integral.

Flux and Fluid Flow

If $\vec{v}$ is the velocity vector field of a fluid, we have

$$
\boxed{\begin{array}{ccc} \text{Rate fluid flows} \\ \text{through surface } S \end{array} = \begin{array}{c} \text{Flux of } \vec{v} \\ \text{through } S \end{array} = \int_S \vec{v} \cdot d\vec{A}}
$$

The rate of fluid flow is measured in units of volume per unit time.

Example 2   Find the flux of the vector field shown in Figure 19.8, and given by

$$
\vec{B}(x, y, z) = \frac{-y\vec{i} + x\vec{j}}{x^2 + y^2},
$$

through the square S of side 2 shown in Figure 19.9, oriented in the $\vec{j}$ direction.

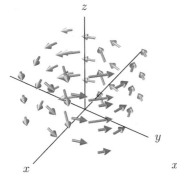
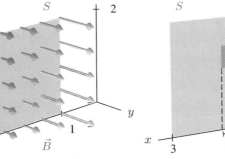
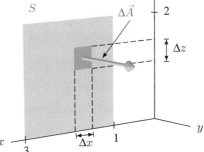

Figure 19.8: The vector field $\vec{B}(x, y, z) = \frac{-y\vec{i} + x\vec{j}}{x^2 + y^2}$

Figure 19.9: Flux of $\vec{B}$ through the square S of side 2 in xy-plane and oriented in $\vec{j}$ direction.

Figure 19.10: A small patch of surface with area $\|\Delta\vec{A}\| = \Delta x \Delta z$

Solution   Consider a small rectangular patch with area vector $\Delta\vec{A}$ in S, with sides Δx and Δz so that $\|\Delta\vec{A}\| = \Delta x\, \Delta z$. Since $\Delta\vec{A}$ points in the $\vec{j}$ direction we have $\Delta\vec{A} = \vec{j}\,\Delta x \Delta z$. (See Figure 19.10.)

At the point $(x, 0, z)$ in S, substituting $y = 0$ into $\vec{B}$ gives $\vec{B}(x, 0, z) = (1/x)\vec{j}$. Thus, we have

$$
\text{Flux through small patch} \approx \vec{B} \cdot \Delta\vec{A} = \left(\frac{1}{x}\vec{j}\right) \cdot (\vec{j}\,\Delta x \Delta z) = \frac{1}{x}\Delta x\, \Delta z.
$$

Therefore,

$$
\text{Flux through surface} = \int_S \vec{B} \cdot d\vec{A} = \lim_{\|\Delta\vec{A}\| \to 0} \sum \vec{B} \cdot \Delta\vec{A} = \lim_{\substack{\Delta x \to 0 \\ \Delta z \to 0}} \sum \frac{1}{x} \Delta x\, \Delta z.
$$

This last expression is a Riemann sum for the double integral $\int_R \frac{1}{x}\, dA$, where R is the square $1 \le x \le 3$, $0 \le z \le 2$. Thus,

$$
\text{Flux through surface} = \int_S \vec{B} \cdot d\vec{A} = \int_R \frac{1}{x}\, dA = \int_0^2 \int_1^3 \frac{1}{x}\, dx\, dz = 2\ln 3.
$$

The result is positive since the vector field is passing through the surface in the positive direction.

Example 3   Each of the vector fields in Figure 19.11 consists entirely of vectors parallel to the xy-plane, and is constant in the z direction (that is, the vector field looks the same in any plane parallel to the xy-plane). For each one, say whether you expect the flux through a closed surface surrounding the origin to be positive, negative, or zero. In part (a) the surface is a closed cube with faces parallel to the axes; in parts (b) and (c) the surface is a closed cylinder. In each case we choose the outward orientation. (See Figure 19.12.)

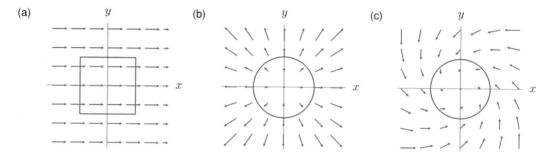

Figure 19.11: Flux of a vector field through the closed surfaces whose cross-sections are shown in the xy-plane

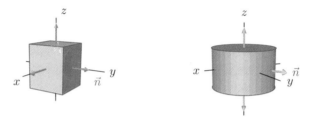

Figure 19.12: The closed cube and closed cylinder, both oriented outward

Solution   (a)   Since the vector field appears to be parallel to the faces of the cube which are perpendicular to the y- and z-axes, we expect the flux through these faces to be zero. The fluxes through the two faces perpendicular to the x-axis appear to be equal in magnitude and opposite in sign, so we expect the net flux to be zero.

(b)   Since the top and bottom of the cylinder are parallel to the flow, the flux through them is zero. On the round surface of the cylinder, $\vec{v}$ and $\Delta\vec{A}$ appear to be everywhere parallel and in the same direction, so we expect each term $\vec{v} \cdot \Delta\vec{A}$ to be positive, and therefore the flux integral $\int_S \vec{v} \cdot d\vec{A}$ to be positive.

(c)   As in part (b), the flux through the top and bottom of the cylinder is zero. In this case $\vec{v}$ and $\Delta\vec{A}$ are not parallel on the round surface of the cylinder, but since the fluid appears to be flowing inwards as well as swirling, we expect each term $\vec{v} \cdot \Delta\vec{A}$ to be negative, and therefore the flux integral to be negative.

Calculating Flux Integrals Using $d\vec{A} = \vec{n}\,dA$

For a small patch of surface ΔS with normal $\vec{n}$ and area ΔA, the area vector is $\Delta\vec{A} = \vec{n}\,\Delta A$. The next example shows how we can use this relationship to compute a flux integral.

Example 4 An electric charge q is placed at the origin in 3-space. The resulting electric field $\vec{E}(\vec{r})$ at the point with position vector $\vec{r}$ is given by

$$\vec{E}(\vec{r}) = q\frac{\vec{r}}{\|\vec{r}\|^3}, \qquad \vec{r} \neq \vec{0}.$$

Find the flux of $\vec{E}$ out of the sphere of radius R centered at the origin. (See Figure 19.13.)

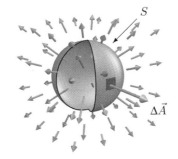

Figure 19.13: Flux of $\vec{E} = q\vec{r}/\|\vec{r}\|^3$ through the surface of a sphere of radius R centered at the origin

Solution This vector field points radially outward from the origin in the same direction as $\vec{n}$. Thus, since $\vec{n}$ is a unit vector,

$$\vec{E} \cdot \Delta\vec{A} = \vec{E} \cdot \vec{n}\,\Delta A = \|\vec{E}\|\,\Delta A.$$

On the sphere, $\|\vec{E}\| = q/R^2$, so

$$\int_S \vec{E} \cdot d\vec{A} = \lim_{\|\Delta\vec{A}\| \to 0} \sum \vec{E} \cdot \Delta\vec{A} = \lim_{\Delta A \to 0} \sum \frac{q}{R^2}\,\Delta A = \frac{q}{R^2}\lim_{\Delta A \to 0}\sum \Delta A.$$

The last sum approximates the surface area of the sphere. In the limit as the subdivisions get finer we have

$$\lim_{\Delta A \to 0}\sum \Delta A = \text{Surface area of sphere.}$$

Thus, the flux is given by

$$\int_S \vec{E} \cdot d\vec{A} = \frac{q}{R^2}\lim_{\Delta A \to 0}\sum \Delta A = \frac{q}{R^2} \cdot \text{Surface area of sphere} = \frac{q}{R^2}(4\pi R^2) = 4\pi q.$$

This result is known as Gauss's law.

Instead of using Riemann sums, we often write $d\vec{A} = \vec{n}\,dA$, as in the next example.

Example 5 Suppose S is the surface of the cube bounded by the six planes $x = \pm 1$, $y = \pm 1$, and $z = \pm 1$. Compute the flux of the electric field $\vec{E}$ of the previous example outward through S.

Solution It is enough to compute the flux of $\vec{E}$ through a single face, say the top face S_1 defined by $z = 1$, where $-1 \leq x \leq 1$ and $-1 \leq y \leq 1$. By symmetry, the flux of $\vec{E}$ through the other five faces of S must be the same.

On the top face, S_1, we have $d\vec{A} = \vec{k}\,dx\,dy$ and

$$\vec{E}(x, y, 1) = q\frac{x\vec{i} + y\vec{j} + \vec{k}}{(x^2 + y^2 + 1)^{3/2}}.$$

The corresponding flux integral is given by

$$\int_{S_1} \vec{E} \cdot d\vec{A} = q \int_{-1}^{1} \int_{-1}^{1} \frac{x\vec{i} + y\vec{j} + \vec{k}}{(x^2 + y^2 + 1)^{3/2}} \cdot \vec{k} \, dx \, dy = q \int_{-1}^{1} \int_{-1}^{1} \frac{1}{(x^2 + y^2 + 1)^{3/2}} \, dx \, dy.$$

Computing this integral numerically shows that

$$\text{Flux through top face} = \int_{S_1} \vec{E} \cdot d\vec{A} \approx 2.0944q.$$

Thus,

$$\text{Total flux of } \vec{E} \text{ out of cube} = \int_{S} \vec{E} \cdot d\vec{A} \approx 6(2.0944q) = 12.5664q.$$

Example 4 on page 391 showed that the flux of $\vec{E}$ through a sphere of radius R centered at the origin is $4\pi q$. Since $4\pi \approx 12.5664$, Example 5 suggests that

$$\text{Total flux of } \vec{E} \text{ out of cube} = 4\pi q.$$

By computing the flux integral in Example 5 exactly, it is possible to verify that the flux of $\vec{E}$ through the cube and the sphere are exactly equal. When we encounter the Divergence Theorem in Chapter 20 we will see why this is so.

Notes on Orientation

Two difficulties can occur in choosing an orientation. The first is that if the surface is not smooth, it may not have a normal vector at every point. For example, a cube does not have a normal vector along its edges. When we have a surface, such as a cube, which is made of a finite number of smooth pieces, we choose an orientation for each piece separately. The best way to do this is usually clear. For example, on the cube we choose the outward orientation on each face. (See Figure 19.14.)

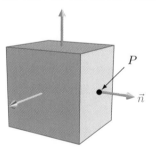

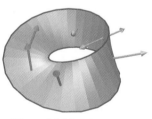

Figure 19.14: The orientation vector field $\vec{n}$ on the cube surface determined by the choice of unit normal vector at the point P

Figure 19.15: The Möbius strip is an example of a non-orientable surface

The second difficulty is that there are some surfaces which cannot be oriented at all, such as the *Möbius strip* in Figure 19.15.

Problems for Section 19.1

1.   Let $\vec{F}(x, y, z) = z\vec{i}$. For each of the surfaces in (a)–(e), say whether the flux of $\vec{F}$ through the surface is positive, negative, or zero. In each case, the orientation of the surface is indicated by the given normal vector.

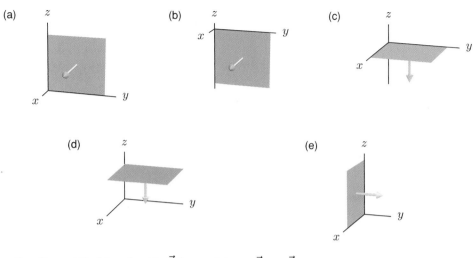

2. Repeat Problem 1 with $\vec{F}(x, y, z) = -z\vec{i} + x\vec{k}$.

3. Repeat Problem 1 with the vector field $\vec{F}(\vec{r}) = \vec{r}$.

4. Arrange the following flux integrals,

$$\int_{S_i} \vec{F} \cdot d\vec{A},$$

with $i = 1, 2, 3, 4$, in ascending order if $\vec{F} = -\vec{i} - \vec{j} + \vec{k}$ and S_i are the following surfaces:

- S_1 is a horizontal square of side 1 with one corner at $(0, 0, 2)$, above the first quadrant of the xy-plane, oriented upward.
- S_2 is a horizontal square of side 1 with one corner at $(0, 0, 3)$, above the third quadrant of the xy-plane, oriented upward.
- S_3 is a square of side $\sqrt{2}$ in the xz-plane with one corner at the origin, one edge along the positive x-axis, one along the negative z-axis, oriented in the negative y- direction.
- S_4 is a square of side $\sqrt{2}$ with one corner at the origin, one edge along the positive y-axis, one corner at $(1, 0, 1)$, oriented upward.

5. Compute the flux of the vector field $\vec{v} = 2\vec{i} + 3\vec{j} + 5\vec{k}$ through each of the rectangular regions in (a)–(d), assuming each is oriented as shown.

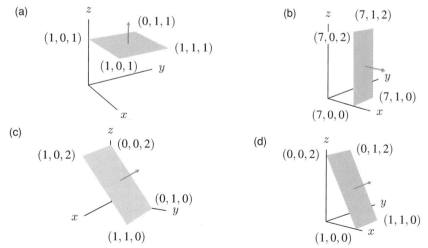

6. Figure 19.16 shows a cross-section of the earth's magnetic field. Say whether the magnetic flux through a horizontal plate, oriented skyward, is positive, negative, or zero if the plate is
    (a)   At the north pole.   (b)   At the south pole.   (c)   On the equator.
    [Note: You may assume that the earth's magnetic and geographic poles coincide.]

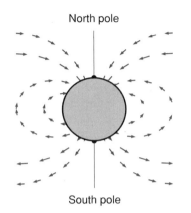

Figure 19.16

7. (a)   What do you think will be the electric flux through the cylindrical surface that is placed as shown in the constant electric field in Figure 19.17? Why?
    (b)   What if the cylinder is placed upright, as shown in Figure 19.18? Explain.

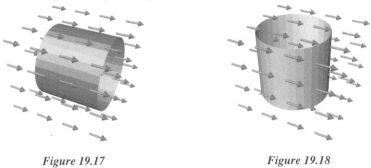

Figure 19.17   *Figure 19.18*

For Problems 8–14, compute the flux integral of the given vector field through the given surface S.

8. $\vec{F} = 2\vec{i}$ and S is a disk of radius 2 on the plane $x + y + z = 2$, oriented upward.

9. $\vec{F} = -y\vec{i} + x\vec{j}$ and S is the square plate in the yz-plane with corners at $(0, 1, 1)$, $(0, -1, 1)$, $(0, 1, -1)$, and $(0, -1, -1)$, oriented in the positive x-direction.

10. $\vec{F} = -y\vec{i} + x\vec{j}$ and S is the disk in the xy-plane with radius 2, oriented upwards and centered at the origin.

11. $\vec{F} = \vec{r}$ and S is the disk of radius 2 parallel to the xy-plane oriented upwards and centered at $(0, 0, 2)$.

12. $\vec{F} = (2 - x)\vec{i}$ and S is the cube whose vertices include the points $(0, 0, 0)$, $(3, 0, 0)$, $(0, 3, 0)$, $(0, 0, 3)$, and oriented outward.

13. $\vec{F} = (x^2 + y^2)\vec{i} + xy\vec{j}$ and S is the square in the xy-plane with corners at $(1, 1, 0)$, $(-1, 1, 0)$, $(1, -1, 0)$, $(-1, -1, 0)$, and oriented upward.

14. $\vec{F} = \vec{r}/r^2$ and S is the sphere of radius R centered at the origin, oriented outwards.

15. Let S be the cube with side length 2, faces parallel to the coordinate planes, and centered at the origin.

    (a) Calculate the total flux of the constant vector field $\vec{v} = -\vec{i} + 2\vec{j} + \vec{k}$ out of S by computing the flux through each face separately.
    (b) Calculate the flux out of S for any constant vector field $\vec{v} = a\vec{i} + b\vec{j} + c\vec{k}$.
    (c) Do your answers in parts (a) and (b) make sense? Explain.

16. Let S be the tetrahedron with vertices at the origin and at $(1, 0, 0)$, $(0, 1, 0)$ and $(0, 0, 1)$.

    (a) Calculate the total flux of the constant vector field $\vec{v} = -\vec{i} + 2\vec{j} + \vec{k}$ out of S by computing the flux through each face separately.
    (b) Calculate the flux out of S in part (a) for any constant vector field $\vec{v}$.
    (c) Do your answers in parts (a) and (b) make sense? Explain.

17. Suppose the z-axis carries a constant electric charge density of λ units of charge per unit length, with $\lambda > 0$, and that $\vec{E}$ is the resulting electric field.

    (a) Sketch the electric field, $\vec{E}$, in the xy-plane, given that

    $$\vec{E}(x, y, z) = 2\lambda \frac{x\vec{i} + y\vec{j}}{x^2 + y^2}.$$

    (b) Compute the flux of $\vec{E}$ outward through the cylinder $x^2 + y^2 = R^2$, for $0 \le z \le h$.

18. Explain why if $\vec{F}$ has constant magnitude on S and is everywhere normal to S and in the direction of orientation, then

    $$\int_S \vec{F} \cdot d\vec{A} = \|\vec{F}\| \cdot \text{Area of } S.$$

19. Let $P(x, y, z)$ be the pressure at the point (x, y, z) in a fluid. Let $\vec{F}(x, y, z) = P(x, y, z)\vec{k}$. Let S be the surface of a body submerged in the fluid. If S is oriented inward, show that $\int_S \vec{F} \cdot d\vec{A}$ is the buoyant force on the body, that is, the force upwards on the body due to the pressure of the fluid surrounding it. [Hint: $\vec{F} \cdot d\vec{A} = P(x, y, z)\vec{k} \cdot d\vec{A} = (P(x, y, z) d\vec{A}) \cdot \vec{k}$.]

20. Consider the function $\rho(x, y, z)$ which gives the electrical charge density at points in space. The vector field $\vec{J}(x, y, z)$ gives the electric current density at any point in space and is defined so that the current through a small area $d\vec{A}$ is given by

    $$\text{Current through small area} \approx \vec{J} \cdot d\vec{A}.$$

    Suppose S is a closed surface enclosing a volume W.

    (a) What do the following integrals represent, in terms of electricity?

    (i) $\displaystyle\int_W \rho \, dV$ (ii) $\displaystyle\int_S \vec{J} \cdot d\vec{A}$

    (b) Using the fact that an electric current through a surface is the rate at which electric charge passes through the surface per unit time, explain why

    $$\int_S \vec{J} \cdot d\vec{A} = -\frac{\partial}{\partial t}\left(\int_W \rho \, dV\right).$$

21. A fluid is flowing along a cylindrical pipe of radius a in the $\vec{i}$ direction. The velocity of the fluid at a radial distance r from the center of the pipe is $\vec{v} = u(1 - r^2/a^2)\vec{i}$.

    (a) What is the significance of the constant u?

(b) What is the velocity of the fluid at the wall of the pipe?

(c) Find the flux through a circular cross-section of the pipe.

22. Suppose a region of 3-space has a temperature which varies from point to point. Let $T(x, y, z)$ be the temperature at a point (x, y, z). Newton's law of cooling says that grad T is proportional to the heat flow vector field, $\vec{F}$, where $\vec{F}$ points in the direction in which heat is flowing and has magnitude equal to the rate of flow of heat.

(a) Suppose $\vec{F} = k \operatorname{grad} T$ for some constant k. What is the sign of k?

(b) Explain why this form of Newton's law of cooling makes sense.

(c) Let W be a region of space bounded by the surface S. Explain why

$$\begin{array}{c} \text{Rate of heat} \\ \text{loss from } W \end{array} = k \int_S (\operatorname{grad} T) \cdot d\vec{A}.$$

23. This problem investigates the behavior of the electric field produced by an infinitely long, straight, uniformly charged wire. (There is no current running through the wire — all charges are fixed.) Assuming that the wire is infinitely long means that we can assume that the electric field is normal to any cylinder that has the wire as an axis, and that the magnitude of the field is constant on any such cylinder. Denote by E_r the magnitude of the electric field due to the wire on a cylinder of radius r. (See Figure 19.19.)

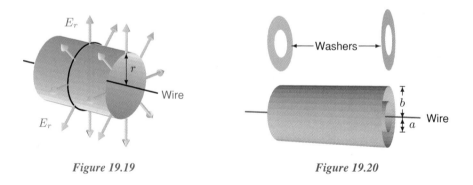

Figure 19.19   *Figure 19.20*

Imagine a closed surface S made up of two cylinders, one of radius a and one of larger radius b, both coaxial with the wire, and the two washers that cap the ends. (See Figure 19.20.) Note that the outward orientation of S means that a normal on the outer cylinder points away from the wire and a normal on the inner cylinder points towards the wire.

(a) Explain why the flux of $\vec{E}$, the electric field, through the washers is 0.

(b) Gauss's Law states that the flux of an electric field through a closed surface S is proportional to the amount of electric charge inside S. Explain why Gauss's Law implies that the flux through the inner cylinder is the same as the flux through the outer cylinder. (Note that the charge on the wire is *not* inside the surface S).

(c) Use part (b) to show that $E_b/E_a = a/b$.

(d) Explain why part (c) shows that the strength of the field due to an infinitely long uniformly charged wire is proportional to $1/r$.

24. Consider an infinite flat sheet uniformly covered with charge. As in the case of the charged wire from Problem 23, symmetry shows that the electric field $\vec{E}$ is perpendicular to the sheet and has the same magnitude at all points that are at the same distance from the sheet. Use Gauss's Law (described in Problem 23) to explain why the field due to the charged sheet is the same at all points in space off the sheet, on any given side of the sheet. [Hint: Consider the flux through the box with sides parallel to the sheet shown in Figure 19.21.]

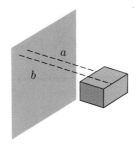

Figure 19.21

19.2 FLUX INTEGRALS FOR GRAPHS, CYLINDERS, AND SPHERES

In Section 19.1 we computed flux integrals in certain simple cases. In this section we see how to compute flux through surfaces that are graphs of functions, through cylinders, and through spheres. In Section 19.3 we see how to compute flux through more general surfaces.

Flux of a Vector Field through the Graph of $z = f(x, y)$

Suppose S is the graph of the differentiable function $z = f(x, y)$, oriented upward, and that $\vec{F}$ is a smooth vector field. In Section 19.1 we subdivided the surface into small pieces with area vector $\Delta \vec{A}$ and defined the flux of $\vec{F}$ through S as follows:

$$\int_S \vec{F} \cdot d\vec{A} = \lim_{\|\Delta \vec{A}\| \to 0} \sum \vec{F} \cdot \Delta \vec{A}.$$

How do we divide S into small pieces? One way is to use the cross-sections of f with x or y constant and take the patches in a wire frame representation of the surface. So we must calculate the area vector of one of these patches, which is approximately a parallelogram.

The Area Vector of a Parallelogram-Shaped Patch

According to the geometric definition of the cross product on page 86, the vector $\vec{v} \times \vec{w}$ has magnitude equal to the area of the parallelogram formed by $\vec{v}$ and $\vec{w}$ and direction perpendicular to this parallelogram and determined by the right-hand rule. Thus, we have

$$\boxed{\text{Area vector of parallelogram} = \vec{A} = \vec{v} \times \vec{w}.}$$

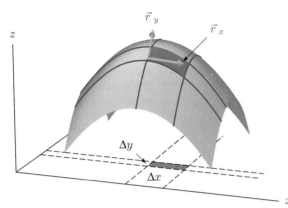

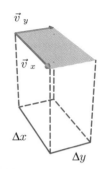

Figure 19.22: Surface showing parameter rectangle and tangent vectors $\vec{r}_x$ and $\vec{r}_y$

Figure 19.23: Parallelogram-shaped patch in the tangent plane to the surface

Consider the patch of surface above the rectangular region with sides Δx and Δy in the xy-plane shown in Figure 19.22. We approximate the area vector, $\Delta\vec{A}$, of this patch by the area vector of the corresponding patch on the tangent plane to the surface. See Figure 19.23 This patch is the parallelogram determined by the vectors $\vec{v}_x$ and $\vec{v}_y$, so its area vector is given by

$$\Delta\vec{A} \approx \vec{v}_x \times \vec{v}_y.$$

To find $\vec{v}_x$ and $\vec{v}_y$, notice that a point on the surface has position vector $\vec{r} = x\vec{i} + y\vec{j} + f(x,y)\vec{k}$. Thus, a cross-section of S with y constant has tangent vector

$$\vec{r}_x = \frac{\partial\vec{r}}{\partial x} = \vec{i} + f_x\vec{k},$$

and a cross-section with x constant has tangent vector

$$\vec{r}_y = \frac{\partial\vec{r}}{\partial y} = \vec{j} + f_y\vec{k}.$$

The vectors $\vec{r}_x$ and $\vec{v}_x$ are parallel because they are both tangent to the surface and in the xz-plane. Since the x-component of $\vec{r}_x$ is $\vec{i}$ and the x-component of $\vec{v}_x$ is $(\Delta x)\vec{i}$, we have $\vec{v}_x = (\Delta x)\vec{r}_x$. Similarly, we have $\vec{v}_y = (\Delta y)\vec{r}_y$. So the upward pointing area vector of the parallelogram is

$$\Delta\vec{A} \approx \vec{v}_x \times \vec{v}_y = \left(\vec{r}_x \times \vec{r}_y\right)\Delta x\,\Delta y = \left(-f_x\vec{i} - f_y\vec{j} + \vec{k}\right)\Delta x\,\Delta y.$$

This is our approximation for the area vector $\Delta\vec{A}$ on the surface. Replacing $\Delta\vec{A}$, Δx, and Δy by $d\vec{A}$, dx and dy, we write

$$d\vec{A} = \left(-f_x\vec{i} - f_y\vec{j} + \vec{k}\right)dx\,dy.$$

> ## The Flux of $\vec{F}$ through a Surface given by a Graph of $z = f(x, y)$
>
> Suppose the surface S is the part of the graph of $z = f(x, y)$ above a region R in the xy-plane, and suppose S is oriented upward. The flux of $\vec{F}$ through S is
>
> $$\int_S \vec{F} \cdot d\vec{A} = \int_R \vec{F}(x, y, f(x, y)) \cdot \left(-f_x \vec{i} - f_y \vec{j} + \vec{k}\right) dx \, dy.$$

Example 1 Compute $\int_S \vec{F} \cdot d\vec{A}$ where $\vec{F}(x, y, z) = z\vec{k}$ and S is the rectangular plate with corners $(0, 0, 0)$, $(1, 0, 0)$, $(0, 1, 3)$, $(1, 1, 3)$, oriented upwards.

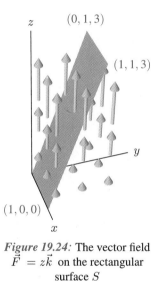

Figure 19.24: The vector field $\vec{F} = z\vec{k}$ on the rectangular surface S

Solution We find the equation for the plane S in the form $z = f(x, y)$. Since f is linear, with x-slope equal to 0 and y-slope equal to 3, and $f(0, 0) = 0$, we have

$$z = f(x, y) = 0 + 0x + 3y = 3y.$$

Thus, we have

$$d\vec{A} = (-f_x \vec{i} - f_y \vec{j} + \vec{k}) \, dx \, dy = (0\vec{i} - 3\vec{j} + \vec{k}) \, dx \, dy = (-3\vec{j} + \vec{k}) \, dx \, dy.$$

The flux integral is therefore

$$\int_S \vec{F} \cdot d\vec{A} = \int_0^1 \int_0^1 3y\vec{k} \cdot (-3\vec{j} + \vec{k}) \, dx \, dy = \int_0^1 \int_0^1 3y \, dx \, dy = 1.5.$$

Flux of a Vector Field through a Cylindrical Surface

Consider the cylinder of radius R centered on the z-axis illustrated in Figure 19.25 and oriented away from the z-axis. The small patch of surface, or parameter rectangle, in Figure 19.26

has surface area given by

$$\Delta A \approx R \, \Delta \theta \, \Delta z.$$

The outward unit normal $\vec{n}$ points in the direction of $x\vec{i} + y\vec{j}$, so

$$\vec{n} = \frac{x\vec{i} + y\vec{j}}{\|x\vec{i} + y\vec{j}\|} = \frac{R\cos\theta\vec{i} + R\sin\theta\vec{j}}{R} = \cos\theta\vec{i} + \sin\theta\vec{j}.$$

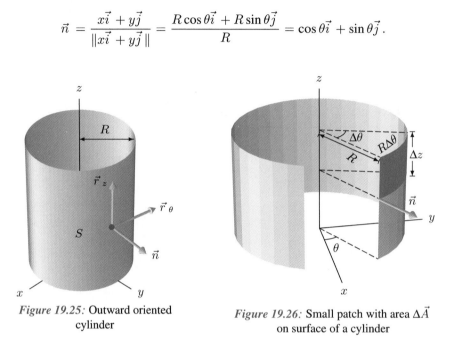

Figure 19.25: Outward oriented cylinder

Figure 19.26: Small patch with area $\Delta\vec{A}$ on surface of a cylinder

Therefore, the area vector of the parameter rectangle is approximated by

$$\Delta\vec{A} = \vec{n}\,\Delta A \approx \left(\cos\theta\vec{i} + \sin\theta\vec{j}\right)R\,\Delta z\,\Delta\theta.$$

Replacing $\Delta\vec{A}$, Δz, and $\Delta\theta$ by $d\vec{A}$, dz, and $d\theta$, we write

$$d\vec{A} = \left(\cos\theta\vec{i} + \sin\theta\vec{j}\right)R\,dz\,d\theta.$$

This gives the following result:

The Flux of a Vector Field through a Cylinder

The flux of $\vec{F}$ through the cylindrical surface S, of radius R and oriented away from the z-axis, is given by

$$\int_S \vec{F} \cdot d\vec{A} = \int_T \vec{F}(R,\theta,z) \cdot \left(\cos\theta\vec{i} + \sin\theta\vec{j}\right)R\,dz\,d\theta,$$

where T is the θz-region corresponding to S.

Example 2   Compute $\int_S \vec{F} \cdot d\vec{A}$ where $\vec{F}(x, y, z) = y\vec{j}$ and S is the part of the cylinder of radius 2 centered on the z-axis with $x \geq 0, y \geq 0$, and $0 \leq z \leq 3$. The surface is oriented towards the z-axis.

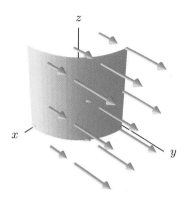

Figure 19.27: The vector field
$\vec{F} = y\vec{j}$ on the surface S

Solution   In cylindrical coordinates, we have $R = 2$ and $\vec{F} = y\vec{j} = 2\sin\theta\vec{j}$. Since the orientation of S is toward the z-axis, the flux across S is given by

$$\int_S \vec{F} \cdot d\vec{A} = -\int_T 2\sin\theta\vec{j} \cdot (\cos\theta\vec{i} + \sin\theta\vec{j})2 \, dz \, d\theta = -4\int_0^{\pi/2}\int_0^3 \sin^2\theta \, dz \, d\theta = -3\pi.$$

Flux of a Vector Field through a Spherical Surface

Consider the piece of the sphere of radius R centered at the origin, oriented outward illustrated in Figure 19.28. The small parameter rectangle in Figure 19.28 has surface area given by

$$\Delta A \approx R^2 \sin\phi \, \Delta\phi \, \Delta\theta.$$

The outward unit normal $\vec{n}$ points in the direction of $\vec{r} = x\vec{i} + y\vec{j} + z\vec{k}$, so

$$\vec{n} = \frac{\vec{r}}{\|\vec{r}\|} = \sin\phi\cos\theta\vec{i} + \sin\phi\sin\theta\vec{j} + \cos\phi\vec{k}.$$

Therefore, the area vector of the parameter rectangle is approximated by

$$\Delta\vec{A} \approx \vec{n}\,\Delta A = \frac{\vec{r}}{\|\vec{r}\|}\Delta A = \left(\sin\phi\cos\theta\vec{i} + \sin\phi\sin\theta\vec{j} + \cos\phi\vec{k}\right)R^2\sin\phi\,\Delta\phi\,\Delta\theta.$$

Replacing $\Delta\vec{A}$, $\Delta\phi$, and $\Delta\theta$ by $d\vec{A}$, $d\phi$, and $d\theta$, we write

$$d\vec{A} = \frac{\vec{r}}{\|\vec{r}\|}\,dA = \left(\sin\phi\cos\theta\vec{i} + \sin\phi\sin\theta\vec{j} + \cos\phi\vec{k}\right)R^2\sin\phi\,d\phi\,d\theta.$$

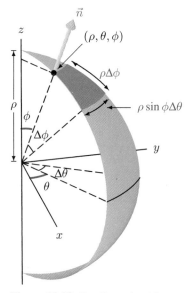

Figure 19.28: Small patch with area
$\Delta \vec{A}$ on surface of a sphere

Thus, we obtain the following result:

The Flux of a Vector Field through a Sphere

The flux of $\vec{F}$ through the spherical surface S, with radius R and oriented away from the origin, is given by

$$\int_S \vec{F} \cdot d\vec{A} = \int_S \vec{F} \cdot \frac{\vec{r}}{\|\vec{r}\|} \, dA$$

$$= \int_T \vec{F}(R, \theta, \phi) \cdot \left(\sin \phi \cos \theta \vec{i} + \sin \phi \sin \theta \vec{j} + \cos \phi \vec{k} \right) R^2 \sin \phi \, d\phi \, d\theta,$$

where T is the $\theta \phi$-region corresponding to S.

Example 3   Find the flux of $\vec{F} = z\vec{k}$ through S, the upper hemisphere of radius 2 centered at the origin, oriented outward.

Solution   The hemisphere S is parameterized by spherical coordinates θ and ϕ, with $0 \le \theta \le 2\pi$ and $0 \le \phi \le \pi/2$. Since $R = 2$ and $\vec{F} = z\vec{k} = 2\cos\phi\vec{k}$, the flux is

$$\int_S \vec{F} \cdot d\vec{A} = \int_S 2\cos\phi\vec{k} \cdot (\sin\phi\cos\theta\vec{i} + \sin\phi\sin\theta\vec{j} + \cos\phi\vec{k})4\sin\phi \, d\phi \, d\theta$$

$$= \int_0^{2\pi} \int_0^{\pi/2} 8\sin\phi\cos^2\phi \, d\phi \, d\theta = 2\pi \left(8 \left(\frac{-\cos^3\phi}{3} \right) \Big|_{\phi=0}^{\pi/2} \right) = \frac{16\pi}{3}.$$

Example 4   The magnetic field $\vec{B}$ due to an *ideal magnetic dipole*, $\vec{\mu}$, located at the origin is defined to be

$$\vec{B}(\vec{r}) = -\frac{\vec{\mu}}{\|\vec{r}\|^3} + \frac{3(\vec{\mu} \cdot \vec{r})\vec{r}}{\|\vec{r}\|^5}.$$

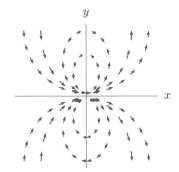

Figure 19.29: The magnetic field of a dipole, $\vec{i}$, at the origin:

$$\vec{B} = \frac{-\vec{i}}{\|\vec{r}\|^3} + \frac{3(\vec{i} \cdot \vec{r})\vec{r}}{\|\vec{r}\|^5}$$

Figure 19.29 shows a sketch of $\vec{B}$ in the plane $z = 0$ for the dipole $\vec{\mu} = \vec{i}$. Notice that $\vec{B}$ is similar to the magnetic field of a bar magnet with its north pole at the tip of the vector $\vec{i}$ and its south pole at the tail of the vector $\vec{i}$.

Compute the flux of $\vec{B}$ outward through the sphere S with center at the origin and radius R.

Solution   Since $\vec{i} \cdot \vec{r} = x$ and $\|\vec{r}\| = R$ on the sphere of radius R, we have

$$\int_S \vec{B} \cdot d\vec{A} = \int_S \left(-\frac{\vec{i}}{\|\vec{r}\|^3} + \frac{3(\vec{i} \cdot \vec{r})\vec{r}}{\|\vec{r}\|^5} \right) \cdot \frac{\vec{r}}{\|\vec{r}\|} \, dA = \int_S \left(-\frac{\vec{i} \cdot \vec{r}}{\|\vec{r}\|^4} + \frac{3(\vec{i} \cdot \vec{r})\|\vec{r}\|^2}{\|\vec{r}\|^6} \right) dA$$

$$= \int_S \frac{2\vec{i} \cdot \vec{r}}{\|\vec{r}\|^4} \, dA = \int_S \frac{2x}{\|\vec{r}\|^4} \, dA = \frac{2}{R^4} \int_S x \, dA,$$

But the sphere S is centered at the origin. Thus, the contribution to the integral from each positive x- value is canceled by the contribution from the corresponding negative x-value; so $\int_S x \, dA = 0$. Therefore,

$$\int_S \vec{B} \cdot d\vec{A} = \frac{2}{R^4} \int_S x \, dA = 0.$$

Problems for Section 19.2

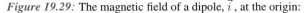

In Problems 1–12 compute the flux of the vector field, $\vec{F}$, through the surface, S.

1. $\vec{F} = (x - y)\vec{i} + z\vec{j} + 3x\vec{k}$ and S is the part of the plane $z = x + y$ above the rectangle $0 \le x \le 2, 0 \le y \le 3$, oriented upward.

2. $\vec{F} = \vec{r}$ and S is the part of the plane $x + y + z = 1$ above the rectangle $0 \le x \le 2, 0 \le y \le 3$, oriented downward.

3. $\vec{F} = \vec{r}$ and S is the part of the surface $z = x^2 + y^2$ above the disk $x^2 + y^2 \le 1$, oriented downward.

4. $\vec{F}(x, y, z) = 2x\vec{j} + y\vec{k}$ and S is the part of the surface $z = -y + 1$ above the square $0 \le x \le 1, 0 \le y \le 1$, oriented upward.

5. $\vec{F} = 3x\vec{i} + y\vec{j} + z\vec{k}$ and S is the part of the surface $z = -2x - 4y + 1$ above the triangle R in the xy-plane with vertices $(0, 0)$, $(0, 2)$, $(1, 0)$, oriented upward.

6. $\vec{F} = x\vec{i} + y\vec{j}$ and S is the part of the surface $z = 25 - (x^2 + y^2)$ above the disk of radius 5 centered at the origin, oriented upward.

7. $\vec{F} = \cos(x^2 + y^2)\vec{k}$ and S is as in Problem 6.

8. $\vec{F} = -y\vec{j} + z\vec{k}$ and S is the part of the surface $z = y^2 + 5$ over the rectangle $-2 \leq x \leq 1$, $0 \leq y \leq 1$, oriented upward.

9. $\vec{F}(x, y, z) = -xz\vec{i} - yz\vec{j} + z^2\vec{k}$ and S is the cone $z = \sqrt{x^2 + y^2}$ for $0 \leq z \leq 6$, oriented upward.

10. $\vec{F} = y\vec{i} + \vec{j} - xz\vec{k}$ and S is the surface $y = x^2 + z^2$, with $x^2 + z^2 \leq 1$, oriented in the positive y-direction.

11. $\vec{F} = xz\vec{i} + y\vec{k}$ and S is the hemisphere $x^2 + y^2 + z^2 = 9, z \geq 0$, oriented upward.

12. $\vec{F} = x^2\vec{i} + y^2\vec{j} + z^2\vec{k}$ and S is the oriented triangular surface shown in Figure 19.30.

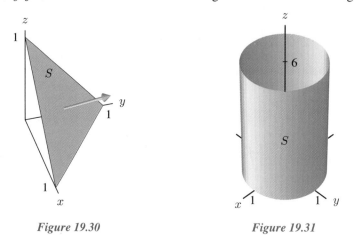

Figure 19.30   *Figure 19.31*

In Problems 13–14, compute the flux of the vector field, $\vec{F}$, through the cylindrical surface shown in Figure 19.31, oriented away from the z-axis.

13. $\vec{F} = x\vec{i} + y\vec{j}$ 14. $\vec{F} = xz\vec{i} + yz\vec{j} + z^3\vec{k}$

In Problems 15–16 compute the flux of the vector field, $\vec{F}$, through the given spherical surface, S.

15. $\vec{F} = z^2\vec{k}$ and S is the upper hemisphere of the sphere $x^2 + y^2 + z^2 = 25$, oriented away from the origin.

16. $\vec{F} = x\vec{i} + y\vec{j} + z\vec{k}$ and S is the surface of the sphere $x^2 + y^2 + z^2 = a^2$, oriented outward.

For Problems 17–18, an electric charge q is placed at the origin in 3-space. The induced electric field $\vec{E}(\vec{r})$ at the point with position vector $\vec{r}$ is given by

$$\vec{E}(\vec{r}) = q\frac{\vec{r}}{\|\vec{r}\|^3}, \qquad \vec{r} \neq \vec{0}.$$

17. Let S be the open cylinder of height $2H$ and radius R given by $x^2 + y^2 = R^2, -H \leq z \leq H$, oriented outward.

(a) Show that the flux of $\vec{E}$, the electric field, through S is given by

$$\int_S \vec{E} \cdot d\vec{A} = 4\pi q\frac{H}{\sqrt{H^2 + R^2}}.$$

(b) What are the limits of the flux $\int_S \vec{E} \cdot d\vec{A}$ if
    (i) $H \to 0$ or $H \to \infty$ when R is fixed?
    (ii) $R \to 0$ or $R \to \infty$ when H is fixed?

18. Let S be the outward oriented, closed cylinder of height $2H$ and radius R whose curved surface is given by $x^2 + y^2 = R^2$, $-H \leq z \leq H$, whose top is given by $z = H$, $x^2 + y^2 \leq R^2$, and bottom by $z = -H$, $x^2 + y^2 \leq R^2$. Use the result of Problem 17 to show that the flux of the electric field, $\vec{E}$, through S is given by

$$\int_S \vec{E} \cdot d\vec{A} = 4\pi q.$$

Notice that the flux is independent of both the height, H, and radius, R, of the cylinder.

19. Calculate the flux of
$$\vec{F} = (xze^{yz})\vec{i} + xz\vec{j} + (5 + x^2 + y^2)\vec{k}$$

through the disk $x^2 + y^2 \leq 1$ in the xy-plane, oriented upward.

20. Calculate the flux of

$$\vec{H} = (e^{xy} + 3z + 5)\vec{i} + (e^{xy} + 5z + 3)\vec{j} + (3z + e^{xy})\vec{k}$$

through the square of side 2 with one vertex at the origin, one edge along the positive y-axis, one edge in the xz-plane with $x > 0$, $z > 0$, and the normal $\vec{n} = \vec{i} - \vec{k}$.

19.3 NOTES ON FLUX INTEGRALS OVER PARAMETERIZED SURFACES

Most of the flux integrals we are likely to encounter can be computed using the methods of Sections 19.1 and 19.2. In this section we briefly consider the general case: how to compute the flux of a smooth vector field $\vec{F}$ through a smooth oriented surface, S, parameterized by

$$\vec{r} = \vec{r}(s, t),$$

for (s, t) in some region R of the parameter space. The method is similar to the one used for graphs in Section 19.2. We consider a parameter rectangle on the surface S corresponding to a rectangular region with sides Δs and Δt in the parameter space. (See Figure 19.32.)

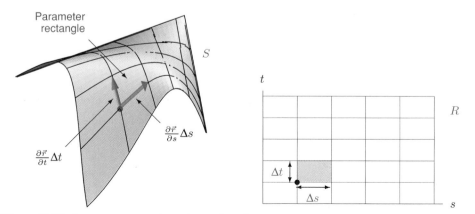

Figure 19.32: Parameter rectangle on the surface S corresponding to a small rectangular region in the parameter space, R

If Δs and Δt are small, the area vector, $\Delta \vec{A}$, of the patch is approximately the area vector of the parallelogram defined by the vectors

$$\vec{r}(s + \Delta s, t) - \vec{r}(s, t) \approx \frac{\partial \vec{r}}{\partial s} \Delta s, \qquad \text{and} \qquad \vec{r}(s, t + \Delta t) - \vec{r}(s, t) \approx \frac{\partial \vec{r}}{\partial t} \Delta t.$$

Thus

$$\Delta \vec{A} \approx \frac{\partial \vec{r}}{\partial s} \times \frac{\partial \vec{r}}{\partial t} \Delta s \, \Delta t.$$

We assume that the vector $\partial \vec{r} / \partial s \times \partial \vec{r} / \partial t$ is never zero and points in the direction of the unit normal orientation vector $\vec{n}$. If the vector $\partial \vec{r} / \partial s \times \partial \vec{r} / \partial t$ points in the opposite direction to $\vec{n}$, we reverse the order of the cross-product. Replacing $\Delta \vec{A}$, Δs, and Δt by $d \vec{A}$, ds, and dt, we write

$$d \vec{A} = \left(\frac{\partial \vec{r}}{\partial s} \times \frac{\partial \vec{r}}{\partial t} \right) ds \, dt.$$

The Flux of a Vector Field through a Parameterized Surface

The flux of a smooth vector field $\vec{F}$ through a smooth oriented surface S parameterized by $\vec{r} = \vec{r}(s, t)$, where (s, t) varies in a parameter region R, is given by

$$\int_S \vec{F} \cdot d \vec{A} = \int_R \vec{F}(\vec{r}(s, t)) \cdot \left(\frac{\partial \vec{r}}{\partial s} \times \frac{\partial \vec{r}}{\partial t} \right) ds \, dt.$$

We choose the parameterization so that $\partial \vec{r} / \partial s \times \partial \vec{r} / \partial t$ is never zero and points in the direction of $\vec{n}$ everywhere.

Example 1   Find the flux of the vector field $\vec{F} = x \vec{i} + y \vec{j}$ through the surface S, oriented downward and given by

$$x = 2s, \quad y = s + t, \quad z = 1 + s - t, \qquad \text{where } 0 \le s \le 1, \quad 0 \le t \le 1.$$

Solution   Since S is parameterized by

$$\vec{r}(s, t) = 2s \vec{i} + (s + t) \vec{j} + (1 + s - t) \vec{k},$$

we have

$$\frac{\partial \vec{r}}{\partial s} = 2 \vec{i} + \vec{j} + \vec{k} \quad \text{and} \quad \frac{\partial r}{\partial t} = \vec{j} - \vec{k},$$

so

$$\frac{\partial \vec{r}}{\partial s} \times \frac{\partial \vec{r}}{\partial t} = \begin{vmatrix} \vec{i} & \vec{j} & \vec{k} \\ 2 & 1 & 1 \\ 0 & 1 & -1 \end{vmatrix} = -2 \vec{i} + 2 \vec{j} + 2 \vec{k}.$$

Since the vector $-2 \vec{i} + 2 \vec{j} + 2 \vec{k}$ points upward, we use $2 \vec{i} - 2 \vec{j} - 2 \vec{k}$ for downward orientation. Thus, the flux integral is given by

$$\int_S \vec{F} \cdot d \vec{A} = \int_0^1 \int_0^1 (2s \vec{i} + (s + t) \vec{j}) \cdot (2 \vec{i} - 2 \vec{j} - 2 \vec{k}) \, ds \, dt$$

$$= \int_0^1 \int_0^1 (4s - 2s - 2t) \, ds \, dt = \int_0^1 \int_0^1 (2s - 2t) \, ds \, dt$$

$$= \int_0^1 \left(s^2 - 2st \Big|_{s=0}^{s=1} \right) dt = \int_0^1 (1 - 2t) \, dt = t - t^2 \Big|_0^1 = 0.$$

Area of a Parameterized Surface

The area ΔA of a small parameter rectangle is the magnitude of its area vector $\Delta \vec{A}$. Therefore,

$$\text{Area of } S = \sum \Delta A = \sum \|\Delta \vec{A}\| \approx \sum \left\| \frac{\partial \vec{r}}{\partial s} \times \frac{\partial \vec{r}}{\partial t} \right\| \Delta s \, \Delta t.$$

Taking the limit as the area of the parameter rectangles tends to zero, we are led to the following expression for the area of S.

The Area of a Parameterized Surface

The area of a surface S which is parameterized by $\vec{r} = \vec{r}(s,t)$, where (s,t) varies in a parameter region R, is given by

$$\int_S dA = \int_R \left\| \frac{\partial \vec{r}}{\partial s} \times \frac{\partial \vec{r}}{\partial t} \right\| ds \, dt.$$

Example 2   Compute the surface area of a sphere of radius a.

Solution   We take the sphere S of radius a centered at the origin and parameterize it with the spherical coordinates ϕ and θ. The parameterization is

$$x = a \sin \phi \cos \theta, \quad y = a \sin \phi \sin \theta, \quad z = a \cos \phi, \qquad \text{for} \quad 0 \leq \theta \leq 2\pi, \quad 0 \leq \phi \leq \pi.$$

We compute

$$\frac{\partial \vec{r}}{\partial \phi} \times \frac{\partial \vec{r}}{\partial \theta} = (a \cos \phi \cos \theta \vec{i} + a \cos \phi \sin \theta \vec{j} - a \sin \phi \vec{k}) \times (-a \sin \phi \sin \theta \vec{i} + a \sin \phi \cos \theta \vec{j})$$

$$= a^2 (\sin^2 \phi \cos \theta \vec{i} + \sin^2 \phi \sin \theta \vec{j} + \sin \phi \cos \phi \vec{k})$$

and so

$$\left\| \frac{\partial \vec{r}}{\partial \phi} \times \frac{\partial \vec{r}}{\partial \theta} \right\| = a^2 \sin \phi.$$

Thus, we see that the surface area of the sphere S is given by

$$\text{Surface area} = \int_S dA = \int_R \left\| \frac{\partial \vec{r}}{\partial \phi} \times \frac{\partial \vec{r}}{\partial \theta} \right\| d\phi \, d\theta = \int_{\phi=0}^{\pi} \int_{\theta=0}^{2\pi} a^2 \sin \phi \, d\theta \, d\phi = 4\pi a^2.$$

Problems for Section 19.3

In Problems 1–5 compute the flux of the vector field, $\vec{F}$, through the parameterized surface, S.

1. $\vec{F} = z\vec{k}$ and S is oriented toward the z-axis and given by

$$x = s + t, \quad y = s - t, \quad z = s^2 + t^2, \qquad 0 \leq s \leq 1, \ 0 \leq t \leq 1.$$

2. $\vec{F} = y\vec{i} + x\vec{j}$ and S is oriented away from the z-axis and given by

$$x = 3 \sin s, \quad y = 3 \cos s, \quad z = t + 1, \qquad 0 \leq s \leq \pi, \ 0 \leq t \leq 1.$$

3. $\vec{F} = z\vec{i} + x\vec{j}$ and S is oriented upward and given by

$$x = s^2, \quad y = 2s + t^2, \quad z = 5t, \quad 0 \le s \le 1, \ 1 \le t \le 3.$$

4. $\vec{F} = -\dfrac{2}{x}\vec{i} + \dfrac{2}{y}\vec{j}$ and S is oriented upward and parameterized by a and θ, where

$$x = a\cos\theta, \quad y = a\sin\theta, \quad z = \sin a^2, \quad 1 \le a \le 3, \ 0 \le \theta \le \pi.$$

5. $\vec{F} = x^2 y^2 z\vec{k}$ and S is the cone $\sqrt{x^2 + y^2} = z$, with $0 \le z \le R$, oriented downward. Parametrize the cone using cylindrical coordinates. (See Figure 19.33.)

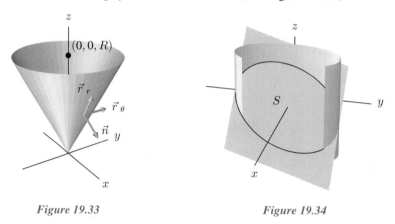

Figure 19.33   *Figure 19.34*

6. Find the area of the ellipse S on the plane $2x + y + z = 2$ cut out by the circular cylinder $x^2 + y^2 = 2x$. (See Figure 19.34.)

7. Evaluate $\int_S \vec{F} \cdot d\vec{A}$, where $\vec{F} = (bx/a)\vec{i} + (ay/b)\vec{j}$ and S is the elliptic cylinder oriented away from the z-axis, and given by $x^2/a^2 + y^2/b^2 = 1$, $|z| \le c$, where a, b, c are positive constants.

8. Consider the surface S formed by rotating the graph of $y = f(x)$ around the x-axis between $x = a$ and $x = b$. Assume that $f(x) \ge 0$ for $a \le x \le b$. Show that the surface area of S is $2\pi \int_a^b f(x)\sqrt{1 + f'(x)^2}\, dx$.

As we remarked in Section 19.1, the limit defining a flux integral might not exist if we subdivide the surface in the wrong way. One way to get around this is to take the formula for a flux integral over a parameterized surface that we have developed in Section 19.3, and use it as the *definition* of the flux integral. In Problems 9–12 we explore how this works.

9. Use a parameterization to verify the formula for a flux integral over a surface graph on page 399.

10. Use a parameterization to verify the formula for a flux integral over a cylindrical surface on page 400.

11. Use a parameterization to verify the formula for a flux integral over a spherical surface on page 402.

12. One problem with defining the flux integral using a parameterization is that the integral appears to depend on the choice of parameterization. However, the flux through a surface ought not to depend on how the surface is parameterized. Suppose that the surface S has two

parameterizations, $\vec{r} = \vec{r}(s,t)$ for (s,t) in the region R of st-space, and also $\vec{r} = r(u,v)$ for (u,v) in the region T in uv-space, and suppose that the two parameterizations are related by the change of variables

$$u = u(s,t) \quad v = v(s,t).$$

Suppose that the Jacobian determinant $\partial(u,v)/\partial(s,t)$ is positive at every point (s,t) in R. Use the change of variables formula for double integrals on page 267 to show that computing the flux integral using either parameterization gives the same result.

REVIEW PROBLEMS FOR CHAPTER NINETEEN

For Problems 1–2, let $\vec{F}(\vec{r}) = \vec{r}$ and let S be a square plate perpendicular to the z-axis and centered on the z-axis. Sketch as a function of time the flux of $\vec{F}$ through S as S moves in the given manner.

1. S moves from far up the positive z-axis to far down the negative z-axis. Assume S is oriented upward.

2. S rotates about an axis parallel to the x-axis, through the center of S. Assume that S is far up the z-axis so that $\vec{r}$ is approximately constant on S as S rotates, and that S is initially oriented upward.

3. Repeat Problems 1–2 with $\vec{F}(\vec{r}) = \vec{r}/r^3$, where $r = \|\vec{r}\|$.

For Problems 4–7 find the flux of the constant vector field $\vec{v} = \vec{i} - \vec{j} + 3\vec{k}$ through the given surfaces.

4. A disk of radius 2 in the xy-plane oriented upward.

5. A triangular plate of area 4 in the yz-plane oriented in the positive x-direction.

6. A square plate of area 4 in the yz-plane oriented in the positive x-direction.

7. The triangular plate with vertices $(1,0,0)$, $(0,1,0)$, $(0,0,1)$, oriented away from the origin.

In Problems 8–15 compute the flux of the given vector field, $\vec{F}$, through the given surface, S.

8. $\vec{F} = x\vec{i} + y\vec{j} + (z^2 + 3)\vec{k}$ and S is the rectangle $z = 4, 0 \le x \le 2, 0 \le y \le 3$, oriented in the positive z-direction.

9. $\vec{F} = z\vec{i} + y\vec{j} + 2x\vec{k}$ and S is the rectangle $z = 4, 0 \le x \le 2, 0 \le y \le 3$, oriented in the positive z-direction.

10. $\vec{F} = (x + \cos z)\vec{i} + y\vec{j} + 2x\vec{k}$ and S is the rectangle $x = 2, 0 \le y \le 3, 0 \le z \le 4$, oriented in the positive x-direction.

11. $\vec{F} = x^2\vec{i} + (x + e^y)\vec{j} - \vec{k}$, and S is the rectangle $y = -1, 0 \le x \le 2, 0 \le z \le 4$, oriented in the negative y-direction.

12. $\vec{F} = (5 + xy)\vec{i} + z\vec{j} + yz\vec{k}$ and S is the 2×2 square plate in the yz-plane centered at the origin, oriented in the positive x-direction.

13. $\vec{F} = x\vec{i} + y\vec{j}$ and S is the surface of a closed cylinder of radius 2 and height 3 centered on the z-axis with its base in the xy-plane.

14. $\vec{F} = -y\vec{i} + x\vec{j} + z\vec{k}$ and S is the surface of a closed cylinder of radius 1 centered on the z-axis with base in the plane $z = -1$ and top in the plane $z = 1$.

15. $\vec{F} = x^2\vec{i} + y^2\vec{j} + z\vec{k}$ and S is the cone $z = \sqrt{x^2 + y^2}$, oriented upward with $x^2 + y^2 \le 1$, $x \ge 0, y \ge 0$.

16. Suppose water is flowing down a cylindrical pipe of radius 2 cm and that the speed is $(3 - (3/4)r^2)$ cm/sec at a distance r cm from the center of the pipe. Find the flux through the circular cross-section of the pipe, oriented so that the flow is positive.

17. Suppose that $\vec{E}$ is a *uniform* electric field on 3-space, so $\vec{E}(x, y, z) = a\vec{i} + b\vec{j} + c\vec{k}$, for all points (x, y, z), where a, b, c are constants. Show, with the aid of symmetry, that the flux of $\vec{E}$ through each of the following closed surfaces S is zero:

    (a) S is the cube bounded by the planes $x = \pm 1$, $y = \pm 1$, and $z = \pm 1$
    (b) S is the sphere $x^2 + y^2 + z^2 = 1$
    (c) S is the cylinder bounded by $x^2 + y^2 = 1$, $z = 0$, and $z = 2$

18. According to Coulomb's Law the electrostatic field $\vec{E}$, at the point with position vector $\vec{r}$ in 3-space, due to a charge q at the origin is given by

$$\vec{E}(\vec{r}) = q\frac{\vec{r}}{\|\vec{r}\|^3}.$$

    Let S_a be the outward oriented sphere in 3-space of radius $a > 0$ and center at the origin. Show that the flux of the resulting electric field $\vec{E}$ through the surface S_a is equal to $4\pi q$, for any radius a. This is Gauss's Law for a single point charge.

19. An infinitely long straight wire lying along the z-axis carries an electric current I flowing in the $\vec{k}$ direction. Ampère's Law in magnetostatics says that the current gives rise to a magnetic field $\vec{B}$ given by

$$\vec{B}(x, y, z) = \frac{I}{2\pi}\frac{-y\vec{i} + x\vec{j}}{x^2 + y^2}.$$

    (a) Sketch the field $\vec{B}$ in the xy-plane.
    (b) Suppose S_1 is a disk with center at $(0, 0, h)$, radius a, and parallel to the xy-plane, oriented in the $\vec{k}$ direction. What is the flux of $\vec{B}$ through S_1? Is your answer reasonable?
    (c) Suppose S_2 is the rectangle given by $x = 0$, $a \leq y \leq b$, $0 \leq z \leq h$, and oriented in the $-\vec{i}$ direction. What is the flux of $\vec{B}$ through S_2? Does your answer seem reasonable?

20. An *ideal electric dipole* in electrostatics is characterized by its position in 3-space and its dipole moment vector $\vec{p}$. The electric field $\vec{D}$, at the point with position vector $\vec{r}$, of an ideal electric dipole located at the origin with dipole moment $\vec{p}$ is given by

$$\vec{D}(\vec{r}) = 3\frac{(\vec{r} \cdot \vec{p})\vec{r}}{\|\vec{r}\|^5} - \frac{\vec{p}}{\|\vec{r}\|^3}.$$

    Assume $\vec{p} = p\vec{k}$, so the dipole points in the $\vec{k}$ direction and has magnitude p.

    (a) What is the flux of $\vec{D}$ through a sphere S with center at the origin and radius $a > 0$?
    (b) The field $\vec{D}$ is a useful approximation to the electric field $\vec{E}$ produced by two "equal and opposite" charges, q at $\vec{r}_2$ and $-q$ at $\vec{r}_1$, where the distance $\|\vec{r}_2 - \vec{r}_1\|$ is small. The dipole moment of this configuration of charges is defined to be $q(\vec{r}_2 - \vec{r}_1)$. Gauss's Law in electrostatics says that the flux of $\vec{E}$ through S is equal to 4π times the total charge enclosed by S. What is the flux of $\vec{E}$ through S if the charges at $\vec{r}_1$ and $\vec{r}_2$ are enclosed by S? How does this compare with your answer for the flux of $\vec{D}$ through S if $\vec{p} = q(\vec{r}_2 - \vec{r}_1)$?

CHAPTER TWENTY

CALCULUS OF VECTOR FIELDS

We have seen two ways of integrating vector fields in three dimensions: along curves and over surfaces. Now we will look at two ways of differentiating them. If we view the vector field as the velocity field of a fluid flow, then one method of differentiation (the divergence) tells us about the net strength of outflow from a point, and the other method (the curl) tells us about the strength of rotation around a point. Each method fits together with one of the ways of integrating to form a vector analogue of the Fundamental Theorem of Calculus: the Divergence Theorem relating the divergence to flux, and Stokes' theorem relating the curl to circulation around a closed path.

20.1 THE DIVERGENCE OF A VECTOR FIELD

Imagine that the vector fields in Figures 20.1 and 20.2 are velocity vector fields describing the flow of a fluid.[1] Figure 20.1 suggests outflow from the origin; for example, it could represent the expanding cloud of matter in the big bang theory of the origin of the universe. We say that the origin is a *source*. Figure 20.2 suggests flow into the origin; in this case we say that the origin is a *sink*.

In this section we will use the flux out of a closed surface surrounding a point to measure the outflow per unit volume there, also called the *divergence*, or *flux density*.

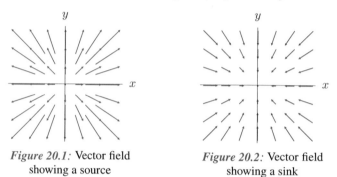

Figure 20.1: Vector field
showing a source

Figure 20.2: Vector field
showing a sink

Definition of Divergence

To measure the outflow per unit volume of a vector field at a point, we calculate the flux out of a small sphere centered at the point, divide by the volume enclosed by the sphere, then take the limit of this flux to volume ratio as the sphere contracts around the point.

> **Geometric Definition of Divergence**
>
> The **divergence**, or **flux density**, of a smooth vector field $\vec{F}$, written **div** $\vec{F}$, is a scalar-valued function defined by
>
> $$\operatorname{div} \vec{F}(x, y, z) = \lim_{\text{Volume}\to 0} \frac{\int_S \vec{F} \cdot d\vec{A}}{\text{Volume of } S}.$$
>
> Here S is a sphere centered at (x, y, z), oriented outwards, that contracts down to (x, y, z) in the limit.

The limit can be computed using other shapes as well, such as the cubes in Example 2. Problem 2 on page 453 explains why this is justified.

> **Cartesian Coordinate Definition of Divergence**
>
> If $\vec{F} = F_1\vec{i} + F_2\vec{j} + F_3\vec{k}$, then
>
> $$\operatorname{div} \vec{F} = \frac{\partial F_1}{\partial x} + \frac{\partial F_2}{\partial y} + \frac{\partial F_3}{\partial z}.$$

---

[1]Although not all vector fields represent physically realistic fluid flows, it is useful to think of them this way.

Example 1   Calculate the divergence of $\vec{F}(\vec{r}) = \vec{r}$ at the origin

(a) Using the geometric definition.

(b) Using the Cartesian coordinate definition.

Solution   (a) Using the method of Example 4 on page 391, you can calculate the flux of $\vec{F}$ out of the sphere of radius a, centered at the origin; it is $4\pi a^3$. So we have

$$\operatorname{div} \vec{F}(0,0,0) = \lim_{a \to 0} \frac{\text{Flux}}{\text{Volume}} = \lim_{a \to 0} \frac{4\pi a^3}{\frac{4}{3}\pi a^3} = \lim_{a \to 0} 3 = 3.$$

(b) In coordinates, $\vec{F}(x, y, z) = x\vec{i} + y\vec{j} + z\vec{k}$, so

$$\operatorname{div} \vec{F} = \frac{\partial}{\partial x}(x) + \frac{\partial}{\partial y}(y) + \frac{\partial}{\partial z}(z) = 1 + 1 + 1 = 3.$$

The next example shows that the divergence can be negative if there is net inflow to a point.

Example 2   (a) Using the geometric definition, find the divergence of $\vec{v} = -x\vec{i}$ at: (i) $(0, 0, 0)$ (ii) $(2, 2, 0)$.

(b) Confirm that the coordinate definition gives the same results.

Solution   (a) (i) The vector field $\vec{v} = -x\vec{i}$ is parallel to the x axis, as shown in Figure 20.3. To compute the flux density, we use a cube S_1, centered at the origin with edges parallel to the axes, of length $2c$. Then the flux through the faces perpendicular to the y- and z-axes is zero (because the vector field is parallel to these faces). On the faces perpendicular to the x-axis, the vector field and the outward normal are parallel but point in opposite directions. On the face at $x = c$, we have

$$\vec{v} \cdot \Delta\vec{A} = -c\,\|\Delta\vec{A}\|.$$

On the face at $x = -c$, the dot product is still negative, and

$$\vec{v} \cdot \Delta\vec{A} = -c\,\|\Delta\vec{A}\|.$$

Therefore, the flux through the cube is given by

$$\int_{S_1} \vec{v} \cdot d\vec{A} = \int_{\text{Face } x=-c} \vec{v} \cdot d\vec{A} + \int_{\text{Face } x=c} \vec{v} \cdot d\vec{A}$$

$$= -c \cdot \text{Area of one face} + (-c) \cdot \text{Area of other face} = -2c(2c)^2 = -8c^3.$$

Thus,

$$\operatorname{div} \vec{v}(0,0,0) = \lim_{\text{Volume} \to 0} \frac{\displaystyle\int_S \vec{v} \cdot d\vec{A}}{\text{Volume of cube}} = \lim_{c \to 0} \left(\frac{-8c^3}{(2c)^3} \right) = -1.$$

Since the vector field points inward towards the yz-plane, it makes sense that the divergence is negative at the origin.

(ii) Take S_2 to be a cube as before, but centered this time at the point $(2, 2, 0)$. See Figure 20.3. As before, the flux through the faces perpendicular to the y- and z-axes is zero. On the face at $x = 2 + c$,

$$\vec{v} \cdot \Delta\vec{A} = -(2 + c)\,\|\Delta\vec{A}\|.$$

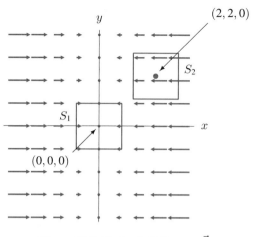

Figure 20.3: Vector field $\vec{v} = -x\vec{i}$

On the face at $x = 2 - c$ with outward normal, the dot product is positive, and

$$\vec{v} \cdot \Delta\vec{A} = (2 - c)\,\|\Delta\vec{A}\|.$$

Therefore, the flux through the cube is given by

$$\int_{S_2} \vec{v} \cdot d\vec{A} = \int_{\text{Face } x=2-c} \vec{v} \cdot d\vec{A} + \int_{\text{Face } x=2+c} \vec{v} \cdot d\vec{A}$$

$$= (2 - c) \cdot \text{Area of one face} - (2 + c) \cdot \text{Area of other face} = -2c(2c)^2 = -8c^3.$$

Then, as before,

$$\text{div}\,\vec{v}\,(2, 2, 0) = \lim_{\text{Volume}\to 0} \frac{\int_S \vec{v} \cdot d\vec{A}}{\text{Volume of cube}} = \lim_{c\to 0} \left(\frac{-8c^3}{(2c)^3} \right) = -1.$$

Although the vector field is flowing away from the point $(2, 2, 0)$ on the left, this outflow is smaller in magnitude than the inflow on the right, so the the net outflow is negative.

(b)   Since $\vec{v} = -x\vec{i} + 0\vec{j} + 0\vec{k}$, the formula gives

$$\text{div}\,\vec{v} = \frac{\partial}{\partial x}(-x) + \frac{\partial}{\partial y}(0) + \frac{\partial}{\partial z}(0) = -1 + 0 + 0 = -1.$$

Why Do the Two Definitions of Divergence Give the Same Result?

The geometric definition defines $\text{div}\,\vec{F}$ as the flux density of $\vec{F}$. To see why the coordinate definition is also the flux density, imagine computing the flux out of a small box-shaped surface S at (x_0, y_0, z_0), with sides of length Δx, Δy, and Δz parallel to the axes. On S_1 (the back face of the box shown in Figure 20.4, where $x = x_0$), the outward normal is in the negative x-direction, so $d\vec{A} = -dy\,dz\,\vec{i}$. Assuming $\vec{F}$ is approximately constant on S_1, we have

$$\int_{S_1} \vec{F} \cdot d\vec{A} = \int_{S_1} \vec{F} \cdot (-\vec{i})\,dy\,dz \approx -F_1(x_0, y_0, z_0) \int_{S_1} dy\,dz$$

$$= -F_1(x_0, y_0, z_0) \cdot \text{Area of } S_1 = -F_1(x_0, y_0, z_0)\,\Delta y\,\Delta z.$$

On S_2, the face where $x = x_0 + \Delta x$, the outward normal points in the positive x-direction, so $d\vec{A} = dy\,dz\,\vec{i}$. Therefore,

$$\int_{S_2} \vec{F} \cdot d\vec{A} = \int_{S_2} \vec{F} \cdot \vec{i} \, dy \, dz \approx F_1(x_0 + \Delta x, y_0, z_0) \int_{S_2} dy \, dz$$
$$= F_1(x_0 + \Delta x, y_0, z_0) \cdot \text{Area of } S_2 = F_1(x_0 + \Delta x, y_0, z_0) \, \Delta y \, \Delta z.$$

Thus

$$\int_{S_1} \vec{F} \cdot d\vec{A} + \int_{S_2} \vec{F} \cdot d\vec{A} \approx F_1(x_0 + \Delta x, y_0, z_0)\Delta y \Delta z - F_1(x_0, y_0, z_0)\Delta y \Delta z$$
$$= \frac{F_1(x_0 + \Delta x, y_0, z_0) - F_1(x_0, y_0, z_0)}{\Delta x} \Delta x \Delta y \Delta z$$
$$\approx \frac{\partial F_1}{\partial x} \Delta x \Delta y \Delta z.$$

By an analogous argument, the contribution to the flux from S_3 and S_4 (the surfaces perpendicular to the y-axis) is approximately

$$\frac{\partial F_2}{\partial y} \Delta x \, \Delta y \, \Delta z,$$

and the contribution to the flux from S_5 and S_6 is approximately

$$\frac{\partial F_3}{\partial z} \Delta x \, \Delta y \, \Delta z.$$

Thus, adding these contributions, we have

$$\text{Total flux through } S \approx \frac{\partial F_1}{\partial x} \Delta x \, \Delta y \, \Delta z + \frac{\partial F_2}{\partial y} \Delta x \, \Delta y \, \Delta z + \frac{\partial F_3}{\partial z} \Delta x \, \Delta y \, \Delta z.$$

Since the volume of the box is $\Delta x \, \Delta y \, \Delta z$, the flux density is

$$\frac{\text{Total flux through } S}{\text{Volume of box}} \approx \frac{\dfrac{\partial F_1}{\partial x}\Delta x \Delta y \Delta z + \dfrac{\partial F_2}{\partial y}\Delta x \Delta y \Delta z + \dfrac{\partial F_3}{\partial z}\Delta x \Delta y \Delta z}{\Delta x \Delta y \Delta z}$$
$$= \frac{\partial F_1}{\partial x} + \frac{\partial F_2}{\partial y} + \frac{\partial F_3}{\partial z}.$$

Problem 2 on page 453 gives a more detailed justification that the two definitions give the same result.

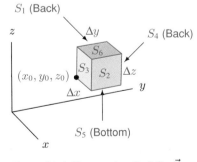

Figure 20.4: Box used to find div $\vec{F}$ at (x_0, y_0, z_0)

Divergence Free Vector Fields

A vector field $\vec{F}$ is said to be *divergence free* or *solenoidal* if div$\vec{F} = 0$ everywhere that $\vec{F}$ is defined.

Example 3   Figure 20.5 shows, for three values of the constant p, the vector field

$$\vec{E} = \frac{\vec{r}}{\|\vec{r}\|^p} \qquad \vec{r} \neq \vec{0}.$$

(a) Find a formula for div $\vec{E}$.
(b) Is there a value of p for which $\vec{E}$ is divergence free? If so, find it.

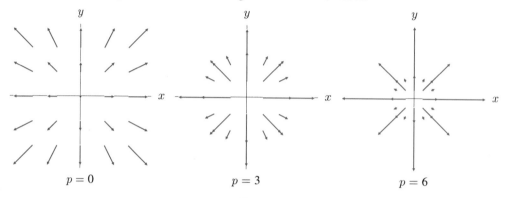

Figure 20.5: The vector field $\vec{E}(\vec{r}) = \vec{r}/\|\vec{r}\|^p$ for $p = 0, 3$, and 6

Solution   (a) The components of $\vec{E}$ are

$$\vec{E} = \frac{x}{(x^2 + y^2 + z^2)^{p/2}}\vec{i} + \frac{y}{(x^2 + y^2 + z^2)^{p/2}}\vec{j} + \frac{z}{(x^2 + y^2 + z^2)^{p/2}}\vec{k}.$$

We compute the partial derivatives

$$\frac{\partial}{\partial x}\left(\frac{x}{(x^2 + y^2 + z^2)^{p/2}}\right) = \frac{1}{(x^2 + y^2 + z^2)^{p/2}} - \frac{px^2}{(x^2 + y^2 + z^2)^{(p/2)+1}}$$

$$\frac{\partial}{\partial y}\left(\frac{y}{(x^2 + y^2 + z^2)^{p/2}}\right) = \frac{1}{(x^2 + y^2 + z^2)^{p/2}} - \frac{py^2}{(x^2 + y^2 + z^2)^{(p/2)+1}}$$

$$\frac{\partial}{\partial z}\left(\frac{z}{(x^2 + y^2 + z^2)^{p/2}}\right) = \frac{1}{(x^2 + y^2 + z^2)^{p/2}} - \frac{pz^2}{(x^2 + y^2 + z^2)^{(p/2)+1}}.$$

So

$$\begin{aligned}
\text{div } \vec{E} &= \frac{3}{(x^2 + y^2 + z^2)^{p/2}} - \frac{p(x^2 + y^2 + z^2)}{(x^2 + y^2 + z^2)^{(p/2)+1}} \\
&= \frac{3 - p}{(x^2 + y^2 + z^2)^{p/2}} = \frac{3 - p}{\|\vec{r}\|^p}.
\end{aligned}$$

(b) The divergence is zero when $p = 3$, so $\vec{F}(\vec{r}) = \vec{r}/\|\vec{r}\|^3$ is a divergence free vector field.

Magnetic Fields

An important class of divergence free vector fields is the magnetic fields. One of Maxwell's Laws of Electromagnetism is that the magnetic field $\vec{B}$ satisfies

$$\text{div } \vec{B} = 0.$$

Example 4 An infinitesimal current loop, similar to that shown in Figure 20.6, is called a *magnetic dipole*. Its magnitude is described by a constant vector $\vec{\mu}$, called the dipole moment. The magnetic field due to a magnetic dipole with moment $\vec{\mu}$ is

$$\vec{B} = -\frac{\vec{\mu}}{\|\vec{r}\|^3} + \frac{3(\vec{\mu} \cdot \vec{r})\vec{r}}{\|\vec{r}\|^5}, \qquad \vec{r} \neq \vec{0}.$$

Show that div $\vec{B} = 0$.

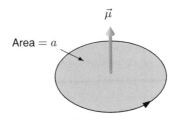

Area $= a$

Figure 20.6: A current loop

Solution To show that div $\vec{B} = 0$ we can use the following version of the product rule for the divergence: if g is a scalar function and $\vec{F}$ is a vector field, then

$$\text{div}(g\vec{F}) = (\text{grad } g) \cdot \vec{F} + g \, \text{div } \vec{F}.$$

(See Problem 12 on page 418.) Thus, since div $\vec{\mu} = 0$, we have

$$\text{div}\left(\frac{\vec{\mu}}{\|\vec{r}\|^3}\right) = \text{div}\left(\frac{1}{\|\vec{r}\|^3}\vec{\mu}\right) = \text{grad}\left(\frac{1}{\|\vec{r}\|^3}\right) \cdot \vec{\mu} + \frac{1}{\|\vec{r}\|^3} \cdot 0$$

and

$$\text{div}\left(\frac{(\vec{\mu} \cdot \vec{r})\vec{r}}{\|\vec{r}\|^5}\right) = \text{grad}(\vec{\mu} \cdot \vec{r}) \cdot \frac{\vec{r}}{\|\vec{r}\|^5} + (\vec{\mu} \cdot \vec{r}) \, \text{div}\left(\frac{\vec{r}}{\|\vec{r}\|^5}\right).$$

From Problems 18 and 19 on page 134 and Example 3 on page 416 we have

$$\text{grad}\left(\frac{1}{\|\vec{r}\|^3}\right) = \frac{-3\vec{r}}{\|\vec{r}\|^5}, \qquad \text{grad}(\vec{\mu} \cdot \vec{r}) = \vec{\mu}, \qquad \text{div}\left(\frac{\vec{r}}{\|\vec{r}\|^5}\right) = \frac{-2}{\|\vec{r}\|^5}.$$

Putting these results together gives

$$\text{div } \vec{B} = -\text{grad}\left(\frac{1}{\|\vec{r}\|^3}\right) \cdot \vec{\mu} + 3 \, \text{grad}(\vec{\mu} \cdot \vec{r}) \cdot \frac{\vec{r}}{\|\vec{r}\|^5} + 3(\vec{\mu} \cdot \vec{r}) \, \text{div}\left(\frac{\vec{r}}{\|\vec{r}\|^5}\right)$$

$$= \frac{3\vec{r} \cdot \vec{\mu}}{\|\vec{r}\|^5} + \frac{3\vec{\mu} \cdot \vec{r}}{\|\vec{r}\|^5} - \frac{6\vec{\mu} \cdot \vec{r}}{\|\vec{r}\|^5}$$

$$= 0.$$

Alternative Notation for Divergence

Using $\nabla = \dfrac{\partial}{\partial x}\vec{i} + \dfrac{\partial}{\partial y}\vec{j} + \dfrac{\partial}{\partial z}\vec{k}$, we can write

$$\text{div } \vec{F} = \nabla \cdot \vec{F} = \left(\frac{\partial}{\partial x}\vec{i} + \frac{\partial}{\partial y}\vec{j} + \frac{\partial}{\partial z}\vec{k}\right) \cdot (F_1\vec{i} + F_2\vec{j} + F_3\vec{k}) = \frac{\partial F_1}{\partial x} + \frac{\partial F_2}{\partial y} + \frac{\partial F_3}{\partial z}.$$

Problems for Section 20.1

1. Draw two vector fields that have positive divergence everywhere.
2. Draw two vector fields that have negative divergence everywhere.

3. Draw two vector fields that have zero divergence everywhere.

In Problems 4–10, find the divergence of the given vector field. (Note: $\vec{r} = x\vec{i} + y\vec{j} + z\vec{k}$.)

4. $\vec{F}(x, y) = -x\vec{i} + y\vec{j}$

5. $\vec{F}(x, y) = -y\vec{i} + x\vec{j}$

6. $\vec{F}(x, y) = (x^2 - y^2)\vec{i} + 2xy\vec{j}$

7. $\vec{F}(\vec{r}) = \vec{a} \times \vec{r}$

8. $\vec{F}(x, y) = \dfrac{-y\vec{i} + x\vec{j}}{x^2 + y^2}$

9. $\vec{F}(\vec{r}) = \dfrac{\vec{r} - \vec{r}_0}{\|\vec{r} - \vec{r}_0\|}$, $\vec{r} \neq \vec{r}_0$

10. $\vec{F}(x, y, z) = (-x + y)\vec{i} + (y + z)\vec{j} + (-z + x)\vec{k}$

11. Show that if $\vec{a}$ is a constant vector and $f(x, y, z)$ is a function, then $\text{div}(f\vec{a}) = (\text{grad } f) \cdot \vec{a}$.

12. Show that if $g(x, y, z)$ is a scalar valued function and $\vec{F}(x, y, z)$ is a vector field, then

$$\text{div}(g\vec{F}) = (\text{grad } g) \cdot \vec{F} + g \,\text{div}\, \vec{F}.$$

In Problems 13–15, use Problem 12 with $\vec{r} = x\vec{i} + y\vec{j} + z\vec{k}$ to find the divergence of the given vector field.

13. $\vec{F}(\vec{r}) = \dfrac{1}{\|\vec{r}\|^p}\vec{a} \times \vec{r}$

14. $\vec{B} = \dfrac{1}{x^a}\vec{r}$

15. $\vec{G}(\vec{r}) = (\vec{b} \cdot \vec{r})\vec{a} \times \vec{r}$

16. Which of the following two vector fields has the greater divergence at the origin? Assume the scales are the same on each.

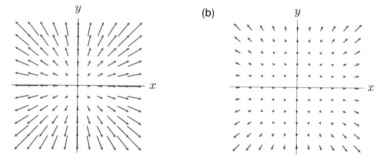

17. For each of the following vector fields, say whether the divergence is positive, zero, or negative at the indicated point.

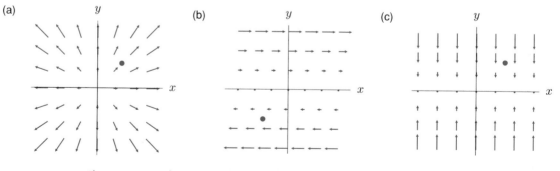

18. Let $\vec{F}(x, y, z) = z\vec{k}$.

    (a) Calculate div $\vec{F}$.

    (b) Sketch $\vec{F}$. Does it appear to be diverging? Does this agree with your answer to part (a)?

19. Let $\vec{F}(\vec{r}) = \vec{r}/\|\vec{r}\|^3$ (in 3-space), $\vec{r} \neq \vec{0}$.

   (a) Calculate div $\vec{F}$.
   (b) Sketch $\vec{F}$. Does it appear to be diverging? Does this agree with your answer to part (a)?

For Problems 20–22,

(a) Find the flux of the given vector field through a cube in the first octant with edge length c, one corner at the origin and edges along the axes.
(b) Use your answer to part (a) to find div $\vec{F}$ at the origin using the geometric definition.
(c) Compute div $\vec{F}$ at the origin using partial derivatives.

20. $\vec{F} = x\vec{i}$ 21. $\vec{F} = 2\vec{i} + y\vec{j} + 3\vec{k}$ 22. $\vec{F} = x\vec{i} + y\vec{j}$

23. (a) Find the flux of the vector field $\vec{F} = x\vec{i} + y\vec{j}$ through the surface of the closed cylinder of radius c and height c, centered on the z-axis with base in the xy-plane. (See Figure 20.7.)
   (b) Use your answer to part (a) to find div $\vec{F}$ at the origin using the geometric definition.
   (c) Compute div $\vec{F}$ at the origin using partial derivatives.

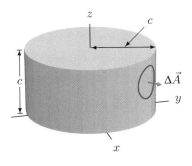

Figure 20.7

Problems 24–25 involve electric fields. Electric charge produces a vector field $\vec{E}$, called the electric field, which represents the force on a unit positive charge placed at the point. Two positive or two negative charges repel one another, whereas two charges of opposite sign attract one another. The divergence of $\vec{E}$ is proportional to the density of the electric charge (that is, the charge per unit volume), with a positive constant of proportionality.

24. Suppose a certain distribution of electric charge produces the electric field shown in Figure 20.8. Where are the charges that produced this electric field concentrated? Which concentrations are positive and which are negative?

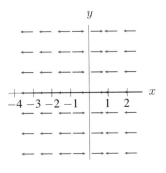

Figure 20.8

25.   The electric field at the point $\vec{r}$ as a result of a point charge at the origin is $\vec{E}(\vec{r}) = k\vec{r}/\|\vec{r}\|^3$.

   (a)   Calculate div $\vec{E}$ for $\vec{r} \neq \vec{0}$.
   (b)   Calculate the limit suggested by the geometric definition of div $\vec{E}$ at the point $(0, 0, 0)$.
   (c)   Explain what your answers mean in terms of charge density.

26.   The divergence of a magnetic vector field $\vec{B}$ must be zero everywhere. Which of the following vector fields cannot be a magnetic vector field?

   (a) $\vec{B}(x, y, z) = -y\vec{i} + x\vec{j} + (x + y)\vec{k}$
   (b) $\vec{B}(x, y, z) = -z\vec{i} + y\vec{j} + x\vec{k}$
   (c) $\vec{B}(x, y, z) = (x^2 - y^2 - x)\vec{i} + (y - 2xy)\vec{j}$

27.   If $f(x, y, z)$ and $g(x, y, z)$ are functions with continuous second partial derivatives, show that

$$\text{div}(\text{grad } f \times \text{grad } g) = 0.$$

28.   In Problem 22 on page 396 it was shown that the rate of heat loss from a volume V in a region of non-uniform temperature equals $k \int_S (\text{grad } T) \cdot d\vec{A}$, where k is a constant, S is the surface bounding V, and $T(x, y, z)$ is the temperature at the point (x, y, z) in space. By taking the limit as V contracts to a point, show that $\partial T/\partial t = B \text{ div grad } T$ at that point, where B is a constant with respect to x, y, z, but may depend on time, t.

29.   A vector field in the plane is a *point source* at the origin if its direction is away from the origin at every point, its magnitude depends only on the distance from the origin, and its divergence is zero away from the origin.

   (a)   Explain why a point source at the origin must be of the form $\vec{v} = [f(x^2 + y^2)](x\vec{i} + y\vec{j})$ for some positive function f.
   (b)   Show that $\vec{v} = K(x^2 + y^2)^{-1}(x\vec{i} + y\vec{j})$ is a point source at the origin if $K > 0$.
   (c)   Determine the magnitude $\|\vec{v}\|$ of the source in part (b) as a function of the distance from its center.
   (d)   Sketch the vector field $\vec{v} = (x^2 + y^2)^{-1}(x\vec{i} + y\vec{j})$.
   (e)   Show that $\phi = \frac{K}{2} \log(x^2 + y^2)$ is a potential function for the source in part (b).

30.   A vector field in the plane is a *point sink* at the origin if its direction is toward the origin at every point, its magnitude depends only on the distance from the origin, and its divergence is zero away from the origin.

   (a)   Explain why a point sink at the origin must be of the form $\vec{v} = [f(x^2 + y^2)](x\vec{i} + y\vec{j})$ for some negative function f.
   (b)   Show that $\vec{v} = K(x^2 + y^2)^{-1}(x\vec{i} + y\vec{j})$ is a point sink at the origin if $K < 0$.
   (c)   Determine the magnitude $\|\vec{v}\|$ of the sink in part (b) as a function of the distance from its center.
   (d)   Sketch the vector field $\vec{v} = -(x^2 + y^2)^{-1}(x\vec{i} + y\vec{j})$.
   (e)   Show that $\phi = \frac{K}{2} \log(x^2 + y^2)$ is a potential function for the sink in part (b).

20.2   THE DIVERGENCE THEOREM

The Divergence Theorem is a multivariable analogue of the Fundamental Theorem of Calculus; it says that the integral of the flux density over a solid region equals the flux integral through the boundary of the region.

The Boundary of a Solid Region

The boundary of a solid region may be thought of as the skin between the interior of the region and the space around it. For example, the boundary of a solid ball is a spherical surface, the boundary of a solid cube is its six faces, and the boundary of a solid cylinder is a tube sealed at both ends by disks. (See Figure 20.9). A surface which is the boundary of a solid region is called a *closed surface*.

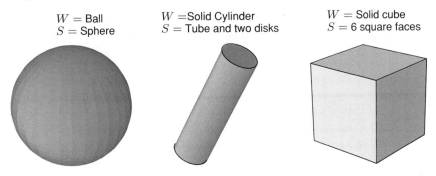

W = Ball
S = Sphere

W = Solid Cylinder
S = Tube and two disks

W = Solid cube
S = 6 square faces

Figure 20.9: Several solid regions and their boundaries

Calculating the Flux from the Flux Density

Consider a solid region W in 3-space whose boundary is the closed surface S. There are two ways to find the total flux of a vector field $\vec{F}$ out of W. One is to calculate the flux of $\vec{F}$ through S:

$$\text{Flux out of } W = \int_S \vec{F} \cdot d\vec{A}.$$

Another way is to use div $\vec{F}$, which gives the flux density at any point in W. We subdivide W into small boxes, as shown in Figure 20.10. Then, for a small box of volume ΔV,

$$\text{Flux out of box} \approx \text{Flux density} \cdot \text{Volume} = \text{div } \vec{F} \, \Delta V.$$

What happens when we add the fluxes out of all the boxes? Consider two adjacent boxes, as shown in Figure 20.11. The flux through the shared wall is counted twice, once out of the box on each side. When we add the fluxes, these two contributions cancel, so we get the flux out of the solid region formed by joining the two boxes. Continuing in this way, we find that

$$\text{Flux out of } W = \sum \text{Flux out of small boxes} \approx \sum \text{div } \vec{F} \, \Delta V.$$

We have approximated the flux by a Riemann sum. As the subdivision gets finer, the sum approaches an integral, so

$$\text{Flux out of } W = \int_W \text{div } \vec{F} \, dV.$$

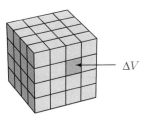

Figure 20.10: Subdivision
of region into small boxes

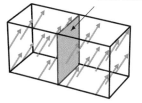

Fluxes through
inner wall cancel

Figure 20.11: Adding the flux out of
adjacent boxes

We have calculated the flux in two ways, as a flux integral and as a volume integral. Therefore, these two integrals must be equal. This result holds even if W is not a rectangular solid, as shown in Figure 20.10. Thus, we have the following result.

The Divergence Theorem

If W is a solid region whose boundary S is a piecewise smooth surface, and if $\vec{F}$ is a smooth vector field which is defined everywhere in W and on S, then

$$\int_S \vec{F} \cdot d\vec{A} = \int_W \operatorname{div} \vec{F} \, dV,$$

where S is given the outward orientation.

We will give a proof of the Divergence Theorem using the coordinate definition of the divergence in Section 20.6.

Example 1   Use the Divergence Theorem to calculate the flux of the vector field $\vec{F}(\vec{r}) = \vec{r}$ through the sphere of radius a centered at the origin.

Solution   In Example 4 on page 391 we computed the flux integral directly:

$$\int_S \vec{r} \cdot d\vec{A} = 4\pi a^3.$$

Now we use $\operatorname{div} \vec{F} = 3$ and the Divergence Theorem:

$$\int_S \vec{r} \cdot d\vec{A} = \int_W \operatorname{div} \vec{F} \, dV = \int_W 3 \, dV = 3\left(\frac{4}{3}\pi a^3\right) = 4\pi a^3.$$

Example 2   Use the Divergence Theorem to calculate the flux of the vector field

$$\vec{F}(x, y, z) = (x^2 + y^2)\vec{i} + (y^2 + z^2)\vec{j} + (x^2 + z^2)\vec{k}$$

through the cube in Figure 20.12.

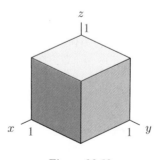

Figure 20.12

Solution   The divergence of $\vec{F}$ is $\operatorname{div} \vec{F} = 2x + 2y + 2z$. Since $\operatorname{div} \vec{F}$ is positive everywhere in the first quadrant, the flux through S is positive. By the Divergence Theorem,

$$\int_S \vec{F} \cdot d\vec{A} = \int_0^1 \int_0^1 \int_0^1 2(x + y + z)\, dx\, dy\, dz = \int_0^1 \int_0^1 x^2 + 2x(y + z) \Big|_0^1 \, dy\, dz$$

$$= \int_0^1 \int_0^1 1 + 2(y + z)\, dy\, dz = \int_0^1 y + y^2 + 2yz \Big|_0^1 \, dz$$

$$= \int_0^1 (2 + 2z)\, dz = 2z + z^2 \Big|_0^1 = 3.$$

The Divergence Theorem and Divergence Free Vector Fields

An important application of the Divergence Theorem is the study of divergence free vector fields.

Example 3   In Example 3 on page 416 we saw that the following vector field is divergence free:

$$\vec{F}(\vec{r}) = \frac{\vec{r}}{\|\vec{r}\|^3}, \qquad \vec{r} \neq \vec{0}.$$

Calculate $\int_S \vec{F} \cdot d\vec{A}$, using the Divergence Theorem if possible, for the following surfaces:

(a) S_1 is the sphere of radius a centered at the origin.
(b) S_2 is the sphere of radius a centered at the point $(2a, 0, 0)$.

Solution   (a)   We cannot use the Divergence Theorem directly because $\vec{F}$ is not defined everywhere inside the sphere (it is not defined at the origin). Since $\vec{F}$ points outward everywhere on S_1, the flux out of S_1 is positive. On S_1,

$$\vec{F} \cdot d\vec{A} = \|\vec{F}\| dA = \frac{a}{a^3} dA,$$

so

$$\int_{S_1} \vec{F} \cdot d\vec{A} = \frac{1}{a^2} \int_{S_1} dA = \frac{1}{a^2} (\text{Area of } S_1) = \frac{1}{a^2} 4\pi a^2 = 4\pi.$$

Notice that the flux is not zero, although div $\vec{F}$ is zero everywhere it is defined.

(b)   Suppose W is the solid region enclosed by S_2. Since div $\vec{F} = 0$ everywhere in W, we can use the Divergence Theorem in this case, giving

$$\int_{S_2} \vec{F} \cdot d\vec{A} = \int_W \text{div } \vec{F} \, dV = \int_W 0\, dV = 0.$$

The Divergence Theorem applies to any solid region W and its boundary S, even in cases where the boundary consists of two or more surfaces. For example, if W is the solid region between the sphere S_1 of radius 1 and the sphere S_2 of radius 2, both centered at the same point, then the boundary of W consists of both S_1 and S_2. The Divergence Theorem requires the outward orientation, which on S_2 points away from the center and on S_1 points towards the center (see Figure 20.13).

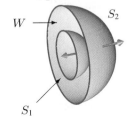

Figure 20.13: Cut-away view of the region W between two spheres, showing orientation vectors

Example 4   Let S_1 be the sphere of radius 1 centered at the origin and let S_2 be the ellipsoid $x^2 + y^2 + 4z^2 = 16$, both oriented outward. For

$$\vec{F}(\vec{r}) = \frac{\vec{r}}{\|\vec{r}\|^3}, \qquad \vec{r} \neq \vec{0},$$

show that

$$\int_{S_1} \vec{F} \cdot d\vec{A} = \int_{S_2} \vec{F} \cdot d\vec{A}.$$

Solution   The ellipsoid contains the sphere; let W be the solid region between them. Since W does not contain the origin, div $\vec{F}$ is defined and equal to zero everywhere in W. Thus, if S is the boundary of W, then

$$\int_S \vec{F} \cdot d\vec{A} = \int_W \text{div}\, \vec{F}\, dV = 0.$$

But S consists of S_2 oriented outwards and S_1 oriented inwards, so

$$0 = \int_S \vec{F} \cdot d\vec{A} = \int_{S_2} \vec{F} \cdot d\vec{A} - \int_{S_1} \vec{F} \cdot d\vec{A},$$

and thus

$$\int_{S_2} \vec{F} \cdot d\vec{A} = \int_{S_1} \vec{F} \cdot d\vec{A}.$$

In Example 3 we showed that $\int_{S_1} \vec{F} \cdot d\vec{A} = 4\pi$, so $\int_{S_2} \vec{F} \cdot d\vec{A} = 4\pi$ also. Note that it would have been more difficult to compute the integral over the ellipsoid directly.

Electric Fields

The electric field produced by a positive point charge q placed at the origin is

$$\vec{E} = q\frac{\vec{r}}{\|\vec{r}\|^3}.$$

Using Example 3 we see that the flux of the electric field through any sphere centered at the origin is $4\pi q$. In fact, using the idea of Example 4, we can show that the flux of $\vec{E}$ through any simple closed surface containing the origin is $4\pi q$. See Problems 20 and 21 on page 457. This is a special case of Gauss's Law, which states that the flux of an electric field through any closed surface is proportional to the total charge enclosed by the surface. Carl Friedrich Gauss (1777-1855) also discovered the Divergence Theorem, which is sometimes called Gauss's Theorem.

Harmonic Functions

A function ϕ of three variables x, y, and z is said to be *harmonic* in a region if $\text{div}(\text{grad}\,\phi) = 0$ at every point in the region. This equation is also written $\nabla^2 \phi = 0$, because $\text{div}(\text{grad}\,\phi) = \nabla \cdot (\nabla \phi)$. For example, the steady-state temperature in a region of space is harmonic, as is the electric potential in a charge-free region of space. Several of the basic properties of harmonic functions can be deduced from the Divergence Theorem (see Problems 19, 23, and 25).

Example 5   It is a fact that a nonconstant harmonic function ϕ cannot have a local maximum. Using the Divergence Theorem, explain why this makes sense.

Solution   Suppose that a nonconstant function ϕ has a local maximum at (x, y, z). For most functions you are likely to encounter, this means that, near (x, y, z), the vector field grad ϕ points approximately towards (x, y, z), because it points in the direction of increasing ϕ. Taking a small sphere S centered at (x, y, z), oriented outwards, we therefore have

$$\int_S \text{grad}\,\phi \cdot d\vec{A} < 0.$$

But this is impossible if ϕ is harmonic, since by the Divergence Theorem

$$\int_S \operatorname{grad}\phi \cdot d\vec{A} = \int_W \operatorname{div}(\operatorname{grad}\phi)dV = 0,$$

where W is the ball enclosed by S. Therefore a harmonic function cannot have a local maximum.

An important property of harmonic functions is their mean value property, discovered by Gauss. If ϕ is a harmonic function in the region enclosed by a sphere, then the value of ϕ at the center of the sphere equals the average value of ϕ on the sphere. For example, at equilibrium, the temperature at a point in space equals the average value of the temperature on any sphere centered at the point.

Problems for Section 20.2

For Problems 1–3, compute the flux integral $\int_S \vec{F} \cdot d\vec{A}$ in two ways, if possible, directly and using the Divergence Theorem. In each case, S is closed and oriented outwards.

1. $\vec{F}(\vec{r}) = \vec{r}$ and S is the cube enclosing the volume $0 \leq x \leq 2, 0 \leq y \leq 2$, and $0 \leq z \leq 2$.

2. $\vec{F}(x, y, z) = y\vec{j}$ and S is a vertical cylinder of height 2, with its base a circle of radius 1 on the xy-plane, centered at the origin. S includes the disks that close it off at top and bottom.

3. $\vec{F}(x, y, z) = -z\vec{i} + x\vec{k}$ and S is a square pyramid with height 3 and base on the xy-plane of side length 1.

4. Suppose V_1 and V_2 are the rectangular solids in the first quadrant shown in Figure 20.14. Both have sides of length 1 parallel to the axes; V_1 has one corner at the origin, while V_2 has the corresponding corner at the point $(1, 0, 0)$. Suppose S_1 and S_2 are the six-faced surfaces of V_1 and V_2, respectively. Suppose V is the box-shaped volume consisting of V_1 and V_2 together, and having outside surface S. Are the following true or false? Give reasons.

   (a) If $\vec{F}$ is a constant vector field, $\int_S \vec{F} \cdot d\vec{A} = 0$

   (b) If S_1, S_2 and S are all oriented outward and $\vec{F}$ is any vector field

$$\int_S \vec{F} \cdot d\vec{A} = \int_{S_1} \vec{F} \cdot d\vec{A} + \int_{S_2} \vec{F} \cdot d\vec{A}.$$

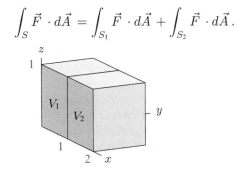

Figure 20.14

5. Calculate $\int_{S_2} \vec{F} \cdot d\vec{A}$ where $\vec{F} = x^2\vec{i} + 2y^2\vec{j} + 3z^2\vec{k}$ and S_2 is as in Problem 4. Do this directly and using the Divergence Theorem.

6. Use the Divergence Theorem to evaluate the flux integral $\int_S (x^2\vec{i} + (y - 2xy)\vec{j} + 10z\vec{k}) \cdot d\vec{A}$, where S is the sphere of radius 5 centered at the origin, oriented outward.

7. Use the Divergence Theorem to calculate the flux of the vector field $\vec{F}(x, y, z) = -z\vec{i} + x\vec{k}$ through the sphere of radius a centered at the origin. Give a geometric explanation for your answer.

8. Suppose $\vec{F}$ is a vector field with div $\vec{F} = 10$. Find the flux of $\vec{F}$ out of a cylinder of height a and radius a, centered on the z-axis and with base in the xy-plane.

9. Consider the vector field $\vec{F} = \vec{r} / \|\vec{r}\|^3$.

    (a) Calculate div $\vec{F}$ for $\vec{r} \neq \vec{0}$.
    (b) Find the flux of $\vec{F}$ out of a box of side a centered at the origin and with edges parallel to the axes.

10. Suppose $\vec{G}$ is a vector field with the property that $\vec{G} = \vec{r}$ for $2 \leq \|\vec{r}\| \leq 7$ and that the flux of $\vec{G}$ through the sphere of radius 3 centered at the origin is 8π. Find the flux of $\vec{G}$ through the sphere of radius 5 centered at the origin.

11. Suppose that a vector field $\vec{F}$ satisfies div$\vec{F} = 0$ everywhere. Show that $\int_S \vec{F} \cdot d\vec{A} = 0$ for every closed surface S.

12. The gravitational field, $\vec{F}$, of a planet of mass m at the origin is given by

$$\vec{F} = -Gm\frac{\vec{r}}{\|\vec{r}\|^3}.$$

Use the Divergence Theorem to show that the flux of the gravitational field through the sphere of radius a is independent of a. [Hint: Consider the region bounded by two concentric spheres.]

13. A basic property of the electric field $\vec{E}$ is that its divergence is zero at points where there is no charge. Suppose that the only charge is along the z-axis, and that the electric field $\vec{E}$ points radially out from the z-axis and its magnitude depends only on the distance r from the z-axis. Use the Divergence Theorem to show that the magnitude of the field is proportional to $1/r$. [Hint: Consider a solid region consisting of a cylinder of finite length whose axis is the z-axis, and with a smaller concentric cylinder removed.]

14. If a surface S is submerged in an incompressible fluid, a force $\vec{F}$ is exerted on one side of the surface by the pressure in the fluid. If we choose a coordinate system where the z-axis is vertical, with the positive direction upward and the fluid level at $z = 0$, then the component of force in the direction of a unit vector $\vec{u}$ is given by the following:

$$\vec{F} \cdot \vec{u} = -\int_S z\rho g\vec{u} \cdot d\vec{A},$$

where ρ is the density of the fluid (mass/volume), g is the acceleration due to gravity, and the surface is oriented away from the side on which the force is exerted. In this problem we will consider a totally submerged closed surface enclosing a volume V. We are interested in the force of the liquid on the external surface, so S is oriented inward.

    (a) Use the Divergence Theorem to show that the force in the $\vec{i}$ and $\vec{j}$ directions is zero.
    (b) Use the Divergence Theorem to show that the force in the $\vec{k}$ direction is $\rho g V$, the weight of the volume of fluid with the same volume as V. This is *Archimedes' Principle.*

15. Heat is generated inside the earth by radioactive decay. Assume it is generated uniformly throughout the earth at a rate of 30 watts per cubic kilometer. (A watt is a rate of heat production.) The heat then flows to the earth's surface where it is lost to space. Let $\vec{F}(x, y, z)$ denote the flow of heat measured in watts per square kilometer. By definition, the flux of $\vec{F}$ across a surface is the amount of heat flowing through the surface per unit of time.

    (a) What is the value of div $\vec{F}$? Include units.

(b)   Assume the heat flows outward symmetrically. Verify that $\vec{F} = \alpha\vec{r}$, where $\vec{r} = x\vec{i} + y\vec{j} + z\vec{k}$ and α is a suitable constant, satisfies the given conditions. Find α.

(c)   Let $T(x, y, z)$ denote the temperature inside the earth. Heat flows according to the equation $\vec{F} = -k\operatorname{grad}T$, where k is a constant. Explain why this makes sense physically.

(d)   If T is in $°C$, then $k = 30,000$ watts/km$°C$. Assuming the earth is a sphere with radius 6400 km and surface temperature $20°C$, what is the temperature at the center?

16.   Show that $\nabla^2\phi(x, y, z) = \partial^2\phi/\partial x^2 + \partial^2\phi/\partial y^2 + \partial^2\phi/\partial z^2$.

17.   Show that linear functions are harmonic.

18.   What is the condition on the constant coefficients a, b, c, d, e, f such that the function $ax^2 + by^2 + cz^2 + dxy + exz + fyz$ is harmonic?

19.   Use the Divergence Theorem to show that if ϕ is harmonic in a region W, then $\int_S \nabla\phi \cdot d\vec{A} = 0$ for every closed surface S in W such that the region enclosed by S lies wholly within W.

20.   Let $\phi = 1/(x^2 + y^2 + z^2)^{1/2} = 1/\|\vec{r}\|$.

(a)   Show that ϕ is harmonic everywhere except the origin.

(b)   Give a geometric explanation for the fact that ϕ has no local maximum or minimum.

(c)   Compute $\int_S \nabla\phi \cdot d\vec{A}$ where S is the sphere of radius 1 centered at the origin. Does your answer contradict the assertion of Problem 19?

21.   Show that a nonconstant harmonic function cannot have a local minimum and that it can achieve a minimum value in a closed region only on the boundary.

22.   Show that if ϕ is a harmonic function, then $\operatorname{div}(\phi\operatorname{grad}\phi) = \|\operatorname{grad}\phi\|^2$.

23.   Suppose that ϕ is a harmonic function in the region enclosed by a closed surface S and suppose that $\phi = 0$ at all points of S. Show that $\phi = 0$ at all points of the region enclosed by S. [Hint: Apply the Divergence Theorem to $\int_S \phi\operatorname{grad}\phi \cdot d\vec{A}$ and use Problem 22.]

24.   Suppose that ϕ_1 and ϕ_2 are harmonic functions in the region enclosed by a closed surface S and suppose that $\phi_1 = \phi_2$ at all points of S. Show that $\phi_1 = \phi_2$ at all points of the region enclosed by S. [Hint: Use Problem 23.]

25.   Suppose u and v are harmonic functions in a region W. Use the Divergence Theorem to show that for every closed surface S in W such that the volume enclosed by S lies completely within W

$$\int_S u\operatorname{grad}v \cdot d\vec{A} = \int_S v\operatorname{grad}u \cdot d\vec{A}.$$

20.3   THE CURL OF A VECTOR FIELD

The divergence of a vector field is a sort of scalar derivative which measures its outflow per unit volume. Now we introduce a vector derivative, the curl, which measures the circulation of a vector field. Imagine holding the paddle-wheel in Figure 20.15 in the flow shown by Figure 20.16. The speed at which the paddle-wheel spins (its angular velocity) measures the strength of circulation. Note that the angular velocity depends on the direction in which the stick is pointing. The paddle-wheel spins one way if the stick is pointing up and the opposite way if it is pointing down. If the stick is pointing horizontally it doesn't spin at all, because the velocity field strikes opposite vanes of the paddle with equal force.

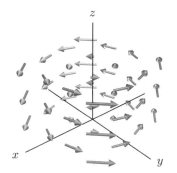

Figure 20.15: A device for
measuring circulation

Figure 20.16: A vector field with circulation
about the z-axis

Circulation Density

We measure the strength of the circulation using a closed curve. Suppose C is a circle with center (x, y, z) in the plane perpendicular to $\vec{n}$, traversed in the direction determined from $\vec{n}$ by the right-hand rule. (See Figures 20.17 and 20.18.)

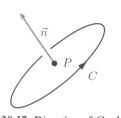

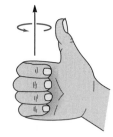

Figure 20.17: Direction of C relates to
direction of $\vec{n}$ by the right-hand rule

Figure 20.18: When the thumb points in the direction of $\vec{n}$,
the fingers curl in the forward direction around C

We make the following definition:

The **circulation density** of a smooth vector field $\vec{F}$ at (x, y, z) around the direction of the unit vector $\vec{n}$ is defined to be

$$\text{circ}_{\vec{n}} \, \vec{F}\,(x, y, z) = \lim_{\text{Area} \to 0} \frac{\text{Circulation around } C}{\text{Area inside } C} = \lim_{\text{Area} \to 0} \frac{\displaystyle\int_C \vec{F} \cdot d\vec{r}}{\text{Area inside } C},$$

provided the limit exists.

The circulation density determines the angular velocity[3] of the paddle-wheel in Figure 20.15 provided you could make one sufficiently small and light and insert it without disturbing the flow.

Example 1   Consider the vector field $\vec{F}$ in Figure 20.16. Suppose that $\vec{F}$ is parallel to the xy-plane and that at a distance r from the z-axis it has magnitude $2r$. Calculate $\text{circ}_{\vec{n}} \, \vec{F}$ at the origin for
(a) $\vec{n} = \vec{k}$ (b) $\vec{n} = -\vec{k}$ (c) $\vec{n} = \vec{i}$.

[3]In fact it is twice the angular velocity. See Example 3 on page 431.

Solution   (a)   Take a circle C of radius a in the xy-plane, centered at the origin, traversed in a direction determined from $\vec{k}$ by the right hand rule. Then, since $\vec{F}$ is tangent to C everywhere and points in the forward direction around C, we have

$$\text{Circulation around } C = \int_C \vec{F} \cdot d\vec{r} = \|\vec{F}\| \cdot \text{Circumference of } C = 2a(2\pi a) = 4\pi a^2.$$

Thus, the circulation density is

$$\text{circ}_{\vec{k}}\, \vec{F} = \lim_{a \to 0} \frac{\text{Circulation around } C}{\text{Area inside } C} = \lim_{a \to 0} \frac{4\pi a^2}{\pi a^2} = 4.$$

(b)   If $\vec{n} = -\vec{k}$ the circle is traversed in the opposite direction, so the line integral changes sign. Thus

$$\text{circ}_{-\vec{k}}\, \vec{F} = -4.$$

(c)   The circulation around $\vec{i}$ is calculated using circles in the yz-plane. Since $\vec{F}$ is everywhere perpendicular to such a circle C,

$$\int_C \vec{F} \cdot d\vec{r} = 0.$$

Thus, we have

$$\text{circ}_{\vec{i}}\, \vec{F} = \lim_{a \to 0} \frac{\int_C \vec{F} \cdot d\vec{r}}{\pi a^2} = \lim_{a \to 0} \frac{0}{\pi a^2} = 0.$$

Definition of the Curl

Example 1 shows that the circulation density of a vector field can be positive, negative, or zero, depending on the direction. We assume that there is one direction in which the circulation density is greatest. Now we define a single vector quantity that incorporates all these different circulation densities.

Geometric Definition of Curl

The curl of a smooth vector field $\vec{F}$, written curl $\vec{F}$, is the vector field with the following properties
- The direction of curl $\vec{F}(x, y, z)$ is the direction $\vec{n}$ for which $\text{circ}_{\vec{n}}(x, y, z)$ is the greatest.
- The magnitude of curl $\vec{F}(x, y, z)$ is the circulation density of $\vec{F}$ around that direction.

If the circulation density is zero around every direction, then we define the curl to be $\vec{0}$.

Cartesian Coordinate Definition of Curl

If $\vec{F} = F_1\vec{i} + F_2\vec{j} + F_3\vec{k}$, then

$$\text{curl}\, \vec{F} = \left(\frac{\partial F_3}{\partial y} - \frac{\partial F_2}{\partial z}\right)\vec{i} + \left(\frac{\partial F_1}{\partial z} - \frac{\partial F_3}{\partial x}\right)\vec{j} + \left(\frac{\partial F_2}{\partial x} - \frac{\partial F_1}{\partial y}\right)\vec{k}.$$

Example 2   For each field in Figure 20.19, use the sketch and the geometric definition to decide whether the curl at the origin points up, down, or is the zero vector. Then check your answer using the coordinate definition of curl. Note that the vector fields have no z-components and are independent of z.

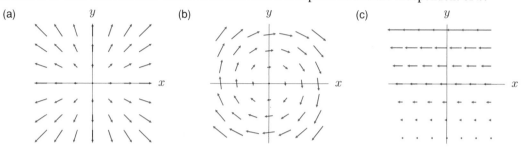

Figure 20.19: Sketches in the xy-plane of (a) $\vec{F} = x\vec{i} + y\vec{j}$ (b) $\vec{F} = y\vec{i} - x\vec{j}$ (c) $\vec{F} = -(y+1)\vec{i}$

Solution   (a)   This vector field shows no rotation, and the circulation around any closed curve appears to be zero, so we suspect that the curl is the zero vector. The coordinate definition of curl gives

$$\text{curl } \vec{F} = \left(\frac{\partial(0)}{\partial y} - \frac{\partial y}{\partial z}\right)\vec{i} + \left(\frac{\partial x}{\partial z} - \frac{\partial(0)}{\partial x}\right)\vec{j} + \left(\frac{\partial y}{\partial x} - \frac{\partial x}{\partial y}\right)\vec{k} = \vec{0}.$$

(b)   This vector field is rotating around the z-axis. By the right-hand rule, the circulation density around $\vec{k}$ is negative, so we expect the z-component of the curl points down. The coordinate definition gives

$$\text{curl } \vec{F} = \left(\frac{\partial(0)}{\partial y} - \frac{\partial(-x)}{\partial z}\right)\vec{i} + \left(\frac{\partial y}{\partial z} - \frac{\partial(0)}{\partial x}\right)\vec{j} + \left(\frac{\partial(-x)}{\partial x} - \frac{\partial y}{\partial y}\right)\vec{k} = -2\vec{k}.$$

(c)   At first glance, you might expect this vector field to have zero curl, as all the vectors are parallel to the x-axis. However, if you find the circulation around the curve C in Figure 20.20, the sides contribute nothing (they are perpendicular to the vector field), the bottom contributes a negative quantity (the curve is in the opposite direction to the vector field), and the top contributes a larger positive quantity (the curve is in the same direction as the vector field and the magnitude of the vector field is larger at the top than at the bottom). Thus, the circulation around C is positive and hence we expect the curl to be nonzero and point up. The coordinate definition gives

$$\text{curl } \vec{F} = \left(\frac{\partial(0)}{\partial y} - \frac{\partial(0)}{\partial z}\right)\vec{i} + \left(\frac{\partial(-(y+1))}{\partial z} - \frac{\partial(0)}{\partial x}\right)\vec{j} + \left(\frac{\partial(0)}{\partial x} - \frac{\partial(-(y+1))}{\partial y}\right)\vec{k} = \vec{k}.$$

Another way to see that the curl is nonzero in this case is to imagine the vector field representing the velocity of moving water. A boat sitting in the water tends to rotate, as the water moves faster on one side than the other.

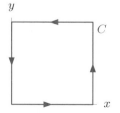

Figure 20.20

Alternative Notation for Curl

Using $\nabla = \frac{\partial}{\partial x}\vec{i} + \frac{\partial}{\partial y}\vec{j} + \frac{\partial}{\partial z}\vec{k}$, we can write

$$\text{curl } \vec{F} = \nabla \times \vec{F} = \begin{vmatrix} \vec{i} & \vec{j} & \vec{k} \\ \frac{\partial}{\partial x} & \frac{\partial}{\partial y} & \frac{\partial}{\partial z} \\ F_1 & F_2 & F_3 \end{vmatrix}.$$

Example 3   A flywheel is rotating with angular velocity $\vec{\omega}$ and the velocity of a point P with position vector $\vec{r}$ is given by $\vec{v} = \vec{\omega} \times \vec{r}$. (See Figure 20.21.) Calculate curl $\vec{v}$.

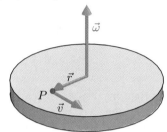

Figure 20.21: Rotating flywheel

Solution   If $\vec{\omega} = \omega_1\vec{i} + \omega_2\vec{j} + \omega_3\vec{k}$, we have

$$\vec{v} = \vec{\omega} \times \vec{r} = \begin{vmatrix} \vec{i} & \vec{j} & \vec{k} \\ \omega_1 & \omega_2 & \omega_3 \\ x & y & z \end{vmatrix} = (\omega_2 z - \omega_3 y)\vec{i} + (\omega_3 x - \omega_1 z)\vec{j} + (\omega_1 y - \omega_2 x)\vec{k}.$$

Thus,

$$\text{curl } \vec{v} = \begin{vmatrix} \vec{i} & \vec{j} & \vec{k} \\ \frac{\partial}{\partial x} & \frac{\partial}{\partial y} & \frac{\partial}{\partial z} \\ \omega_2 z - \omega_3 y & \omega_3 x - \omega_1 z & \omega_1 y - \omega_2 x \end{vmatrix}$$

$$= \left(\frac{\partial}{\partial y}(\omega_1 y - \omega_2 x) - \frac{\partial}{\partial z}(\omega_3 x - \omega_1 z) \right)\vec{i} + \left(\frac{\partial}{\partial z}(\omega_2 z - \omega_3 y) - \frac{\partial}{\partial x}(\omega_1 y - \omega_2 x) \right)\vec{j}$$

$$+ \left(\frac{\partial}{\partial x}(\omega_3 x - \omega_1 z) - \frac{\partial}{\partial y}(\omega_2 z - \omega_3 y) \right)\vec{k}$$

$$= 2\omega_1\vec{i} + 2\omega_2\vec{j} + 2\omega_3\vec{k} = 2\vec{\omega}.$$

Thus, as we would expect, curl $\vec{v}$ is parallel to the axis of rotation of the flywheel (namely, the direction of $\vec{\omega}$) and the magnitude of curl $\vec{v}$ is larger the faster the flywheel is rotating (that is, the larger the magnitude of $\vec{\omega}$).

Why Do the Two Definitions of Curl Give the Same Result?

Using Green's Theorem in Cartesian coordinates, we can show that for curl $\vec{F}$ defined in Cartesian coordinates

$$\boxed{\text{curl } \vec{F} \cdot \vec{n} = \text{circ}_{\vec{n}} \, \vec{F}.}$$

This shows that curl $\vec{F}$ defined in Cartesian coordinates satisfies the geometric definition, since the left hand side takes its maximum value when $\vec{n}$ points in the same direction as curl $\vec{F}$, and in that case its value is $\| \text{curl } \vec{F} \|$.

The following example justifies this formula in a specific case. Problems 27 and 28 on page 436 show how to prove curl $\vec{F} \cdot \vec{n} = \text{circ}_{\vec{n}} \; \vec{F}$ in general.

Example 4   Use the definition of curl in Cartesian coordinates and Green's Theorem to show that

$$\left(\text{curl } \vec{F} \right) \cdot \vec{k} = \text{circ}_{\vec{k}} \; \vec{F}.$$

Solution   Using the definition of curl in Cartesian coordinates, the left hand side of the formula is

$$\left(\text{curl } \vec{F} \right) \cdot \vec{k} = \frac{\partial F_2}{\partial x} - \frac{\partial F_1}{\partial y}.$$

Now let's look at the right hand side. The circulation density around $\vec{k}$ is calculated using circles perpendicular to $\vec{k}$; hence the $\vec{k}$-component of $\vec{F}$ does not contribute to it, that is, the circulation density of $\vec{F}$ around $\vec{k}$ is the same as the circulation density of $F_1 \vec{i} + F_2 \vec{j}$ around $\vec{k}$. But in any plane perpendicular to $\vec{k}$, z is constant, so in that plane F_1 and F_2 are functions of x and y alone. Thus $F_1 \vec{i} + F_2 \vec{j}$ can be thought of as a two-dimensional vector field on the horizontal plane through the point (x, y, z) where the circulation density is being calculated. Let C be a circle in this plane, with radius a and centered at (x, y, z), and let R be the region enclosed by C. Green's Theorem says that

$$\int_C (F_1 \vec{i} + F_2 \vec{j}) \cdot d\vec{r} = \int_R \left(\frac{\partial F_2}{\partial x} - \frac{\partial F_1}{\partial y} \right) dA.$$

When the circle is small, $\partial F_2 / \partial x - \partial F_1 / \partial y$ is approximately constant on R, so

$$\int_R \left(\frac{\partial F_2}{\partial x} - \frac{\partial F_1}{\partial y} \right) dA \approx \left(\frac{\partial F_2}{\partial x} - \frac{\partial F_1}{\partial y} \right) \cdot \text{Area of } R = \left(\frac{\partial F_2}{\partial x} - \frac{\partial F_1}{\partial y} \right) \pi a^2.$$

Thus, taking a limit as the radius of the circle goes to zero, we have

$$\text{circ}_{\vec{k}} \; \vec{F}(x, y, z) = \lim_{a \to 0} \frac{\int_C (F_1 \vec{i} + F_2 \vec{j}) \cdot d\vec{r}}{\pi a^2} = \lim_{a \to 0} \frac{\int_R \left(\frac{\partial F_2}{\partial x} - \frac{\partial F_1}{\partial y} \right) dA}{\pi a^2} = \frac{\partial F_2}{\partial x} - \frac{\partial F_1}{\partial y}.$$

Curl Free Vector Fields

A vector field is said to be *curl free* or *irrotational* if curl $\vec{F} = \vec{0}$ everywhere that $\vec{F}$ is defined.

Example 5   Figure 20.22 shows the vector field $\vec{B}$ for three values of the constant p, where $\vec{B}$ is defined on 3-space by

$$\vec{B} = \frac{-y \vec{i} + x \vec{j}}{(x^2 + y^2)^{p/2}}.$$

(a)   Find a formula for curl $\vec{B}$.

(b)   Is there a value of p for which $\vec{B}$ is curl free? If so, find it.

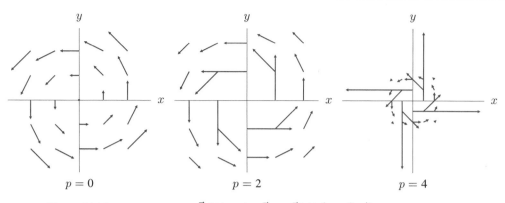

Figure 20.22: The vector field $\vec{B}(\vec{r}) = (-y\vec{i} + x\vec{j})/(x^2 + y^2)^{p/2}$ for $p = 0, 2,$ and 4

Solution   (a)   We can use the following version of the product rule for curl. If ϕ is a scalar function and $\vec{F}$ is a vector field, then

$$\operatorname{curl}(\phi\vec{F}) = \phi \operatorname{curl} \vec{F} + (\operatorname{grad} \phi) \times \vec{F}.$$

(See Problem 17 on page 434.) We write $\vec{B} = \phi\vec{F} = \dfrac{1}{(x^2 + y^2)^{p/2}}(-y\vec{i} + x\vec{j})$. Then

$$\operatorname{curl} \vec{F} = \operatorname{curl}(-y\vec{i} + x\vec{j}) = 2\vec{k}$$

$$\operatorname{grad} \phi = \operatorname{grad}\left(\frac{1}{(x^2 + y^2)^{p/2}}\right) = \frac{-p}{(x^2 + y^2)^{(p/2)+1}}(x\vec{i} + y\vec{j}).$$

Thus, we have

$$\operatorname{curl} \vec{B} = \frac{1}{(x^2 + y^2)^{p/2}}\operatorname{curl}(-y\vec{i} + x\vec{j}) + \operatorname{grad}\left(\frac{1}{(x^2 + y^2)^{p/2}}\right) \times (-y\vec{i} + x\vec{j})$$

$$= \frac{1}{(x^2 + y^2)^{p/2}}2\vec{k} + \frac{-p}{(x^2 + y^2)^{(p/2)+1}}(x\vec{i} + y\vec{j}) \times (-y\vec{i} + x\vec{j})$$

$$= \frac{1}{(x^2 + y^2)^{p/2}}2\vec{k} + \frac{-p}{(x^2 + y^2)^{(p/2)+1}}(x^2 + y^2)\vec{k}$$

$$= \frac{2 - p}{(x^2 + y^2)^{p/2}}\vec{k}.$$

(b)   The curl is zero when $p = 2$. Thus, when $p = 2$ the vector field is curl free:

$$\vec{B} = \frac{-y\vec{i} + x\vec{j}}{x^2 + y^2}.$$

Problems for Section 20.3

Compute the curl of the vector fields in Problems 1–7.

1. $\vec{F} = (x^2 - y^2)\vec{i} + 2xy\vec{j}$

2. $\vec{F}(\vec{r}) = \vec{r}/\|\vec{r}\|$

3. $\vec{F} = x^2\vec{i} + y^3\vec{j} + z^4\vec{k}$

4. $\vec{F} = e^x\vec{i} + \cos y\vec{j} + e^{z^2}\vec{k}$

5. $\vec{F} = 2yz\vec{i} + 3xz\vec{j} + 7xy\vec{k}$

6. $\vec{F} = (-x+y)\vec{i} + (y+z)\vec{j} + (-z+x)\vec{k}$

7. $\vec{F} = (x+yz)\vec{i} + (y^2 + xzy)\vec{j} + (zx^3y^2 + x^7y^6)\vec{k}$

8. Use the geometric definition to find the curl of the vector field $\vec{F}(\vec{r}) = \vec{r}$. Check your answer using the coordinate definition.

9. Using your answers to Problems 3–4, make a conjecture about the value of curl $\vec{F}$ when the vector field $\vec{F}$ has a certain form. (What form?) Show why your conjecture is true.

10. Let $\vec{F}$ be the vector field in Figure 20.16 on page 428. It is rotating counterclockwise around the z-axis when viewed from above. Suppose that at a distance r from the z-axis $\vec{F}$ has magnitude $2r$.

    (a) Find a formula for $\vec{F}$.
    (b) Find curl $\vec{F}$ using the coordinate definition and relate your answer to circulation density.

11. Decide whether each of the following vector fields has a nonzero curl at the origin. In each case, the vector field is shown in the xy-plane; assume it has no z-component and is independent of z.

12. A large fire becomes a fire-storm when the nearby air acquires a circulatory motion. The associated updraft has the effect of bringing more air to the fire, causing it to burn faster. Records show that a fire-storm developed during the Chicago Fire of 1871 and during the Second World War bombing of Hamburg, Germany, but there was no fire-storm during the Great Fire of London in 1666. Explain how a fire-storm could be identified using the curl of a vector field.

13. Show that curl $(\vec{F} + \vec{C}) = \text{curl } \vec{F}$ for a constant vector field $\vec{C}$.

14. For any constant vector field $\vec{c}$, and any vector field, $\vec{F}$, show that $\text{div}(\vec{F} \times \vec{c}) = \vec{c} \cdot \text{curl } \vec{F}$.

15. In Chapter 18 we saw how the Fundamental Theorem of Calculus for Line Integrals implies $\int_C \text{grad } f \cdot d\vec{r} = 0$ for any smooth closed path C and any smooth function f.

    (a) Use the geometric definition of curl to deduce that curl grad $f = \vec{0}$.
    (b) Verify that curl grad $f = \vec{0}$ using the coordinate definition.

16. If $\vec{F}$ is any vector field whose components have continuous second partial derivatives, show that div curl $\vec{F} = 0$.

17. Show that curl $(\phi\vec{F}) = \phi \text{ curl } \vec{F} + (\text{grad } \phi) \times \vec{F}$ for a scalar function ϕ and a vector field $\vec{F}$.

18. A vortex that rotates at constant angular velocity ω about the z-axis has velocity vector field $\vec{v} = \omega(-y\vec{i} + x\vec{j})$.

    (a) Sketch the vector field with $\omega = 1$ and the vector field with $\omega = -1$.
    (b) Determine the speed $\|\vec{v}\|$ of the vortex as a function of the distance from its center.
    (c) Compute div $\vec{v}$ and curl $\vec{v}$.
    (d) Compute the circulation of $\vec{v}$ counterclockwise about the circle of radius R in the xy-plane, centered at the origin.

Problems 19–21 concern the vector fields in Figure 20.23. In each case, assume that the cross-section is the same in all other planes parallel to the given cross-section.

19. Three of the vector fields have zero curl at each point shown. Which are they? How do you know?

20. Three of the vector fields have zero divergence at each point shown. Which are they? How do you know?

21. Four of the line integrals $\int_{C_i} \vec{F} \cdot d\vec{r}$ are zero. Which are they? How do you know?

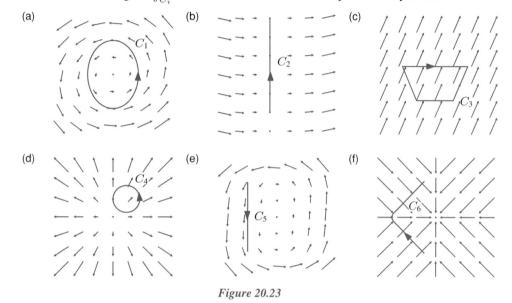

Figure 20.23

22. Show that if ϕ is a harmonic function, then grad ϕ is both curl free and divergence free.

23. It is a theorem of Helmholtz that every vector field $\vec{F}$ equals the sum of a curl free vector field and a divergence free vector field. Show how to do this, assuming that there is a function ϕ such that $\nabla^2\phi = \text{div}\vec{F}$.

24. Express $(3x + 2y)\vec{i} + (4x + 9y)\vec{j}$ as the sum of a curl free vector field and a divergence free vector field.

25. Find a vector field $\vec{F}$ such that curl $\vec{F} = 2\vec{i} - 3\vec{j} + 4\vec{k}$. [Hint: Try $\vec{F} = \vec{v} \times \vec{r}$ for some vector $\vec{v}$.]

26. Figure 20.24 gives a sketch of a velocity vector field $\vec{F} = y\vec{i} + x\vec{j}$ in the xy-plane.
     (a) What is the direction of rotation of a thin twig placed at the origin along the x-axis?
     (b) What is the direction of rotation of a thin twig placed at the origin along the y-axis?
     (c) Compute curl $\vec{F}$.

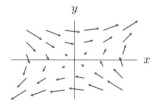

Figure 20.24

27. Let $\vec{F}$ be a smooth vector field and let $\vec{u}$ and $\vec{v}$ be constant vectors. Using the definition of curl $\vec{F}$ in Cartesian coordinates, show that

$$\operatorname{grad}(\vec{F} \cdot \vec{v}) \cdot \vec{u} - \operatorname{grad}(\vec{F} \cdot \vec{u}) \cdot \vec{v} = (\operatorname{curl} \vec{F}) \cdot \vec{u} \times \vec{v}.$$

28. Let $\vec{F}$ be a smooth vector field in 3-space. Consider a plane L in 3-space parameterized by

$$\vec{r}(s, t) = \vec{r}_0 + s\vec{u} + t\vec{v},$$

where $\vec{u}$ and $\vec{v}$ are orthogonal unit vectors. We can think of this plane as a copy of the Cartesian plane with coordinates s and t, sitting in 3-space. We define a 2-dimensional vector field, $\vec{G}$, on this plane by

$$\vec{G}(s, t) = \text{component of } \vec{F}(\vec{r}(s, t)) \text{ parallel to } L.$$

(a) Show that $\vec{G} = G_1 \vec{u} + G_2 \vec{v}$, where $G_1 = \vec{F} \cdot \vec{u}$, and $G_2 = \vec{F} \cdot \vec{v}$.

(b) Show that

$$\frac{\partial G_2}{\partial s} - \frac{\partial G_1}{\partial t} = \operatorname{grad}(\vec{F} \cdot \vec{v}) \cdot \vec{u} - \operatorname{grad}(\vec{F} \cdot \vec{u}) \cdot \vec{v}.$$

(c) Let $\vec{n} = \vec{u} \times \vec{v}$. Use part (b) and Problem 27 to deduce that

$$\operatorname{curl} \vec{F} \cdot \vec{n} = \frac{\partial G_2}{\partial s} - \frac{\partial G_1}{\partial t}.$$

(d) Use the method of Example 4 on page 432 and Green's Theorem to conclude that

$$\operatorname{curl} \vec{F}(\vec{r}_0) \cdot \vec{n} = \operatorname{circ}_{\vec{n}} \vec{F}(\vec{r}_0).$$

Since $\vec{r}_0$, $\vec{u}$, and $\vec{v}$ can be chosen to be any vectors, this proves that, in general,

$$(\operatorname{curl} \vec{F}) \cdot \vec{n} = \operatorname{circ}_{\vec{n}} \vec{F}.$$

20.4 STOKES' THEOREM

The Divergence Theorem says that the integral of the flux density over a solid region is equal to the flux through the surface bounding the region. Similarly, Stokes' Theorem says that the integral of the circulation density over a surface is equal to the circulation around the boundary of the surface.

The Boundary of a Surface

The *boundary* of a surface S is the curve running around the edge of S (like the hem around the edge of a piece of cloth). An orientation of S determines an orientation for its boundary, C, as follows. Pick a positive normal vector $\vec{n}$ on S, near C, and use the right hand rule to determine a direction of travel around $\vec{n}$. This in turn determines a direction of travel around the boundary C. See Figure 20.25. Another way of describing the orientation on C is that someone walking along C in the forward direction, body upright in the direction of the positive normal on S, would have the surface on their left. Notice that the boundary can consist of two or more curves, as the surface on the right in Figure 20.25 shows.

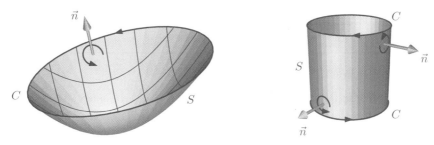

Figure 20.25: Two oriented surfaces and their boundaries

Calculating the Circulation from the Circulation Density

Consider a closed, oriented curve C in 3-space. We can find the circulation of a vector field $\vec{F}$ around C by calculating the line integral:

$$\begin{array}{c} \text{Circulation} \\ \text{around } C \end{array} = \int_C \vec{F} \cdot d\vec{r} \,.$$

If C is the boundary of an oriented surface S, there is another way to calculate the circulation using curl $\vec{F}$. We subdivide S into pieces as shown on the surface on the left in Figure 20.25. If $\vec{n}$ is a positive unit normal vector to a piece of surface with area ΔA, then $\Delta \vec{A} = \vec{n} \, \Delta A$. In addition, $\mathrm{circ}_{\vec{n}} \, \vec{F}$ is the circulation density of $\vec{F}$ around $\vec{n}$, so

$$\begin{array}{c} \text{Circulation of } \vec{F} \text{ around} \\ \text{boundary of the piece} \end{array} \approx \left(\mathrm{circ}_{\vec{n}} \, \vec{F} \right) \Delta A = ((\mathrm{curl} \, \vec{F}) \cdot \vec{n}) \Delta A = (\mathrm{curl} \vec{F}) \cdot \Delta \vec{A} \,.$$

Next we add up the circulations around all the small pieces. The line integral along the common edge of a pair of adjacent pieces appears with opposite sign in each piece, so it cancels out. (See Figure 20.26.) When we add up all the pieces the internal edges cancel and we are left with the circulation around C, the boundary of the entire surface. Thus,

$$\begin{array}{c} \text{Circulation} \\ \text{around } C \end{array} = \sum \begin{array}{c} \text{Circulation around} \\ \text{boundary of pieces} \end{array} \approx \sum \mathrm{curl} \, \vec{F} \cdot \Delta \vec{A} \,.$$

Taking the limit as $\Delta A \to 0$, we get

$$\begin{array}{c} \text{Circulation} \\ \text{around } C \end{array} = \int_S \mathrm{curl} \, \vec{F} \cdot d\vec{A} \,.$$

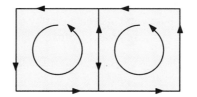

Figure 20.26: Two adjacent pieces of
the surface

We have expressed the circulation as a line integral around C and as a flux integral over S; thus, the two integrals must be equal. Hence we have

---

Stokes' Theorem

If S is a smooth oriented surface with piecewise smooth, oriented boundary C, and if $\vec{F}$ is a smooth vector field which is defined on S and C, then

$$\int_C \vec{F} \cdot d\vec{r} = \int_S \operatorname{curl} \vec{F} \cdot d\vec{A}.$$

The orientation of C is determined from the orientation of S according to the right hand rule.

A proof of Stokes' Theorem using the coordinate definition of curl is given in Section 20.6.

Example 1   Let $\vec{F}(x, y, z) = -2y\vec{i} + 2x\vec{j}$. Use Stokes' Theorem to find $\int_C \vec{F} \cdot d\vec{r}$, where C is a circle

(a)   Parallel to the yz-plane, of radius a, centered at a point on the x-axis, with either orientation.
(b)   Parallel to the xy-plane, of radius a, centered at a point on the z-axis, oriented counterclockwise as viewed from a point on the z-axis above the circle.

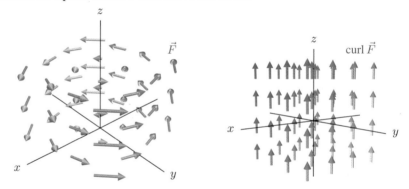

Figure 20.27: The vector fields $\vec{F}$ and curl $\vec{F}$

Solution   We have curl $\vec{F} = 4\vec{k}$. Figure 20.27 shows sketches of $\vec{F}$ and curl $\vec{F}$.

(a)   Let S be the disk enclosed by C. Since S lies in a vertical plane and curl $\vec{F}$ points vertically everywhere, the flux of curl $\vec{F}$ through S is zero. Hence, by Stokes' Theorem,

$$\int_C \vec{F} \cdot d\vec{r} = \int_S \operatorname{curl} \vec{F} \cdot d\vec{A} = 0.$$

It makes sense that the line integral is zero. If C is parallel to the yz-plane (even if it is not lying in the plane), the symmetry of the vector field means that the line integral of $\vec{F}$ over the top half of the circle cancels the line integral over the bottom half.

(b)   Let S be the horizontal disk enclosed by C. Since curl $\vec{F}$ is a constant vector field pointing in the direction of $\vec{k}$, we have, by Stokes' Theorem,

$$\int_C \vec{F} \cdot d\vec{r} = \int_S \operatorname{curl} \vec{F} \cdot d\vec{A} = \| \operatorname{curl} \vec{F} \| \cdot \text{Area of } S = 4\pi a^2.$$

Since $\vec{F}$ is circling around the z-axis in the same direction as C, we expect the line integral to be positive. In fact, in Example 1 on page 428, we computed this line integral directly.

Curl Free Vector Fields

Stokes' Theorem applies to any oriented surface S and its boundary C, even in cases where the boundary consists of two or more curves. This is particularly useful in studying curl free vector fields.

Example 2   A current I flows along the z-axis in the $\vec{k}$ direction. The induced magnetic field $\vec{B}\,(x, y, z)$ is

$$\vec{B}\,(x, y, z) = \frac{2I}{c}\left(\frac{-y\vec{i} + x\vec{j}}{x^2 + y^2}\right),$$

where c is the speed of light. In Example 5 on page 432 we showed that curl $\vec{B} = \vec{0}$.

(a) Compute the circulation of $\vec{B}$ around the circle C_1 in the xy-plane of radius a, centered at the origin, and oriented counterclockwise when viewed from above.

(b) Use part (a) and Stokes' Theorem to compute $\int_{C_2} \vec{B} \cdot d\vec{r}$, where C_2 is the ellipse $x^2 + 9y^2 = 9$ in the plane $z = 2$, oriented counterclockwise when viewed from above.

Solution   (a) On the circle C_1, we have $\|\vec{B}\| = 2I/(ca)$. Since $\vec{B}$ is tangent to C_1 everywhere and points in the forward direction around C_1,

$$\int_{C_1} \vec{B} \cdot d\vec{r} = \int_{C_1} \|\vec{B}\|\, dr = \frac{2I}{ca} \cdot \text{Length of } C_1 = \frac{2I}{ca} \cdot 2\pi a = \frac{4\pi I}{c}.$$

(b) Let S be the conical surface extending from C_1 to C_2 in Figure 20.28. The boundary of this surface has two pieces, $-C_2$ and C_1. The orientation of C_1 leads to the outward normal on S, which forces us to choose the clockwise orientation on C_2. By Stokes' Theorem,

$$\int_S \text{curl}\,\vec{B} \cdot d\vec{A} = \int_{-C_2} \vec{B} \cdot d\vec{r} + \int_{C_1} \vec{B} \cdot d\vec{r} = -\int_{C_2} \vec{B} \cdot d\vec{r} + \int_{C_1} \vec{B} \cdot d\vec{r}.$$

Since curl $\vec{B} = \vec{0}$, we have $\int_S \text{curl}\,\vec{B} \cdot d\vec{A} = 0$, so the two line integrals must be equal:

$$\int_{C_2} \vec{B} \cdot d\vec{r} = \int_{C_1} \vec{B} \cdot d\vec{r} = \frac{4\pi I}{c}.$$

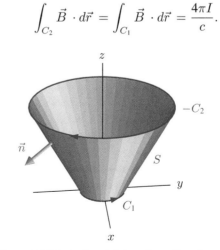

Figure 20.28: Surface joining C_1 to C_2, oriented to satisfy the conditions of Stoke's Theorem

Curl Fields

A vector field $\vec{F}$ is called a *curl field* if $\vec{F} = \operatorname{curl}\vec{G}$ for some vector field $\vec{G}$. Recall that if $\vec{F} = \operatorname{grad} f$, then f is called a potential function. By analogy, if a vector field $\vec{F} = \operatorname{curl}\vec{G}$, then $\vec{G}$ is called a *vector potential* for $\vec{F}$. The following example shows that the flux of a curl field through a surface depends only on the boundary of the surface. This is analogous to the fact that the line integral of a gradient field depends only on the endpoints of the path.

Example 3   Suppose $\vec{F} = \operatorname{curl}\vec{G}$. Suppose that S_1 and S_2 are two oriented surfaces with the same boundary C. Show that, if S_1 and S_2 determine the same orientation on C (as in Figure 20.29), then

$$\int_{S_1} \vec{F}\cdot d\vec{A} = \int_{S_2}\vec{F}\cdot d\vec{A}.$$

If S_1 and S_2 determine opposite orientations on C, then

$$\int_{S_1} \vec{F}\cdot d\vec{A} = -\int_{S_2}\vec{F}\cdot d\vec{A}.$$

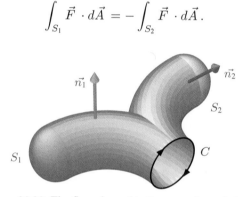

Figure 20.29: The flux of a curl is the same through the two surfaces S_1 and S_2 if they determine the same orientation on the boundary, C

Solution   Since $\vec{F} = \operatorname{curl}\vec{G}$, by Stokes' Theorem we have

$$\int_{S_1}\vec{F}\cdot d\vec{A} = \int_{S_1}\operatorname{curl}\vec{G}\cdot d\vec{A} = \int_C \vec{G}\cdot d\vec{r}$$

and

$$\int_{S_2}\vec{F}\cdot d\vec{A} = \int_{S_2}\operatorname{curl}\vec{G}\cdot d\vec{A} = \int_C \vec{G}\cdot d\vec{r}.$$

In each case the line integral on the right must be computed using the orientation determined by the surface. Thus, the two flux integrals of $\vec{F}$ are the same if the orientations are the same and they are opposite if the orientations are opposite.

Problems for Section 20.4

1. Can you use Stokes' Theorem to compute the line integral $\int_C (2x\vec{i} + 2y\vec{j} + 2z\vec{k})\cdot d\vec{r}$ where C is the straight line from the point $(1,2,3)$ to the point $(4,5,6)$? Why or why not?

In Problems 2–5 compute the given line integral using Stokes' Theorem.

2. $\int_C \vec{F} \cdot d\vec{r}$, where $\vec{F} = (z - 2y)\vec{i} + (3x - 4y)\vec{j} + (z + 3y)\vec{k}$ and C is the circle $x^2 + y^2 = 4$, $z = 1$, oriented counterclockwise when viewed from above.

3. $\int_C \vec{F} \cdot d\vec{r}$ where $\vec{F} = (2x - y)\vec{i} + (x + 4y)\vec{j}$ and C is a circle of radius 10, centered at the origin

    (a) In the xy-plane, oriented clockwise as viewed from the positive z-axis.
    (b) In the yz-plane, oriented clockwise as viewed from the positive x-axis.

4. $\int_C \vec{F} \cdot d\vec{r}$, with $\vec{F} = \vec{r}/\|\vec{r}\|^3$ where C is the path consisting of line segments from $(1, 0, 1)$ to $(1, 0, 0)$ to $(0, 0, 1)$ back to $(1, 0, 1)$.

5. Find the circulation of the vector field $\vec{F} = xz\vec{i} + (x + yz)\vec{j} + x^2\vec{k}$ around the circle $x^2 + y^2 = 1$, $z = 2$, oriented counterclockwise when viewed from above.

6. Compute the line integral $\int_C ((yz^2 - y)\vec{i} + (xz^2 + x)\vec{j} + 2xyz\vec{k}) \cdot d\vec{r}$ where C is the circle of radius 3 in the xy-plane, centered at the origin, oriented counterclockwise as viewed from the positive z-axis. Do it two ways: (a) Directly   (b) Using Stokes' Theorem

7. Let S be the surface given by $z = 1 - x^2$ for $0 \le x \le 1$ and $-2 \le y \le 2$, oriented upward. Verify Stokes' Theorem for $\vec{F} = xy\vec{i} + yz\vec{j} + xz\vec{k}$. Sketch the surface S and the curve C that bounds S.

8. Verify Stokes' Theorem for $\vec{F} = y\vec{i} + z\vec{j} + x\vec{k}$ and S, the paraboloid $z = 1 - (x^2 + y^2)$, $z \ge 0$ oriented upward. [Hint: Use polar coordinates.]

9. Suppose that C is a closed curve in the xy-plane, oriented counterclockwise when viewed from above. Show that $\frac{1}{2} \int_C (-y\vec{i} + x\vec{j}) \cdot d\vec{r}$ equals the area of the region R in the xy-plane enclosed by C.

10. The vector fields $\vec{F}$ and $\vec{G}$ are sketched in Figures 20.30 and 20.31. Each vector field has no z-component and is independent of z. Assume all the axes have the same scales.

    (a) What can you say about div $\vec{F}$ and div $\vec{G}$ at the origin?
    (b) What can you say about curl $\vec{F}$ and curl $\vec{G}$ at the origin?
    (c) Is there a closed surface around the origin such that $\vec{F}$ has a nonzero flux through it?
    (d) Repeat part (c) for $\vec{G}$.
    (e) Is there a closed curve around the origin such that $\vec{F}$ has a nonzero circulation around it?
    (f) Repeat part (e) for $\vec{G}$.

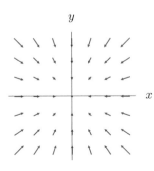

Figure 20.30: Cross-section of $\vec{F}$

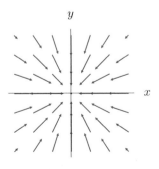

Figure 20.31: Cross-section of $\vec{G}$

11. Use Stokes' Theorem to show that $\int_S \text{curl } \vec{F} \cdot d\vec{A} = 0$ for any surface S which is the boundary surface of a solid region W. Use the geometric definition of the divergence to deduce that $\text{div curl } \vec{F} = 0$.

12. Show that Green's Theorem is a special case of Stokes' Theorem.

13. A vector field $\vec{F}$ is defined everywhere except on the z-axis and $\text{curl } \vec{F} = \vec{0}$ everywhere where $\vec{F}$ is defined. What can you say about $\int_C \vec{F} \cdot d\vec{r}$ if C is a circle of radius 1 in the xy-plane, and if the center of C is at   (a)   the origin,   (b)   the point $(2, 0)$?

14. Evaluate $\int_C (-z\vec{i} + y\vec{j} + x\vec{k}) \cdot d\vec{r}$, where C is a circle of radius 2 around the y-axis with orientation indicated in Figure 20.32.

15. Let $\vec{F} = -z\vec{j} + y\vec{k}$, let C be the circle of radius a in the yz-plane oriented clockwise as viewed from the positive x-axis, and let S be the disk in the yz-plane enclosed by C, oriented in the positive x-direction. See Figure 20.33.

(a) Evaluate directly $\int_C \vec{F} \cdot d\vec{r}$.

(b) Evaluate directly $\int_S \text{curl } \vec{F} \cdot d\vec{A}$.

(c) The answers in parts (a) and (b) are not equal. Explain why this does not contradict Stokes' Theorem.

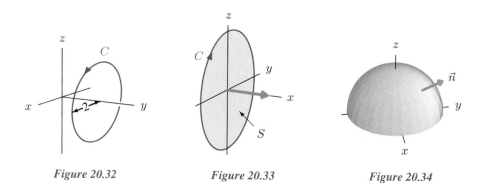

Figure 20.32   *Figure 20.33*   *Figure 20.34*

16. Let $\vec{F} = (8yz - z)\vec{j} + (3 - 4z^2)\vec{k}$.

(a) Show that $\vec{G} = 4yz^2\vec{i} + 3x\vec{j} + xz\vec{k}$ is a vector potential for $\vec{F}$.

(b) Evaluate $\int_S \vec{F} \cdot d\vec{A}$ where S is the hemisphere of radius 5 shown in Figure 20.34, oriented upwards. [Hint: Use Example 3 on page 440 to simplify the calculation.]

17. Let $\vec{F} = -y\vec{i} + x\vec{j} + \cos(xy)z\vec{k}$ and let S be the surface of the lower unit hemisphere $x^2 + y^2 + z^2 = 1, z \leq 0$, oriented with outward pointing normal. Find $\int_S \text{curl } \vec{F} \cdot d\vec{A}$.

18. Water in a bathtub has velocity vector field near the drain given, for x, y, z in cm, by

$$\vec{F} = -\frac{y + xz}{(z^2 + 1)^2}\vec{i} - \frac{yz - x}{(z^2 + 1)^2}\vec{j} - \frac{1}{z^2 + 1}\vec{k} \qquad \text{cm/sec}.$$

(a) The drain in the bathtub is a disk in the xy-plane with center at the origin and radius 1 cm. Find the rate at which the water is leaving the bathtub (that is, the rate at which water is flowing through the disk). Give units for your answer.

(b) Find the divergence of $\vec{F}$.

(c) Find the flux of the water through the hemisphere of radius 1, centered at the origin, lying below the xy-plane and oriented downward.

(d) Find $\int_C \vec{G} \cdot d\vec{r}$ where C is the edge of the drain, oriented clockwise when viewed from above, and where

$$\vec{G} = \frac{1}{2}\left(\frac{y}{z^2+1}\vec{i} - \frac{x}{z^2+1}\vec{j} - \frac{x^2+y^2}{(z^2+1)^2}\vec{k}\right).$$

(e) Calculate curl $\vec{G}$.

(f) Explain why your answers to parts (c) and (d) are equal.

20.5 THE THREE FUNDAMENTAL THEOREMS

We have now seen three multivariable versions of the Fundamental Theorem of Calculus. In this section we will examine some consequences of these theorems.

Fundamental Theorem of Calculus for Line Integrals

$$\int_C \operatorname{grad} f \cdot d\vec{r} = f(Q) - f(P).$$

Stokes' Theorem

$$\int_S \operatorname{curl} \vec{F} \cdot d\vec{A} = \int_C \vec{F} \cdot d\vec{r}.$$

Divergence Theorem

$$\int_W \operatorname{div} \vec{F} \, dV = \int_S \vec{F} \cdot d\vec{A}.$$

Notice that, in each case, the region of integration on the right is the boundary of the region on the left (except that for the first theorem we simply evaluate f at the boundary points); the integrand on the left is a sort of derivative of the integrand on the right; see Figure 20.35.

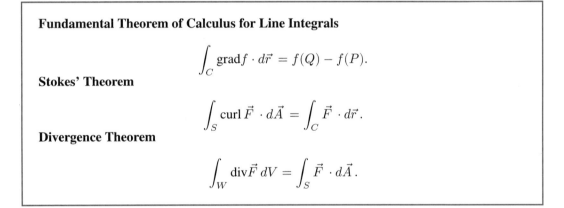
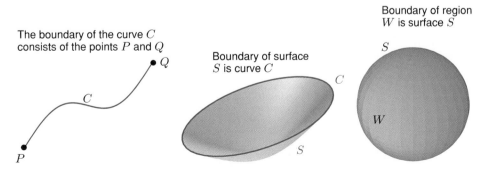

The boundary of the curve C consists of the points P and Q

Boundary of surface S is curve C

Boundary of region W is surface S

Figure 20.35: Regions and their boundaries for the three fundamental theorems

The Gradient and the Curl

Suppose that $\vec{F}$ is a smooth gradient field, so $\vec{F} = \text{grad } f$ for some function f. Using the Fundamental Theorem for Line Integrals, we saw in Chapter 18 that

$$\int_C \vec{F} \cdot d\vec{r} = 0$$

for any closed curve C. Thus, for any unit vector $\vec{n}$

$$\text{circ}_{\vec{n}} \vec{F} = \lim_{\text{Area} \to 0} \frac{\displaystyle\int_C \vec{F} \cdot d\vec{r}}{\text{Area of } C} = \lim_{\text{Area} \to 0} \frac{0}{\text{Area}} = 0,$$

where the limit is taken over circles C in a plane perpendicular to $\vec{n}$, and oriented by the right hand rule. Thus the circulation density of $\vec{F}$ is zero in every direction, so curl $\vec{F} = \vec{0}$, that is,

$$\boxed{\text{curl grad } f = \vec{0}.}$$

(This formula can also be verified using the coordinate definition of curl. See Problem 15 on page 434.)

Is the converse true? Is any vector field whose curl is zero a gradient field? Suppose that curl $\vec{F} = \vec{0}$ and let us consider the line integral $\int_C \vec{F} \cdot d\vec{A}$ for a closed curve C contained in the domain of $\vec{F}$. If C is the boundary curve of an oriented surface S that lies wholly in the domain of curl $\vec{F}$, then Stokes' Theorem asserts that

$$\int_C \vec{F} \cdot d\vec{r} = \int_S \text{curl } \vec{F} \cdot d\vec{A} = \int_S \vec{0} \cdot d\vec{A} = 0.$$

If we knew that $\int_C \vec{F} \cdot d\vec{r} = 0$ for every closed curve C, then $\vec{F}$ would be path-independent, and hence a gradient field. Thus we need to know whether every closed curve in the domain of $\vec{F}$ is the boundary of an oriented surface contained in the domain. It can be quite difficult to determine if a given curve is the boundary of a surface (suppose, for example, that the curve is knotted in a complicated way). However, if the curve can be contracted smoothly to a point, remaining all the time in the domain of $\vec{F}$, then it is the boundary of a surface, namely, the surface it sweeps through as it contracts.[3] Thus, we have proved the test for a gradient field that we stated in Chapter 18.

The Curl Test for Vector Fields in 3-Space

Suppose $\vec{F}$ is a smooth vector field on 3-space such that
- The domain of $\vec{F}$ has the property that every closed curve in it can be contracted to a point in a smooth way, staying at all times within the domain.
- curl $\vec{F} = \vec{0}$.

Then $\vec{F}$ is path-independent, and thus is a gradient field.

Example 6 on page 375 shows how the curl test is applied. Problems 14 – 16 on pages 447 – 448, are about contractibility.

The Curl and The Divergence

In this section we will use the second two fundamental theorems to get a test for a vector field to be a curl field, that is, a field of the form $\vec{F} = \text{curl } \vec{G}$ for some $\vec{G}$.

---

[3]The surface might intersect itself, but this doesn't matter for the proof of Stokes' Theorem that we will give in Section 20.6

Example 1   Suppose that $\vec{F}$ is a smooth curl field. Use Stokes' Theorem to show that for any closed surface, S, contained in the domain of $\vec{F}$

$$\int_S \vec{F} \cdot d\vec{A} = 0.$$

Solution   Suppose $\vec{F} = \text{curl}\,\vec{G}$. Draw a closed curve C on the surface S, thus dividing S into two surfaces S_1 and S_2 as shown in Figure 20.36. Pick the orientation for C corresponding to S_1; then the orientation of C corresponding to S_2 is the opposite. Thus, using Stokes' Theorem,

$$\int_{S_1} \vec{F} \cdot d\vec{A} = \int_{S_1} \text{curl}\,\vec{G} \cdot d\vec{A} = \int_C \vec{G} \cdot d\vec{r} = -\int_{S_2} \text{curl}\,\vec{G} \cdot d\vec{A} = -\int_{S_2} \vec{F} \cdot d\vec{A}.$$

Thus, for any closed surface S, we have

$$\int_S \vec{F} \cdot d\vec{A} = \int_{S_1} \vec{F} \cdot d\vec{A} + \int_{S_2} \vec{F} \cdot d\vec{A} = 0.$$

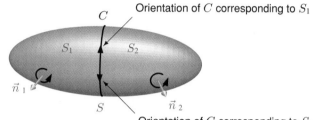

Figure 20.36: The closed surface S divided into two surfaces S_1 and S_2

Thus, if $\vec{F} = \text{curl}\,\vec{G}$, we use the result of Example 1 to see that

$$\text{div}\,\vec{F} = \lim_{\text{Volume}\to 0} \frac{\displaystyle\int_S \vec{F} \cdot d\vec{A}}{\text{Volume enclosed by } S} = \lim_{\text{Volume}\to 0} \frac{0}{\text{Volume}} = 0,$$

where the limit is taken over spheres S contracting down to a point. So we conclude that:

$$\boxed{\text{div}\,\text{curl}\,\vec{G} = 0.}$$

(This formula can also be verified using coordinates. See Problem 16 on page 434.)

Is every vector field whose divergence is zero a curl field? It turns out that we have the following analogue of the curl test, though we will not prove it.

The Divergence Test for Vector Fields in 3-Space

Suppose $\vec{F}$ is a smooth vector field on 3-space such that
- The domain of $\vec{F}$ has the property that every closed surface in it is the boundary of a solid region completely contained in the domain.
- $\text{div}\,\vec{F} = 0$.

Then $\vec{F}$ is a curl field.

Example 2   Consider the vector fields $\vec{E} = q\dfrac{\vec{r}}{\|\vec{r}\|^3}$ and $\vec{B} = \dfrac{2I}{c}\left(\dfrac{-y\vec{i} + x\vec{j}}{x^2 + y^2}\right)$.

(a) Calculate div $\vec{E}$ and div $\vec{B}$.
(b) Do $\vec{E}$ and $\vec{B}$ satisfy the divergence test?
(c) Is either $\vec{E}$ or $\vec{B}$ a curl field?

Solution (a) Example 3 on page 416 shows that div $\vec{E} = 0$. The following calculation shows div $\vec{B} = 0$ also:

$$\text{div } \vec{B} = \frac{2I}{c}\left(\frac{\partial}{\partial x}\left(\frac{-y}{x^2 + y^2}\right) + \frac{\partial}{\partial y}\left(\frac{x}{x^2 + y^2}\right) + \frac{\partial}{\partial z}(0)\right)$$

$$= \frac{2I}{c}\left(\frac{2xy}{(x^2 + y^2)^2} + \frac{-2yx}{(x^2 + y^2)^2}\right) = 0.$$

(b) The domain of $\vec{E}$ is 3-space minus the origin, so a region is contained in the domain if it misses the origin. Thus the surface of a sphere centered at the origin is contained in the domain of E, but the solid ball inside is not. Hence $\vec{E}$ does not satisfy the divergence test.

   The domain of $\vec{B}$ is 3-space minus the z-axis, so a region is contained in the domain if it avoids the z-axis. If S is a surface bounding a solid region W, then the z-axis cannot pierce W without piercing S as well. Hence, if S avoids the z-axis, so does W. Thus $\vec{B}$ satisfies the divergence test.

(c) In Example 3 on page 423 we computed the flux of $\vec{r}/\|\vec{r}\|^3$ through a sphere centered at the origin, and found it was 4π, so the flux of $\vec{E}$ through this sphere is $4\pi q$. Thus, $\vec{E}$ cannot be a curl field, because by Example 1, the flux of a curl field through a closed surface is zero.

   On the other hand, $\vec{B}$ satisfies the divergence test, so it must be a curl field. In fact, Problem 5 below shows that

$$\vec{B} = \text{curl}\left(\frac{-I}{c}\ln(x^2 + y^2)\vec{k}\right).$$

Problems for Section 20.5

Which of the vector fields in Problems 1–2 is a gradient field?

1. $\vec{F} = yz\vec{i} + (xz + z^2)\vec{j} + (xy + 2yz)\vec{k}$ 2. $\vec{G} = -y\vec{i} + x\vec{j}$

3. Let $\vec{B} = b\vec{k}$, for some constant b. Show that the following are all possible vector potentials for $\vec{B}$: (a) $\vec{A} = -by\vec{i}$ (b) $\vec{A} = bx\vec{j}$ (c) $\vec{A} = \frac{1}{2}\vec{B} \times \vec{r}$.

4. Find a vector potential for the constant vector field $\vec{B}$ whose value at every point is $\vec{b}$.

5. Show that $\vec{A} = \dfrac{-I}{c}\ln(x^2 + y^2)\vec{k}$ is a vector potential for $\vec{B} = \dfrac{2I}{c}\left(\dfrac{-y\vec{i} + x\vec{j}}{x^2 + y^2}\right)$.

6. Is there a vector field $\vec{G}$ such that curl $\vec{G} = y\vec{i} + x\vec{j}$? How do you know?

For each vector field in Problems 7–8, determine whether a vector potential exists. If so, find one.

7. $\vec{F} = 2x\vec{i} + (3y - z^2)\vec{j} + (x - 5z)\vec{k}$ 8. $\vec{G} = x^2\vec{i} + y^2\vec{j} + z^2\vec{k}$

9. An electric charge q at the origin produces an electric field $\vec{E} = q\vec{r}/\|\vec{r}\|^3$.

    (a) Does curl $\vec{E} = \vec{0}$?
    (b) Does $\vec{E}$ satisfy the curl test?
    (c) Is $\vec{E}$ a gradient field?

10. Suppose c is the speed of light. A thin wire along the z-axis carrying a current I produces a magnetic field

$$\vec{B} = \frac{2I}{c}\left(\frac{-y\vec{i} + x\vec{j}}{x^2 + y^2}\right),$$

    (a) Does curl $\vec{B} = \vec{0}$?
    (b) Does $\vec{B}$ satisfy the curl test?
    (c) Is $\vec{B}$ a gradient field?

11. For constant p, consider the vector field $\vec{E} = \dfrac{\vec{r}}{\|\vec{r}\|^p}$.

    (a) Find curl $\vec{E}$.
    (b) Find the domain of $\vec{E}$.
    (c) For which values of p does $\vec{E}$ satisfy the curl test? For those values of p, find a potential function for $\vec{E}$.

12. The magnetic field, $\vec{B}$, due to a magnetic dipole with moment $\vec{\mu}$ satisfies div $\vec{B} = 0$ and is given by

$$\vec{B} = -\frac{\vec{\mu}}{\|\vec{r}\|^3} + \frac{3(\vec{\mu}\cdot\vec{r})\vec{r}}{\|\vec{r}\|^5}, \qquad \vec{r} \neq \vec{0}.$$

    (a) Does $\vec{B}$ satisfy the divergence test?
    (b) Show that a vector potential for $\vec{B}$ is given by $\vec{A} = \dfrac{\vec{\mu} \times \vec{r}}{\|\vec{r}\|^3}$.

    [Hint: Use Problem 17 on page 434. The identities in Example 3 on page 431, Problem 19 on page 134, and Problem 24 on page 92 might also be useful.]

    (c) Does your answer to part (a) contradict your answer to part (b)? Explain.

13. Suppose that $\vec{A}$ is a vector potential for $\vec{B}$.

    (a) Show that $\vec{A} + \operatorname{grad}\psi$ is also a vector potential for $\vec{B}$, for any function ψ with continuous second-order partial derivatives. (The vector potentials $\vec{A}$ and $\vec{A} + \operatorname{grad}\psi$ are called *gauge equivalent* and the transformation, for any ψ, from $\vec{A}$ to $\vec{A} + \operatorname{grad}\psi$ is called a *gauge transformation*.)
    (b) What is the divergence of $\vec{A} + \operatorname{grad}\psi$? How should ψ be chosen such that $\vec{A} + \operatorname{grad}\psi$ has zero divergence? (If div $\vec{A} = 0$, the magnetic vector potential $\vec{A}$ is said to be in *Coulomb gauge*.)

The condition in the 3-dimensional curl test about contracting curves can be stated more precisely as follows. A curve C is said to be *smoothly contractible* to a point P if there is a family of parameterized closed curves $C_s, 0 \leq s \leq 1$, with parameterizations

$$\vec{r} = \vec{r}_s(t), \qquad a \leq t \leq b,$$

such that C_0 is the original curve C and C_1 is the point P. (Thus, as s moves from 0 to 1, the curve C_s shrinks from C to P; imagine an animated picture, which at time s shows C_s.) We require that $\vec{r}_s(t)$ be smooth as a function of the two variables s and t. Notice that, since the curves C_s are closed, we must have $\vec{r}_s(a) = \vec{r}_s(b)$ for each s. Then the condition in the curl test is that every closed curve in the domain of $\vec{F}$ be smoothly contractible to a point in such a way that C_s is in the domain of $\vec{F}$ for every s. Problems 14–16 use these ideas.

14. Let C be the circle of radius 1 in the xy-plane, centered at the origin. Write down a family of curves C_s that smoothly contracts C to the origin.

15. Show that any smoothly parameterized curve C in 3-space can be smoothly contracted to any point P. [Hint: Contract along straight lines joining the curve to P.]

16. If C is a closed curve which is smoothly contractible to a point P, show that C is the boundary of a surface S which is smoothly parameterized by a rectangle (S may intersect itself). [Hint: Use the two variables s and t to parameterize the surface.]

20.6 PROOF OF THE DIVERGENCE THEOREM AND STOKES' THEOREM

In this section we give proofs of the Divergence Theorem and Stokes' Theorem using the definitions in Cartesian coordinates.

Proof of the Divergence Theorem

For the Divergence Theorem, we use the same approach as we used for Green's Theorem; first prove the theorem for rectangular regions, then use the change of variables formula to prove it for regions parameterized by rectangular regions, and finally paste such regions together to form general regions.

Proof for Rectangular Solids with Sides Parallel to the Axes

Consider a smooth vector field $\vec{F}$ defined on the rectangular solid V: $a \leq x \leq b$, $c \leq y \leq d$, $e \leq z \leq f$. (See Figure 20.37). We start by computing the flux of $\vec{F}$ through the two faces of V perpendicular to the x-axis, A_1 and A_2, both oriented outward:

$$\int_{A_1} \vec{F} \cdot d\vec{A} + \int_{A_2} \vec{F} \cdot d\vec{A} = -\int_e^f \int_c^d F_1(a, y, z)\, dy\, dz + \int_e^f \int_c^d F_1(b, y, z)\, dy\, dz$$
$$= \int_e^f \int_c^d (F_1(b, y, z) - F_1(a, y, z))\, dy\, dz.$$

By the Fundamental Theorem of Calculus,

$$F_1(b, y, z) - F_1(a, y, z) = \int_a^b \frac{\partial F_1}{\partial x}\, dx,$$

so

$$\int_{A_1} \vec{F} \cdot d\vec{A} + \int_{A_2} \vec{F} \cdot d\vec{A} = \int_e^f \int_c^d \int_a^b \frac{\partial F_1}{\partial x}\, dx\, dy\, dz = \int_V \frac{\partial F_1}{\partial x}\, dV.$$

By a similar argument, we can show

$$\int_{A_3} \vec{F} \cdot d\vec{A} + \int_{A_4} \vec{F} \cdot d\vec{A} = \int_V \frac{\partial F_2}{\partial y}\, dV \quad \text{and} \quad \int_{A_5} \vec{F} \cdot d\vec{A} + \int_{A_6} \vec{F} \cdot d\vec{A} = \int_V \frac{\partial F_3}{\partial z}\, dV.$$

Adding these, we get

$$\int_A \vec{F} \cdot d\vec{A} = \int_V \left(\frac{\partial F_1}{\partial x} + \frac{\partial F_2}{\partial y} + \frac{\partial F_3}{\partial z} \right) dV = \int_V \operatorname{div} \vec{F} \, dV.$$

This is the Divergence Theorem for the region V.

Proof for Regions Parameterized by Rectangular Solids

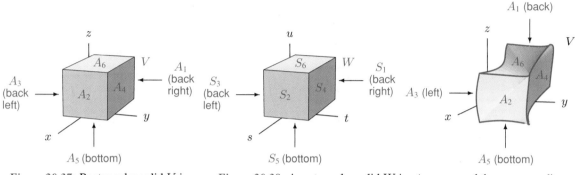

Figure 20.37: Rectangular solid V in
xyz-space

Figure 20.38: A rectangular solid W in stu-space and the corresponding
curved solid V in xyz-space

Now suppose we have a smooth change of coordinates

$$x = x(s,t,u), \qquad y = y(s,t,u), \qquad z = z(s,t,u).$$

Consider a curved solid V in xyz-space corresponding to a rectangular solid W in stu-space. See
Figure 20.38. We suppose that the change of coordinates is one-to-one on the interior of W, and
that its Jacobian determinant is positive on W. We prove the Divergence Theorem for V using the
Divergence Theorem for W.

Let A be the boundary of V. To prove the Divergence Theorem for V, we must show that

$$\int_A \vec{F} \cdot d\vec{A} = \int_V \operatorname{div} \vec{F} \, dV.$$

First we express the flux through A as a flux integral in stu-space over S, the boundary of the
rectangular region W. In vector notation the change of coordinates is

$$\vec{r} = \vec{r}(s,t,u) = x(s,t,u)\vec{i} + y(s,t,u)\vec{j} + z(s,t,u)\vec{k}.$$

The face A_1 of V is parameterized by

$$\vec{r} = \vec{r}(a,t,u), \qquad c \le t \le d, \quad e \le u \le f,$$

so on this face

$$d\vec{A} = \pm \frac{\partial \vec{r}}{\partial t} \times \frac{\partial \vec{r}}{\partial u}.$$

In fact, in order to make $d\vec{A}$ point outward, we must choose the negative sign. (Problem 3 on
page 454 shows how this follows from the fact that the Jacobian determinant is positive.) Thus, if
S_1 is the face $s = a$ of W,

$$\int_{A_1} \vec{F} \cdot d\vec{A} = -\int_{S_1} \vec{F} \cdot \frac{\partial \vec{r}}{\partial t} \times \frac{\partial \vec{r}}{\partial u} \, dt \, du,$$

The outward pointing area element on S_1 is $d\vec{S} = -\vec{i}\, dt\, du$. Therefore, if we choose a vector field $\vec{G}$ on stu-space whose component in the s-direction is

$$G_1 = \vec{F} \cdot \frac{\partial \vec{r}}{\partial t} \times \frac{\partial \vec{r}}{\partial u},$$

we have

$$\int_{A_1} \vec{F} \cdot d\vec{A} = \int_{S_1} \vec{G} \cdot d\vec{S}.$$

Similarly, if we define the t and u components of $\vec{G}$ by

$$G_2 = \vec{F} \cdot \frac{\partial \vec{r}}{\partial u} \times \frac{\partial \vec{r}}{\partial s} \quad \text{and} \quad G_3 = \vec{F} \cdot \frac{\partial \vec{r}}{\partial s} \times \frac{\partial \vec{r}}{\partial t},$$

then

$$\int_{A_i} \vec{F} \cdot d\vec{A} = \int_{S_i} \vec{G} \cdot d\vec{S}, \quad i = 2, \ldots, 6.$$

(See Problem 4.) Adding the integrals for all the faces, we find that

$$\int_A \vec{F} \cdot d\vec{A} = \int_S \vec{G} \cdot d\vec{S}.$$

Since we have already proved the Divergence Theorem for the rectangular region W, we have

$$\int_S \vec{G} \cdot d\vec{S} = \int_W \operatorname{div} \vec{G}\, dW,$$

where

$$\operatorname{div} \vec{G} = \frac{\partial G_1}{\partial s} + \frac{\partial G_2}{\partial t} + \frac{\partial G_3}{\partial u}.$$

Problems 5 and 6 on page 454 show that

$$\frac{\partial G_1}{\partial s} + \frac{\partial G_2}{\partial t} + \frac{\partial G_3}{\partial u} = \left| \frac{\partial(x, y, z)}{\partial(s, t, u)} \right| \left(\frac{\partial F_1}{\partial x} + \frac{\partial F_2}{\partial y} + \frac{\partial F_3}{\partial z} \right).$$

So, by the three-variable change of variables formula on page 268,

$$\begin{aligned}
\int_V \operatorname{div} \vec{F}\, dV &= \int_V \left(\frac{\partial F_1}{\partial x} + \frac{\partial F_2}{\partial y} + \frac{\partial F_3}{\partial z} \right) dx\, dy\, dz \\
&= \int_W \left(\frac{\partial F_1}{\partial x} + \frac{\partial F_2}{\partial y} + \frac{\partial F_3}{\partial z} \right) \left| \frac{\partial(x, y, z)}{\partial(s, t, u)} \right| ds\, dt\, du \\
&= \int_W \left(\frac{\partial G_1}{\partial s} + \frac{\partial G_2}{\partial t} + \frac{\partial G_3}{\partial u} \right) ds\, dt\, du \\
&= \int_W \operatorname{div} \vec{G}\, dW.
\end{aligned}$$

In summary, we have shown that

$$\int_A \vec{F} \cdot d\vec{A} = \int_S \vec{G} \cdot d\vec{S}$$

and

$$\int_V \operatorname{div} \vec{F}\, dV = \int_W \operatorname{div} \vec{G}\, dW.$$

By the Divergence Theorem for rectangular solids, the right hand sides of these equations are equal, so the left hand sides are equal also. This proves the Divergence Theorem for the curved region V.

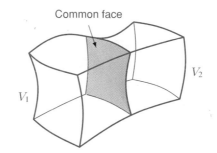

Figure 20.39: Region V formed by
pasting together V_1 and V_2

Pasting Regions Together

As in the proof of Green's Theorem, we prove the Divergence Theorem for more general regions by pasting smaller regions together along common faces. Suppose the solid region V is formed by pasting together solids V_1 and V_2 along a common face, as in Figure 20.39.

The surface A which bounds V is formed by joining the surfaces A_1 and A_2 which bound V_1 and V_2, and then deleting the common face. The outward flux integral of a vector field $\vec{F}$ through A_1 includes the integral across the common face, and the outward flux integral of $\vec{F}$ through A_2 includes the integral over the same face, but oriented in the opposite direction. Thus, when we add the integrals together, the contributions from the common face cancel, and we get the flux integral through A. Thus we have

$$\int_A \vec{F} \cdot d\vec{A} = \int_{A_1} \vec{F} \cdot d\vec{A} + \int_{A_2} \vec{F} \cdot d\vec{A}.$$

But we also have

$$\int_V \text{div}\, \vec{F}\, dV = \int_{V_1} \text{div}\, \vec{F}\, dV + \int_{V_2} \text{div}\, \vec{F}\, dV.$$

So the Divergence Theorem for V follows from the Divergence Theorem for V_1 and V_2. Hence we have proved the Divergence Theorem for any region formed by pasting together regions that can be smoothly parameterized by rectangular solids.

Example 1   Let V be a spherical ball of radius 2, centered at the origin, with a concentric ball of radius 1 removed. Using spherical coordinates, show that the proof of the Divergence Theorem we have given applies to V.

Solution   We cut V into two hollowed hemispheres like the one shown in Figure 20.40, W. In spherical coordinates, W is the rectangle $1 \leq \rho \leq 2$, $0 \leq \phi \leq \pi$, $0 \leq \theta \leq \pi$. Each face of this rectangle becomes part of the boundary of W. The faces $\rho = 1$ and $\rho = 2$ become the inner and outer hemispherical surfaces that form part of the boundary of W. The faces $\theta = 0$ and $\theta = \pi$ become the two halves of the flat part of the boundary of W. The faces $\phi = 0$ and $\phi = \pi$ become line segments along the z-axis. We can form V by pasting together two solid regions like W along the flat surfaces where $\theta = $ constant.

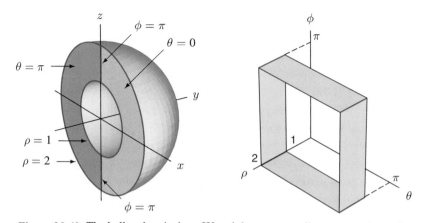

Figure 20.40: The hollow hemisphere W and the corresponding rectangular region in $\rho\theta\phi$-space

Proof of Stokes' Theorem

Consider an oriented surface A, bounded by the curve B. We want to prove Stokes' Theorem:

$$\int_A \operatorname{curl} \vec{F} \cdot d\vec{A} = \int_B \vec{F} \cdot d\vec{r}.$$

We suppose that A has a smooth parameterization $\vec{r} = \vec{r}(s, t)$, so that A corresponds to a region R in the st-plane, and B corresponds to the boundary C of R. See Figure 20.41. We prove Stokes' Theorem for the surface A and a vector field $\vec{F}$ by expressing the integrals on both sides of the theorem in terms of s and t, and using Green's Theorem in the st-plane.

First, we convert the line integral $\int_B \vec{F} \cdot d\vec{r}$ into a line integral around C:

$$\int_B \vec{F} \cdot d\vec{r} = \int_C \vec{F} \cdot \frac{\partial \vec{r}}{\partial s}\, ds + \vec{F} \cdot \frac{\partial \vec{r}}{\partial t}\, dt.$$

So if we define a 2-dimensional vector field $\vec{G} = (G_1, G_2)$ on the st-plane by

$$G_1 = \vec{F} \cdot \frac{\partial \vec{r}}{\partial s} \quad \text{and} \quad G_2 = \vec{F} \cdot \frac{\partial \vec{r}}{\partial t},$$

then

$$\int_B \vec{F} \cdot d\vec{r} = \int_C \vec{G} \cdot d\vec{s},$$

using $\vec{s}$ to denote the position vector of a point in the st-plane.

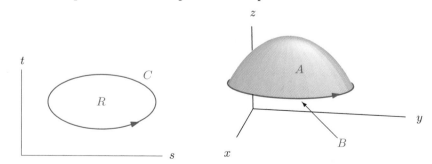

Figure 20.41: A region R in the st-plane and the corresponding surface A in xyz-space; the curve C corresponds to the boundary of B

What about the flux integral $\int_A \text{curl}\, \vec{F} \cdot d\vec{A}$ that occurs on the other side of Stokes' Theorem? In terms of the parameterization,

$$\int_A \text{curl}\, \vec{F} \cdot d\vec{A} = \int_R \text{curl}\, \vec{F} \cdot \frac{\partial \vec{r}}{\partial s} \times \frac{\partial \vec{r}}{\partial t}\, ds\, dt.$$

In Problem 7 on page 455 we show that

$$\text{curl}\, \vec{F} \cdot \frac{\partial \vec{r}}{\partial s} \times \frac{\partial \vec{r}}{\partial t} = \frac{\partial G_2}{\partial s} - \frac{\partial G_1}{\partial t}.$$

Hence

$$\int_A \text{curl}\, \vec{F} \cdot d\vec{A} = \int_R \left(\frac{\partial G_2}{\partial s} - \frac{\partial G_1}{\partial t} \right) ds\, dt.$$

We have already seen that

$$\int_B \vec{F} \cdot d\vec{r} = \int_C \vec{G} \cdot d\vec{s}.$$

By Green's Theorem, the right-hand sides of the last two equations are equal. Hence the left-hand sides are equal as well, which is what we had to prove for Stokes' Theorem.

Problems for Section 20.6

1. Let W be a solid circular cylinder along the z-axis, with a smaller concentric cylinder removed. Parameterize W by a rectangular solid in $r\theta z$-space, where r, θ, and z are cylindrical coordinates.

2. In this section we proved the Divergence Theorem using the coordinate definition of divergence. Now we use the Divergence Theorem to show that the coordinate definition is the same as the geometric definition. Suppose $\vec{F}$ is smooth in a neighborhood of (x_0, y_0, z_0), and let U_R be the ball of radius R with center (x_0, y_0, z_0). Let m_R be the minimum value of div $\vec{F}$ on U_R and let M_R be the maximum value.

   (a) Let S_R be the sphere bounding U_R. Show that

   $$m_R \leq \frac{\int_{S_R} \vec{F} \cdot d\vec{A}}{\text{Volume of } U_R} \leq M_R.$$

   (b) Explain why we can conclude that

   $$\lim_{R \to 0} \frac{\int_{S_R} \vec{F} \cdot d\vec{A}}{\text{Volume of } U_R} = \text{div}\, \vec{F}(x_0, y_0, z_0).$$

   (c) Explain why the statement in part (b) remains true if we replace U_R with a cube of side R, centered at (x_0, y_0, z_0).

Problems 3–6 fill in the details of the proof of the Divergence Theorem.

3. Figure 20.38 on page 449 shows the solid region V in xyz-space parameterized by a rectangular solid W in stu-space using the change of coordinates

$$\vec{r} = \vec{r}(s,t,u), \qquad a \le s \le b, \quad c \le t \le d, \quad e \le u \le f.$$

Suppose that $\dfrac{\partial \vec{r}}{\partial s} \cdot \left(\dfrac{\partial \vec{r}}{\partial t} \times \dfrac{\partial \vec{r}}{\partial u} \right)$ is positive.

(a) Let A_1 be the face of V corresponding to the face $s = a$ of W. Show that $\dfrac{\partial \vec{r}}{\partial s}$, if it is not zero, points into W.

(b) Show that $-\dfrac{\partial \vec{r}}{\partial t} \times \dfrac{\partial \vec{r}}{\partial u}$ is an outward pointing normal on A_1.

(c) Find an outward pointing normal on A_2, the face of V where $s = b$.

4. Show that for the other five faces of the solid V in the proof of the Divergence Theorem (see page 450):

$$\int_{A_i} \vec{F} \cdot d\vec{A} = \int_{S_i} \vec{G} \cdot d\vec{S}, \quad i = 2, 3, 4, 5, 6.$$

5. Suppose that $\vec{F}$ is a vector field and that $\vec{a}$, $\vec{b}$, and $\vec{c}$ are vectors. In this problem we prove the formula

$$\operatorname{grad}(\vec{F} \cdot \vec{b} \times \vec{c}) \cdot \vec{a} + \operatorname{grad}(\vec{F} \cdot \vec{c} \times \vec{a}) \cdot \vec{b} + \operatorname{grad}(\vec{F} \cdot \vec{a} \times \vec{b}) \cdot \vec{c} = (\vec{a} \cdot \vec{b} \times \vec{c}) \operatorname{div} \vec{F}.$$

(a) Interpretating the divergence as flux density, explain why the formula makes sense. [Hint: Consider the flux out of a small parallelepiped with edges parallel to $\vec{a}$, $\vec{b}$, $\vec{c}$.]

(b) Say how many terms there are in the expansion of the left hand side of the formula in Cartesian coordinates, without actually doing the expansion.

(c) Write down all the terms on the left hand side that contain $\partial F_1/\partial x$. Show that these terms add up to $\vec{a} \cdot \vec{b} \times \vec{c} \, \dfrac{\partial F_1}{\partial x}$.

(d) Write down all the terms that contain $\partial F_1/\partial y$. Show that these add to zero.

(e) Explain how the expressions involving the other seven partial derivatives will work out, and how this verifies that the formula holds.

6. Let $\vec{F}$ be a smooth vector field in 3-space, and let

$$x = x(s,t,u), \quad y = y(s,t,u), \quad z = z(s,t,u)$$

be a smooth change of variables, which we will write in vector form as

$$\vec{r} = \vec{r}(s,t,u) = x(s,t,u)\vec{i} + y(s,t,u)\vec{j} + z(s,t,u)\vec{k}.$$

Define a vector field $\vec{G} = (G_1, G_2, G_3)$ on stu-space by

$$G_1 = \vec{F} \cdot \frac{\partial \vec{r}}{\partial t} \times \frac{\partial \vec{r}}{\partial u} \qquad G_2 = \vec{F} \cdot \frac{\partial \vec{r}}{\partial u} \times \frac{\partial \vec{r}}{\partial s} \qquad G_3 = \vec{F} \cdot \frac{\partial \vec{r}}{\partial s} \times \frac{\partial \vec{r}}{\partial t}.$$

(a) Show that

$$\frac{\partial G_1}{\partial s} + \frac{\partial G_2}{\partial t} + \frac{\partial G_3}{\partial u} = \frac{\partial \vec{F}}{\partial s} \cdot \frac{\partial \vec{r}}{\partial t} \times \frac{\partial \vec{r}}{\partial u} + \frac{\partial \vec{F}}{\partial t} \cdot \frac{\partial \vec{r}}{\partial u} \times \frac{\partial \vec{r}}{\partial s} + \frac{\partial \vec{F}}{\partial u} \cdot \frac{\partial \vec{r}}{\partial s} \times \frac{\partial \vec{r}}{\partial t}.$$

(b) Let $\vec{r}_0 = \vec{r}(s_0, t_0, u_0)$, and let

$$\vec{a} = \frac{\partial \vec{r}}{\partial s}(\vec{r}_0), \quad \vec{b} = \frac{\partial \vec{r}}{\partial t}(\vec{r}_0), \quad \vec{c} = \frac{\partial \vec{r}}{\partial u}(\vec{r}_0).$$

Use the chain rule to show that

$$\left(\frac{\partial G_1}{\partial s} + \frac{\partial G_2}{\partial t} + \frac{\partial G_3}{\partial u} \right) \bigg|_{\vec{r} = \vec{r}_0} = \text{grad}(\vec{F} \cdot \vec{b} \times \vec{c}) \cdot \vec{a} + \text{grad}(\vec{F} \cdot \vec{c} \times \vec{a}) \cdot \vec{b} + \text{grad}(\vec{F} \cdot \vec{a} \times \vec{b}) \cdot \vec{c}.$$

(c) Use Problem 5 to show that

$$\frac{\partial G_1}{\partial s} + \frac{\partial G_2}{\partial t} + \frac{\partial G_3}{\partial u} = \left| \frac{\partial(x, y, z)}{\partial(s, t, u)} \right| \left(\frac{\partial F_1}{\partial x} + \frac{\partial F_2}{\partial y} + \frac{\partial F_3}{\partial z} \right).$$

7. This problem completes the proof of Stokes' Theorem. Let $\vec{F}$ be a smooth vector field in 3-space, and let S be a surface parameterized by $\vec{r} = \vec{r}(s, t)$. Let $\vec{r}_0 = \vec{r}(s_0, t_0)$ be a fixed point on S. We define a vector field in st-space as on page 452:

$$G_1 = \vec{F} \cdot \frac{\partial \vec{r}}{\partial s} \qquad G_2 = \vec{F} \cdot \frac{\partial \vec{r}}{\partial t}.$$

(a) Let $\vec{a} = \frac{\partial \vec{r}}{\partial s}(\vec{r}_0), \quad \vec{b} = \frac{\partial \vec{r}}{\partial t}(\vec{r}_0)$. Show that

$$\frac{\partial G_1}{\partial t}(\vec{r}_0) - \frac{\partial G_2}{\partial s}(\vec{r}_0) = \text{grad}(\vec{F} \cdot \vec{a}) \cdot \vec{b} - \text{grad}(\vec{F} \cdot \vec{b}) \cdot \vec{a}.$$

(b) Use Problem 27 on page 436 to show

$$\text{curl}\,\vec{F} \cdot \frac{\partial \vec{r}}{\partial s} \times \frac{\partial \vec{r}}{\partial t} = \frac{\partial G_2}{\partial s} - \frac{\partial G_1}{\partial t}.$$

REVIEW PROBLEMS FOR CHAPTER TWENTY

1. Use the geometric definition of divergence to find div $\vec{v}$ at the origin, where $\vec{v} = -2\vec{r}$. Check that you get the same result using the definition in Cartesian coordinates.

2. Can you evaluate the flux integral in Problem 12 on page 409 by application of the Divergence Theorem? Why or why not?

3. If V is a volume surrounded by a closed surface S, show that $\frac{1}{3} \int_S \vec{r} \cdot d\vec{A} = V$.

4. Use Problem 3 to compute the volume of a sphere of radius R given that its surface area is $4\pi R^2$.

5. Use Problem 3 to compute the volume of a cone of base radius b and height h.
   [Hint: Stand the cone with its point downward and its axis along the positive z-axis.]

6. (a) Find the flux of the vector field $\vec{F} = 2x\vec{i} - 3y\vec{j} + 5z\vec{k}$ through a box with four of its corners at the points $(a, b, c), (a + w, b, c), (a, b + w, c), (a, b, c + w)$ and edge length w. See Figure 20.42.

(b) Use the geometric definition and part (a) to find div $\vec{F}$ at the point (a, b, c).

(c) Find div $\vec{F}$ using partial derivatives.

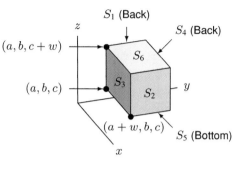

Figure 20.42

7. Suppose $\vec{F} = (3x + 2)\vec{i} + 4x\vec{j} + (5x + 1)\vec{k}$. Use the method of Problem 6 to find div $\vec{F}$ at the point (a, b, c) by two different methods.

8. Compute the flux integral $\int_S (x^3\vec{i} + 2y\vec{j} + 3\vec{k}) \cdot d\vec{A}$, where S is the $2 \times 2 \times 2$ rectangular surface centered at the origin, oriented outward. Do this in two ways:

(a) Directly   (b) By means of the Divergence Theorem

Are the statements in Problems 9–16 true or false? Assume $\vec{F}$ and $\vec{G}$ are smooth vector fields in 3-space. Explain your answer.

9. curl $\vec{F}$ is a vector field.   10. $\operatorname{grad}(fg) = (\operatorname{grad} f) \cdot (\operatorname{grad} g)$

11. $\operatorname{div}(\vec{F} + \vec{G}) = \operatorname{div} \vec{F} + \operatorname{div} \vec{G}$ 12. $\operatorname{grad}(\vec{F} \cdot \vec{G}) = \vec{F}(\operatorname{div} \vec{G}) + (\operatorname{div} \vec{F})\vec{G}$

13. $\operatorname{curl}(f\vec{G}) = (\operatorname{grad} f) \times \vec{G} + f(\operatorname{curl} \vec{G})$

14. div $\vec{F}$ is a scalar whose value can vary from point to point.

15. If $\int_S \vec{F} \cdot d\vec{A} = 12$ and S is a flat disk of area 4π, then div $\vec{F} = 3/\pi$.

16. If $\vec{F}$ is as shown in Figure 20.43, curl $\vec{F} \cdot \vec{j} > 0$.

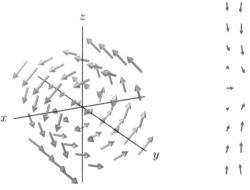

Figure 20.43

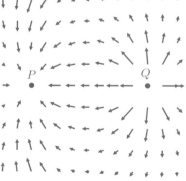

Figure 20.44

17. Figure 20.44 shows the part of a vector field $\vec{E}$ that lies in the xy-plane. Assume that the vector field is independent of z, so that any horizontal cross section looks the same. What can you say about div $\vec{E}$ at the points marked P and Q, assuming it is defined?

18. Suppose $\vec{F} = \vec{r}/\|\vec{r}\|^3$. Find $\int_S \vec{F} \cdot d\vec{A}$ where S is the ellipsoid $x^2 + 2y^2 + 3z^2 = 6$. Give reasons for your computation.

19. A central vector field is one of the form $\vec{F} = f(r)\vec{r}$ where f is any function of $r = \|\vec{r}\|$. Show that any central vector field is irrotational.

20. According to Coulomb's Law, the electrostatic field $\vec{E}$ at the point $\vec{r}$ due to a charge q at the origin is given by

$$\vec{E}(\vec{r}) = q\frac{\vec{r}}{\|\vec{r}\|^3}.$$

  (a) Compute div $\vec{E}$.
  (b) Let S_a be the sphere of radius a centered at the origin and oriented outwards. Show that the flux of $\vec{E}$ through S_a is $4\pi q$.
  (c) Could you have used the Divergence Theorem in part (b)? Explain why or why not.
  (d) Let S be an arbitrary, closed, outward-oriented surface surrounding the origin. Show that the flux of $\vec{E}$ through S is again $4\pi q$. [Hint: Apply the Divergence Theorem to the solid region lying between a small sphere S_a and the surface S.]

21. According to Coulomb's Law, the electric field $\vec{E}$ at the point $\vec{r}$ due to a charge q at the point $\vec{r}_0$ is given by

$$\vec{E}(\vec{r}) = q\frac{(\vec{r} - \vec{r}_0)}{\|\vec{r} - \vec{r}_0\|^3}.$$

Suppose S is a closed, outward-oriented surface and that $\vec{r}_0$ does not lie on S. Use Problem 20 to show that

$$\int_S \vec{E} \cdot d\vec{A} = \begin{cases} 4\pi q & \text{if } q \text{ lies inside } S, \\ 0 & \text{if } q \text{ lies outside } S. \end{cases}$$

22. At the point with position vector $\vec{r}$, the electric field $\vec{E}$ of an ideal electric dipole with moment $\vec{p}$ located at the origin is given by

$$\vec{E}(\vec{r}) = 3\frac{(\vec{r} \cdot \vec{p})\vec{r}}{\|\vec{r}\|^5} - \frac{\vec{p}}{\|\vec{r}\|^3}.$$

  (a) What is div $\vec{E}$?
  (b) Suppose S is an outward oriented, closed smooth surface surrounding the origin. Compute the flux of $\vec{E}$ through S. Can you use the Divergence Theorem directly to compute the flux? Explain why or why not. [Hint: First compute the flux of the dipole field $\vec{E}$ through an outward-oriented sphere S_a with center at the origin and radius a. Then apply the Divergence Theorem to the region W lying between S and a small sphere S_a.]

23. Due to roadwork ahead, the traffic on a highway slows linearly from 55 miles/hour to 15 miles/hour over a 2000 foot stretch of road, then crawls along at 15 miles/hour for 5000 feet, then speeds back up linearly to 55 miles/hour in the next 1000 feet, after which it moves steadily at 55 miles/hour.

  (a) Sketch a velocity vector field for the traffic flow.
  (b) Write a formula for the velocity vector field $\vec{v}$ (miles/hour) as a function of the distance x feet from the initial point of slowdown. (Take the direction of motion to be $\vec{i}$ and consider the various sections of the road separately.)
  (c) Compute div $\vec{v}$ at $x = 1000, 5000, 7500, 10,000$. Be sure to include the proper units.

24. The velocity field $\vec{v}$ in Problem 23 does not give a complete description of the traffic flow, for it takes no account of the spacing between vehicles. Let ρ be the density (cars/mile) of highway, where we assume that ρ depends only on x.

    (a) Using your highway experience, arrange in ascending order: $\rho(0), \rho(1000), \rho(5000)$.

    (b) What are the units and interpretation of the vector field $\rho\vec{v}$?

    (c) Would you expect $\rho\vec{v}$ to be constant? Why? What does this mean for $\text{div}(\rho\vec{v})$?

    (d) Determine $\rho(x)$ if $\rho(0) = 75$ cars/mile and $\rho\vec{v}$ is constant.

    (e) If the highway has two lanes, find the approximate number of feet between cars at $x = 0, 1000$, and 5000.

25. (a) A river flows across the xy-plane in the positive x-direction and around a circular rock of radius 1 centered at the origin. The velocity of the river can be modeled using the potential function $\phi = x + (x/(x^2 + y^2))$. Compute the velocity vector field, $\vec{v} = \text{grad}\,\phi$.

    (b) Show that $\text{div}\,\vec{v} = 0$.

    (c) Show that the flow of $\vec{v}$ is tangent to the circle $x^2 + y^2 = 1$. This means that no water crosses the circle. The water on the outside must therefore all flow around the circle.

    (d) Use a computer to sketch the vector field $\vec{v}$ in the region outside the unit circle.

26. Evaluate

$$\vec{F} = (\text{grad}\,\phi) + \vec{v} \times \vec{r}$$

where

$$\phi(x, y, z) = \tfrac{1}{2}(a_1 x^2 + b_2 y^2 + c_3 z^2 + (a_2 + b_1)xy + (a_3 + c_1)xz + (b_3 + c_2)yz)$$

and

$$\vec{v} = \tfrac{1}{2}((c_2 - b_3)\vec{i} + (a_3 - c_1)\vec{j} + (b_1 - a_2)\vec{k}).$$

Explain why every linear vector field can be written in the form $(\text{grad}\,\phi) + \vec{v} \times \vec{r}$.

Problems 27–28 use the fact that the electric field, $\vec{E}$, is related to the charge density, $\rho(x, y, z)$, in units of charge/volume, by the equation

$$\text{div}\,\vec{E} = 4\pi\rho.$$

In addition, there is an electric potential, ϕ, whose gradient gives the electric field:

$$\vec{E} = -\,\text{grad}\,\phi.$$

27. Calculate and describe in words the electric field and the charge distribution corresponding to the potential function defined as follows:

$$\phi = \begin{cases} x^2 + y^2 + z^2 & \text{for } x^2 + y^2 + z^2 \le \frac{b^2}{4} \\ \dfrac{b^2}{4} - \dfrac{b^3}{4(x^2 + y^2 + z^2)^{1/2}} & \text{for } \frac{b^2}{4} \le x^2 + y^2 + z^2 \end{cases}$$

28. A vector field which could possibly represent an electric field is given by

$$\vec{E} = 10xy\vec{i} + (5x^2 - 5y^2)\vec{j}.$$

    (a) Calculate the line integral of $\vec{E}$ from the origin to the point (a, b) along the path which runs straight from the origin to the point $(a, 0)$ and then straight from $(a, 0)$ to the point (a, b).

(b) Calculate the line integral of $\vec{E}$ between the same points as in part (a) but via the point $(0, b)$.

(c) Why do your answers to part (a) and (b) suggest that $\vec{E}$ could indeed be an electric field?

(d) Find the electric potential, ϕ, and calculate grad ϕ to confirm that $\vec{E} = -\text{grad } \phi$.

29. The relations between the electric field, $\vec{E}$, the magnetic field, $\vec{B}$, the charge density, ρ, and the current density, $\vec{J}$, at a point in space are described by the equations

$$\text{div } \vec{E} = 4\pi\rho,$$

$$\text{curl } \vec{B} - \frac{1}{c}\frac{\partial \vec{E}}{\partial t} = \frac{4\pi}{c}\vec{J},$$

where c is a constant (the speed of light).

(a) Using the results of Problem 16 on page 434, show that

$$\frac{\partial \rho}{\partial t} + \text{div } \vec{J} = 0.$$

(b) What does the equation in part (a) say about charge and current density? Explain in intuitive terms why this is reasonable.

(c) Why do you think the equation in part (a) is called the charge conservation equation?

30. A vector field is a *point source* at the origin in 3-space if its direction is away from the origin at every point, its magnitude depends only on the distance from the origin, and its divergence is zero except at the origin. (Such a vector field might be used to model the photon flow out of a star or the neutrino flow out of a supernova.)

(a) Show that $\vec{v} = K(x^2 + y^2 + z^2)^{-3/2}(x\vec{i} + y\vec{j} + z\vec{k})$ is a point source at the origin if $K > 0$.

(b) Determine the magnitude $\|\vec{v}\|$ of the source in part (a) as a function of the distance from its center.

(c) Compute the flux of $\vec{v}$ through a sphere of radius r centered at the origin.

(d) Compute the flux of $\vec{v}$ through a closed surface that does not contain the origin.

31. A basic property of the magnetic field $\vec{B}$ is that curl $\vec{B} = \vec{0}$ in a region where there is no current. Consider the magnetic field around a long thin wire carrying a constant current. The magnitude of the magnetic field depends only on the distance from the wire and its direction is always tangent to the circle around the wire traversed in a direction related to the direction of the current by the right hand rule. Use Stokes' Theorem to deduce that the magnitude of the magnetic field is proportional to the reciprocal of the distance from the wire. [Hint: Consider an annulus (ring) around the wire. Its boundary has two pieces: an inner and an outer circle.]

32. The speed of a naturally occurring vortex (tornado, waterspout, whirlpool) is a decreasing function of the distance from its center, so the constant angular velocity model of Problem 18 on page 434 is inappropriate. A *free vortex* circulating about the z-axis has vector field $\vec{v} = K(x^2 + y^2)^{-1}(-y\vec{i} + x\vec{j})$ where K is a constant.

(a) Sketch the vector field with $K = 1$ and the vector field with $K = -1$.

(b) Determine the speed $\|\vec{v}\|$ of the vortex as a function of the distance from its center.

(c) Compute div $\vec{v}$.

(d) Show that curl $\vec{v} = \vec{0}$.

(e) Compute the circulation of $\vec{v}$ counterclockwise about the circle of radius R at the origin.

(f) The computations in parts (d) and (e) show that $\vec{v}$ has curl $\vec{0}$, but has nonzero circulation around the closed curve in part (e). Explain why this does not contradict Stokes' Theorem.

APPENDICES

A REVIEW OF LOCAL LINEARITY FOR ONE VARIABLE

If you zoom in on the graph of a smooth function of one variable, $y = f(x)$, around the point $x = a$, the graph looks more and more like a straight line, and thus becomes indistinguishable from its tangent line at that point. (See Figure A.1.)

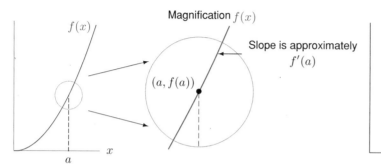

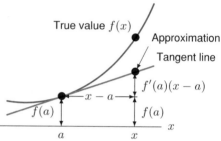

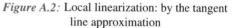

Figure A.1: Zooming in on a portion of a function of one variable until the graph is almost straight

Figure A.2: Local linearization: by the tangent line approximation

The slope of the tangent line is the derivative $f'(a)$, and the line passes through the point $(a, f(a))$, so its equation is

$$y = f(a) + f'(a)(x - a).$$

(See Figure A.2.) Now we approximate the values of f by the y-values from the tangent line, giving the following result.

The Tangent Line Approximation for values of x near a

$$f(x) \approx f(a) + f'(a)(x - a)$$

We are thinking of a as fixed, so that $f(a)$ and $f'(a)$ are constants and the expression on the right-hand side is linear in x. The fact that f is approximately a linear function of x near a is expressed by saying f is *locally linear* near $x = a$.

Example 1   Find the local linearization at $x = 2$ of the one-variable function u, given that $u(2) = 135$ and $u'(2) = 16$.

Solution   Since $u(2) = 135$ and $u'(2) = 16$, the tangent line approximation to $u(x)$ at $x = 2$ is

$$u(x) \approx u(2) + u'(2)(x - 2) = 135 + 16(x - 2) \quad \text{for } x \text{ near } 2.$$

B MAXIMA AND MINIMA OF FUNCTIONS OF ONE VARIABLE

If f is a function of one variable and x is a point in its domain, we say

- p is a *critical point* of f if $f'(p) = 0$ or $f'(p)$ is undefined
- f has a *local maximum* at a critical point, x_0, if $f(x) \leq f(x_0)$ for all x near x_0
- f has a *local minimum* at a critical point, x_0, if $f(x) \geq f(x_0)$ for all x near x_0
- f has a *global maximum* at x_0 if $f(x) \leq f(x_0)$ for all x
- f has a *global minimum* at x_0 if $f(x) \geq f(x_0)$ for all x

Global extrema (that is, global maxima and minima) can only occur at local extrema or at the endpoints of an interval. (See Figure B.3.) To find local extrema, first find the critical points. To find global extrema, evaluate the function at the critical points and the endpoints of the interval (if they are included).

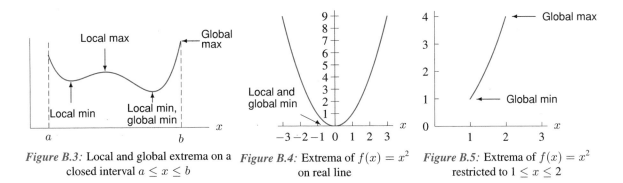

Figure B.3: Local and global extrema on a closed interval $a \leq x \leq b$

Figure B.4: Extrema of $f(x) = x^2$ on real line

Figure B.5: Extrema of $f(x) = x^2$ restricted to $1 \leq x \leq 2$

Functions do not necessarily have local or global extrema — it depends on the function and on the domain under consideration. For example, $f(x) = x^2$ has a local minimum at $x = 0$, and this local minimum is also the global minimum, but it has no local or global maxima. (See Figure B.4.) If, on the other hand, we look at the same function on the domain $1 \leq x \leq 2$ then f has a global minimum at $x = 1$ and a global maximum at $x = 2$ (see Figure B.5).

To find the critical points, we solve the equation

$$f' = 0 \qquad \text{(or } f' \text{ undefined).}$$

To decide if a critical point is a local maximum, local minimum, or neither, use the Second Derivative Test:

- If $f'(p) = 0$ and $f''(p) < 0$, then f has a local maximum at p.
- If $f'(p) = 0$ and $f''(p) > 0$, then f has a local minimum at p.

C DETERMINANTS

We introduce the determinant of an array of numbers. Each 2 by 2 array of numbers has another number associated with it, called its determinant, which is given by

$$\begin{vmatrix} a_1 & a_2 \\ b_1 & b_2 \end{vmatrix} = a_1 b_2 - a_2 b_1.$$

For example

$$\begin{vmatrix} 2 & 5 \\ -4 & -6 \end{vmatrix} = 2(-6) - 5(-4) = 8.$$

Each 3 by 3 array of numbers also has a number associated with it, also called a determinant, which is defined in terms of 2 by 2 determinants as follows:

$$\begin{vmatrix} a_1 & a_2 & a_3 \\ b_1 & b_2 & b_3 \\ c_1 & c_2 & c_3 \end{vmatrix} = a_1 \begin{vmatrix} b_2 & b_3 \\ c_2 & c_3 \end{vmatrix} - a_2 \begin{vmatrix} b_1 & b_3 \\ c_1 & c_3 \end{vmatrix} + a_3 \begin{vmatrix} b_1 & b_2 \\ c_1 & c_2 \end{vmatrix}.$$

Notice that the determinant of the 2 by 2 array multiplied by a_i is the determinant of the array found by removing the row and column containing a_i. Also, note the minus sign in the second term. An example is given by

$$\begin{vmatrix} 2 & 1 & -3 \\ 0 & 3 & -1 \\ 4 & 0 & 5 \end{vmatrix} = 2 \begin{vmatrix} 3 & -1 \\ 0 & 5 \end{vmatrix} - 1 \begin{vmatrix} 0 & -1 \\ 4 & 5 \end{vmatrix} + (-3) \begin{vmatrix} 0 & 3 \\ 4 & 0 \end{vmatrix} = 2(15 + 0) - 1(0 - (-4)) + (-3)(0 - 12) = 62.$$

Suppose the vectors $\vec{a}$ and $\vec{b}$ have components $\vec{a} = a_1 \vec{i} + a_2 \vec{j} + a_3 \vec{k}$ and $\vec{b} = b_1 \vec{i} + b_2 \vec{j} + b_3 \vec{k}$. Recall that the cross product $\vec{a} \times \vec{b}$ is given by the expression

$$\vec{a} \times \vec{b} = (a_2 b_3 - a_3 b_2) \vec{i} + (a_3 b_1 - a_1 b_3) \vec{j} + (a_1 b_2 - a_2 b_1) \vec{k}.$$

Notice that if we expand the following determinant, we get the cross product:

$$\begin{vmatrix} \vec{i} & \vec{j} & \vec{k} \\ a_1 & a_2 & a_3 \\ b_1 & b_2 & b_3 \end{vmatrix} = \vec{i} (a_2 b_3 - a_3 b_2) - \vec{j} (a_1 b_3 - a_3 b_1) + \vec{k} (a_1 b_2 - a_2 b_1) = \vec{a} \times \vec{b}.$$

Determinants give a useful way of computing cross products.

D   REVIEW OF ONE-VARIABLE INTEGRATION

Definition of the One-Variable Integral

The one-variable integral

$$\int_a^b f(x)\, dx$$

is defined to be a limit of *Riemann sums*, which may be constructed as follows. We divide the interval $a \le x \le b$ into n equal subdivisions, each of width Δx. Thus, $\Delta x = (b - a)/n$.

Suppose $x_0, x_1, x_2, \ldots, x_n$ are the endpoints of the subdivisions, as in Figures D.6 and D.7. We construct two special Riemann sums:

$$\text{Left-hand sum} = f(x_0)\Delta x + f(x_1)\Delta x + \cdots + f(x_{n-1})\Delta x$$

and

$$\text{Right-hand sum} = f(x_1)\Delta x + f(x_2)\Delta x + \cdots + f(x_n)\Delta x.$$

To define the definite integral, we take the limit of these sums as n goes to infinity.

The **definite integral** of f from a to b, written

$$\int_a^b f(x)\, dx,$$

is the limit of the left-hand or right-hand sums with n subdivisions as n gets arbitrarily large. In other words,

$$\int_a^b f(x)\, dx = \lim_{n\to\infty} (\text{left-hand sum}) = \lim_{n\to\infty} \left(\sum_{i=0}^{n-1} f(x_i)\Delta x \right)$$

and

$$\int_a^b f(x)\, dx = \lim_{n\to\infty} (\text{right-hand sum}) = \lim_{n\to\infty} \left(\sum_{i=1}^{n} f(x_i)\Delta x \right).$$

Each of these sums is called a *Riemann sum*, f is called the *integrand*, and a and b are called the *limits of integration*.

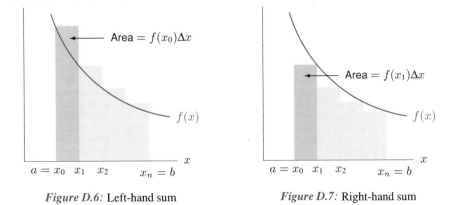

Figure D.6: Left-hand sum   *Figure D.7:* Right-hand sum

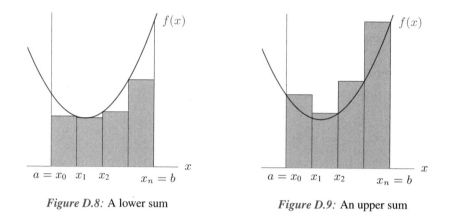

Figure D.8: A lower sum   *Figure D.9:* An upper sum

Other Riemann sums are obtained by evaluating the function at other points in each subinterval, not just the left and right endpoints. Graphically, this means that the graph of the function can intersect the top of each rectangle at any point, not just at the endpoints. Figure D.8 shows all rectangles sitting below the curve, giving an underestimate for the integral called a *lower sum*. Figure D.9 shows an *upper sum* for the same integral.

Interpretations of the Definite Integral

As an Area

If $f(x)$ is positive we can interpret each term $f(x_0)\Delta x$, $f(x_1)\Delta x$, ... in a left- or right-hand Riemann sum as the area of a rectangle. As the width Δx of the rectangles approaches zero, the rectangles fit the curve of the graph more exactly, and the sum of their areas gets closer and closer to the area under the curve, shaded in Figure D.10. Thus, we conclude that

---

If $f(x) \geq 0$ and $a < b$:

$$\text{Area under graph of } f \atop \text{between } a \text{ and } b = \int_a^b f(x)\,dx.$$

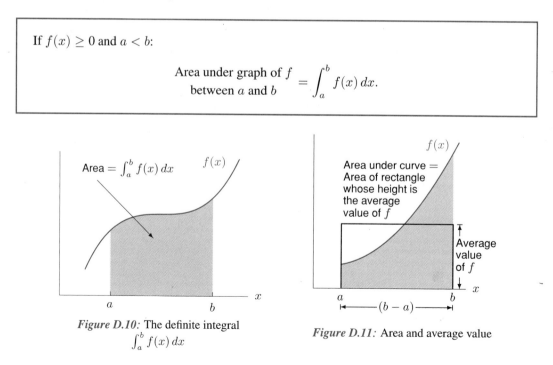

Figure D.10: The definite integral $\int_a^b f(x)\,dx$

Figure D.11: Area and average value

As an Average Value

The definite integral can be used to compute the average value of a function:

$$\text{Average value of } f \text{ from } a \text{ to } b = \frac{1}{b-a} \int_a^b f(x)\,dx.$$

If $f(x) \geq 0$, we can think of the average value of f as the height of the rectangle with base $(b-a)$ and area equal to the area under the graph of f for $a \leq x \leq b$. (See Figure D.11.)

When $f(x)$ Represents a Density

If $f(x)$ represents a density, say a population density or the density of a substance, then we can calculate the total mass or total population using a definite integral. We divide the region into small pieces and find the population or mass of each one by multiplying the density by the size of that piece. Taking a limit, we obtain

If $f(x)$ represents the density of a substance along the interval $a \leq x \leq b$, the

$$\text{Total mass} = \int_a^b f(x)\,dx.$$

The Fundamental Theorem of Calculus

Suppose $f = F'$. Since $F'(t)$ is the rate of change of $F(t)$ with respect to t, and since the definite integral of a rate of change of some quantity is the total change in that quantity, we find that

$$\int_a^b F'(t)\,dt = \int_a^b \left(\text{Rate of change of } F(t)\right)\,dt = \text{Change in } F(t) \text{ between } a \text{ and } b = F(b)-F(a).$$

Thus we have

The Fundamental Theorem of Calculus

If $f = F'$, then

$$\int_a^b f(t)\,dt = F(b) - F(a).$$

The Fundamental Theorem can be used to evaluate definite integrals whenever we can find an antiderivative, or indefinite integral, for the integrand. Appendix E gives a brief table of indefinite integrals.

Problems for Section D

In Problems 1–20, find an antiderivative for each of the functions given.

1. $\displaystyle\int \left(x^2 + 2x + \frac{1}{x}\right) dx$

2. $\displaystyle\int \frac{t+1}{t^2} \, dt$

3. $\displaystyle\int \frac{(t+2)^2}{t^3} \, dt$

4. $\displaystyle\int \sin t \, dt$

5. $\displaystyle\int \cos 2t \, dt$

6. $\displaystyle\int \frac{x}{x^2+1} \, dx$

7. $\displaystyle\int \tan \theta \, d\theta$

8. $\displaystyle\int e^{5z} \, dz$

9. $\displaystyle\int t e^{t^2+1} \, dt$

10. $\displaystyle\int \frac{dz}{1+z^2}$

11. $\displaystyle\int \frac{dz}{1+4z^2}$

12. $\displaystyle\int \sin^2 \theta \cos \theta \, d\theta$

13. $\displaystyle\int \sin 5\theta \cos^3 5\theta \, d\theta$

14. $\displaystyle\int \sin^3 z \cos^3 z \, dz$

15. $\displaystyle\int \frac{(\ln x)^2}{x} \, dx$

16. $\displaystyle\int \cos \theta \sqrt{1 + \sin \theta} \, d\theta$

17. $\displaystyle\int x e^x \, dx$

18. $\displaystyle\int t^3 e^t \, dt$

19. $\displaystyle\int x \ln x \, dx$

20. $\displaystyle\int \frac{1}{\cos^2 \theta} \, d\theta$

In Problems 21–25, find the definite integral by two methods (Fundamental Theorem and numerically).

21. $\displaystyle\int_1^3 x(x^2+1)^{70} \, dx$

22. $\displaystyle\int_0^1 \frac{dx}{x^2+1}$

23. $\displaystyle\int_0^{10} z e^{-z} \, dz$

24. $\displaystyle\int_{-\pi/3}^{\pi/4} \sin^3 \theta \cos \theta \, d\theta$

25. $\displaystyle\int_1^4 \frac{e^{\sqrt{x}}}{\sqrt{x}} \, dx$

26. The graph of dy/dt against t is in Figure D.12. Suppose the three shaded regions each have area 2. Given that $y = 0$ when $t = 0$, draw the graph of y against t, indicating all special features the graph might have (known heights, maxima and minima, inflection points, etc.). Pay particular attention to the relationship between the graphs. Mark $t_1, t_2, \ldots, t_5$ on the t axis.[1]

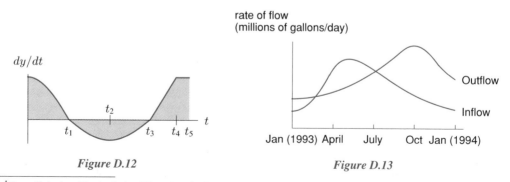

Figure D.12   Figure D.13

---
[1] From *Calculus: The Analysis of Functions*, by Peter D. Taylor (Toronto: Wall & Emerson, Inc., 1992)

27. The Quabbin Reservoir in the western part of Massachusetts provides most of Boston's water. The graph in Figure D.13 represents the flow of water in and out of the Quabbin Reservoir throughout 1993.

(a) Sketch a possible graph for the quantity of water in the reservoir, as a function of time.

(b) When, in the course of 1993, was the quantity of water in the reservoir largest? Smallest? Mark and label these points on the graph you drew in part (a).

(c) When was the quantity of water decreasing most rapidly? Again, mark and label this time on Figure D.13 and the graph you drew in part (a).

(d) By July 1994 the quantity of water in the reservoir was about the same as in January 1993. Draw plausible graphs for the flow into and the flow out of the reservoir for the first half of 1994. Explain your graph.

28. The rate at which the world's oil is being consumed is continuously increasing. Suppose the rate (in billions of barrels per year) is given by the function $r = f(t)$, where t is measured in years and $t = 0$ is the start of 1990.

(a) Write a definite integral which represents the total quantity of oil used between the start of 1990 and the start of 1995.

(b) Suppose $r = 32e^{0.05t}$. Using a left-hand sum with five subdivisions, find an approximate value for the total quantity of oil used between the start of 1990 and the start of 1995.

(c) Interpret each of the five terms in the sum from part (b) in terms of oil consumption.

29. A rod has length 2 meters. At a distance x meters from its left end, the density of the rod is given by

$$\rho(x) = 2 + 6x \text{ g/m}.$$

(a) Write a Riemann sum approximating the total mass of the rod.

(b) Find the exact mass by converting the sum into an integral.

30. The density of cars (in cars per mile) down a 20-mile stretch of the Pennsylvania Turnpike can be approximated by

$$\rho(x) = 300 \left(2 + \sin \left(4\sqrt{x + 0.15} \right) \right),$$

where x is the distance in miles from the Breezewood toll plaza.

(a) Sketch a graph of this function for $0 \leq x \leq 20$.

(b) Write a sum that approximates the total number of cars on this 20-mile stretch.

(c) Find the total number of cars on the 20-mile stretch.

31. Circle City, a typical metropolis, is very densely populated near its center, and its population gradually thins out toward the city limits. In fact, its population density is $10,000(3 - r)$ people/square mile at a distance r miles from the center.

(a) Assuming that the population density at the city limits is zero, find the radius of the city.

(b) What is the total population of the city?

32. The density of oil in a circular oil slick on the surface of the ocean at a distance r meters from the center of the slick is given by $\rho(r) = 50/(1 + r)$ kg/m².

(a) If the slick extends from $r = 0$ to $r = 10,000$ m, find a Riemann sum approximating the total mass of oil in the slick.

(b) Find the exact value of the mass of oil in the slick by turning your sum into an integral and evaluating it.

(c) Within what distance r is half the oil of the slick contained?

33. An exponential model for the density of the earth's atmosphere says that if the temperature of the atmosphere were constant, then the density of the atmosphere as a function of height, h (in meters), above the surface of the earth would be given by

$$\rho(h) = 1.28e^{-0.000124h} \text{ kg/m}^3.$$

   (a) Write (but do not evaluate) a sum that approximates the mass of the portion of the atmosphere from $h = 0$ to $h = 100$ m (i.e., the first 100 meters above sea level). Assume the radius of the earth is 6370 km.

   (b) Find the exact answer by turning your sum in part (a) into an integral. Evaluate the integral.

34. Water is flowing in a cylindrical pipe of radius 1 inch. Because water is viscous and sticks to the pipe, the rate of flow varies with the distance from the center. The speed of the water at a distance r inches from the center is $10(1 - r^2)$ inches per second. What is the rate (in cubic inches per second) at which water is flowing through the pipe?

35. The reflector behind a car headlight is made in the shape of the parabola, $x = \frac{4}{9}y^2$, with a circular cross-section, as shown in Figure D.14.

   (a) Find a Riemann sum approximating the volume contained by this headlight.
   (b) Find the volume exactly.

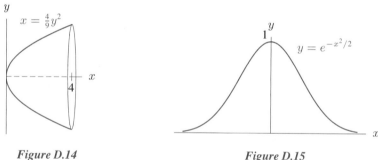

Figure D.14   Figure D.15

36. Rotate the bell-shaped curve $y = e^{-x^2/2}$ shown in Figure D.15 around the y-axis, forming a hill-shaped solid of revolution. By slicing horizontally, find the volume of this hill.

37. The circumference of the trunk of a certain tree at different heights above the ground is given in the following table.

Height (feet)	0	20	40	60	80	100	120
Circumference (feet)	26	22	19	14	6	3	1

Assume all horizontal cross-sections of the trunk are circles. Estimate the volume of the tree trunk using the trapezoid rule.

38. Most states expect to run out of space for their garbage soon. In New York, solid garbage is packed into pyramid-shaped dumps with square bases. (The largest such dump is on Staten Island.) A small community has a dump with a base length of 100 yards. One yard vertically above the base, the length of the side parallel to the base is 99 yards; the dump can be built up to a vertical height of 20 yards. (The top of the pyramid is never reached.) If 65 cubic yards of garbage arrive at the dump every day, how long will it be before the dump is full?

E TABLE OF INTEGRALS

A Short Table of Indefinite Integrals

I. Basic Functions

1. $\int x^n \, dx = \dfrac{1}{n+1} x^{n+1} + C, \quad n \neq -1$

5. $\int \sin x \, dx = -\cos x + C$

2. $\int \dfrac{1}{x} \, dx = \ln |x| + C$

6. $\int \cos x \, dx = \sin x + C$

3. $\int a^x \, dx = \dfrac{1}{\ln a} a^x + C$

7. $\int \tan x \, dx = -\ln |\cos x| + C$

4. $\int \ln x \, dx = x \ln x - x + C, \quad x > 0$

II. Products of e^x, $\cos x$, and $\sin x$

8. $\int e^{ax} \sin(bx) \, dx = \dfrac{1}{a^2 + b^2} e^{ax} [a \sin(bx) - b \cos(bx)] + C$

9. $\int e^{ax} \cos(bx) \, dx = \dfrac{1}{a^2 + b^2} e^{ax} [a \cos(bx) + b \sin(bx)] + C$

10. $\int \sin(ax) \sin(bx) \, dx = \dfrac{1}{b^2 - a^2} [a \cos(ax) \sin(bx) - b \sin(ax) \cos(bx)] + C, \quad a \neq b$

11. $\int \cos(ax) \cos(bx) \, dx = \dfrac{1}{b^2 - a^2} [b \cos(ax) \sin(bx) - a \sin(ax) \cos(bx)] + C, \quad a \neq b$

12. $\int \sin(ax) \cos(bx) \, dx = \dfrac{1}{b^2 - a^2} [b \sin(ax) \sin(bx) + a \cos(ax) \cos(bx)] + C, \quad a \neq b$

III. Product of Polynomial $p(x)$ with $\ln x$, e^x, $\cos x$, $\sin x$

13. $\int x^n \ln x \, dx = \dfrac{1}{n+1} x^{n+1} \ln x - \dfrac{1}{(n+1)^2} x^{n+1} + C, \quad n \neq -1, \quad x > 0$

14. $\int p(x) e^{ax} \, dx = \dfrac{1}{a} p(x) e^{ax} - \dfrac{1}{a} \int p'(x) e^{ax} \, dx$

$\qquad = \dfrac{1}{a} p(x) e^{ax} - \dfrac{1}{a^2} p'(x) e^{ax} + \dfrac{1}{a^3} p''(x) e^{ax} - \cdots$

$\qquad (+ - + - \ldots) \quad \text{(signs alternate)}$

15. $\int p(x) \sin ax \, dx = -\dfrac{1}{a} p(x) \cos ax + \dfrac{1}{a} \int p'(x) \cos ax \, dx$

$\qquad = -\dfrac{1}{a} p(x) \cos ax + \dfrac{1}{a^2} p'(x) \sin ax + \dfrac{1}{a^3} p''(x) \cos ax - \cdots$

$\qquad (- + + - - + + \ldots) \quad \text{(signs alternate in pairs after first term)}$

16. $\int p(x) \cos ax \, dx = \dfrac{1}{a} p(x) \sin ax - \dfrac{1}{a} \int p'(x) \sin ax \, dx$

$\qquad = \dfrac{1}{a} p(x) \sin ax + \dfrac{1}{a^2} p'(x) \cos ax - \dfrac{1}{a^3} p''(x) \sin ax - \cdots$

$\qquad (+ + - - + + - - \ldots) \quad \text{(signs alternate in pairs)}$

IV. Integer Powers of $\sin x$ and $\cos x$

17. $\displaystyle\int \sin^n x \, dx = -\frac{1}{n}\sin^{n-1} x \cos x + \frac{n-1}{n}\int \sin^{n-2} x \, dx, \quad n \text{ positive}$

18. $\displaystyle\int \cos^n x \, dx = \frac{1}{n}\cos^{n-1} x \sin x + \frac{n-1}{n}\int \cos^{n-2} x \, dx, \quad n \text{ positive}$

19. $\displaystyle\int \frac{1}{\sin^m x}\, dx = \frac{-1}{m-1}\frac{\cos x}{\sin^{m-1} x} + \frac{m-2}{m-1}\int \frac{1}{\sin^{m-2} x}\, dx, \quad m \neq 1, m \text{ positive}$

20. $\displaystyle\int \frac{1}{\sin x}\, dx = \frac{1}{2}\ln\left|\frac{(\cos x)-1}{(\cos x)+1}\right| + C$

21. $\displaystyle\int \frac{1}{\cos^m x}\, dx = \frac{1}{m-1}\frac{\sin x}{\cos^{m-1} x} + \frac{m-2}{m-1}\int \frac{1}{\cos^{m-2} x}\, dx, \quad m \neq 1, m \text{ positive}$

22. $\displaystyle\int \frac{1}{\cos x}\, dx = \frac{1}{2}\ln\left|\frac{(\sin x)+1}{(\sin x)-1}\right| + C$

23. $\displaystyle\int \sin^m x \cos^n x \, dx$: If m is odd, let $w = \cos x$. If n is odd, let $w = \sin x$. If both m and n are even and non-negative, convert all to $\sin x$ or all to $\cos x$ (using $\sin^2 x + \cos^2 x = 1$), and use IV-17 or IV-18. If m and n are even and one of them is negative, convert to whichever function is in the denominator and use IV-19 or IV-21. The case in which both m and n are even and negative is omitted.

V. Quadratic in the Denominator

24. $\displaystyle\int \frac{1}{x^2 + a^2}\, dx = \frac{1}{a}\arctan\frac{x}{a} + C, \quad a \neq 0$

25. $\displaystyle\int \frac{bx + c}{x^2 + a^2}\, dx = \frac{b}{2}\ln|x^2 + a^2| + \frac{c}{a}\arctan\frac{x}{a} + C, \quad a \neq 0$

26. $\displaystyle\int \frac{1}{(x-a)(x-b)}\, dx = \frac{1}{a-b}(\ln|x-a| - \ln|x-b|) + C, \quad a \neq b$

27. $\displaystyle\int \frac{cx + d}{(x-a)(x-b)}\, dx = \frac{1}{a-b}\big[(ac+d)\ln|x-a| - (bc+d)\ln|x-b|\big] + C, \quad a \neq b$

VI. Integrands Involving $\sqrt{a^2 + x^2}$, $\sqrt{a^2 - x^2}$, $\sqrt{x^2 - a^2}$, $a > 0$

28. $\displaystyle\int \frac{1}{\sqrt{a^2 - x^2}}\, dx = \arcsin\frac{x}{a} + C$

29. $\displaystyle\int \frac{1}{\sqrt{x^2 \pm a^2}}\, dx = \ln\left|x + \sqrt{x^2 \pm a^2}\right| + C$

30. $\displaystyle\int \sqrt{a^2 \pm x^2}\, dx = \frac{1}{2}\left(x\sqrt{a^2 \pm x^2} + a^2 \int \frac{1}{\sqrt{a^2 \pm x^2}}\, dx\right) + C$

31. $\displaystyle\int \sqrt{x^2 - a^2}\, dx = \frac{1}{2}\left(x\sqrt{x^2 - a^2} - a^2 \int \frac{1}{\sqrt{x^2 - a^2}}\, dx\right) + C$

F   REVIEW OF DENSITY FUNCTIONS AND PROBABILITY

Understanding the distribution of various quantities through the population can be important to decision makers. For example, the income distribution gives useful information about the economic structure of a society. In this section we will look at the distribution of ages in the US. To allocate funding for education, health care, and social security, the government needs to know how many people are in each age group. We will see how to represent such information by a density function.

US Age Distribution

TABLE F.1   *Distribution of ages in the US in 1990*

Age group	Percentage of total population
0–20	30%
20–40	31%
40–60	24%
60–80	14%
Over 80	1%

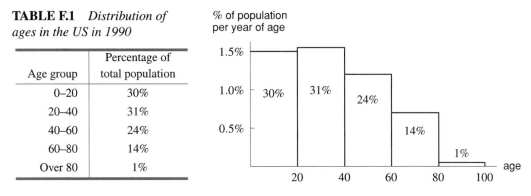

Figure F.16: How ages were distributed in the US in 1990

Suppose we have the data in Table F.1 showing how the ages of the US population were distributed in 1990. To represent this information graphically we use a *histogram*, putting a vertical bar above each age group in such a way that the *area* of each bar represents the percentage in that age group. The total area of all the rectangles is 100% = 1. We will assume that there is nobody over 100 years old, so that the last age group is 80–100. For the 0–20 age group, the base of the rectangle is 20, and we want the area to be 30%, so the height must be 30%/20 = 1.5%. Notice that the vertical axis is measured in percent/year. (See Figure F.16.)

Example 1   In 1990, what percentage of the US population was:
(a)   Between 20 and 60 years old?
(b)   Less than 10 years old?
(c)   Between 75 and 80 or between 80 and 85 years old?

Solution   (a)   We add the percentages, so 31% + 24% = 55%.
(b)   To find the percentage less than 10 years old, we could assume, for example, that the population was distributed evenly over the 0–20 group. (This means we are assuming that babies were born at a fairly constant rate over the last 20 years, which is probably reasonable.) If we make this assumption, then we can say that the population less than 10 years old was about half that in the 0–20 group, that is, 15%. Notice that we get the same result by computing the area of the rectangle from 0 to 10. (See Figure F.17.)
(c)   To find the population between 75 and 80 years old, since 14% of Americans in 1990 were in the 60-80 group, we might apply the same reasoning and say that $\frac{1}{4}(14\%) = 3.5\%$ of the population was in this age group. This result is represented as an area in Figure F.17. The

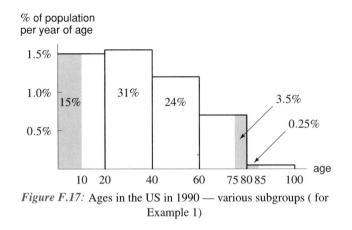

Figure F.17: Ages in the US in 1990 — various subgroups (for Example 1)

assumption that the population was evenly distributed is not a good one here; certainly there were more people between the ages of 60 and 65 than between 75 and 80. Thus, the estimate of 3.5% is certainly too high.

Again using the (faulty) assumption that ages in each group were distributed uniformly, we would find that the percentage between 80 and 85 was $\frac{1}{4}(1\%) = 0.25\%$. (See Figure F.17.) This estimate is also poor — there were certainly more people in the 80–85 group than, say, the 95–100 group, and so the 0.25% estimate is too low. In addition, although the percentage of 80–85-year-olds was certainly smaller than the percentage of 75–80-year-olds, the difference between 0.25% and 3.5% (a factor of 14) is unreasonably large. We can expect the transition from one age group to the next to be smoother and more gradual.

Smoothing Out the Histogram

We could get better estimates if we had smaller age groups (each age group in Figure F.16 is 20 years, which is quite large) or if the histogram were smoother. Suppose we have the more detailed data in Table F.2, which leads to the new histogram in Figure F.18.

As we get more detailed information, the upper silhouette of the histogram becomes smoother, but the area of any of the bars still represents the percentage of the population in that age group. Imagine, in the limit, replacing the upper silhouette of the histogram by a smooth curve in such a way that the area under the curve above one age group is the same as the area in the corresponding rectangle. The total area under the whole curve is again 100% = 1. (See Figure F.18.)

The Age Density Function

If t is age in years, we define $p(t)$, the age *density function*, to be a function which "smooths out" the age histogram. This function has the property that

$$\begin{array}{c}\text{Fraction of population} \\ \text{between ages } a \text{ and } b\end{array} = \begin{array}{c}\text{Area under} \\ \text{graph of } p \\ \text{between } a \text{ and } b\end{array} = \int_a^b p(t)\,dt.$$

If a and b are the smallest and largest possible ages (say, $a = 0$ and $b = 100$), so that the ages of all of the population are between a and b, then

$$\int_a^b p(t)\,dt = \int_0^{100} p(t)\,dt = 1.$$

TABLE F.2 *Ages in the US in 1990 (more detailed)*

Age group	Percentage of total population
0–10	15%
10–20	15%
20–30	16%
30–40	15%
40–50	13%
50–60	11%
60–70	9%
70–80	5%
80–90	1%

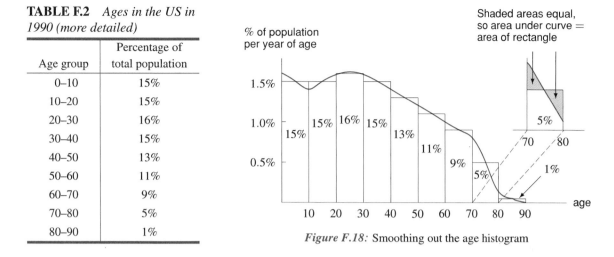

Figure F.18: Smoothing out the age histogram

What does the age density function p tell us? Notice that we have not talked about the meaning of $p(t)$ itself, but *only* of the integral $\int_a^b p(t)\, dt$. Let's look at this in a bit more detail. Suppose, for example, that $p(10) = 0.015 = 1.5\%$ per year. This is *not* telling us that 1.5% of the population is precisely 10 years old (where 10 years old means exactly 10, not $10\frac{1}{2}$, not $10\frac{1}{4}$, not 10.1). However, $p(10) = 0.015$ does tell us that for some small interval Δt around 10, the fraction of the population with ages in this interval is approximately $p(10)\, \Delta t = 0.015\, \Delta t$. Notice also that the units of $p(t)$ are *% per year*, so $p(t)$ must be multiplied by years to give a percentage of the population.

The Density Function

In order to generalize the idea of the age distribution, let us look at a general density function. Suppose we are interested in how a certain characteristic, x, is distributed through a population. For example, x might be height, age, or wattage, and the population might be people, or any set of objects such as light bulbs. Then we define a general density function with the following properties:

> The function, $p(x)$, is a **density function** if
>
> $$\begin{array}{c} \text{Fraction of population} \\ \text{for which } x \text{ is} \\ \text{between } a \text{ and } b \end{array} = \begin{array}{c} \text{Area under} \\ \text{graph of } p \\ \text{between } a \text{ and } b \end{array} = \int_a^b p(x)\,dx.$$
>
> $$\int_{-\infty}^{\infty} p(x)\, dx = 1 \quad \text{and} \quad p(x) \geq 0 \quad \text{for all } x.$$

The density function must be nonnegative if its integral always gives a fraction of the population. Also, the fraction of the population with x between $-\infty$ and ∞ is 1 because the entire population has the characteristic x between $-\infty$ and ∞. The function $p(t)$ used to smooth out the age histogram satisfies this definition of a density function. We do not assign a meaning to the value of $p(x)$ alone, but rather interpret $p(x)\, \Delta x$ as the fraction of the population with the characteristic in a short interval of length Δx around x.

Example 2   The graph in Figure F.19 shows the distribution of the number of years of education completed by adults in a population. What does the graph tell us?

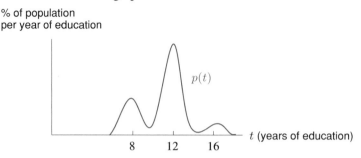

Figure F.19: Distribution of years of education

Solution   The fact that most of the area under the graph of the density function is concentrated in two humps, centered at 8 and 12 years, indicates that most of the population belongs to one of two groups, those who leave school after finishing approximately 8 years and those who finish about 12 years. There is a smaller group of people who finish approximately 16 years of school.

The density function is often approximated by formulas, as in the next example.

Example 3   Find reasonable formulas representing the density function for the US age distribution, assuming the function is constant at 1.5% up to age 40 and then drops linearly.

Solution   We need to construct a linear function sloping downward from age 40 in such a way that $p(40) = 1.5\%$ per year $= 0.015$ and that $\int_0^{100} p(t)dt = 1$. Suppose b is as in Figure F.20. Since

$$\int_0^{100} p(t)dt = \int_0^{40} p(t)dt + \int_{40}^{100} p(t)dt = 40(0.015) + \frac{1}{2}(0.015)b = 1,$$

we have

$$\frac{0.015}{2}b = 0.4, \quad \text{giving} \quad b \approx 53.3.$$

Thus the slope of the line is $-0.015/53.3 \approx -0.00028$, so for $40 \le t \le 40 + 53.3 = 93.3$,

$$p(t) - 0.015 = -0.00028(t - 40),$$
$$p(t) = 0.0262 - 0.00028t.$$

According to this way of smoothing the data, there is no one over 93.3 years old.

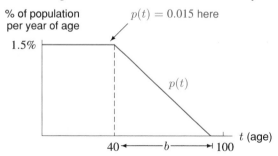

Figure F.20: Age density function

Probability

Suppose we pick a member of the US population at random and ask what is the probability that the person is between, say, the ages of 60 and 65. The probability, or chance, that the person is in a certain age group is equal to the fraction of the population in that age group. Consider the density function $p(t)$ defined on page 474 to describe the distribution of ages in the US. We can use the density function to calculate probabilities as follows:

$$\begin{matrix} \text{Probability that} \\ \text{a person is between} \\ \text{ages } a \text{ and } b \end{matrix} \quad = \quad \begin{matrix} \text{Fraction of population} \\ \text{between ages } a \text{ and } b \end{matrix} \quad = \int_a^b p(t)\, dt.$$

The Median and Mean

It is often useful to be able to give an "average" value for a distribution. Two measures that are in common use are the *median* and the *mean*.

The Median

A **median** is a value T such that half the population has values of x less than (or equal to) T, and half the population has values of x greater than (or equal to) T. Thus, a median T satisfies

$$\int_{-\infty}^{T} p(x)\, dx = 0.5,$$

where p is the density function. In other words, half the area under the graph of p lies to the left of T.

Example 4   Find the median age in the US in 1990, using the age density function given by

$$p(t) = \begin{cases} 0.015 & \text{for } 0 \leq t \leq 40 \\ 0.0262 - 0.00028t & \text{for } 40 < t \leq 93.3. \end{cases}$$

Solution   We want to find the value of T such that

$$\int_{-\infty}^{T} p(t)\, dt = \int_{0}^{T} p(t)\, dt = 0.5.$$

Since $p(t) = 1.5\%$ up to age 40, we have

$$\text{Median} = T = \frac{50\%}{1.5\%} \approx 33 \text{ years}.$$

(See Figure F.21.)

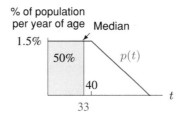

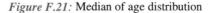

Figure F.21: Median of age distribution

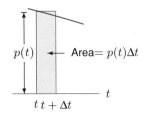

Figure F.22: Shaded area is percentage of population with age between t and $t + \Delta t$

The Mean

Another commonly used average value is the *mean*. To find the mean of N numbers, you add the numbers and divide the sum by N. For example, the mean of the numbers 1, 2, 7, and 10 is $(1 + 2 + 7 + 10)/4 = 5$. The mean age of the entire US population is therefore defined as

$$\frac{\sum \text{Ages of all people in the US}}{\text{Total number of people in the US}}.$$

Calculating the sum of all the ages directly would be an enormous task; we will approximate the sum by an integral. The idea is to "slice up" the age axis and consider the people whose age is between t and $t + \Delta t$. How many are there?

The percentage of the population between t and $t + \Delta t$ is the area under the graph of p between these points, which is well approximated by the area of the rectangle, $p(t)\Delta t$. (See Figure F.22.) If the total number of people in the population is N, then

$$\begin{array}{c}\text{Number of people with age} \\ \text{between } t \text{ and } t + \Delta t\end{array} \approx p(t)\Delta t N.$$

The age of all of these people is approximately t:

$$\begin{array}{c}\text{Sum of ages of people} \\ \text{between age } t \text{ and } t + \Delta t\end{array} \approx t p(t)\Delta t N.$$

Therefore, adding and factoring out an N gives us

$$\text{Sum of ages of all people} \approx \left(\sum t p(t)\Delta t \right) N.$$

In the limit, as we allow Δt to shrink to 0, the sum becomes an integral, so as an approximation

$$\text{Sum of ages of all people} = \left(\int_0^{100} t p(t)dt \right) N.$$

Therefore, with N equal to the total number of people in the US, and assuming no person is over 100 years old,

$$\text{Mean age} = \frac{\text{Sum of ages of all people in US}}{N} = \int_0^{100} t p(t)dt.$$

We can give the same argument for any[2] density function $p(x)$.

If a quantity has density function $p(x)$,

$$\textbf{Mean value} \text{ of the quantity} = \int_{-\infty}^{\infty} x p(x)\, dx.$$

---

[2]Provided all the relevant improper integrals converge.

It can be shown that the mean is the point on the horizontal axis where the region under the graph of the density function, if it were made out of cardboard, would balance.

Example 5   Find the mean age of the US population, using the density function of Example 4.

Solution   The approximate formulas for p are

$$p(t) = \begin{cases} 0.015 & \text{for } 0 \leq t \leq 40 \\ 0.0262 - 0.00028t & \text{for } 40 < t \leq 93.3. \end{cases}$$

Using these formulas, we compute

$$\text{Mean age} = \int_0^{100} tp(t)dt = \int_0^{40} t(0.015)dt + \int_{40}^{93.3} t(0.0262 - 0.00028t)dt$$

$$= 0.015\frac{t^2}{2}\Big|_0^{40} + 0.0262\frac{t^2}{2}\Big|_{40}^{93.3} - 0.00028\frac{t^3}{3}\Big|_{40}^{93.3} \approx 35 \text{ years.}$$

The mean is shown is Figure F.23..

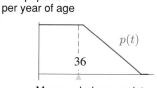

% of population
per year of age

$p(t)$

36

Mean = balance point

Figure F.23: Mean of age distribution

Normal Distributions

How much rain do you expect will fall in your home town this year? If you live in Anchorage, Alaska, the answer would be something close to 15 inches (including the snow). Of course, you don't expect exactly 15 inches. Some years there will be more than 15 inches, and some years there will be less. Most years, however, the amount of rainfall will be close to 15 inches; only rarely will it be well above or well below 15 inches. What does the density function for the rainfall look like? To answer this question, we look at rainfall data over many years. It lies on a bell-shaped curve which peaks at 15 inches and slopes downward approximately symmetrically on either side. This is an example of a normal distribution.

Normal distributions are frequently used to model real phenomena, from grades on an exam to the number of airline passengers on a particular flight. A normal distribution is characterized by its *mean*, μ, and its *standard deviation*, σ. The mean tells us where the data is clustered: the location of the central peak. The standard deviation tells us how closely the data is clustered around the mean. A small value of σ tells us that the data is close to the mean; a large σ tells us the data is spread out. The formula for a normal distribution is as follows.

A **normal distribution** has a density function of the form

$$p(x) = \frac{1}{\sigma\sqrt{2\pi}}e^{-(x-\mu)^2/(2\sigma^2)},$$

where μ is the mean of the distribution and σ is the standard deviation, with $\sigma > 0$.

The factor of $1/(\sigma\sqrt{2\pi})$ in front of the function is there to make the area under its graph equal to 1. That the factor $\sqrt{2\pi}$ is involved is one of the truly remarkable discoveries of mathematics.

To model the rainfall in Anchorage, we use a normal distribution with $\mu = 15$. The standard deviation can be estimated by looking at the data; we will take it to be 1. (See Figure F.24.)

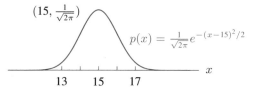

Figure F.24: Normal distribution with $\mu = 15$ and $\sigma = 1$

In the following example, we will verify that for a normal distribution a certain percentage of the data always lies within a certain number of standard deviations from the mean.

Example 6   For Anchorage's rainfall, use the normal distribution with the density function

$$p(x) = \frac{1}{\sqrt{2\pi}}e^{-(x-15)^2/2},$$

to compute the fraction of the years with rainfall between
(a)   14 and 16 inches,   (b)   13 and 17 inches,   (c)   12 and 18 inches.

Solution   (a)   The fraction of the years with annual rainfall between 14 and 16 inches is $\int_{14}^{16} \frac{1}{\sqrt{2\pi}}e^{-(x-15)^2/2}\, dx$.

Since there is no elementary antiderivative for $e^{-(x-15)^2/2}$, we find the integral numerically. Its value is about 0.68.

$$\begin{array}{c}\text{Fraction of years with rainfall} \\ \text{between 14 and 16 inches}\end{array} = \int_{14}^{16} \frac{1}{\sqrt{2\pi}}e^{-(x-15)^2/2}\, dx \approx 0.68.$$

(b)   Finding the integral numerically again, we get

$$\begin{array}{c}\text{Fraction of years with rainfall} \\ \text{between 13 and 17 inches}\end{array} = \int_{13}^{17} \frac{1}{\sqrt{2\pi}}e^{-(x-15)^2/2}\, dx \approx 0.95.$$

(c)   Similarly,

$$\begin{array}{c}\text{Fraction of years with rainfall} \\ \text{between 12 and 18 inches}\end{array} = \int_{12}^{18} \frac{1}{\sqrt{2\pi}}e^{-(x-15)^2/2}\, dx \approx 0.997.$$

Since 0.95 is so close to 1, we expect that most of the time the rainfall will be between 13 and 17 inches a year.

Notice that in the preceding example, the standard deviation is 1 inch, so rainfall between 14 and 16 inches a year is within one standard deviation of the mean. Similarly, rainfall between 13 and 17 inches is within 2 standard deviations of the mean, and rainfall between 12 and 18 inches is within three standard deviations of the mean. The fractions of the observations within one, two, and three standard deviations of the mean calculated in the previous example hold for any normal distribution.

> # Rules of Thumb for Any Normal Distribution
>
> - About 68% of the observations are within one standard deviation of the mean.
> - About 95% of the observations are within two standard deviations of the mean.
> - Over 99% of the observations are within three standard deviations of the mean.

Problems for Section F

In Problems 1–3, sketch graphs of a density function which could represent the distribution of income through a population with the given characteristics.

1. A large middle class.

2. Small middle and upper classes and many poor people.

3. Small middle class, many poor and many rich people.

4. A large number of people take a standardized test, receiving scores described by the density function p graphed in Figure F.25. Does the density function imply that most people receive a score near 50? Explain why or why not.

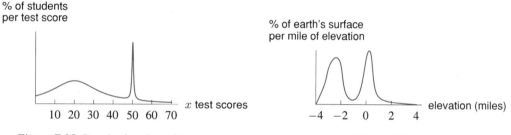

Figure F.25: Density function of test scores   *Figure F.26*

5. Figure F.26 shows the distribution of elevation, in miles, across the earth's surface. Positive elevation denotes land above sea level; negative elevation shows land below sea level (i.e., the ocean floor).

    (a) Describe in words the elevation of most of the earth's surface.
    (b) Approximately what fraction of the earth's surface is below sea level?

6. Consider a pendulum swinging through a small angle. The x-coordinate of the bob moves between $-a$ and a, as shown in Figure F.27.

    (a) Draw the density function for the location of the x-coordinate of the pendulum bob (i.e., neglect up-and-down motion). In order to do this, imagine a camera taking pictures of the pendulum at random instants. Where is the bob most likely to be found? Least likely? [Hint: Consider the speed of the pendulum at different points in its path. Is the camera more likely to take a photograph of the bob at a point on its path where the bob is moving quickly, or where it is moving slowly?]

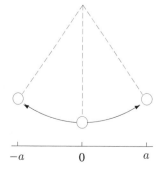

Figure F.27

(b) Now sketch the function

$$f(x) = \begin{cases} \dfrac{1}{\pi\sqrt{a^2 - x^2}} & -a < x < a; \\ 0 & |x| \geq a. \end{cases}$$

How does this graph compare with the one you drew in part (a)?

(c) Assuming that the function given in part (b) is the density function for the pendulum, what do you expect

$$\int_{-a}^{a} \frac{1}{\pi\sqrt{a^2 - x^2}} \, dx$$

to be? Check this by computing the integral.

(d) Does it seem reasonable, physically speaking, that $f(x)$ "blows up" at a and $-a$? Explain your answer.

7. IQ scores are believed to be normally distributed with mean 100 and standard deviation 15.

  (a) Write a formula for the density distribution of IQ scores.
  (b) Estimate the fraction of the population with IQ between 115 and 120.

8. Show that the area under the graph of the density function of the normal distribution

$$p(x) = \frac{1}{\sqrt{2\pi}} e^{-(x-15)^2/2}$$

is 1. This function has no elementary antiderivative, so you must do it numerically. Make it clear in your solution what limits of integration you used.

9. (a) Using a calculator or computer, sketch graphs of the density function of the normal distribution

$$p(x) = \frac{1}{\sigma\sqrt{2\pi}} e^{-(x-\mu)^2/(2\sigma^2)}$$

  (i) For fixed μ (say, $\mu = 5$) and varying σ (say, $\sigma = 1, 2, 3$).
  (ii) For varying μ (say, $\mu = 4, 5, 6$) and fixed σ (say, $\sigma = 1$).

  (b) Explain how the graphs confirm that μ is the mean of the distribution and that σ shows how closely the data is clustered around the mean.

10. Let v be the speed, in meters/second, of an oxygen molecule, and let $p(v)$ be the density function of the speed distribution of oxygen molecules at room temperature. Maxwell showed that

$$p(v) = av^2 e^{-mv^2/(2kT)},$$

where $k = 1.4 \times 10^{-23}$ is the Boltzmann constant, T is the temperature in Kelvin (at room temperature, $T = 293$), and $m = 5 \times 10^{-26}$ is the mass of the oxygen molecule in kilograms.

  (a) Find the value of a.
  (b) Estimate the median and the mean speed. Find the maximum of $p(v)$.
  (c) How do your answers in part (b) for the mean and the maximum of $p(v)$ change as T changes?

G REVIEW OF POLAR COORDINATES

Polar coordinates are another way of describing points in an xy-plane. The x- and y-coordinates can be thought of as instructions on how to get to the point. To get to the point $(1, 2)$ go 1 unit horizontally and 2 units vertically. Polar coordinates can be thought of the same way. There is an r-coordinate, which tells you how far to go along the ray extending to the point from the origin, and there is the θ-coordinate, which is an angle, and it tells you the angle the ray makes with the positive x-axis. (See Figure G.28.)

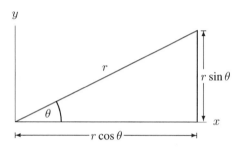

Figure G.28: Polar coordinates

Example 1   Give the polar coordinates of the points $(1, 0)$, $(0, 1)$, $(-1, 0)$, and $(1, 1)$.

Solution   To get to the point $(1, 0)$, you go 1 unit along the horizontal axis, and there you are. So its r-coordinate is 1 and its θ-coordinate is 0, since you go along the x-axis.

The point $(0, 1)$ is also one unit from the origin, so you start out the same way as before, by going 1 unit out along the ray. Then you have to go around the circle of radius 1 through an arc of $\pi/2$ to get to the point $(0, 1)$. So $r = 1$ and $\theta = \pi/2$.

The point $(-1, 0)$ also has r-coordinate equal to 1, but this time you have to go halfway around the circle to get there, so its θ-coordinate is π.

The point $(1, 1)$ is a distance of $\sqrt{2}$ from the origin, and the ray from the origin to it makes an angle of $\pi/4$ with the horizontal ray. Thus, it has r-coordinate equal to $\sqrt{2}$ and θ-coordinate equal to $\pi/4$. See Figure G.29.

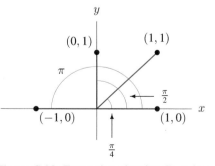

Figure G.29: Four points showing Cartesian
and polar coordinates

Conversion Between Polar and Cartesian Coordinates

Suppose a point has Cartesian coordinates (x, y). Look at Figure G.28. The distance r of the point from the origin is the length of the hypotenuse, which is $\sqrt{x^2 + y^2}$ by Pythagoras' theorem. The angle θ with the positive half of the x-axis satisfies $\tan \theta = y/x$.

On the other hand, if you are given r and θ, then from trigonometry you can see that $x = r \cos \theta$ and $y = r \sin \theta$.

Relation Between Polar and Cartesian Coordinates

$$x = r \cos \theta \qquad r = \sqrt{x^2 + y^2}$$

$$y = r \sin \theta \qquad \tan \theta = y/x.$$

Example 2   Give the Cartesian coordinates of the points with polar coordinates $(2, 3\pi/2)$ and $(2, 1)$.

Solution   Both points have r-coordinate equal to 2, so they are 2 units from the origin. The first one has θ-coordinate equal to $3\pi/2$, which is three quarters of a full revolution, so it is three quarters of the way around the circle of radius 2, at $(0, -2)$.

The second point has θ-coordinate equal to 1. From the formulas above, we see that

$$x = 2 \cos 1 = 1.0806 \quad \text{and} \quad y = 2 \sin 1 = 1.6830.$$

Problems for Section G

For Problems 1–7, give Cartesian coordinates for the points with the following polar coordinates (r, θ). The angles are measured in radians.

1. $(1, 0)$ 2. $(0, 1)$ 3. $(2, \pi)$ 4. $(\sqrt{2}, 5\pi/4)$

5. $(5, -\pi/6)$ 6. $(3, \pi/2)$ 7. $(1, 1)$

For Problems 8–15, give polar coordinates for the points with the following Cartesian coordinates. Choose $0 \leq \theta < 2\pi$.

8. $(1, 0)$ 9. $(0, 2)$ 10. $(1, 1)$ 11. $(-1, 1)$

12. $(-3, -3)$ 13. $(0.2, -0.2)$ 14. $(3, 4)$ 15. $(-3, 1)$

16. Every point in the plane can be represented by some pair of polar coordinates, but are the polar coordinates (r, θ) uniquely determined by the Cartesian coordinates (x, y)? In other words, for each pair of Cartesian coordinates, is there one and only one pair of polar coordinates for that point? Why or why not?

ANSWERS TO ODD NUMBERED PROBLEMS

Section 11.1

1 (a) 80-90°F
   (b) 60-72°F
   (c) 60-100°F
11 (a) Decreasing
   (b) Increasing
15 Half wave-length of original
   Speed = 2 seats/sec

Section 11.2

1 B, B, B
3 $(1, -1, -3)$; Front, left, below
9 Cylinder radius 2 along x-axis
11 Q
13 $(1.5, 0.5, -0.5)$
15 $(-4, 2, 7)$
19 $(x - 1)^2 + (y - 2)^2$
   $+ (z - 3)^2 = 25$

Section 11.3

1 (a) Decreases
   (b) Increases
3 (a) Bowl
   (b) Neither
   (c) Plate
   (d) Bowl
   (e) Plate
5 (a) I
   (b) V
   (c) IV
   (d) II
   (e) III
9 (b) Increasing x

Section 11.4

11 (a) A
   (b) B
   (c) A
23 (a) (III)
   (b) (I)
   (c) (V)
   (d) (II)

(e) (IV)
27 (a) (II) (E)
   (b) (I) (D)
   (c) (III) (G)
29 $\alpha + \beta > 1$: increasing
   $\alpha + \beta = 1$: constant
   $\alpha + \beta < 1$: decreasing

Section 11.5

1 $\Delta z = 0.4$; $z = 2.4$
3 $f(x, y) = 2 - \dfrac{1}{2} \cdot x - \dfrac{2}{3} \cdot y$
5 $z = 4 + 3x + y$
7 No
9 $f(x, y) = 2x - 0.5y + 1$
11 -1.0
13 $f(x, y) = 3 - 2x + 3y$
19 (a) no
   (b) no
   (c) no
   (e) 0.5 of GPA

Section 11.6

3 (a) I
   (b) II
7 Spheres
9 $9x - \frac{5}{2}y + \frac{1}{3}z + \frac{67}{6}$
11 $-(3/2)x - 3y + 3z - 4$
13 Hyperboloid of two sheets
15 Elliptic paraboloid
17 Ellipsoid
19 Sphere

Section 11.7

3 (b) Yes
5 0
7 0
9 1
13 $\pm\sqrt{(1 - c)/c}$.
15 No

Chapter 11 Review

1 Vertical line through $(2, 1, 0)$
3 $(x/5) + (y/3) + (z/2) = 1$
11 Could not be true
13 Might be true
15 True
17 $3x - 5y + 1 = c$
19 $2x^2 + y^2 = k$
23 $g(x, y) = 3x + y$

Section 12.1

1 $\vec{p} = 2\vec{w}$
   $\vec{q} = -\vec{u}$
   $\vec{r} = \vec{u} + \vec{w}$
   $\vec{s} = 2\vec{w} - \vec{u}$
   $\vec{t} = \vec{u} - \vec{w}$
3 $\sqrt{11}$
5 $-15\vec{i} + 25\vec{j} + 20\vec{k}$
7 $\sqrt{65}$
9 $\vec{i} + 3\vec{j}$
11 $-4.5\vec{i} + 8\vec{j} + 0.5\vec{k}$
13 $\sqrt{11}$
15 5.6
17 $\vec{i} + 4\vec{j}$
19 $-\vec{i}$
23 $3\vec{i} + 4\vec{j}$
25 $\vec{a} = \vec{b} = \vec{c} = 3\vec{k}$
   $\vec{d} = 2\vec{i} + 3\vec{k}$
   $\vec{e} = \vec{j}$
   $\vec{f} = -2\vec{i}$
27 $\|\vec{u}\| = \sqrt{6}$
   $\|\vec{v}\| = \sqrt{5}$
29 (a) $(3/5)\vec{i} + (4/5)\vec{j}$
   (b) $6\vec{i} + 8\vec{j}$

Section 12.2

1 Scalar
3 Vector
5 (a) $50\vec{i}$

(b) $-50\vec{j}$
(c) $25\sqrt{2}\vec{i} - 25\sqrt{2}\vec{j}$
(d) $-25\sqrt{2}\vec{i} + 25\sqrt{2}\vec{j}$

7 $180\vec{i} + 72\vec{j}$

9 P
Towards the center

11 $(79.00, 79.33, 89.00, 68.33, 89.33)$

13 $48.3°$ east of north
744 km/hr

15 548.6 km/hr

Section 12.3

1 -38

3 14

5 238

7 1.91 radians $(109.5°)$

9 $\vec{u} \perp \vec{v}$ for $t = 2$ or -1.
No values of t make $\vec{u}$ parallel
to $\vec{v}$.

11 $3\vec{i} + 4\vec{j} - \vec{k}$
(multiples of)

13 $-x + 2y + z = 1$

15 $2x - 3y + 7z = 19$

17 $3x + y + z = -1$

19 $\vec{a} = -\frac{8}{21}\vec{d} + (\frac{79}{21}\vec{i} + \frac{10}{21}\vec{j} - \frac{118}{21}\vec{k})$

21 $38.7°$

25 Bantus

Section 12.4

1 $-\vec{i}$

3 $-\vec{i} + \vec{j} + \vec{k}$

5 $-2\vec{i} + 2\vec{j}$

7 $\vec{a} \times \vec{b} = -2\vec{i} - 7\vec{j} - 13\vec{k}$
$\vec{a} \cdot (\vec{a} \times \vec{b}) = 0$
$\vec{b} \cdot (\vec{a} \times \vec{b}) = 0$

9 Max: 6
Min: 0

11 $x + y + z = 1$

13 (a) 1.5
(b) $y = 1$

15 $4x + 26y + 14z = 0$

19 (b) $\vec{c} \perp \vec{a}$ and $\|\vec{c}\| = \|\vec{a}\|$
(c) $-a_2 b_1 + a_1 b_2$

Chapter 12 Review

3 $\vec{p} = -\frac{4\sqrt{5}}{5}\vec{i} - \frac{2\sqrt{5}}{5}\vec{j}$

5 $-\vec{u}, \vec{v}, \vec{v} - \vec{u}, \vec{u} - \vec{v}$

7 (a) $t = 1$
(b) no t values
(c) any t values

9 $-2\vec{k}$

11 $\vec{i} - \vec{j}$

13 $\sqrt{6}/2$

15 $\vec{n} = 4\vec{i} + 6\vec{k}$

17 (a) $(21/5, 0, 0)$
(b) $(0, -21, 0)$ and $(0, 0, 3)$
(for example)
(c) $\vec{n} = 5\vec{i} - \vec{j} + 7\vec{k}$
(for example)
(d) $21\vec{j} + 3\vec{k}$
(for example)

19 $-\vec{i}/2 + \sqrt{3}\vec{j}/2$
$-\vec{i}/2 - \sqrt{3}\vec{j}/2$

21 Vector

25 (a) $30\vec{i} - 2\vec{k}$
$20\vec{i} + 15\vec{j} - 5\vec{k}$
$12\vec{i} + 30\vec{j} + 3\vec{k}$
(b) $30\vec{i} + \vec{j}$
$20\vec{i} + 16\vec{j} - 3\vec{k}$
$12\vec{i} + 31\vec{j} + 5\vec{k}$

27 $38.7°$ south of east.

33 $9x - 16y + 12z = 5$
0.23

Section 13.1

1 $f_x(3, 2) \approx 2$
$f_y(3, 2) \approx -1$

3 (c) Positive
(d) Negative

5 (a) Both negative
(b) Both negative

7 (a) $\partial P / \partial t$:
dollars/month
Rate of change in pay-
ments with time.

negative
(b) $\partial P / \partial r$:
dollars/percentage point
Rate of change in pay-
ments with interest rate.
positive

9 (a) Negative
(b) Positive

11 $f_T(5, 20) \approx 1$

13 (a) $2.5, 0.02$
(b) $3.33, 0.02$
(c) $3.33, 0.02$

15 -1

Section 13.2

1 -1

3 $2xy + 10x^4 y$

5 y

7 $a/(2\sqrt{x})$

9 $21x^5 y^6 - 96x^4 y^2 + 5x$

11 $(a + b)/2$

13 $2B/u_0$

15 $2mv/r$

17 Gm_1/r^2

19 $-2\pi r/T^2$

21 $\epsilon_0 E$

23 $c\cos(ct - 5x)$

25 $(15a^2 bcx^7 - 1)/(ax^3 y)$

27 $[x^2 y(-3\lambda + 10) - 3\lambda^4(8\lambda^2 - 27\lambda + 50)]/2(\lambda^2 - 3\lambda + 5)^{3/2}$

29 $\pi xy/\sqrt{2\pi xyw - 13x^7 y^3 v}$

31 $z_x = 7x^6 + yx^{y-1}$
$z_y = 2^y \ln 2 + x^y \ln x$

33 13.6

35 (a) $3.3, 2.5$
(b) $4.1, 2.1$
(c) $4, 2$

37 (a) Pre^{rt}
(b) e^{rt}

39 $h_x(2, 5) = -0.38$ ft/seat
$h_t(2, 5) = 0.76$ ft/second

Section 13.3

1   $z = 6 + 3x + y$

3   $z = -4 + 2x + 4y$

5   $z = 9 + 6(x-3) + 9(y-1)$

7   $P(r,L) \approx 80 + 2.5(r-8) +$ $0.02(L - 4000)$, $P(r,L) \approx$ $120 + 3.33(r-8) + 0.02(L - 6000)$, $P(r,L) \approx 160 + 3.33(r-13) + 0.02(L-7000)$.

11   $df = y\cos(xy)\,dx$ $+ x\cos(xy)\,dy$

13   $dg = (2u + v)\,du + u\,dv$

15   $df = dx - dy$

17   $dP \approx 2.395\,dK + 0.008\,dL$

19   $df = \frac{1}{3}dx + 2dy$ $f(1.04, 1.98) \approx 2.973$

21   (a)   Increases
      (b)   Increases
      (c)   55 joules

23   l is 1%; g is 2%

Section 13.4

1   (a)   1.01
      (b)   0.98

3   1

5   2.12

7   $x > 2$

9   (a)   Negative
      (b)   Negative

11   $\nabla z = \frac{1}{y}\cos\left(\frac{x}{y}\right)\vec{i}$ $- \frac{x}{y^2}\cos\left(\frac{x}{y}\right)\vec{j}$

13   $\nabla z = e^y\vec{i} + e^y(1 + x + y)\vec{j}$

15   $\nabla z = 2x\cos(x^2 + y^2)\vec{i}$ $+ 2y\cos(x^2 + y^2)\vec{j}$

17   $2m\vec{i} + 2n\vec{j}$

19   $-\frac{(t^2 - 2t + 4)}{(2s\sqrt{s})}\vec{i} + \frac{(2t-2)}{\sqrt{s}}\vec{j}$

21   $\left(\frac{5\alpha}{\sqrt{5\alpha^2 + \beta}}\right)\vec{i} + \left(\frac{1}{2\sqrt{5\alpha^2 + \beta}}\right)\vec{j}$

23   $50\vec{i} + 96\vec{j}$

25   $(1/2)\vec{i} + (1/2)\vec{j}$

27   (a) $2/\sqrt{13}$

      (b) $1/\sqrt{17}$
      (c) $\vec{i} + \frac{1}{2}\vec{j}$

29   (a) $-\sqrt{2}/2$
      (b) $\sqrt{3} + 1/2$

31   4.4

33   $(3\sqrt{5} - 2\sqrt{2})\vec{i}$ $+ (4\sqrt{2} - 3\sqrt{5})\vec{j}$

35   $y = 2x - 7$

37   P

39   (a) P, Q
      (c) $\|\operatorname{grad} f\|$
          $f_{\vec{u}} = \|\operatorname{grad} f\|\cos\theta$

Section 13.5

1   $10/3$

3   $2z + 3x + 2y = 17$

5   $x + 3y + 7z = -9$ $\vec{i} + 3\vec{j} + 7\vec{k}$

7   (a) $6.33\vec{i} + 0.76\vec{j}$
      (b) -34.69

9   (b)   Valley

11   (a) $x = y = 0$ and $z \neq 0$.
      (b) $y = 0$; $2x - y + 2z = 3$
      (c) $-\vec{j}$, $\frac{2}{3}\vec{i} - \frac{1}{3}\vec{j} + \frac{2}{3}\vec{k}$

13   $2x - y - z = 4$

Section 13.6

1   $\frac{dz}{dt} = e^{-t}\sin(t)(2\cos t - \sin t)$

3   $(t^3 - 2)/(t + t^4)$

5   $2e^{1-t^2}(1 - 2t^2)$

7   $\frac{\partial z}{\partial u} = (e^{-v\cos u} - v\cos(u)e^{-u\sin v})\sin v$ $- (-u\sin(v)e^{-v\cos u} + e^{-u\sin v})v\sin u$
     $\frac{\partial z}{\partial v} = (e^{-v\cos u} - v\cos(u)e^{-u\sin v})u\cos v$ $+ (-u\sin(v)e^{-v\cos u} + e^{-u\sin v})\cos u$

9   $\frac{\partial z}{\partial u} = e^v/u$ $\frac{\partial z}{\partial v} = e^v\ln u$

11   $\frac{\partial z}{\partial u} = 2ue^{(u^2 - v^2)}(1 + u^2 + v^2)$ $\frac{\partial z}{\partial v} = 2ve^{(u^2 - v^2)}(1 - u^2 - v^2)$

13   $\frac{\partial z}{\partial u} = \frac{1}{vu}\cos\left(\frac{\ln u}{v}\right)$ $\frac{\partial z}{\partial v} = -\frac{\ln u}{v^2}\cos\left(\frac{\ln u}{v}\right)$

15   $\frac{\partial w}{\partial u} = \frac{\partial w}{\partial x}\frac{\partial x}{\partial u} + \frac{\partial w}{\partial y}\frac{\partial y}{\partial u} + \frac{\partial w}{\partial z}\frac{\partial z}{\partial u}$ $\frac{\partial w}{\partial v} = \frac{\partial w}{\partial x}\frac{\partial x}{\partial v} + \frac{\partial w}{\partial y}\frac{\partial y}{\partial v} + \frac{\partial w}{\partial z}\frac{\partial z}{\partial v}$

17   -0.6

21   $\left(\frac{\partial U}{\partial P}\right)_T = \left(\frac{\partial U}{\partial V}\right)_T \left(\frac{\partial V}{\partial P}\right)_T$

23   (a) $\frac{\partial z}{\partial r} = \cos\theta\frac{\partial z}{\partial x} + \sin\theta\frac{\partial z}{\partial y}$
          $\frac{\partial z}{\partial\theta} = r(\cos\theta\frac{\partial z}{\partial y} - \sin\theta\frac{\partial z}{\partial x})$
      (b) $\frac{\partial z}{\partial y} = \sin\theta\frac{\partial z}{\partial r} + \frac{\cos\theta}{r}\frac{\partial z}{\partial\theta}$
          $\frac{\partial z}{\partial x} = \cos\theta\frac{\partial z}{\partial r} - \frac{\sin\theta}{r}\frac{\partial z}{\partial\theta}$

Section 13.7

1   $f_{xx} = 2$, $f_{yy} = 2$ $f_{yx} = 2$, $f_{xy} = 2$

3   $f_{xx} = 0$ $f_{xy} = e^y = f_{yx}$ $f_{yy} = xe^y$

5   $f_{xx} = -(\sin(x^2 + y^2))4x^2$ $+ 2\cos(x^2 + y^2)$
     $f_{xy} = -(\sin(x^2 + y^2))4xy$ $= f_{yx}$
     $f_{yy} = -(\sin(x^2 + y^2))4y^2$ $+ 2\cos(x^2 + y^2)$

7   $f_{xx} = -(\sin\left(\frac{x}{y}\right))(\frac{1}{y^2})$
     $f_{xy} = -(\sin\left(\frac{x}{y}\right))(\frac{-x}{y^2})(\frac{1}{y})$ $+ (\cos\left(\frac{x}{y}\right))(\frac{-1}{y^2}) = f_{yx}$
     $f_{yy} = -(\sin\left(\frac{x}{y}\right))(\frac{-x}{y^2})^2$ $+ (\cos\left(\frac{x}{y}\right))(\frac{2x}{y^3})$

9   $z_{yy} = 0$

11   (a)   Positive
      (b)   Zero
      (c)   Positive
      (d)   Zero
      (e)   Zero

13   (a)   Negative
      (b)   Zero
      (c)   Negative

(d) Zero

(e) Zero

15 (a) Zero

(b) Negative

(c) Zero

(d) Negative

(e) Zero

17 (a) Negative

(b) Negative

(c) Zero

(d) Zero

(e) Zero

19 (a) Negative

(b) Positive

(c) Positive

(d) Positive

(e) Negative

Section 13.8

1 (a) $u(4, 1) \approx 56.05°C$
$u(8, 1) \approx 70.05°C$

(b) $u(6, 2) \approx 62.1°C$

3 $c = D(a^2 + b^2)$

13 $a = -b^2$

15 (a) $u(0, t) = 0$
$u(1, t) = 0$

(b) $a = -b^2 = -(\pi k)^2$
for any integer k

17 $A = a/(b^2 + c^2), a > 0$

Section 13.9

1 $Q(x, y) = 1 - 2x^2 - y^2$

3 $Q(x, y) = -y + x^2 - y^2/2$

5 $L(x, y) = 2e + e(x - 1) + 3e(y - 1)$
$Q(x, y) = 2e + e(x - 1) + 3e(y - 1) + e(x - 1)(y - 1) + 2e(y - 1)^2$

7 $L(x, y) = \sqrt{2} + \frac{1}{\sqrt{2}}(x - 1) + \frac{1}{\sqrt{2}}(y - 1)$
$Q(x, y) = \sqrt{2} + \frac{1}{\sqrt{2}}(x - 1) + \frac{1}{\sqrt{2}}(y - 1) + \frac{1}{4\sqrt{2}}(x - 1)^2 - \frac{1}{2\sqrt{2}}(x - 1)(y - 1) + \frac{1}{4\sqrt{2}}(y - 1)^2$

9 $L(x, y) = \frac{e}{2} + \frac{e}{4}(x - 1) + \frac{e}{4}(y - 1)$
$Q(x, y) = \frac{e}{2} + \frac{e}{4}(x - 1) + \frac{e}{4}(y - 1) - \frac{e}{8}(x - 1)^2 + \frac{e}{4}(x - 1)(y - 1) + \frac{e}{8}(y - 1)^2$

11 $L(x, y) = \frac{\pi}{4} + \frac{1}{2}(x - 1) - \frac{1}{2}(y - 1)$
$Q(x, y) = \frac{\pi}{4} + \frac{1}{2}(x - 1) - \frac{1}{2}(y - 1) - \frac{1}{4}(x - 1)^2 + \frac{1}{4}(y - 1)^2$

15 (a) xy
$1 - \frac{1}{2}(x - \frac{\pi}{2})^2 - \frac{1}{2}(y - \frac{\pi}{2})^2$

17 (a) $L(x, y) = 1,$
$|E_L(x, y)| \leq 0.047$

(b) $Q(x, y) = 1 + (1/2)x^2 - (1/2)y^2,$
$|E_Q(x, y)| \leq 0.0047$

19 (a) $L(x, y) = 0,$
$|E_L(x, y)| \leq 0.14$

(b) $Q(x, y) = x^2 + y^2,$
$|E_Q(x, y)| \leq 0.036$

Section 13.10

1 (b) No

(c) No

(d) No

(e) Exist, not continuous.

3 (b) Yes

(c) Yes

(d) No

(e) Exist, not continuous

5 (b) Yes

(d) No

(f) No

7 (c) No

(e) No

9 (a) No

Chapter 13 Review

1 $\partial z/\partial x = \frac{14x+7}{(x^2+x-y)^{-6}}$
$\partial z/\partial y = -7(x^2 + x - y)^6$

3 $\partial f/\partial p = (1/q)e^{p/q}$
$\partial f/\partial q = -(p/q^2)e^{p/q}$

5 $\partial z/\partial x = 4x^3 - 7x^6 y^3 + 5y^2$
$\partial z/\partial y = -3x^7 y^2 + 10xy$

7 $\partial w/\partial s = \ln(s + t) + \frac{s}{(s+t)}$
$\partial w/\partial t = \frac{s}{(s+t)}$

9 $84/5$

11 False

13 True

15 False

17 (a) $f_w(2, 2) \approx 2.78$
$f_z(2, 2) \approx 4.01$

(b) $f_w(2, 2) \approx 2.773$
$f_z(2, 2) = 4$

19 (a) negative, positive, up if positive, down if negative

(b) $\pi < t < 2\pi$

(c) $0 < x < 3\pi/2$ and $0 < t < \pi/2$ or $3\pi/2 < t < 5\pi/2$.

23 (a) $\partial g/\partial m = G/r^2$
$\partial g/\partial r = -2Gm/r^3$

29 $3e/\sqrt{5}$

31 $f_{\vec{u}}(3, 1) \approx -1.64$

33 $\pm 4\sqrt{\frac{2}{11}} \left(\frac{1}{2}, -\frac{1}{2}, \frac{3}{2}\right)$

35 (a) $-14, 2.5$

(b) -2.055

(c) 14.221 in direction of $-14\vec{i} + 2.5\vec{j}$.

(d) $f(x, y) = f(2, 3) = 7.56.$

(e) For example, $\vec{v} = 2.5\vec{i} + 14\vec{j},$

(f) -0.32

37 (a) $\left.\frac{\partial w}{\partial u}\right|_{(1,\pi)} = 3 + \pi$
$\left.\frac{\partial w}{\partial v}\right|_{(1,\pi)} = 4 - \pi$

(b) $\left.\frac{dw}{dt}\right|_{t=1} = 5\pi - 3\pi^2$

41 $x - y$

43 (b) $x^2 + y^2 + z^2 = c^2$

(c) Outward, exponentially decreasing

Section 14.1

1 A: no B: yes, max
   C: yes, saddle
5 Saddle pts: $(1, -1), (-1, 1)$
   local max $(-1, -1)$
   local min $(1, 1)$.
7 $(1, -1)$ and $(-1, 1)$, both
   are saddle points.
9 Critical points: $(0, 0), (\pm\pi, 0)$,
      $(\pm 2\pi, 0), (\pm 3\pi, 0), \cdots$
   Local minima: $(0, 0)$,
      $(\pm 2\pi, 0), \pm 4\pi, 0), \cdots$
   Saddle points: $(\pm\pi, 0)$,
      $(\pm 3\pi, 0), (\pm 5\pi, 0), \cdots$
11 Local max: $(1, 5)$
13 Local minimum
15 Local maximum
17 (a) $(1, 3)$ is a minimum

Section 14.2

1 Mississippi:
      87−88(max), 83−87(min)
   Alabama:
      88−89(max), 83−87(min)
   Pennsylvania:
      89 − 90 (max), 80 (min)
   New York:
      81 − 84 (max),
      74 − 76 (min)
   California:
      100 − 101 (max),
      65 − 68 (min)
   Arizona:
      102 − 107 (max),
      85 − 87 (min)
   Massachusetts:
      81 − 84 (max), 70 (min)
3 Neither.
5 Min = 0 at $(0, 0)$
      (not on boundary)
   Max = 2 at $(1, 1), (1, -1)$,
      $(-1, -1)$ and $(-1, 1)$
      (on boundary)
7 Max = 0 at $(0, 0)$
      (not on boundary)
   Min = −2 at $(1, -1), (-1, -1)$,
      $(-1, 1)$ and $(1, 1)$
      (on boundary)

9 $q_1 = 300, q_2 = 225$.
11 $h = 25\%, t = 25°C$
15 $l = w = h = 45$ cm
17 $y = 2/3 - x/2$
19 (a) 255.2 million
   (c) 320.6 million
21 (b) 0.2575

Section 14.3

1 Min $= -\sqrt{2}$,
   max $= \sqrt{2}$
3 Min $= \frac{3}{4}$, no max
5 Min $= \sqrt{2}$, max $= 2$
7 Min $= -\sqrt{35}$,
   max $= \sqrt{35}$
9 max: 0, no min
11 max: $\frac{3}{\sqrt{6}}$, min: $-\frac{3}{\sqrt{6}}$
13 Max: $\frac{\sqrt{2}}{4}$, min: $-\frac{\sqrt{2}}{4}$.
15 Max: $f(\frac{1}{\sqrt{5}}, \frac{3}{\sqrt{5}}) = 2\sqrt{5}$
   Min: $f(-\frac{1}{\sqrt{5}}, -\frac{3}{\sqrt{5}}) = -2\sqrt{5}$
17 Max: 1
   Min: −1.
19 $q_1 = 50$ units
   $q_2 = 150$ units
21 (b) $S = 1000 - 10l$
23 $r = \sqrt[3]{\frac{50}{\pi}}$
   $h = 2\sqrt[3]{\frac{50}{\pi}}$
25 Along $x = 2y$ line
27 (c) $D = 10, N = 20$,
      $V \approx 9{,}779$
   (d) $\lambda = 14.67$
   (e) $68, rise

Chapter 14 Review

1 Local maximum: $(\pi/3, \pi/3)$
3 $(\sqrt{2}, -\sqrt{2}/2)$ saddle point
7 Maxima: $(-1, 1)$ and
   $(1, -1)$
   Minimum: $(0, 0)$
9 $p_1 = 110, p_2 = 115$.
11 $K = 20$
   $L = 30$
   $C = \$7{,}000$
13 (a) Reduce K by 1/2
      unit, increase L by
      1 unit.

15 $-\frac{1}{2m^2}(1 + \frac{2(v_1 v_2)^{1/2}}{v_1 + v_2})$ weeks^2
17 $d \approx 5.37$m, $w \approx 6.21$m,
    $\theta = \pi/3$ radians
19 (b) − grad d
21 (a) $\frac{a}{(v_1 \cos\theta_1)} + \frac{b}{(v_2 \cos\theta_2)}$
23 $d \approx 0.9148$.

Section 15.1

1 Lower sum : 0.34
   Upper sum : 0.62
3 Upper sum = 46.63
   Lower sum = 8
   Average $\approx$ 27.3
5 Positive
7 40/3
9 $\int_R w(x, y) \, dx \, dy \approx 2700$
   cubic feet,
   where R is the region
   $0 \leq x \leq 60$,
   $0 \leq y \leq 8$
11 (a) About 148 tornados
   (b) About 56 tornados
   (c) About 2 tornados
13 (a) positive
   (b) positive
   (c) positive
   (d) zero
   (e) zero
   (f) zero
   (g) negative
   (h) zero
   (i) negative
   (j) zero
   (k) zero
   (l) positive
   (m) positive
   (n) positive
   (o) zero
   (p) zero

Section 15.2

1 $\frac{4}{15}(9\sqrt{3} - 4\sqrt{2} - 1) = $
   2.38176

3 32/9

5 $\int_1^4 \int_1^2 f \, dy \, dx$
   or $\int_1^2 \int_1^4 f \, dx \, dy$

7 $\int_1^4 \int_{(x-1)/3}^2 f \, dy \, dx$

9 $(e^4 - 1)(e^3 - e)$

11 ≈ -2.68

13 14

15 $\frac{e-1}{2}$

17 $\frac{2}{9}(3\sqrt{3} - 2\sqrt{2})$

19 $\int_{-5}^5 \int_{-\sqrt{25-y^2}}^{\sqrt{25-y^2}} (25-x^2-y^2) \, dx \, dy$

21 $\int_0^4 \int_{y-4}^{(4-y)/2} (4-2x-y) \, dx \, dy$

23 Volume = 6

25 1/10

27 $\frac{1}{2}(1 - \cos 1) = 0.23$

Section 15.3

1 2

3 $a + b + 2c$

7 Limits do not make sense.

9 Limits do not make sense.

13 $\frac{15}{2}$

15 $\int_{-1}^1 \int_{-\sqrt{1-x^2}}^{\sqrt{1-x^2}} \int_{-\sqrt{1-z^2}}^{\sqrt{1-z^2}} dy \, dz \, dx$

17 $m = 1/36g$; $(\bar{x}, \bar{y}, \bar{z}) = (1/4, 1/8, 1/12)$

19 $m(b^2 + c^2)/3$

Section 15.4

1 0.7854

3 0.7966

5 4

7 0.79

9 4

Section 15.5

1 $\int_{\pi/4}^{3\pi/4} \int_0^2 f \, r \, dr \, d\theta$

3 $\int_0^{2\pi} \int_0^{\sqrt{2}} f \, r \, dr \, d\theta$

13 0

15 $-2/3$

17 6

19 $32\pi(\sqrt{2} - 1)/3$

21 (a) $\int_{\pi/2}^{3\pi/2} \int_1^4 \delta(r, \theta) \, r \, dr \, d\theta$
    (b) (i)
    (c) About 39,000

Section 15.6

1 $200\pi/3$

3 25π

5 $\int_0^1 \int_0^{2\pi} \int_0^4 \delta \cdot r \, dr \, d\theta \, dz$

7 $\int_0^{2\pi} \int_0^{\pi/6} \int_0^3 \delta \cdot \rho^2 \sin\phi \, d\rho \, d\phi \, d\theta$

9 $\int_0^3 \int_0^1 \int_0^5 \delta \, dz \, dy \, dx$

11 π

13 (a) Positive
    (b) Zero

15 $25\pi/6$

17 27π

19 $3/\sqrt{2}$

21 3/4

25 $3I = \frac{6}{5}a^2$; $I = \frac{2}{5}a^2$.

Section 15.7

1 (a) 20/27
    (b) 199/243

3 (a) $k = 8$
    (b) 1/3

5 $\int_{65}^{100} \int_{0.8}^1 f(x, y) \, dx \, dy$

7 $f(x, y) = \frac{30}{\pi} e^{-50(x-5)^2 - 18(y-15)^2}$

9 (a) $\int_\theta^{\pi/6} \int_{\frac{1}{\cos\theta}}^4 p(r, \theta) r \, dr \, d\theta$
    (b) $\int_{\frac{\pi}{6}}^{\frac{\pi}{6}+\frac{\pi}{12}} \int_{\frac{1}{\cos\theta}}^4 p(r, \theta) r \, dr \, d\theta +$
    $\int_{\frac{\pi}{6}+\frac{\pi}{12}}^{\frac{2\pi}{6}} \int_{\frac{1}{\sin\theta}}^4 p(r, \theta) r \, dr \, d\theta$

Section 15.8

3 $\rho^2 \sin\phi$

5 13.5

7 9

Chapter 15 Review

1 9200 cubic miles

7 85/12

9 $10(e - 2)$

11 $-4\cos 4 + 2\sin 4$
    $+ 3\cos 3 - 2\sin 3 - 1$

13 $\int_0^4 \int_{\frac{y}{2}-2}^{-y+4} f(x, y) \, dx \, dy$
    or $\int_{-2}^0 \int_0^{2x+4} f(x, y) \, dy \, dx +$
    $\int_0^4 \int_0^{-x+4} f(x, y) \, dy \, dx.$

17 $162\pi/5$

19 8π

21 ≈ 183

23 $8\pi R^5/15$

25 $2\pi Gm(r_2 - r_1 - \sqrt{r_2^2 + h^2} + \sqrt{r_1^2 + h^2}$

27 $40\sqrt{2}, 54\sqrt{2}$

Section 16.1

1 The particle moves on straight lines from $(0, 1)$ to $(1, 0)$ to $(0, -1)$ to $(-1, 0)$ and back to $(0, 1)$.

3 The particle moves on straight lines from $(-1, 1)$ to $(1, 1)$ to $(-1, -1)$ to $(1, -1)$ and back to $(-1, 1)$.

5 Clockwise for all t.

7 Clockwise: $t < 0$,
   Counter-clockwise: $t > 0$.

9 Counterclockwise: $t > 0$.

13 $x = -2$, $y = t$

15 $x = -2\cos t$, $y = 2\sin t$,
    $0 \le t \le 2\pi$

17 $x = 5\cos t$, $y = 7\sin t$,
    $0 \le t \le 2\pi$

19　(a)　Right of $(2, 4)$
　　(b)　$(-1, -3)$ to $(2, 4)$
　　(c)　$t < -2/3$

21　(a)　$a = b = 0, k = 5$ or -5
　　(b)　$a = 0, b = 5, k = 5$ or -5
　　(c)　$a = 10, b = -10, k = \sqrt{200}$ or $-\sqrt{200}$

23　$x = 2\cos t, y = 0, z = 2\sin t$

25　$x = 2 + 3t, \quad y = 3 - t, \quad z = -1 + t.$

27　$x = 1, \quad y = 0, \quad z = t.$

29　$x = 1 + 2t, \quad y = 2 + 4t, \quad z = 5 - t$

31　Yes

33　(a)　Straight lines
　　(b)　No
　　(c)　$(1, 2, 3)$

Section 16.2

1　(a)　Both parameterize the line
　　　　$y = 3x - 2$
　　(b)　Slope $= 3,0$
　　　　y-intercept $= -2$

3　(b)　$-\vec{i} - 10\vec{j} - 7\vec{k}$.
　　(c)　$\vec{r} = (1 - t)\vec{i} + (3 - 10t)\vec{j} - 7t\vec{k}$.

5　(a)　Spiral
　　(b)　$\vec{v}(2) = -2.24\vec{i} + 0.08\vec{j}$,
　　　　$\vec{v}(4) = 2.38\vec{i} - 3.37\vec{j}$,
　　　　$\vec{v}(6) = 2.63\vec{i} + 5.48\vec{j}$.
　　(c)　$\vec{v}(2) = -2.235\vec{i} + 0.077\vec{j}$,
　　　　$\vec{v}(4) = 2.374\vec{i} - 3.371\vec{j}$,
　　　　$\vec{v}(6) = 2.637\vec{i} + 5.482\vec{j}$.

7　$\vec{v} = -2t\sin(t^2)\vec{i} + 2t\cos(t^2)\vec{j}$,
　　Speed $= 2|t|$,
　　Particle stops when $t = 0$.

9　$\vec{v} = (2t - 2)\vec{i} + (3t^2 - 3)\vec{j} + (2t^3 - 12t^2)\vec{k}$,
　　Speed $= ((2t - 2)^2 + (3t^2 - 3)^2 + (12t^3 - 12t^2)^2)^{1/2}$,
　　Particle stops at $t = 1$.

11　$\vec{v} = -3\sin t\vec{i} + 4\cos t\vec{j}$,
　　$\vec{a} = -3\cos t\vec{i} - 4\sin t\vec{j}$

13　$\vec{v} = 3\vec{i} + \vec{j} - \vec{k}, \quad \vec{a} = \vec{0}$

15　$D = \sqrt{42}$

17　$D \approx 24.6$

19　$x = 5 + 3(t - 7), y = 4 + 1(t - 7), z = 3 + 2(t - 7)$.

21　(a)　$\vec{v}(2) \approx -4\vec{i} + 5\vec{j}$,
　　　　Speed $\approx \sqrt{41}$
　　(b)　About $t = 1.5$
　　(c)　About $t = 3$

23　(a)　No
　　(b)　$t = 5$
　　(c)　$\vec{v}(5) \approx 0.959\vec{i} + 0.284\vec{j} + 2\vec{k}$
　　(d)　$\vec{r} \approx 0.284\vec{i} - 0.959\vec{j} + 10\vec{k} + (t - 5)(0.959\vec{i} + 0.284\vec{j} + 2\vec{k})$.

25　$\vec{r}(t) = 22.1t\vec{i} + 66.4t\vec{j} + (442.7t - 4.9t^2)\vec{k}$

27　(a)　$x(t) = 5\sin t, y(t) = 5\cos t, z(t) = 8$
　　(b)　$\vec{v} = -5\vec{j}, \vec{a} = -5\vec{i}$
　　(c)　$x_{tt}(t) = y_{tt}(t) = 0$,
　　　　$z_{tt}(t) = -g$,
　　　　$x_t(0) = z_t(0) = 0$,
　　　　$y_t(0) = -5, x_t(0) = 5$,
　　　　$y_t(0) = 0, z_t(0) = 8$

29　(a)　C, E.
　　(b)　E, C.
　　(c)　Yes, when the light beam is tangential to shoreline.
　　(d)　The speed is not defined at corners.

31　$-(2t + 4t^3)/(1 + t^2 + t^4)^2$

Section 16.3

1　A horizontal disk of radius 5 in the plane $z = 7$.

3　A cylinder of radius 5 around the z-axis, $0 \le z \le 7$.

5　A cone of height and radius 5.

7　Cylinder, elliptical cross-section.

9　$x = a\cos\theta, y = a\sin\theta$, $z = z$

11　$x = u, y = v, z = 2u + v - 1$

13　No

15　Horizontal circle

17　$x = 5\sin\phi\cos\theta$
　　$y = 5\sin\phi\sin\theta$
　　$z = 5\cos\phi$

19　$x = a + d\sin\phi\cos\theta, y = b + d\sin\phi\sin\theta, z = c + d\cos\phi$
　　for $0 \le \phi \le \pi$ and $0 \le \theta \le 2\pi$.

21　If $\theta < \pi$, then $(\theta + \pi, \pi/4)$
　　If $\theta \ge \pi$, then $(\theta - \pi, \pi/4)$

23　$x = u\cos v, y = u\sin v$, $z = u$ for $0 \le v \le 2\pi$

25　(a)　$x = r\cos\theta$,
　　　　$0 \le r \le a$,
　　　　$y = r\sin\theta$,
　　　　$0 \le \theta < 2\pi$,
　　　　$z = \frac{hr}{a}$
　　(b)　$x = \frac{az}{h}\cos\theta$,
　　　　$0 \le z \le h$,
　　　　$y = \frac{az}{h}\sin\theta$,
　　　　$0 \le \theta < 2\pi$,
　　　　$z = z$

27　$x = ((\frac{z}{10})^2 + 1)\cos\theta$,
　　$y = ((\frac{z}{10})^2 + 1)\sin\theta$,
　　$z = z$,
　　$0 \le \theta \le 2\pi$,
　　$0 \le z \le 10$.

29　(a)　$z = (x^2/2) + (y^2/2)$
　　　　$0 \le x + y \le 2$
　　　　$0 \le x - y \le 2$

31　(a)　$x^2 + y^2 + z^2 = 1$,
　　　　$x, y, z \ge 0$.

33　$\vec{r}(t) = x_0\vec{i} + y_0\vec{j} + z_0\vec{k} + a\cos t\vec{u} + a\sin t\vec{v}$

35　(a)　$x = (\cos(\frac{\pi}{3}t) + 3)\cos\theta$
　　　　$y = (\cos(\frac{\pi}{3}t) + 3)\sin\theta$
　　　　$z = t \qquad 0 \le \theta \le 2\pi, 0 \le t \le 48$
　　(b)　456π in.³

Section 16.4

1　Circle:
　　$(x - 2)^2 + (y - 2)^2 = 1$

3　Parabola:
　　$y = (x - 2)^2, 1 \le x \le 3$

5　Implicit:
　　$x^2 - 2x + y^2 = 0, y < 0$,

Explicit:
$$y = -\sqrt{-x^2 + 2x},$$
Parametric:
$$x = 1 + \cos t, \ y = \sin t,$$
with $\pi \le t \le 2\pi$

7   $x + 2y = 0$

11   (a) $z = 1.054217$
    (b) $m(x, y, z) \approx 0 + 6x -$
      $7(y - 1) - 4(z - 1)$,
      $f_2(0.01, 0.98) \approx 1.05$
    (c) $\partial f_2/\partial x$ at $(0, 1)$ is $3/2$,
      $\partial f_2/\partial y$ at $(0, 1)$ is $-7/4$

13   (a)   Graph tangent to plane
    (b) $z \approx 7 - (2/5)(x - 3) -$
      $(4/5)(y - 5)$

15   (a) $S = \ln(a^a(1-a)^{(1-a)}) +$
      $\ln b - a \ln p_1 - (1 - a) \ln p_2$
    (b) $b = \dfrac{e^c p_1^a p_2^{(1-a)}}{a^a(1-a)^{(1-a)}}$

Section 16.5

1   250,000 stadia or 46,000 km

5   Not always closed

Chapter 16 Review

1   $x = t, \ y = 5$

3   $x = 4 + 4\sin t, \ y = 4 - 4\cos t$

5   $x = 2 - t, \ y = -1 + 3t, \ z = 4 + t.$

7   $x = 1 + 2t, \ y = 1 - 3t, \ z = 1 + 5t.$

9   $x = 3\cos t$
    $y = 5$
    $z = -3\sin t$

11   (a) $(I) = C_4, (II) = C_1,$
      $(III) = C_2, (IV) = C_6$
    (b) $C_3 : 0.5\cos t\vec{i} - 0.5\sin t\vec{j}$,
      $C_5 : -2\cos(\frac{t}{2})\vec{i}$
      $- 2\sin(\frac{t}{2})\vec{j}$

13   (a)   In the direction given by
      the vector: $\vec{i} - \vec{j}$
    (b)   Directions given by unit
      vectors:
      $\frac{1}{\sqrt{2}}\vec{i} + \frac{1}{\sqrt{2}}\vec{j}$
      $-\frac{1}{\sqrt{2}}\vec{i} - \frac{1}{\sqrt{2}}\vec{j}$

(c) -4

15   Line Equation:
    $x = 1 + 2t$
    $y = 2 + 3t$
    $z = 3 + 4t$
    Shortest distance: $\sqrt{174}/29$

17   The equation of the curve is
    $x = 1 - 2y^2, -1 \le y \le 1.$

19   (a) $(2, 3, 0)$
    (b) 2
    (c)   No; not on line

21   (a) $x = (V\cos A) \cdot t - 15$
      $y = $
      $-16t^2 + (V\sin A)t + 6$
    (c) $A \approx 52°$

23   (a) $(x, y) = (t, 1)$
    (b) $(x, y) = $
      $(t + \cos t, 1 - \sin t)$

25   $x = 2 - 2s - 2t,$
    $y = 3 - 3s - 0.5t,$
    $z = -1.6 + 3.6s + 1.6t$

29   (a) $a = 0.0893, b = 4.48$
    (b) $a = 0.427, b = 8.26$
    (c) $a = 1.43, b = 21.4$

31   (a)   spread out
    (b)   spread out
    (c)   compressed
    (d)   let $c < 0$

Section 17.1

1   (a)   IV
    (b)   III
    (c)   I
    (d)   II

9   $\vec{V} = -y\vec{i}$

11   $\vec{V} = -x\vec{i} - y\vec{j} = -\vec{r}$

13   $\vec{V} = -y\vec{i} + x\vec{j}$

15   $\vec{F}(x, y) = x\vec{i}$
    (for example)

17   $\vec{F}(x, y) = \dfrac{y\vec{i} - x\vec{j}}{\sqrt{x^2 + y^2}}$
    (for example)

Section 17.2

1   $y = $ constant

3   $y = -\frac{2}{3}x + c$

9   (a)   III
    (b)   I
    (c)   II
    (d)   V
    (e)   VI
    (f)   IV

Chapter 17 Review

1   (b)   (i)   yes
      (ii)   no
      (iii)   yes
      (iv)   no

5   (a) $\frac{1}{x^2 + y^2 + z^2}$
    (b) $\frac{1}{\sqrt{x^2 + y^2 + z^2}}$
    (c) $\frac{x}{\sqrt{x^2 + y^2 + z^2}}\vec{i} + \frac{y}{\sqrt{x^2 + y^2 + z^2}}\vec{j}$
      $+ \frac{z}{\sqrt{x^2 + y^y + z^2}}\vec{k}$
    (d) $\frac{-x}{\sqrt{x^+ y^2 + z^2}}\vec{i} + \frac{-y}{\sqrt{x^2 + y^2 + z^2}}\vec{j}$
      $+ \frac{-z}{\sqrt{x^2 + y^y + z^2}}\vec{k}$
    (e) $\frac{\cos t}{2\sqrt{2}}\vec{i} + \frac{\sin t}{2\sqrt{2}}\vec{j} + \frac{1}{2\sqrt{2}}\vec{k}$
    (f) $\frac{1}{\sqrt{2}}$

Section 18.1

1   Positive

3   Positive

5   $\int_{C_3} \vec{F} \cdot d\vec{r} < \int_{C_1} \vec{F} \cdot d\vec{r}$
    $< \int_{C_2} \vec{F} \cdot d\vec{r}$

7   Positive

9   0

11   0

13   16

15   32

19   C_1, C_2

21   (a)   Various values
    (b)   Various values

23   (a)   Various values
    (b)   Various values

27   $-GMm/8000$

Section 18.2

1   116.28

3   $82/3$

5   $e^2 - e$

7   85.32

9   24π

11   $C_1 : (t, \sqrt{2t - t^2}),$
     $0 \le t \le 2$
     $C_2 : (t, -2(t-1)^2),$
     $-1 \le t \le 2$
     $C_3 : (t, \sin t),$
     $-2\pi \le t \le -\pi$

13   (a) $11/6$
     (b) $7/6$

Section 18.3

3   Yes

5   No

7   $9/2$

9   $\frac{3}{\sqrt{2}} \ln(\frac{3}{\sqrt{2}} + 1)$

11   (a) e
     (b) e

13   (b)   No

19   (a) $\pi/2$
     (b)   No

21   (a)   Increases

Section 18.4

3   $f(x,y) = x^2 y + 2y^4 + K$
     $K = $ constant

5   No

7   Yes, $f = x^2 y^3 + xy + C$

9   Yes, $f = \ln A|xyz|$ where A
     is a positive constant.

11   (b) $-\pi$

13   πab

15   $3/2$

Section 18.5

1   $x = -1 + r\cos\theta,$
     $y = 2 + r\sin\theta,$
     $2 \le r \le 3, 0 \le \theta \le 2\pi$

3   $x = s,$
     $y = tg(s) + (1-t)f(s),$
     $a \le s \le b, 0 \le t \le 1$

Chapter 18 Review

1   (a)   Negative
     (b) C_1: Positive
           C_2, C_4: Zero
           C_3: Negative
     (c)   Negative

3   2

5   12

7   False

9   True

11   -58

15   (a) $\omega = 3000\,\text{rad/hr}$
           $K = 3 \cdot 10^7\,\text{m}^2\cdot\text{rad/hr}$
     (c) $r < 100$ m, circulation
           is $2\omega\pi r^2$
           $r \ge 100$ m, circulation
           is $2K\pi$

17   (b)   Circles
     (c)   No

Section 19.1

1   (a)   Positive
     (b)   Negative
     (c)   Zero
     (d)   Zero
     (e)   Zero

3   (a)   Zero
     (b)   Zero
     (c)   Zero
     (d)   Negative
     (e)   Zero

5   (a)   5
     (b)   4
     (c)   11
     (d)   9

7   (a)   Zero
     (b)   Zero

9   Zero

11   8π

13   Zero

15   (a)   Zero
     (b)   Zero

17   (b) $4\pi\lambda h$

21   (a)   Maximum speed
     (b)   0
     (c) $\pi u a^2/2$

Section 19.2

1   6

3   $\pi/2$

5   $7/3$

7   $\pi \sin 25$

9   1296π

11   $-81\pi/4$

13   12π

15   $625\pi/2$

17   (b)   (i) $\lim_{H\to 0} \int_S \vec{E} \cdot d\vec{A} = 0$
                 $\lim_{H\to\infty} \int_S \vec{E} \cdot d\vec{A} = 4\pi q$

           (ii) $\lim_{R\to 0} \int_S \vec{E} \cdot d\vec{A} = 4\pi q$
                 $\lim_{R\to\infty} \int_S \vec{E} \cdot d\vec{A} = 0$

19   $11\pi/2$

Section 19.3

1   $4/3$

3   195

5   $-\pi R^7/28$

7   $2\pi c(a^2 + b^2)$

Chapter 19 Review

5   4

7   1.5

9   12

11   $-8(1 + e^{-1})$

13   24π

15   $(\pi/6) - 1/3$

19   (b)   0
     (c) $Ih \ln|b/a|/2\pi$

Section 20.1

5   0

7   0

9   $2/\|\vec{r} - \vec{r}_0\|$

13 0

15 $\vec{b} \cdot (\vec{a} \times \vec{r})$

17 (a) Positive
    (b) Zero
    (c) Negative

19 (a) 0

21 (a) Flux = c^3
    (b) 1
    (c) 1

23 (a) $2\pi c^3$
    (b) 2
    (c) 2

25 (a) 0
    (b) Undefined.

Section 20.2

1 24

3 Zero

5 $\int_{S_2} \vec{F} \cdot d\vec{A} = 8$

9 (a) 0
    (b) 4π

11 $\int_S \vec{F} \cdot d\vec{A}$
    $= \int_R \operatorname{div} \vec{F} \, dV = 0$

15 (a) 30 watts/km^3
    (b) $\alpha = 10$ watts/km^3
    (d) 6847°C

Section 20.3

1 $4y\vec{k}$

3 $\vec{0}$

5 $4x\vec{i} - 5y\vec{j} + z\vec{k}$

7 $(2x^3yz + 6x^7y^5 - xy)\vec{i}$
    $+ (-3x^2y^2z - 7x^6y^6 + y)\vec{j}$
    $+ (yz - z)\vec{k}$

9 $\operatorname{curl}(F_1(x)\vec{i} + F_2(y)\vec{j} + F_3(z)\vec{k}) = $
    0

11 (a) Zero curl
    (b) Nonzero curl
    (c) Nonzero curl

19 $(c), (d), (f)$

21 $C_2, C_3, C_4, C_6,$

23 $\nabla\phi + (\vec{F} - \nabla\phi)$

25 $\vec{F} = (-\frac{3}{2}z - 2y)\vec{i}$

$+ (2x - z)\vec{j} + (y + \frac{3}{2}x)\vec{k}$

Section 20.4

1 No

3 (a) -200π
    (b) 0

5 π

13 (a) Can't say anything
    (b) 0

15 (a) $-2\pi a^2$
    (b) $2\pi a^2$
    (c) Orientations not related

17 -2π

Section 20.5

1 Yes

7 Yes
    $(-xy + 5yz)\vec{i}$
    $+ (2xy + xz^2)\vec{k}$.

9 (a) Yes
    (b) Yes
    (c) Yes

11 (a) $\operatorname{curl} \vec{E} = \vec{0}$
    (b) 3-space minus a point if
         $p > 0$
         3-space if $p \leq 0$.
    (c) Satisfies test for all p.
         $\phi(r) = r^{2-p}$ if $p \neq 2$.
         $\phi(r) = \ln r$ if $p = 2$.

15 $\vec{r} = (1 - s)\vec{r}(t) + s\vec{r}_0$
    $a \leq t \leq b$

Section 20.6

5 (b) 54

Chapter 20 Review

1 $\operatorname{div} \vec{v} = -6$

5 $\pi b^2 h/3$

7 $\operatorname{div} \vec{F} = 3$

9 True

11 True

13 True

15 False

17 $\operatorname{div} \vec{E}(P) \leq 0$, $\operatorname{div} \vec{E}(Q) \geq$
    0.

23 (b) $\vec{v}(x) = (55 - x/50)\vec{i}$
         mph
         if $0 \leq x < 2000$
         $\vec{v}(x) = 15\vec{i}$ mph
         if $2000 \leq x < 7000$
         $\vec{v}(x) = (15$
         $+ (x-7000)/25)\vec{i}$ mph
         if $7000 \leq x < 8000$
         $\vec{v}(x) = 55\vec{i}$ mph
         if $x \geq 8000$

    (c) $\operatorname{div} \vec{v}(1000) = -1/50$
         $\operatorname{div} \vec{v}(5000) = 0$
         $\operatorname{div} \vec{v}(7500) = 1/25$
         $\operatorname{div} \vec{v}(10,000) = 0$
         mph/ft

25 (a) $\vec{v} = (1 + \frac{y^2-x^2}{(x^2+y^2)^2})\vec{i}$
         $+ \frac{-2xy}{(x^2+y^2)^2}\vec{j}$

Appendix D

1 $(1/3)x^3 + x^2 + \ln|x| + C$,
    C a constant.

3 $\ln|t| - 4/t - 2/t^2 + C$, C a
    constant.

5 $(1/2)\sin 2t + C$, C a constant.

7 $-\ln|\cos\theta| + C$, C a constant.

9 $(1/2)e^{t^2+1} + C$, C a constant.

11 $(1/2)\tan^{-1} 2z + C$, C a
    constant.

13 $(-1/20)\cos^4 5\theta + C$,
    C a constant.

15 $(1/3)(\ln x)^3 + C$, C a con-
    stant.

17 $xe^x - e^x + C$, C a constant.

19 $(1/2)x^2 \ln x - (1/4)x^2 + C$,
    C a constant.

21 $(1/142)(10^{71} - 2^{71})$

23 $-11e^{-10} + 1$

25 $2e(e - 1) \approx 9.34$.

27 (b) Maximum in July 1993
         Minimum in Jan 1994
    (c) Incr. fastest in May 1993
         Decr. fastest in Oct 1993

29 (a) $\sum_{i=1}^{N} \rho(x_i)\Delta x$

(b) 16 grams

31 (a) 3 miles

(b) 282,743

33 (a) $\sum_{i=0}^{N-1} 4\pi(r_e + h_i)^2$ $\times 1.28e^{-0.000124h_i}\Delta h$

(b) 6.48×10^{16}

35 (a) $\sum_{i=1}^{N} \pi \frac{9x_i}{4}\Delta x$

(b) 18π

37 2267.32 cubic feet

Appendix F

5 (b) about $\frac{3}{4}$

7 (a) $p(x) = \dfrac{e^{-\frac{1}{2}\left(\frac{x-100}{15}\right)^2}}{15\sqrt{2\pi}}$

(b) 6.7% of the population

Appendix G

1 (1,0)

3 (−2,0)

5 $(\frac{5\sqrt{3}}{2}, -\frac{5}{2})$

7 $(\cos 1, \sin 1)$

9 $(2, \pi/2)$

11 $(\sqrt{2}, 3\pi/4)$

13 $(0.28, 7\pi/4)$

15 (3.16, 2.82)

INDEX